Recombinant Gene Expression Protocols

METHODS IN MOLECULAR BIOLOGY™

John M. Walker, SERIES EDITOR

73. **Neuropeptide Protocols,** edited by *G. Brent Irvine and Carvell H. Williams, 1997*
72. **Neurotransmitter Methods,** edited by *Richard C. Rayne, 1997*
71. **PRINS and *In Situ* PCR Protocols,** edited by *John R. Gosden, 1997*
70. **Sequence Data Analysis Guidebook,** edited by *Simon R. Swindell, 1997*
69. **cDNA Library Protocols,** edited by *Ian G. Cowell and Caroline A. Austin, 1997*
68. **Gene Isolation and Mapping Protocols,** edited by *Jacqueline Boultwood, 1997*
67. **PCR Cloning Protocols:** *From Molecular Cloning to Genetic Engineering,* edited by *Bruce A. White, 1996*
66. **Epitope Mapping Protocols,** edited by *Glenn E. Morris, 1996*
65. **PCR Sequencing Protocols,** edited by *Ralph Rapley, 1996*
64. **Protein Sequencing Protocols,** edited by *Bryan J. Smith, 1996*
63. **Recombinant Protein Protocols:** *Detection and Isolation,* edited by *Rocky S. Tuan, 1997*
62. **Recombinant Gene Expression Protocols,** edited by *Rocky S. Tuan, 1997*
61. **Protein and Peptide Analysis by Mass Spectrometry,** edited by *John R. Chapman, 1996*
60. **Protein NMR Protocols,** edited by *David G. Reid, 1997*
59. **Protein Purification Protocols,** edited by *Shawn Doonan, 1996*
58. **Basic DNA and RNA Protocols,** edited by *Adrian J. Harwood, 1996*
57. **In Vitro Mutagenesis Protocols,** edited by *Michael K. Trower, 1996*
56. **Crystallographic Methods and Protocols,** edited by *Christopher Jones, Barbara Mulloy, and Mark Sanderson, 1996*
55. **Plant Cell Electroporation and Electrofusion Protocols,** edited by *Jac A. Nickoloff, 1995*
54. **YAC Protocols,** edited by *David Markie, 1995*
53. **Yeast Protocols:** *Methods in Cell and Molecular Biology,* edited by *Ivor H. Evans, 1996*
52. **Capillary Electrophoresis:** *Principles, Instrumentation, and Applications,* edited by *Kevin D. Altria, 1996*
51. **Antibody Engineering Protocols,** edited by *Sudhir Paul, 1995*
50. **Species Diagnostics Protocols:** *PCR and Other Nucleic Acid Methods,* edited by *Justin P. Clapp, 1996*
49. **Plant Gene Transfer and Expression Protocols,** edited by *Heddwyn Jones, 1995*
48. **Animal Cell Electroporation and Electrofusion Protocols,** edited by *Jac A. Nickoloff, 1995*
47. **Electroporation Protocols for Microorganisms,** edited by *Jac A. Nickoloff, 1995*
46. **Diagnostic Bacteriology Protocols,** edited by *Jenny Howard and David M. Whitcombe, 1995*
45. **Monoclonal Antibody Protocols,** edited by *William C. Davis, 1995*
44. ***Agrobacterium* Protocols,** edited by *Kevan M. A. Gartland and Michael R. Davey, 1995*
43. **In Vitro Toxicity Testing Protocols,** edited by *Sheila O'Hare and Chris K. Atterwill, 1995*
42. **ELISA:** *Theory and Practice,* by *John R. Crowther, 1995*
41. **Signal Transduction Protocols,** edited by *David A. Kendall and Stephen J. Hill, 1995*
40. **Protein Stability and Folding:** *Theory and Practice,* edited by *Bret A. Shirley, 1995*
39. **Baculovirus Expression Protocols,** edited by *Christopher D. Richardson, 1995*
38. **Cryopreservation and Freeze-Drying Protocols,** edited by *John G. Day and Mark R. McLellan, 1995*
37. **In Vitro Transcription and Translation Protocols,** edited by *Martin J. Tymms, 1995*
36. **Peptide Analysis Protocols,** edited by *Ben M. Dunn and Michael W. Pennington, 1994*
35. **Peptide Synthesis Protocols,** edited by *Michael W. Pennington and Ben M. Dunn, 1994*
34. **Immunocytochemical Methods and Protocols,** edited by *Lorette C. Javois, 1994*
33. ***In Situ* Hybridization Protocols,** edited by *K. H. Andy Choo, 1994*
32. **Basic Protein and Peptide Protocols,** edited by *John M. Walker, 1994*
31. **Protocols for Gene Analysis,** edited by *Adrian J. Harwood, 1994*
30. **DNA–Protein Interactions,** edited by *G. Geoff Kneale, 1994*
29. **Chromosome Analysis Protocols,** edited by *John R. Gosden, 1994*
28. **Protocols for Nucleic Acid Analysis by Nonradioactive Probes,** edited by *Peter G. Isaac, 1994*
27. **Biomembrane Protocols:** *II. Architecture and Function,* edited by *John M. Graham and Joan A. Higgins, 1994*
26. **Protocols for Oligonucleotide Conjugates:** *Synthesis and Analytical Techniques,* edited by *Sudhir Agrawal, 1994*
25. **Computer Analysis of Sequence Data:** *Part II,* edited by *Annette M. Griffin and Hugh G. Griffin, 1994*
24. **Computer Analysis of Sequence Data:** *Part I,* edited by *Annette M. Griffin and Hugh G. Griffin, 1994*
23. **DNA Sequencing Protocols,** edited by *Hugh G. Griffin and Annette M. Griffin, 1993*
22. **Microscopy, Optical Spectroscopy, and Macroscopic Techniques,** edited by *Christopher Jones, Barbara Mulloy, and Adrian H. Thomas, 1993*
21. **Protocols in Molecular Parasitology,** edited by *John E. Hyde, 1993*
20. **Protocols for Oligonucleotides and Analogs:** *Synthesis and Properties,* edited by *Sudhir Agrawal, 1993*
19. **Biomembrane Protocols:** *I. Isolation and Analysis,* edited by *John M. Graham and Joan A. Higgins, 1993*

METHODS IN MOLECULAR BIOLOGY™

Recombinant Gene Expression Protocols

Edited by

Rocky S. Tuan

Thomas Jefferson University, Philadelphia, PA

Humana Press Totowa, New Jersey

999 Riverview Drive, Suite 208
Totowa, New Jersey 07512

This publication is printed on acid-free paper. ∞
ANSI Z39.48-1984 (American Standards Institute) Permanence of Paper for Printed Library Materials.

Cover illustration: Fig. 3 from Chapter 20, "Expression of Exogenous Genes in *Xenopus* Oocytes, Eggs, and Embryos," by Koichiro Shiokawa, Chie Koga, Yuzuru Ito, and Mikihito Shibata.

Cover design by Patricia F. Cleary.

For additional copies, pricing for bulk purchases, and/or information about other Humana titles, contact Humana at the above address or at any of the following numbers: Tel: 201-256-1699; Fax: 201-256-8341; E-mail: humana@mindspring.com

Printed in the United States of America. 10 9 8 7 6 5 4 3 2 1

Library of Congress Cataloging-in-Publication Data

Main entry under title:

Methods in molecular biology™.

Recombinant gene expression protocols / edited by Rocky S. Tuan.
p. cm. — (Methods in molecular biology™; vol. 62)
Includes index.
ISBN 0-89603-333-3 (comb)
ISBN 0-89603-480-1 (hardcover) (alk. paper)
1. Genetic recombination—Laboratory manuals. I. Tuan, Rocky S.
II. Series: Methods in molecular biology (Totowa, NJ); 62
QH443.R36 1997
572.8'77—dc21 96-49600
CIP

To Cecilia, Chuck, and my parents.

Preface

The current explosive progress in molecular biological research can be definitively traced to the development of molecular cloning technology. The ability to insert specific gene sequences into cloning vectors and their subsequent expansion is the cornerstone of modern molecular biology. A direct practical outcome of molecular cloning technology is its application to express specific recombinant genes. Currently, recombinant gene products are used in a wide spectrum of applications, including gene therapy, production of bioactive pharmaceuticals, synthesis of novel biopolymers, in agriculture and animal husbandry, and so on.

A fundamental requirement for successful recombinant gene expression is the design of the cloning vector and the choice of the host organism for expression. *Recombinant Gene Expression Protocols* grows out of the need for a laboratory manual that provides the reader the background and rationale, as well as the practical protocols for the preparation of "expression constructs" and their introduction into appropriate host cells and/or organisms. The chapters in this book are grouped by their expression hosts, including *E. coli*, yeast, mammalian cells, nonmammalian eukaryotes such as plants, *Xenopus*, and insects, as well as in transgenic organisms. In-depth information is presented on the important characteristics of expression cloning vectors and the various methods for efficiently introducing expression constructs into target cells and/or organisms.

Throughout *Recombinant Gene Expression Protocols*, the authors have consistently striven for a balanced presentation of both background information and actual laboratory details. It is anticipated that this highly practical format, a feature of the *Methods in Molecular Biology* series, will permit the reader to bring concepts into their personal research practices in a most efficient manner, i.e., help decide which vector and expression host system to use, and how to develop an efficient protocol for the introduction of the expression construct into the host.

The practice of molecular biology as a means to express recombinant genes continues as a rapidly advancing area in basic biomedical research, as well as the biotechnology and pharmaceutical industries. For this reason, it is anticipated that the subjects covered here will continue to be "under development," and serve as the basis for more sophisticated and efficient methodologies in the future.

The preparation of *Recombinant Gene Expression Protocols* would not have been possible without the outstanding work of the contributing authors, all of whom have been most tolerant of my persistent reminders. Dr. John M. Walker, the mastermind of the *Methods in Molecular Biology* series, was instrumental in initiating and guiding the project. The staff at the Humana Press showed great patience and provided excellent guidance and assistance. My wife, Cecilia, and my newborn son, Chuck, both tolerated my indulgence in the project, and always gave me the necessary emotional support throughout the preparation of the volume. Finally, the excellent secretarial assistance of Margaret Feoli, Susan Lowenstein, and in particular, Lynn Stierle, is gratefully acknowledged.

Rocky S. Tuan

Contents

Contents of the Companion Volume

Recombinant Protein Protocols: *Detection and Isolation*

Contributors

SALLY A. ADLER • *Department of Biological Sciences, Carnegie-Mellon University, Pittsburgh, PA*
NATHALIE ALLIOLI • *Centre de Genetique Moleculaire et Cellularie, Universite Claude Bernard Lyon I, Villcurbanne Cedex, France*
RÉMY A. AUBIN • *Life Sciences Division-Biotechnology, Sir F. G. Banting Research Centre, Health Canada, Ottawa, Ontario, Canada*
PAULINA BALBÁS • *Centro de Investigación en Biotechnologia, Universidad Autónoma del Estado de Morelos, Cuernavaca, Mexico*
STEVEN H. BASS • *Department of Cell Genetics, Genentech, South San Francisco, CA*
CHRISTOPHER C. BRODER • *Laboratory of Viral Diseases, National Institute of Allergy and Infectious Diseases, Bethesda, MD*
ANTHONY R. CASHMORE • *Department of Biology, University of Pennsylvania, Philadelphia, PA*
ALISTAIR CHAMBERS • *Department of Genetics, Queen's Medical Centre, University of Nottingham, UK*
DONALD C. CHANG • *Department of Biology, Hong Kong University of Science and Technology, Clear Water Bay, Kowloon, Hong Kong*
HAIFENG CHEN • *Department of Biochemistry and Molecular Biology, School of Medicine, University of Kansas Medical Center, Kansas City, KS*
SHAORONG CHONG • *College of Pharmacy, University of Michigan, Ann Arbor, MI*
ALAN C. CHRISTENSEN • *School of Biological Sciences, University of Nebraska, Lincoln, NE*
CHRISTOPHER COLECLOUGH • *Department of Immunology, St. Jude Children's Research Hospital, Memphis, TN*
KENINE E. COMSTOCK • *Department of Medicine, CF/Pulmonary Research and Treatment Center, University of North Carolina at Chapel Hill, NC*
HENRY DANIELL • *Molecular Genetics Program, Department of Botany and Microbiology, Auburn University, Auburn, AL*
TERRI G. DUNAHAY • *Applied Biological Sciences Branch, National Renewable Energy Laboratory, Golden, CO*

PATRICIA L. EARL • *Laboratory of Viral Diseases, National Institute of Allergy and Infectious Diseases, Bethesda, MD*
ROBERT J. ETCHES • *Department of Animal and Poultry Science, University of Guelph, Ontario, Canada*
ALFRED I. GELLER • *Endocrinology Department, Children's Hospital, Boston, MA*
GEORGE A. GARCIA • *College of Pharmacy, University of Michigan, Ann Arbor, MI*
RANDY GAUGLER • *Department of Entomology, Rutgers University, New Brunswick, NJ*
ANN M. VERRINDER GIBBINS • *Department of Animal and Poultry Science, University of Guelph, Ontario, Canada*
CHARLES GIRDHAM • *Department of Biochemistry, University of California, San Francisco, CA*
JEAN GOULD • *Department of Forest Science, Texas A&M University, College Station, TX*
IAN R. GRAHAM • *Department of Genetics, Queen's Medical Centre, University of Nottingham, UK*
DEAN HARTLEY • *Department of Endocrinology, Children's Hospital, Boston, MA*
SARWAR HASHMI • *Department of Entomology, Rutgers University, New Brunswick, NJ*
JACK A. HEINEMANN • *Department of Plant and Microbial Sciences, University of Canterbury, Christchurch, New Zealand*
JAMES E. HOPPER • *Department of Biochemistry and Molecular Biology, The Pennsylvania State University College of Medicine, Hershey, PA*
MASAYORI INOUYE • *Department of Biochemistry, UMDNJ Robert Wood Johnson Medical School, Piscataway, NJ*
YUZURU ITO • *Laboratory of Molecular Embryology, Zoological Institute, Faculty of Science, University of Tokyo Bunkyo-ku, Tokyo, Japan*
OLENA JACENKO • *Department of Animal Biology, University of Pennsylvania School of Veterinary Medicine, Philadelphia, PA*
JONATHAN W. JARVIK • *Department of Biological Sciences, Carnegie-Mellon University, Pittsburgh, PA*
SUSAN E. KANE • *Department of Cell and Tumor Biology, City of Hope National Medical Center, Duarte, CA*
ROXANN R. KARKHOFF-SCHWEIZER • *Department of Microbiology, College of Veterinary Medicine and Biomedical Sciences, Colorado State University, Fort Collins, CO*

RANDAL J. KAUFMAN • *Department of Biological Chemistry, Howard Hughes Medical Institute, University of Michigan Medical Center, Ann Arbor, MI*
WAYNE A. KEOWN • *Hybridon, Worcester, MA*
CHIE KOGA • *Laboratory of Molecular Embryology, Zoological Institute, Faculty of Science, University of Tokyo Bunkyo-ku, Tokyo, Japan*
PHUNG LANG • *Department of Endocrinology, Children's Hospital, Boston, MA*
CATHERINE LEGRAS • *Centre de Genetique Moleculaire et Cellularie, Universite Claude Bernard Lyon I, Villcurbanne Cedex, France*
FILIP LIM • *Centro de Biología Molecular, Facultad de Ciencias, Universidad Autonoma, Madrid, Spain*
JENNIE P. MATHER • *Genentech, South San Francisco, CA*
RAZMIK MIRZAYANS • W. W. *Cross Cancer Institute, Molecular Oncology Program, Edmonton, Alberta, Canada*
SALLY A. MOODY • *Department of Anatomy, Programs in Genetics and Neuroscience, The George Washington University Medical Center, Washington, DC*
ALISON MOORE • *Genentech, South San Francisco, CA*
LAWRENCE M. MYLIN • *Department of Microbiology and Immunology, The Pennsylvania State University College of Medicine, Hershey, PA*
PER-ÅKE NYGREN • *Department of Biochemistry and Biotechnology, Royal Institute of Technology, Stockholm, Sweden*
JOHN C. OLSEN • *Department of Medicine, CF/Pulmonary Research and Treatment Center, University of North Carolina at Chapel Hill*
DAVID R. O'REILLY • *Department of Biology, Imperial College of Science, Technology, and Medicine, London, UK*
R. PADMANABHAN • *Department of Biochemistry and Molecular Biology, School of Medicine, University of Kansas Medical Center, Kansas City, KS*
RAJI PADMANABHAN • *Laboratory of Experimental Carcinogenesis, National Cancer Institute, Bethesda, MD*
GREGORY S. PARI • *Hybridon, Worcester, MA*
MALCOLM C. PATERSON • *W. W. Cross Cancer Institute, Molecular Oncology Program, Edmonton, Alberta, Canada*
OPHRY PINES • *Department of Molecular Biology, Hebrew University, Hadassah Medical School, Jerusalem, Israel*
VIRGINIA L. PRICE • *Immunex Corporation, Seattle, WA*
BRIGITTE E. SCHONER • *Technology Core Research, Lily Research Laboratories, Lily Corporate Center, Indianapolis, IN*

Herbert P. Schweizer • *Department of Microbiology and Infectious Diseases, University of Calgary, Alberta, Canada*
Robert Shawley • *Genentech, South San Francisco, CA*
Mikihito Shibata • *Laboratory of Molecular Embryology, Zoological Institute, Faculty of Science, University of Tokyo Bunkyo-ku, Tokyo, Japan*
Koichiro Shiokawa • *Laboratory of Molecular Embryology, Zoological Institute, Faculty of Science, University of Tokyo Bunkyo-ku, Tokyo, Japan*
Keshav K. Singh • *Department of Molecular and Cellular Toxicology, Harvard School of Public Health, Boston, MA*
Song Song • *Department of Endocrinology, Children's Hospital, Boston, MA*
Stefan Ståhl • *Department of Biochemistry and Biotechnology, Royal Institute of Technology, Stockholm, Sweden*
Philip Starr • *Department of Endocrinology, Children's Hospital, Boston, MA*
Marzieh Taghavi • *Life Sciences Division-Biotechnology, Sir F. G. Banting Research Centre, Health Canada, Ottawa, Ontario, Canada*
William B. Terzaghi • *Department of Biology, University of Pennsylvania, Philadelphia, PA*
Snorri S. Thorgeirsson • *Laboratory of Experimental Carcinogenesis, National Cancer Institute, Bethesda, MD*
Rocky S. Tuan • *Department of Orthopaedic Surgery and Department of Biochemistry and Molecular Pharmacology, Thomas Jefferson University, Philadelphia, PA*
Mathias Uhlén • *Department of Biochemistry and Biotechnology, Royal Institute of Technology, Stockholm, Sweden*
Gérard Verdier • *Centre de Genetique Moleculaire et Cellularie, Universite Claude Bernard Lyon I, Villcurbanne Cedex, France*
Jean-Paul Vincent • *Laboratory of Molecular Biology, Cambridge, UK*
Yaming Wang • *Department of Endocrinology, Children's Hospital, Boston, MA*
Nathaniel F. Watson • *Department of Medicine, CF/Pulmonary Research and Treatment Center, University of North Carolina at Chapel Hill, NC*
Michael Weinfeld • *W. W. Cross Cancer Institute, Department of Radiobiology, Edmonton, Alberta, Canada*
Daniel G. Yansura • *Department of Cell Genetics, Genentech, South San Francisco, CA*
Linda Yu • *Department of Endocrinology, Children's Hospital, Boston, MA*

I

Introduction

1

Recombinant Gene Expression Protocols

Overview of Experimental Strategies for the Expression of Recombinant Genes

Rocky S. Tuan

1. Introduction

The direct cloning of DNA fragments, either derived from naturally occurring or artificially designed gene sequences, into various cloning vectors, including bacteriophages, plasmids, and viruses, is the cornerstone of modern molecular biology. Such recombinant constructs represent the basic reagents of molecular biology. One of the major applications utilizing cloned DNA sequences is the expression of the cloned DNA into a protein product, i.e., the expression of recombinant genes. Because the cloned DNA sequences can be modified or altered, recombinant expression technology thus enables the investigator to custom-design the final protein product to suit specific requirements. The production of such recombinant gene products is in fact one of the major success stories of modern molecular biology. Thus, in addition to answering fundamental questions related to regulation of gene expression, gene structure and function, and other basic issues of molecular biology, the technology of recombinant genes has revolutionized modern biotechnology and biomedicine. Applications in gene therapy, biopharmaceutics, biopolymers, and the like, have major impacts on science, medicine, agriculture, and industry.

A major technological requirement in successful expression of recombinant genes is the choice of the proper vector/host system and the use of efficient means to introduce the recombinant gene construct into the host cells. This chapter provides a brief overview of the technologies addressing these aspects as described in this book.

2. *Escherichia coli*

The bacterium *Escherichia coli* is undoubtedly the best understood organism in nature. The extensive genetic and biochemical studies of *E. coli* during

From: *Methods in Molecular Biology, vol. 62: Recombinant Gene Expression Protocols*
Edited by: R. Tuan Humana Press Inc., Totowa, NJ

the 1960s and 1970s is the foundation for modern recombinant DNA technology. Because *E. coli* is easy to manipulate experimentally and grows rapidly in relatively simple medium, it is the most commonly used host organism for recombinant gene expression. Cloning vectors used for *E. coli*, commonly plasmids, generally have the following characteristics:

1. A selectable marker to ensure maintenance of the vector in the cell;
2. A regulatable transcriptional promoter, such as *lac*, *trp*, and so on, that can be induced to produce large quantities of mRNA transcripts from the cloned DNA sequence;
3. Appropriate translational control sequences for ribosome binding and initiation of translation; and
4. Polylinker cloning site to facilitate insertion of the sequence of interest into the vector.

Several chapters in this book (Chapters 2, 6, 7, and 9) present in-depth information on the design of plasmid vectors and promoters. For protein expression, the recombinant construct needs to be introduced into the appropriate *E. coli* host strain by transformation. Chapter 3 in this book addresses the design of bacterial hosts for *lac*-based expression vectors. Another commonly used expression vector for *E. coli* is the bacteriophage λ, which is covered in Chapter 4. Other important aspects of recombinant gene expression in *E. coli* are covered in Chapter 5 (Strategies of Gene Fusion) and Chapter 8 (Expression and Secretion of Proteins in *E. coli*). Cell-free transcription and translation can also be coupled for efficient gene expression (Chapter 10).

3. Yeast

The baker's yeast, *Saccharomyces cerevisiae*, is among one of the simplest eukaryotes. In addition to having been characterized extensively in terms of its genetics, yeast can be grown easily either in the laboratory or in large-scale fermentation cultures. For the purpose of expressing recombinant genes, the yeast genome has been manipulated to reduce proteolysis as well as to ensure proper posttranslational modifications. For these reasons, *S. cerevisiae* is a powerful and commonly used host organism for the production of heterologous recombinant proteins. Chapters in this book cover the following main topics:

1. Chapter 11 provides an overview of the characteristics of yeast cloning vectors;
2. Chapters 12 and 13 describe two highly useful inducible expression cassettes, utilizing the *GAL4* and *ADH* promoters, respectively; and
3. Chapter 14 addresses the specifics of using a highly efficient constitutive expression vector containing the phosphoglycerate kinase (PGK) promoter.

4. Viral Expression Vectors

As carriers of genetic information, viruses offer high infectivity, broad host range, and the capacity to accommodate large gene fragments. Viral vectors, such as adenovirus and retrovirus, are currently being utilized actively in designing gene therapies. Chapters in this book provide background and technical information on several viral vectors currently in popular use for recombinant gene expression, including retroviruses (Chapter 17), vaccinia viruses (Chapters 15 and 16), defective herpes virus (Chapter 18), as well as baculovirus for insect cells (Chapter 19).

5. Eukaryotic Expression: Nonmammalian Systems

A popular eukaryotic expression system is the use of baculoviruses in insect cells (Chapter 19). This system allows posttranslational modification, processing, and transport of the expressed proteins as in most higher eukaryotic cells. In addition, the baculoviral expression system utilizes a helper-independent virus which may be propagated to high titers in insect cells adapted to growth in suspension cultures; thereby facilitating the isolation of expressed proteins in large quantities, since the majority of such proteins remain soluble in insect cells. The system also has the advantage that baculoviruses are noninfectious to vertebrate cells, and that their promoters are inactive in mammalian cells, which are important considerations when oncogenes or potentially toxic proteins are to be expressed. Generally, the level of expressed recombinant proteins in insect cells can reach up to half of the total cellular protein late in the infectious cycle, greatly enhancing the production capability of the baculoviral system.

The seminal work of Gurdon and Brown, who injected various types of RNAs into both fertilized and unfertilized *Xenopus* eggs, was a major technological breakthrough in the molecular biology of development. Such an approach has served well as an experimental system to assay for gene activities and functions. In this book, Chapters 20 and 21 present detailed background and technical information on the expression of heterologous or exogenous genes in *Xenopus* oocytes, eggs, and embryos, including specific blastomeres. These methodologies have allowed many investigators to probe the functional aspects of putatively developmentally important genes, in particular those involved in pattern formation and morphogenesis.

6. Expression in Plants

The agricultural importance of expressing recombinant genes in plants cannot be overestimated. In addition to enhancing crop yield, recent emphasis has been on using plants as bioreactors for the production of bioactive reagents. Introduction of exogenous DNA into plant cells requires overcoming the barrier imposed by the cell wall. Detailed technical information is included in this book on recent developments in the introduction of exogenous genes into plants by electroporation (Chapter 34), microprojectile bombardment (Chapter 35), the transforming bacterium *Agrobacterium tumefaciens* (Chapter 36), and the transformation of microalgae using silicon carbide whiskers (Chapter 37). Among these, the biolistic (bombardment) delivery system has also found application in transgenic animal research as well as human gene therapy.

7. Expression in Mammalian Cells

Expression of foreign genes in mammalian cells probably represents the bulk of current applications in recombinant gene expression work, which has become increasingly crucial for the study of basic biological questions as well as a means for production of recombinant gene products for pharmaceutical, agricultural, or other industrial uses. The distinct advantages of expressing recombinant genes in mammalian cells include the following reasons. Cloned DNA sequences derived from higher eukaryotes, either in the form of cDNAs or genomic fragments, are readily and properly expressed because the full compendium of posttranscriptional activities are functional in the host cell. In a similar manner, the mammalian host cell is also capable of posttranslational modifications essential for the activation and/or processing of the translated product, e.g., glycosylation, subunit assembly, and so on. With the proper selection of the host cell type, delivery or compartmentalization of the final gene product can be optimized, thus allowing, for example, the secretion of recombinant protein into the culture medium for efficient subsequent isolation and purification.

The expression of recombinant genes in mammalian cells is covered in depth in this book, and addresses several aspects of this application:

1. The design of expression vectors;
2. The means to introduce the user-designed expression construct into the mammalian host cells;
3. The identification and selection of the transfected cells; and
4. Optimization of growth and recombinant gene expression of the transfected cells.

Chapter 22 presents the background for vector design and suggestions for expression strategies. Several commonly used methods for efficient DNA transfection of mammalian cells are presented in this book. The classic and widely used methods of calcium phosphate coprecipitation and DEAE-dextran-mediated DNA transfection are discussed in detail in Chapter 23. The use of a pulse of high electric field to transiently permeabilize cell membrane and to allow the entry of macromolecules from the external medium into the cell is known as electroporation, and has recently gained much popularity in DNA transfection. The theory and practice of this technology is the subject of Chapter 24. The cationic detergent, polybrene, is a relatively inexpensive reagent which has been recently shown to be highly efficient in mediating DNA transfection of mammalian cells (Chapter 25).

Since DNA transfection is incomplete and efficiencies range from less than 1% to, at best, 50–60%, the selection and isolation of transfected cells from untransfected cells is crucial for the long-term propagation and homogeneity of the cell population. Probably the most commonly used method is the cotransfection of a gene, aminoglycoside phosphotransferase, thus rendering the transfected cell resistant to the antibiotic neomycin, which is normally toxic to mammalian cells by blocking ribosomal function and protein synthesis. In this manner, by selecting for cells that remain viable in otherwise toxic concentrations of the antibiotic G418, clones of cells stably transfected with the desired gene sequence can be isolated. Although highly useful, the G418 selection protocol is time-consuming and works most efficiently only with rapidly dividing cells, and is thus not applicable for transient expression studies. The experimental principle of the procedure of magnetic affinity cell sorting (Chapter 26) involves the cotransfection of a gene sequence which encodes a cell surface moiety, which may be subsequently recognized and tagged by specific antibodies; in this manner, by applying magnetically tagged immunoselection technology, the transfected cells expressing the recombinant genes may be isolated shortly after transfection (as short as 12–18 hours). Another limitation of the G418 selection scheme is that the neomycin-resistance gene is not amplifiable, and therefore does not promote overexpression of the desired recombinant gene. The use of the human multidrug resistance *(MDR1)* gene is described in Chapter 27 as a selectable marker for gene transfection, drug selection, and gene amplification. *MDR1* encodes a 170 kilodalton glycoprotein, P-glycoprotein (Pgp), which functions as a drug transporter to prevent the intracellular accumulation of cytotoxic agents and thereby confers resistance to those agents. Colchicine is one of the drugs transported by Pgp, and expression of *MDR1* cDNA confers colchicine resistance to the cells. When the cells are cultured in increasing concentrations of colchicine, the expression and copy number of transfected *MDR1* can be amplified, with concomitant amplification and overexpression of the cotransfected, desired foreign gene.

The overall strategy of working with transfected cells to optimize their growth, viability, and specific productivity for expression of recombinant proteins, is the subject of Chapter 28. Methods covered include cell cloning, suspension adaptation, adapting cells to reduced serum growth medium, optimizing medium formulation, and monitoring of cell viability.

8. Expression in Transgenic Organisms

The advent of animal transgenesis is one of the modern technical wonders of molecular biology. Transgenic animals, expressing specific exogenous genes or transgenes, have become an increasingly important system to examine and analyze the biological function(s) of the gene of interest. This book (Chapters

29–33) presents the fundamental experimental principles in generating a number of widely studied species of transgenic animals. These species include the fruitfly *Drosophila* (Chapter 29), the nematodes *Caenorhabditis elegans* and *Heterorhabditis bacteriophora* (Chapter 30), mammals (mouse) (Chapter 31), and avian species (chicken) (Chapter 32 and 33). These chapters provide the readers with wide-ranging, contemporary approaches to generating and using such transgenic animals for gene expression studies and applications.

II

EXPRESSION IN *E. COLI*

2

Designing Expression Plasmid Vectors in *E. coli*

Paulina Balbás

1. Introduction

The production of proteins is one of the main applications of genetic engineering in biotechnology. Even though standard cloning procedures are now routine and a large variety of host-vector systems for gene expression are available, difficulties are encountered when theoretical strategies are put into practice, so gene expression is still quite empirical. *E. coli* remains an important host system for the industrial production of proteins from cloned genes, and considerable lore has accumulated since the pioneering gene expression experiments. The extensive knowledge about *E. coli*'s physiology and genetics accounts for its preferential use as a host for gene expression. The inability of this organism to exert certain posttranslational modifications of proteins that lead to correct folding and activity represents its major drawback as a production organism.

The final amount of accumulated product in a fermentation process of a genetically modified *E. coli* strain is dependent upon the rates of biosynthesis vs the rates of degradation, both of the final product (the target protein) and the template (mRNA). Therefore, the two main concerns when designing an expression system should be the production of high levels of proteins and mRNA, and the stability of both macromolecules. With this simple rule in mind, there are four important components of an expression system that determine the final amount of accumulated protein:

1. The genetics of the expression system, such as: the nucleotide sequence and structure of the target gene, the specific transcription and translation signals that direct the synthesis of high amounts of protein and its mRNA, and the nature of the gene product.
2. In the case of plasmid-based systems, a suitable vector with an adequate copy number control mechanism, stability, a selective method, and a known replication mode is essential.

From: *Methods in Molecular Biology, vol. 62: Recombinant Gene Expression Protocols*
Edited by: R. Tuan Humana Press Inc., Totowa, NJ

3. An appropriate host strain containing the precise genetic traits necessary to work in conjunction with the expression signals of the system, and/or to increase mRNA or protein stability.
4. A proper cultivation process, in which fermentation conditions as well as the induction procedure can be very important for the final yield of recombinant proteins present in the culture at the time of harvesting.

These four elements should be carefully evaluated and must all fit into a processing strategy, which should include not only plans for efficient production, but a system for product recovery and purification. The success or failure of a process often depends on the initial choice of the host organism and the expression system. Whatever information is available on expression vectors, fermentation and induction conditions, nucleotide sequences, and so forth, is vital for decision making. This chapter presents practical recommendations for designing strategies for gene expression or improving existing systems, and emphasizing those particular aspects that can be improved for higher product output. Detailed discussion of some particular aspects of gene expression in plasmid-based systems has been reviewed elsewhere *(1)*. For practical techniques, it is suggested that the reader consult the various books and laboratory manuals available that offer step-by-step instructions for DNA cloning and editing, or that list an assortment of expression vectors with maps, descriptions, and uses *(2–4)*.

2. Experimental Considerations

2.1. High-Level Synthesis of mRNA

Transcription initiation is the rate-limiting step for mRNA synthesis. In *E. coli*, the frequency of transcription is regulated by a promoter sequence, which can be modulated by a variety of mechanisms, namely the interaction of one or more effector molecules with specific sequences in the vicinity of the promoter. Consequently, maximizing the efficiency of transcription initiation with the introduction of a strong promoter is the key step necessary to attain the goal of high-level mRNA synthesis. No mRNA synthesis during the early stages of fermentation is desirable to avoid lower growth rates or lethality, therefore tightly regulated promoters and strategies to completely evade transcription have been developed. Transcription initiation is often controlled by one of the following manners: temperature shift (λ pL and pR), chemical induction (*lac* and hybrid promoters containing its –10 region) or metabolic response (*phoA, trp*). Therefore, the choice of promoter must take into consideration the other three components of the system, namely strain genotype, fermentation conditions, and type of plasmid *(1,5)*.

The occurrence of nonspecific transcription termination can affect the levels of mRNA. This phenomenon is bound to be found when heterologous genes are expressed, and analysis of the mRNA sequence often reveals the presence of fortuitous mRNA-like terminator sequences. Disruption of these sequences or addition of antiterminator sequences in the 5'-end of the mRNA may solve this problem. Finally, the transcriptional isolation of the genetic information that leads to protein synthesis is important in order to avoid interference with plasmid or cellular functions. Cloning of a strong transcription terminator at the 3'-end of the gene in question often renders stability increase *(1,5,6)*.

2.2. Enhancing the Stability of mRNA

Degradation of mRNA serves as an adaptative mechanism of cells to respond rapidly to environmental changes, thus serving an important physiological role for the modulation of gene expression. Polynucleotide phosphorylase, RNases, the presence or absence of ribosomes, the translation rate, secondary structures present in the mRNA, the cell growth rate, and other factors have been shown to influence mRNA decay. For high-level expression purposes, two simple approaches have been used successfully: the introduction of mRNA stabilizing 5'-end enhancing sequences and the modulation of growth rates *(1,5,6,8)*.

2.3. High-Level Protein Biosynthesis

Translation is the mRNA-directed process by which amino acids are assembled into a polypeptide chain. Therefore, the mRNA contains the elements that condition the initiation of translation: an initiation AUG codon, a Shine-Dalgarno sequence or ribosome binding site (RBS), and other ribonucleotides that confer folding and spacing attributes to mRNA *(1,6)*. The efficiency of translation strongly depends on the sequence and structure of the RBS element. At least four nucleotides of the general consensus AGGAGG should be present, and the optimal spacing between the RBS and start codon should be about nine nucleotides. The initiation codon AUG is preferred over GUG and UUG. When the transcription initiation site is located far upstream of the translation initiation site, upstream, out-of-frame initiation can occur that may interfere with translation from the proper start. It is therefore useful in such cases to include stop codons in all frames upstream of the RBS sequence *(4)*.

2.4. Protein Stability

It is still not clear what determinants on a protein, either homologous or heterologous, make it stable or labile when synthesized in *E. coli*. Proteolysis is a highly regulated process that exerts a global impact on cellular functions, and although efforts to diminish degradation have been focused on producing strains with decreased capacity for proteolysis, no mutants completely defective on proteolysis have been obtained. Suggested strain genotypes include *lon, ptr, ompT, htrA, rpoH*, and multiple mutants *(5)*. A preferred strategy to overcome selective degradation of proteins has been the generation of gene fusions (translational fusions) that direct the synthesis of hybrid proteins. However, these hybrids require a cleavage process for the release of the target protein from the carrier. Rapidly synthesized abnormal proteins often accumulate as light-refractile aggregates known as inclusion bodies. The extent of intracellular protein aggregates depends not only on the protein itself, but also on the *E. coli* strain, the rate of protein synthesis and the growth conditions. If the protein of interest can be refolded adequately so that activity can be obtained, the formation of inclusion bodies is desirable because proteolysis is lowered and biosynthesis and purification might be facilitated. When the protein cannot be refolded from inclusion bodies, it must be produced in soluble form, which may not be easy to achieve in *E. coli*. If this is the case, expression in a different host may provide a more favorable environment for correct folding *(6,8)*.

2.5. Choice of Expression Vector

The genetic material of the expression system can be maintained in *E. coli* either extrachromosomally on replicative plasmids, or integrated into the host genome. For years, the first approach has been preferred, although the second is advantageous with respect to stability and undesired copy number effects. There is a considerable amount of information in the literature concerning plasmid design, types, construction, structure, function, stability, and so forth, and this information will not be presented here *(1–6)*. However, it is important to emphasize the fact that replicative plasmids containing the DNA for gene expression work in conjunction with the host strain and the growth conditions. Plasmid instability of a recombinant plasmid in a fermentation is the main cause for reduction of the overall levels of accumulated product, which in turn increases the production costs since the growth substrates are consumed by nonproductive cells. The segregational instability of a plasmid may be affected by the overall genetic background of the host, but also by the copy number of the plasmid, a high growth rate of the culture, and the genes contained in the plasmid. These elements have to be studied separately in case of low product output *(9)*.

Plasmid instability is usually an insidious problem in overexpression systems. Integration of the information for gene expression into the chromosome has proven to be an excellent alternative method to tackle plasmid instability problems. Integration strategies include double- and single-crossover recombination between homologous DNA fragments, transposon insertion, and phage delivery systems. Lower copy number of the target gene is the only drawback so far encountered *(4)*.

2.6. Choice of Host Strain

The host genetic background must contain the appropriate traits for regulation of gene expression and few specific nutrient requirements that might increase the fermentation costs. However, there are usually differences observed in the final yield of accumulated gene product, due at least to plasmid instability or protein degradation rates inherent to the host strain. The mode of induction of a promoter affects the host's metabolism differentially, producing alterations that range from mild to dramatic. Trial of a number of candidate strains in order to scan for the highest productivity yields is often recommended *(1,6,9)*.

2.7. Determination of the Fermentation Conditions

Microorganisms can grow under a variety of nutritional, physical, and chemical conditions. Productivity of an expression system is good when abundant biomass containing a large amount of the target protein is obtained. It is therefore important that the substrates (nucleotides and amino acids), catalytic components (RNA polymerase, DNA polymerase, tRNAs, and cofactors), and energy requirements of the host are not limited. In general, fast growth rates are usually achieved only at the expense of product output, for plasmid and DNA instabilities are provoked and harmful metabolites accumulate in the culture medium. As a consequence, the rate of product formation must be evaluated at different growth rates *(1,9)*.

References

1. Balbás, P. and Bolivar, F. (1990) Design and construction of expression plasmid vectors in *E. coli*. *Methods Enzymol.* **185,** 14–37.
2. Balbás, P., Soberón, X., Merino, M., Zurita, M., Lomelí, H., Valle, F., Flores, N., and Bolivar, F. (1986) Plasmid pBR322 and its Special Purpose Derivatives—a Review. *Gene* **50,** 3–40.
3. Gerhardt, P., Murray, R. G. E., Wood, W. A., and Krieg, N. R. (1994) *Methods for General and Molecular Bacteriology*. American Society for Microbiology. ASM Press, Washington, DC.
4. Schwab, H. (1993) Principles of Genetic Engineering for *E. coli*, in *Genetic Engineering of Microorganisms* (Pulher, A., ed.), VCH Publishers, New York, pp. 1–53.
5. Rosenbaum, V., Klahn, T., Lundberg, U., Holmgren, E., von Gabain, A., and Riesner, D. (1993) Co-existing structures of an mRNA stability determinant. *J. Mol. Biol.* **229,** 656–670.
6. Goeddel, D. V. (1990) Systems for heterologous gene expression. *Methods Enzymol.* **185,** 3–7.
7. Georgiou, G. and Bowden, G. A. (1991) Inclusion body formation and the recovery of aggregated recombinant proteins, in *Recombinant DNA Technology and Applications* (Prokop, A., Bajpai, R. K., and Ho, C., eds.), McGraw Hill, New York, pp. 333–356.

8. Bogosian, G., Kane, J. F., Obukowicz, M. G., and Olins, P. O. (1991) Optimizing protein production in recombinant strains of *E. coli*, in *Recombinant DNA Technology and Applications* (Prokop, A., Bajpai, R. K., and Ho, C., eds.), McGraw Hill, New York, pp. 285–315.
9. Shuler, M. L. and Kargi, F. (1992) *Bioprocess Engineering. Basic Concepts.* Prentice Hall, Englewood Cliffs, NJ.

3

Design of Bacterial Hosts for *lac*-Based Expression Vectors

Herbert P. Schweizer and RoxAnn R. Karkhoff-Schweizer

1. Introduction

Since introduction of the first pUC plasmids *(1)*, a great variety of plasmid vectors that use α-complementation and expression from the *lac* promoter, or its derivatives *tac* and *trc* promoters, have been developed (e.g., *see* refs. *2–14*). In order to maximize utilization of these vectors, various *Escherichia coli* host strains have been designed which contain the *lacZ*ΔM15 allele *(15)* necessary for α-complementation and the $lacI^q$ *(16,17)* gene, which allows for overproduction of the *lac* repressor that is required for regulated expression from the *lac* promoter. The development of F episomes *(7,14,18)* or phages *(1,19)* containing these components facilitated the construction of various host strains, provided they do not express β-galactosidase (e.g., Δ*lac* strains). However, these systems suffer from several short-comings that restrict their use:

1. Unless the episomes contain transposon-encoded antibiotic resistance markers (usually kanamycin or tetracycline), which also excludes their use in Tn*5*- or Tn*10*-containing strains, other commonly used F episomes require minimal medium for their maintenance because passage in rich media leads to their frequent loss *(20)*;
2. Since the episomes and phages have been tailored for use in *E. coli*, they cannot be exploited for establishment of a *lac*-based α-complementation and regulated expression system in other bacteria.

More recently, Haima et al. *(21,22)* developed a *lac*-based α-complementation system for *Bacillus subtilis* demonstrating that similar systems could be developed for other bacteria. However, their system is not easily transferable to other Gram-negative bacteria. To circumvent some of these problems, we have developed systems that allow conversion of Gram-negative bacterial strains,

From: *Methods in Molecular Biology, vol. 62: Recombinant Gene Expression Protocols*
Edited by: R. Tuan Humana Press Inc., Totowa, NJ

which do not express β-galactosidase, to host strains for *lac*-based cloning and expression vectors, allowing both α-complementation and regulated expression *(20,23)*. In this chapter, we will describe the use of two transposable elements that facilitate transfer of *lacZ*ΔM15 and *lacI*^{q} between *E. coli* strains and between *E. coli* and other bacterial species. One of these (mini-Tn*5*Lac4; Fig. 1A) is based on the mini-Tn*5* delivery system *(24,25)*, which was chosen for mainly two reasons:

1. Delivery of these mini-transposons is achieved via a convenient delivery system based on the mobilizable pUT suicide plasmid; and
2. Mini-Tn*5*s have been shown to transpose in various bacteria *(24,25)* and thus are very useful tools for genetic engineering *(26)*. The second (mini-D*lac*; Fig. 1B) is an example of a host-specific system that is based on the *Pseudomonas aeruginosa*-specific transposable bacteriophage D3112 *(27,28)*.

2. Materials

1. The following bacterial strains and plasmids are required: *E. coli* strains: S17-1(λ*pir*) *(25)*, BW20767 *(29)*, and S17-1 (λ*pir*)/pUT-Tn*5*Lac4 *(20)*, plasmids pUC19 *(14)*, and/or pBluescript SK (Stratagene, La Jolla, CA). *P. aeruginosa* strains PAO1 *(30)*, CD10 (D3112 *c*ts)/pTBZQR1 *(23)*, and the pUCP *(9,12)* and pRK415 *(6)* broad-host-range plasmids. Strain BW20767 can be obtained from Barry Wanner, Department of Biological Sciences, Purdue University, West Lafayette, IN 47907, and all other strains and plasmids from the authors in the Department of Microbiology, Colorado State University, Fort Collins, CO 80523.
2. LB medium: 10 g bacto-tryptone, 5 g yeast extract, 5 g NaCl, distilled water to 1 L, adjust pH to 7.0 with 5*M* NaOH (several drops), autoclave to sterilize.
3. LB plates: 10 g bacto-tryptone, 5 g yeast extract, 5 g NaCl, 15 g bacto agar, distilled water to 1 L, autoclave to sterilize.
4. LBM: Add 10 mL 1*M* $MgSO_4$ per L LB medium after autoclaving.
5. LBM top agar: 1 g bacto-tryptone, 0.5 g yeast extract, 0.5 g NaCl, 0.6 g bacto agar, distilled water to 100 mL, adjust pH to 7.0 with 5*M* NaOH, autoclave to sterilize. Add 10 mL 1*M* $MgSO_4$ per L medium after autoclaving. Just prior to use, melt by heating in a microwave oven, dispense in 3 mL aliquots into disposable (13 × 100 mm) glass test tubes and keep at 45–50°C.
6. 5*M* NaOH.
7. 5 × M9 stock solution: 30 g Na_2HPO_4, 15 g KH_2PO_4, 2.5 g NaCl, 5 g NH_4Cl, distilled water to 1 L, autoclave to sterilize.
8. M9 medium: 200 mL 5 × M9 stock solution, 10 mL 10 m*M* $CaCl_2$, 1 mL 1*M* $MgSO_4$, autoclaved distilled water to 1 L.
9. M9G plates: 15 g bacto agar, distilled water to 0.8 L, autoclave to sterilize. Let cool to about 50°C, then add 200 mL 5 × M9 stock solution, 10 mL 1*M* glucose, 10 mL 10 m*M* $CaCl_2$ and 1 mL 1*M* $MgSO_4$.
10. 1*M* glucose, filter sterilized.

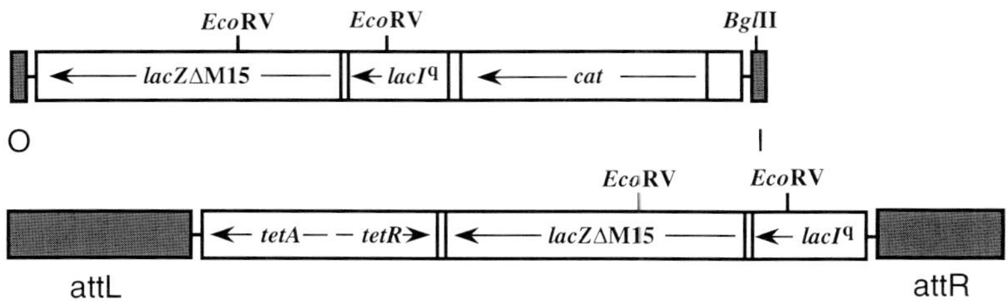

Fig. 1. Physical maps of the mini-transposons mini-Tn*5*Lac4 (7.5 kb) **(A)** and mini-D*lac* (10 kb) **(B)** are shown. The approximate extents of the *lacZ*ΔM15 and *lacI*^{q} genes encoding the ΔM15 β-galactosidase polypeptide and *lac* operon repressor, respectively, the chloramphenicol resistance-encoding gene (*cat*), and the genes specifying tetracycline-resistance (*tet*), as well as their transcriptional orientations, are shown. I and O denote the 19-base pair minimal Tn*5* ends (not drawn to scale) as present on pUT plasmid *(25)*. Symbols *attL* and *attR* indicate the left and right termini of phage D3112, respectively *(28)*. Only selected restriction sites are indicated, notably the *Eco*RV sites flanking a 1531-base pair *lac* operon segment that can be utilized as the probe in genomic Southern analyses.

11. Autoclaved distilled water.
12. Saline: 8.5 g NaCl, distilled water to 1 L, autoclave to sterilize.
13. XGal: in a glass or polypropylene vial prepare a 20 mg/mL stock solution by dissolving 5-bromo-4-chloro-3-indolyl-β-D-thiogalactopyranoside (various suppliers) in dimethylformamide. Use an amber vial or wrap in aluminum foil and store at –20°C.
14. Antibiotics: Antibiotics are purchased from Sigma, St. Louis, MO (or other suppliers) and prepared as stock solutions. Ampicillin (50 mg/mL): dissolve 250 mg in 5 mL distilled water, filter sterilize and store frozen in 1 mL aliquots at –20°C. Carbenicillin (500 mg/mL): Dissolve 2.5 g in 5 mL distilled water, filter sterilize and store frozen in 1 mL aliquots at –20°C. Chloramphenicol (20 mg/mL): dissolve 100 mg in 5 mL ethanol, store at -20°C. Tetracycline (10 mg/mL): dissolve 50 mg in 5 mL 70% ethanol, store in amber or foil-wrapped tube at –20°C.
15. 100% ethanol.
16. 70% ethanol.
17. Chloroform.
18. Dimethylformamide.
19. TG salts: 11 g $CaCl_2 \cdot 2H_2O$, 1.22 g $MgCl_2 \cdot 6H_2O$, 101 g glycerol, distilled water to 1 L, autoclave to sterilize, store at 4°C.
20. 1*M* $MgSO_4$, autoclave to sterilize, store at room temperature.
21. 1*M* $CaCl_2$, autoclave to sterilize, store at room temperature.
22. 0.1*M* $MgCl_2$, autoclave to sterilize, store at 4°C.

23. 0.1*M* $CaCl_2$, autoclave to sterilize, store at 4°C.
24. 0.1*M* $CaCl_2$-10% glycerol, autoclave to sterilize, store at 4°C.
25. 10 m*M* $CaCl_2$, autoclave to sterilize, store at room temperature.
26. Membrane filters: Several kinds of filters can be utilized. We use cellulose acetate or cellulose nitrate filters; 13 mm diameter; 0.45 µm pore size (Sartorius, Göttingen, FRG).

3. Methods

3.1. Construction of E. Coli Host Strains Containing Mini-Tn5Lac4

3.1.1. Conjugal Transfer and Transposition of Mini-Tn5Lac4

1. Preparation of overnight cultures:
   a. For donor: Using a sterile wooden applicator or inoculation loop, pick a colony of BW20767/pUT-Tn5Lac4 and inoculate 4 mL of LB broth (with ampicillin to 100 µg/mL and/or chloramphenicol to 20 µg/mL) and grow overnight at 37°C.
   b. For recipient: Using a sterile wooden applicator or inoculation loop, pick a colony and inoculate 4 mL of LB broth and grow overnight at 37°C.
2. In a sterile microfuge tube containing 1 mL of LB medium, combine 0.1 mL of each overnight culture. Harvest cells in a microcentrifuge at reduced speed (6000*g*, 30 s) at room temperature.
3. Thoroughly drain supernatant and, utilizing a pipet tip, suspend the cells in the residual LB medium.
4. Place 10–20 µL of the cell mixture onto a prewarmed membrane filter, which is placed on a prewarmed (37°C) LB plate. Incubate overnight at 37°C.
5. Transfer filter to a microfuge tube containing 1 mL of M9 medium and vortex vigorously for 1 min to remove the cells from the filter.
6. Harvest cells by centrifugation in a microfuge (13,000*g*; 1 min; room temperature) and wash twice in 1 mL of M9 medium.
7. Suspend cells in 1 mL of M9 medium and prepare a 1:10 dilution in the same medium.
8. Plate 0.1 mL and 0.2 mL aliquots of the undiluted and 1:10-diluted conjugation mixtures on M9G plates supplemented with 15 µg/mL chloramphenicol plus other nutritional requirements as dictated by the genotype of the recipient. Incubate at 37°C for 2–3 d.
9. Utilizing a sterile toothpick, transfer 8–10 colonies onto similar plates and streak for single colonies. Incubate at 37°C.
10. Utilizing a sterile toothpick, transfer single colonies to the following plates: (a) LB containing 15 µg/mL chloramphenicol and (b) LB containing 100 µg/mL ampicillin and 15 µg/mL chloramphenicol. Incubate at 37°C.
11. Pick 4–6 chloramphenicol-resistant and ampicillin-sensitive colonies for the transformation studies.

3.1.2. Test for Functionality of Chromosomally Integrated Mini-Tn5Lac4

3.1.2.1. Preparation of "Turbo"-Competent Cells

1. Utilizing a wooden applicator or inoculation loop, transfer a single colony to 4 mL LB medium (with chloramphenicol to 15 µg/mL) and grow at 37°C either overnight or to early saturation.
2. Transfer 1 mL of this culture to a cold microfuge tube sitting on ice. Briefly centrifuge at room temperature at full speed (13,000*g*; 30 s) (*see* Note 1).
3. Decant medium, immediately place centrifuge tube on ice and suspend cell pellet in 1 mL cold 0.1*M* $MgCl_2$ by quickly pipetting up and down in a sterile 1000 µL pipet tip.
4. Briefly centrifuge at room temperature at full speed (13,000*g*; 30 s).
5. Decant medium, immediately place centrifuge tube on ice and suspend cells in 1 mL cold TG salts by quickly pipetting up and down in a sterile 1000 µL pipet.
6. Keep on ice for 10 min. Briefly centrifuge at room temperature at full speed (13,000*g*; 30 s).
7. Decant medium, immediately place centrifuge tube on ice and suspend cells in 0.25 mL cold TG salts by quickly pipetting up and down in a sterile 1000 µL pipet.
8. At this point, competent cells are ready for immediate use or any residual cells can be frozen on dry ice and stored at –70°C for future use.

3.1.2.2. Transformation of "Turbo"-Competent Cells

1. To a small (13 × 100 mm) disposable borosilicate glass test tube sitting on ice, add 100 µL of competent cells and 2–5 µL (50–100 ng) of DNA from a plasmid containing a *lacZ*α coding sequence (e.g., pUC19 or pBluescript SK).
2. Keep on ice for 15 min; then heat-pulse at 37°C for 2 min.
3. Add 0.5 mL LB medium and shake at 37°C for 1 h.
4. Utilizing a flame-sterilized glass spreader, plate 100 µL and 200 µL aliquots on (a) LB medium (with ampicillin to 100 µg/mL and XGal to 40 µg/mL) and (b) the same medium with IPTG added to 0.5 m*M*. Incubate plates overnight at 37°C.
5. Score the phenotypes of the colonies. Colonies containing a functional mini-Tn5Lac4 element should appear white on the plates containing no IPTG (uninduced) and blue on the plates with IPTG (induced).

*3.2. Transfer and Transposition of Mini-D*lac

The D3112 *c*ts lysates were prepared by thermoinduction, titered, and used for transductions utilizing modifications of the procedures of Darzins and Casadaban *(30,31)*.

3.2.1. Preparation of a D3112 Lysate

1. Utilizing a wooden inoculator or inoculation loop, transfer a single colony of *P. aeruginosa* strain CD10/pTBZQR1 to 4 mL LB medium (with carbenicillin to 500 µg/mL) and shake the culture overnight at 30°C.

2. Dilute 0.4 mL of the overnight culture into 20 mL LB medium without antibiotic and grow for approx 6 h at 30°C.
3. Transfer culture to a 42°C shaker and continue incubation until lysis occurs (*see* Note 2).
4. Add 4 µL 1*M* $CaCl_2$ and 40 µL 1*M* $MgSO_4$, followed by 0.2 mL chloroform.
5. Remove cellular debris by centrifugation (8000*g*, 10 min, 4°C) (*see* Note 3).
6. Carefully remove supernatant (lysate) and stored at 4°C until needed (*see* Note 4).

3.2.2. Titer of a D3112 Lysate

1. Inoculate 4 mL of LBM medium with a single colony of the D3112-sensitive *P. aeruginosa* wild-type strain PAO1. Shake culture overnight at 37°C.
2. Add 0.1 mL of the overnight culture to a disposable (13 × 100 mm) borosilicate glass test tube containing 3 mL melted, 45–50°C LBM top agar. Mix well and pour evenly on a LBM plate. Let top agar harden at room temperature.
3. Prepare a series of dilutions (10^{-2}–10^{-7}) of the D3112 lysate in LBM. Start by addition of 10 µL of undiluted lysate to 990 µL of LBM (10^{-2} dilution), then continue in 10^{-1} increments by addition of 0.1 mL lysate dilutions to 0.9 mL LBM (10^{-3} dilution, and so on).
4. Utilizing a sterile pipet tip and a micropipetor, spot 10 µL of the dilutions on the previously prepared lawn of PAO1 cells. Dry spots at room temperature with lid of plate partially open. Incubate plate overnight at 37°C.
5. Inspect plates and count plaques of the spots where single plaques are clearly discernable. Calculate the titer of the lysate: e.g., if one observes 10 plaque forming units (PFU) in the 10 µL (or 0.010 mL) spot from the 10^{-6} dilution the titer is calculated as 10 PFU × 10^6 per 0.010 mL = 10^9 PFU/mL.

3.2.3. Transduction of Mini-D*lac*

1. Inoculate a single colony of a D3112-sensitive *Pseudomonas* recipient strain into 4 mL LBM medium and grow overnight at 30°C (*see* Note 5).
2. Using a sterile glass pipet or a sterilized pipet tip and a micropipetor, transfer 100 µL of the overnight culture of the recipient cells to a LBM plate equilibrated to 30°C and mix with 100 µL of lysate (titer >10^6) on LBM plates.
3. Utilizing a flame-sterilized glass spreader, distribute mixture evenly on the plate and allow infection to continue by incubating the plate at 30°C for 3 h (*see* Note 6).
4. Wash the cells from the plate with 2 mL of saline by using a disposable flame-sterilized glass spreader. Transfer the cell suspension with a micropipetor to two sterile 1.5-mL microcentrifuge tubes.
5. Harvest the cells by a brief centrifugation at room temperature (13,000*g*, 30 s).
6. Discard supernatant and suspend the cells in a total of 1 mL of saline.
7. Prepare dilutions (up to 10^{-4}) in saline (*see* Section 3.2.2., step 3) and plate 0.1 mL of each on LB plates with 50 µg/mL tetracycline. Incubate at 30°C for 1–2 d.
8. Pick 20 single colonies onto LB-tetracycline plates and streak for single colonies. Incubate at 30°C.

9. Utilizing a sterile toothpick, patch single purified colonies in quadruplicate onto the following plates: (a) two LB plates with carbenicillin to 500 μg/mL and tetracycline to 50 μg/mL; (b) two LB plates with tetracycline to 50 μg/mL. Incubate one set of plates overnight at 30°C and the second set overnight at 42°C.
10. Inspect the plates and retain at least eight tetracycline-resistant, carbenicillin-sensitive and temperature-resistant (growth at 42°C) colonies for the transformation experiment.

*3.2.4. Test for Functionality of Chromosomally Integrated Mini-D*lac

3.2.4.1 Preparation of Competent Cells of *P. aeruginosa*

1. Using a sterile wooden applicator or inoculation loop, pick a single colony of a PAO::mini-D*lac* strain into 4 mL of LB medium and grow overnight at 37°C.
2. With a sterile pipet, transfer 0.4 mL of the overnight culture to 40 mL of fresh LB medium and grow at 37°C until the culture reaches an A_{600nm} of ~1.0.
3. Transfer the culture to a chilled, sterile 45 mL screw-capped centrifuge tube (Nalgene), and keep on ice for 10–20 min.
4. Harvest cells by centrifugation at 4°C (5 min; 6000*g*; Beckman JA-20 rotor or equivalent).
5. Discard supernatant and suspend cell pellet in 20 mL cold, sterile 0.1*M* $MgCl_2$ (*see* Note 1).
6. Pellet the cells as in step 4, discard the supernatant, and suspend cell pellet in 10 mL cold 0.1*M* $CaCl_2$. Keep on ice for 30 min.
7. Pellet cells as in step 4, discard supernatant and suspend the cells in 2 mL cold 0.1*M* $CaCl_2$-10% glycerol.
8. Distribute in 200 μL aliquots into prechilled microfuge tubes sitting on ice, freeze on dry ice, and store at –70°C until use.

3.2.4.2. Transformation of *P. aeruginosa*

1. Thaw an aliquot of frozen competent cells on ice and transfer 0.1 mL to a cold, disposable thin-walled glass or plastic test tube sitting on ice. Add 5–20 μL of DNA isolated from *E. coli* of a broad-host-range plasmid containing a *lacZ*α coding sequence (e.g., the pUCP plasmids or pRK415) prepared by a plasmid mini-preparation method *(31)*. Add 1–10 μL of similarly prepared DNA from *P. aeruginosa (30)*. Incubate on ice for 30–60 min.
2. Quickly transfer tube to a 37°C heating-block or waterbath and continue incubation for 3 min, swirling the tube occasionally.
3. Place the tubes back on ice for 5 min.
4. Add 1 mL of LB and incubate at 37°C for 1–2 h with gentle shaking. Plate 0.1 and 0.2 mL aliquots on selective LB medium with 40 μg/mL XGal +/– 1 m*M* IPTG (e.g., carbenicillin to 500 μg/mL for the pUCP20/21 plasmids *(12)* or tetracycline to 50 μg/mL for the pUCP26/27 *(12)* and pRK415 plasmids). Incubate at the appropriate temperature, usually 37°C, for 24–48 h.
5. Score the phenotypes of the colonies. Colonies containing a functional mini-D*lac* element should appear white on the plates containing no IPTG (uninduced) and blue on the plates with IPTG (induced).

4. Notes

4.1. The Mini-D3112 Based Delivery System

1. Cold means that the solutions are always stored at 4°C and that all exposures to temperatures and materials above 4°C are kept to a minimum.
2. For induction of D3112*c*ts lysogens, the amount of time required for lysis varies widely and can take from a few hours to overnight.
3. Be careful to use a chloroform-resistant centrifuge tube when centrifuging chloroform-treated lysates.
4. D3112 lysates can be stored for at least 6–12 mo without significant loss of titer.
5. D3112 sensitivity can be determined utilizing the spot titer method described in Section 3.2.2.
6. For plating of phage-infected bacterial suspensions, we routinely use a disposable spreader made by flaming shut the thin end of a long Pasteur pipet and bending it in the flame at a 90° angle.

4.2. Problems with pir-Based Mini-Tn5 Delivery Systems

7. Utilizing the described procedure, it was found that mini-Tn5Lac4 transposed at very low, yet detectable frequencies in all *E. coli* strains studied to date.
8. As previously described in a step-by-step protocol *(20)*, this element can also be used with the more widely available S17-1(λ*pir*) or SM10(λ*pir*) *(25)* mobilizer strains. However, the major problems with λ*pir*-containing donors is release of phage and frequent lysogenization of the recipient strain with λ*pir*, thereby circumventing the suicide mechanism. To minimize this problem, it is imperative to adhere to the previously described procedure *(20)*. It is virtually the same as the one described in Section 3.1.1., except that the ion chelator Na_3-citrate is included at high concentrations (20 m*M*) into the selective media during the mating, selection, and subsequent screening steps. But, this problem can entirely be avoided by utilizing the *uidA::pir* *E. coli* donor strains *(29)*, as described in this chapter, or λ-resistant *E. coli* recipient strains.

4.3. Problems with Multiple Mini-Tn Insertions

9. We have previously observed that duplicate insertions of both mini-D*lac* *(23)* and, to a much lesser extent, mini-Tn5Lac4 (H. Schweizer, unpublished observations) can occur. The number of transposons per chromosome can be determined by probing with a *lac*-specific probe as previously described *(23)*. The *lac* operon-specific 1531-base pair *Eco*RV fragment indicated in Fig. 1 is especially useful for this purpose. However, our results indicate that duplicate insertions have neither an effect on blue/white screening nor on regulation of expression from the *lac* promoter by the *lac* repressor *(23)*.
10. Although not explicitly explored by our laboratory, it should be easy in *E. coli* to transduce mini-Tn5Lac4 insertions into other strain backgrounds by P1 transduc-

tion *(32)* using selection of the transposon-encoded chloramphenicol-resistance marker. Thus, transposons present in any given strain could be segregated by genetic means. Similarly, a D3112-mediated transduction *(28)* of the mini-D*lac* element between different *P. aeruginosa* strains could also be envisioned.

4.4. Host Range of Mini-Tn Elements

11. As previously mentioned, the mini-Tn*5*Lac4 element should be widely applicable for genetic engineering of other bacterial strains that are (a) naturally β-galactosidase negative (e.g., *Salmonella typhimurium*), or that have been made β-galactosidase negative by mutagenesis, and are (b) recipients of the *oriT*-mediated pUT transposon delivery system *(24,25)*.
12. One major problem with extension of the host range of the mini-Tn*5*Lac4 element may be that the chloramphenicol-resistance marker may not be applicable in any given strain background (e.g., the commonly used *P. aeruginosa* wild-type strain PAO1 is inherently resistant to chloramphenicol). However, this could easily be circumvented by substitution of the chloramphenicol-resistance marker with other selectable markers, e.g., the readily available and widely used kanamycin-resistance *(33)* or gentamycin-resistance *(34)* markers.

Acknowledgments

This work was performed while both authors were in the Department of Microbiology and Infectious Diseases at the University of Calgary, Calgary, Alberta, Canada. We acknowledge generous financial support by the Medical Research Council of Canada (MRC) and the Canadian Cystic Fibrosis Foundation. H. P. S. supported by a MRC Medical Scholarship.

References

1. Vieira, J. and Messing, J. (1982) The pUC plasmids, an M13mp7-derived system for insertion mutagenesis and sequencing with synthetic universal primers. *Gene* **19,** 259–268.
2. Balbás, P., Soberón, X., Merino, E., Zurita, M., Lomeli, Z., Valle, F., Flores, N., and Bolivar, F. (1986) Plasmid vector pBR322 and its special purpose derivatives—a review. *Gene* **50,** 3–40.
3. Martinez, E., Bartolomé, B., and de la Cruz, F. (1988) pACYC184-derived cloning vectors containing the multiple cloning site and *lacZα* reporter gene of pUC8/9 and pUC18/19 plasmids. *Gene* **68,** 159–162.
4. Bartolomé, B., Jubete, Y., Martinez, E., and de la Cruz, F. (1991) Construction and properties of a family of pACYC184-derived cloning vectors compatible with pBR322. *Gene* **102,** 75–78.
5. Brosius, J. (1992) Compilation of superlinker vectors. *Methods Enzymol.* **216,** 469–483.
6. Keen, N. T., Tamaki, S., Kobayashi, D., and Trollinger, D. (1988) Improved broad-host-range plasmids for DNA cloning in Gram-negative bacteria. *Gene* **70,** 191–197.

7. Messing, J. (1983) New M13 vectors for cloning. *Methods Enzymol.* **101,** 20–78.
8. Vieira, J. and Messing, J. (1991) New pUC-derived cloning vectors with different selectable markers and DNA replication origins. *Gene* **100,** 189–194.
9. Schweizer, H. P. (1991) *Escherichia-Pseudomonas* shuttle vectors derived from pUC18/19. *Gene* **97,** 109–112.
10. Spratt, B. G., Hedge, P. J., te Heesen, S., Edelman, A., and Broome-Smith, J. K. (1986) Kanamycin-resistant vectors that are analogues of plasmids pUC8, pUC9, pEMBL8 and pEMBL9. *Gene* **41,** 337–342.
11. Stewart, G. S. A. B., Lubinsky-Mink, S., and Kuhn, J. (1986) pHG276: a multiple cloning site pBR322 copy number vector expressing a functional *lacZα* peptide from the bacteriophage lambda P_R promoter. *Plasmid* **15,** 182–190.
12. West, S. E. H., Schweizer, H. P., Dall, C., Sample, A. K., and Runyen-Janecky, L. J. (1994) Construction of improved *Escherichia-Pseudomonas* shuttle vectors derived from pUC18/19 and the sequence of the region required for their replication in *Pseudomonas aeruginosa. Gene* **128,** 81–86.
13. Kushner, S. R. and Wang, R. F. (1991) Construction of versatile low-copy-number vectors for cloning, sequencing and gene expression in *Escherichia coli. Gene* **100,** 195–199.
14. Yanisch-Perron, C., Vieira, J., and Messing, J. (1985) Improved M13 cloning vectors and host strains: nucleotide sequences of the M13mp18 and pUC19 vectors. *Gene* **33,** 103–119.
15. Prentki, P. (1992) Nucleotide sequence of the classical *lacZ* deletion ΔM15. *Gene* **122,** 231–232.
16. Müller-Hill, B., Crapo, L., and Gilbert, W. (1968) Mutants that make more *lac* repressor. *Proc. Natl. Acad. Sci. USA* **59,** 1259–1264.
17. Calos, M. (1978) DNA sequence for a low-level promoter of the *lac* repressor gene and an "up" promoter mutation. *Nature* **274,** 762–765.
18. Bullock, W. O., Fernandez, J. M., and Short, J. M. (1987) XL1-Blue: A high efficiency plasmid transforming *recA Escherichia coli* strain with Beta-galactosidase selection. *BioTechniques* **5,** 376–379.
19. Liss, L. (1987) New M13 host: DH5αF' competent cells. *Focus* **9,** 13.
20. Schweizer, H. P. (1994) A method for construction of bacterial hosts for *lac*-based cloning and expression vectors: a complementation and regulated expression. *BioTechniques* **17,** 452–456.
21. Haima, P., van Sinderen, D., Schotting, H., Bron, S., and Venema, G. (1990) Development of a β-galactosidase α-complementation system for molecular cloning in *Bacillus subtilis*. *Gene* **86,** 63–69.
22. Haima, P., van Sinderen, D., Bron, S., and Venema, G. (1990) An improved β-galactosidase α-complementation system for molecular cloning in *Bacillus subtilis*. *Gene* **93,** 41–47.
23. Karkhoff-Schweizer, R. R. and Schweizer, H. P. (1994) Utilization of mini-D*lac* transposable element to create an α-complementation and regulated expression system for molecular cloning in *Pseudomonas aeruginosa*. *Gene* **140,** 7–15.

24. Miller, V. L. and Mekalanos, J. J. (1988) A novel suicide vector and its use in construction of insertion mutations: osmoregulation of outer membrane proteins and virulence determinants in *Vibrio cholerae* requires *toxR*. *J. Bacteriol.* **170,** 2575–2583.
25. De Lorenzo, V. and Timmis, K. N. (1994) Analysis and construction of stable phenotypes in Gram-negative bacteria with Tn*5* and Tn*10*-derived transposons. *Methods Enzymol.* **235,** 386–405.
26. Berg, C. M., Berg, D. E., and Groisman, E. A. (1989) Transposable elements and the genetic engineering of bacteria, in *Mobile DNA* (Berg, D. E. and Howe, M., ed.), American Society for Microbiology, Washington, pp. 879–925.
27. Darzins, A. and Casadaban, M. J. (1989) In vivo cloning of *Pseudomonas aeruginosa* genes with mini-D3112 transposable bacteriophage. *J. Bacteriol.* **171,** 3917–3925.
28. Darzins, A. and Casadaban, M. J. (1989) Mini-D3112 bacteriophage transposable elements for genetic analysis of *Pseudomonas aeruginosa*. *J. Bacteriol.* **171,** 3909–3916.
29. Metcalf, W. W., Jiang, W., and Wanner, B. L. (1994) Use of the *rep* technique for allele replacement to construct new *Escherichia coli* hosts for maintenance of R6Kγ origin plasmids at different copy numbers. *Gene* **138,** 1–7.
30. Schweizer, H. P. and Po, C. (1994) Cloning and nucleotide sequence of the *glpD* gene encoding *sn*-glycerol-3-phosphate dehydrogenase from *Pseudomonas aeruginosa*. *J. Bacteriol.* **176,** 2184–2193.
31. Xiang, C., Wang, H., Shiel, P., Berger, P., and Guerra, D. J. (1994) A modified alkaline lysis miniprep protocol using a single microcentrifuge tube. *BioTechniques* **17,** 30–31.
32. Miller, J. H. (1992) P1 transduction, in *A Short Course in Bacterial Genetics*, Cold Spring Harbor Laboratory Press, Cold Spring Harbor, NY, pp. 263–278.
33. Taylor, L. A. and Rose, R. E. (1988) A correction in the nucleotide sequence of the Tn*903* kanamycin resistance determinant in pUC4K. *Nucleic Acids Res.* **16,** 358.
34. Schweizer, H. P. (1993) Small broad-host-range gentamycin resistance cassettes for site-specific insertion and deletion mutagenesis. *BioTechniques* **15,** 831–833.

4

Bacteriophage λ-Based Expression Vectors

Alan C. Christensen

1. Introduction

Bacteriophage λ has been in use as a cloning vector for over 20 years, and has been extensively used as an expression vector. A historical overview of λ as a cloning vector can be found in Murray *(1)*, and a more recent review of λ vectors is in Chauthaiwale et al. *(2)*. In general, λ is more useful as a tool for expressing foreign genes when a library is to be screened. For high level production and purification of a particular foreign protein in *E. coli*, plasmids are generally preferred, although λ vectors can be used. There are also λ vectors that can readily be converted into plasmids for production of foreign proteins. The purpose of this chapter is to review bacteriophage λ based expression vectors and their features as an aid in choosing a vector or in understanding the features of a particular λ library or clone. Detailed methods on cDNA library construction including mRNA isolation, cDNA synthesis, ligation, plating and screening λ libraries with antibodies or DNA probes can be found in a number of cloning manuals *(3)*. In addition, several λ vectors are commercially available either by themselves or as part of cDNA cloning kits with detailed instructions.

2. Cloning Features of λ Expression Vectors

2.1. Library Screening

Bacteriophage λ has many advantages over plasmids for the construction and screening of recombinant DNA libraries. These include the greater efficiency of packaging λ particles and infecting *E. coli* compared to transformation with plasmids. Lambda libraries are also very conveniently amplified and are very stable and easy to store. Screening plaques by filter lifting is generally easier than screening colonies, and more plaques can be screened per plate than colonies. The disadvantages of using λ are that large-scale DNA preps are

From: *Methods in Molecular Biology, vol. 62: Recombinant Gene Expression Protocols*
Edited by: R. Tuan Humana Press Inc., Totowa, NJ

more time-consuming and yields are lower than for plasmids. Lambda clones also include a much larger proportion of vector DNA, and restriction mapping of λ clones is more difficult than with plasmids. DNA sequencing from λ clones is also more difficult than from plasmids or single-stranded (M13) phage clones. These advantages and disadvantages lead to a strategy of preparing and screening a library in a bacteriophage λ vector, and then subcloning the insert DNA from isolated clones into plasmids for further manipulation, such as DNA preparation, restriction mapping, and DNA sequencing.

2.2. Promoters

For expression of a foreign gene in λ, a suitable promoter needs to be provided upstream of the cloning site. The promoter needs to be regulable, since expression of a foreign gene may be toxic to *E. coli*, which would result in selection against the clone in a library. The promoter should be easily inducible, and a suitable translational initiation site should be provided. Two systems are commonly employed. The first expression vector, λgt11, employed a fusion with the *E. coli* β-galactosidase gene, encoded by *lacZ (4)*. There is a naturally occurring *Eco*RI site near the 3' end of the *lacZ* gene. Foreign DNA inserted into this site (if in frame, *see below*) will result in a fusion protein being produced when the *lacZ* promoter is induced. The lactose analog IPTG (isopropyl-β-D-thiogalactopyranoside) is a convenient and potent inducer of the *lacZ* promoter. The *lacI* gene, which encodes the *lac* repressor, must also be present in the host cell in order to repress the *lacZ* promoter when IPTG is not present. The natural translation start site of the *lacZ* gene drives the expression of the fusion gene. Another vector, λZAP *(5)*, uses a smaller fragment of the *lacZ* gene, called the *lacZα* fragment, for its insertion site *(see below)*. Again, the natural *lacZ* promoter and translation initiation signals drive expression of the fusion, and *lacI* must also be present for repression.

Another expression system, used in the λEXLX and λMOSElox vectors *(6)*, is derived from bacteriophage T7. The T7 late promoter is only recognized by the T7 RNA polymerase, and there is essentially no transcription from this promoter in a cell with only the endogenous *E. coli* RNA polymerase. A fragment of T7 gene 10 is included in the vector, and foreign DNA is ligated into the coding region of the gene. The T7 late promoter and the gene 10 translation initiation signals are used. Expression is controlled by controlling the expression of the T7 RNA polymerase gene.

2.3. Cloning Sites

A variety of restriction sites have been used as cloning sites in λ vectors. The vector, λgt11 *(4)*, uses a single *Eco*RI site in the *lacZ* gene for cloning.

Insert molecules must have *Eco*RI ends and may insert in either orientation. This means that with respect to the reading frame of the foreign DNA insert it can insert in two orientations, sense and antisense, and in three possible reading frames in each orientation. Only one of these six possibilities will result in production of an in-frame fusion to the *lacZ* gene. Other vectors use asymmetric insertion sites, such that one end of the insert has one restriction site and the other end has a different site. For example if a cloning vector has a polylinker that includes unique *Eco*RI and *Xho*I sites, the vector can be cut with both enzymes, and the small oligonucleotide between those sites is removed when the λ arms are purified. The cDNA preparation must then have *Eco*RI sites at one end and *Xho*I sites at the other end. This can be done by using primers for the first and second strand synthesis steps with *Eco*RI and *Xho*I linkers incorporated into them. This directional cloning procedure eliminates the antisense orientation of inserts, with a twofold reduction in the number of clones that need to be screened. A further advantage in this strategy is to use recognition sites for restriction enzymes that are rare-cutters, for example *Not*I or *Sfi*I, which recognize eight base-pair sites and are thus unlikely to cut within the insert. This eliminates the need to protect the internal sites within the cDNA molecules when the linkers are cut before ligation to the vector. A wide variety of the available vectors use this strategy, including λZAP *(5)*, and derivatives of λgt11 such as λgt22, λgt23 *(7)*, and λgt11 Not-Sfi (Promega, Madison, WI).

2.4. Selection of Recombinant Clones

All practical λ expression vectors are insertion vectors rather than replacement vectors. These vectors typically allow up to 8–10 kb cDNA inserts. Since the vector itself is packageable, one of the first considerations in making a library is to distinguish those phages that are merely religated vector arms from those that have incorporated a piece of insert DNA. This can be done in a variety of ways. One way is to insert the foreign DNA into a screenable gene such as the *E. coli lacZ* gene. Phage with an insert are *lacZ*$^-$, whereas the religated vector phage are *lacZ*$^+$, and these can be distinguished by use of the chromogenic substrate for β-galactosidase, X-gal, which turns blue in the presence of the enzyme (X-gal does not induce *lacZ* transcription, so IPTG must also be included). The vectors λZAP and λgt11 (and its derivatives) employ this strategy.

The λZAP vector actually takes advantage of the phenomenon known as α-complementation for *lacZ* expression *(8)*. The vector only includes a fragment of the *lacZ* gene, encompassing the operator, promoter, and the first 190 amino acids of β-galactosidase. This α fragment of β-galactosidase does not have enzymatic activity. The host cell must carry a deletion derivative of the *lacZ* gene, known as *lacZΔM15*, which is a deletion of amino acids 11–41 of the β-galactosidase protein. This fragment is known as the ω fragment and also lacks enzymatic activity. When the α and ω fragments are expressed in the same cell, the two fragments will associate to form an active enzyme. The combination of the vector plus the host will then be *lacZ*$^+$, and will give a blue color if X-gal is included in the plates. If the α peptide in the vector is disrupted by an insertion of foreign DNA, the cells will be *lacZ*$^-$, and will not form blue plaques in the presence of X-gal.

Another strategy involves inserting the foreign DNA into the *cI* gene of λ. The *cI* gene encodes the λ repressor, which is required for lysogenic infection. When plated on an *hfl*$^-$ (high frequency of lysogenization) host, the vector, being *cI*$^+$, forms lysogenic colonies whereas the phage with inserts are *cI*$^-$ and form plaques. When the lytic phage are harvested from such a plate, the phage with inserts are selected. Although not an expression vector, λgt10 *(9)* is a widely used cDNA cloning vector that employs this strategy. The expression vectors λZd39 *(10)* and λPOP6 (Invitrogen, San Diego, CA) also utilize the *cI* gene for selection of inserted DNA.

The third strategy for discriminating against religated vector sequences when making a λ library involves making such ligation events extremely unlikely. One method is to phosphatase treat the vector arms, thus decreasing their likelihood of ligating to one another. A more efficient method is to use an asymmetric cloning site in the vector, as described above. When the inserts are ligated to the vector, the only productive ligations are those that join one insert molecule to one λ left arm and one λ right arm. The two arms will not ligate because the overhanging single-stranded ends left by the two enzymes are not complementary. This procedure has the further advantage of forcing the inserts to always ligate to the vector in one particular orientation, as discussed above. Chimeras with two different cDNA inserts ligated into one vector are also extremely rare when asymmetric cloning sites are used. A variety of vectors, including λZAP *(5)*, λEXLX *(6)*, and some derivatives of λgt11, such as λgt22 and λgt23 *(7)* use this strategy.

2.5. Automatic Subcloning

As previously mentioned, λ is an advantageous vector for use in libraries, however, once a particular clone is isolated, λ has disadvantages compared to plasmids. Subcloning into a plasmid is typically done in order to simplify restriction mapping, DNA sequencing, and large scale preps. Several λ vectors have been engineered to make this process simpler by a scheme referred to as "automatic subcloning." In this strategy, plasmid sequences have been included in the λ vector, and following a suitable manipulation, the plasmid "pops out" of the λ vector, and can be propagated as a plasmid in *E. coli*. One strategy, employed in the vectors λEXLX *(6)*, λMOSElox (Amersham, Arlington

Heights, IL), and λPOP6 (Invitrogen, San Diego, CA), is site-specific recombination. The *loxP* site from bacteriophage P1 is a 34 bp sequence that is recognized by the P1 recombinase, Cre. If the plasmid sequences are flanked by *loxP* sites, site-specific recombination will occur after induction of the Cre protein (or transfer to a host expressing Cre), resulting in a circular DNA molecule being produced. If this molecule has an antibiotic resistance gene and an origin of replication, it can now replicate as a plasmid, thus eliminating the need for subcloning. The other strategy for automatic subcloning is used by λZAP *(5)*. This vector incorporates the Bluescript phagemid into λ. Bluescript is a derivative of M13, the single-stranded DNA phage. When an *E. coli* cell is co-infected by the λZAP clone and an M13 helper phage, the phage origin of replication in Bluescript is activated, and that segment of the λ clone replicates as a circular molecule. Since Bluescript also has all the M13 genes required for single-stranded phage production, these subclones can be used to produce single-stranded DNA for sequencing.

2.6. Eukaryotic Shuttle Sequences

If the λ clone will be used for transformation of eukaryotic cells, it is useful to have appropriate origins of replication and selectable markers for the host cell. The λPOP6 vector (*10*, Invitrogen) has incorporated an origin of replication from the Epstein-Barr virus, and a hygromycin resistance gene, so that after automatic subcloning by the cre/lox system *(see above)*, the plasmid can be used for cell transfections.

3. Expression Features of λ Vectors

3.1. Phage or Plasmids?

A problem with using λ to express recombinant proteins to high levels is that most λ cloning vectors are propagated as lytic phages. This means that after infection of an *E. coli* cell there are only about 50 min during which the foreign protein can be made before the host cell is lysed. Not only does this potentially reduce the yield, it also means that the expressed protein is found in a crude cell lysate, which can cause difficulties in purification. Plasmid-based expression systems will generally be more useful for production and purification of a recombinant protein. However, the λ vectors that include an automatic subcloning feature can be easily and rapidly switched into a plasmid mode of growth. An alternative is λgt11, which can form lysogens for high level production of the fusion protein. A problem with a lysogen versus a plasmid is that the λ lysogen has only one copy of the foreign gene per cell, where a plasmid may have hundreds of copies per cell. Fusion proteins are usually made from λgt11 lysogens by inducing the prophage and the *lacZ* promoter simultaneously *(4,9)*. This means that the cells enter the lytic phase of infection, which increases the copy number, but production of the fusion protein ends with cell lysis again.

3.2. Promoter Induction

The two promoters most often used in λ expression systems are the *E. coli* *lacZ* and bacteriophage T7 gene 10 promoters. The *lacZ* promoter may be leaky, particularly if the host cell has a single copy *lacI* gene and the *lacZ* promoter is multicopy. The *lacI*^{q} mutation overexpresses the *lac* repressor, thus maintaining more efficient repression of the *lacZ* promoter. The *lacI* gene can also be put on a multicopy plasmid to decrease the background level of expression from *lacZ*. The *lacZ* gene is typically derepressed by addition of IPTG. Since IPTG is not hydrolyzed by β-galactosidase, derepression is continuous once the IPTG has been added. For screening libraries, IPTG is often incorporated into the plate, into a top agar layer, or impregnated onto a membrane filter that is laid over the plate. For fusion protein production, IPTG is generally added to liquid cultures.

The T7 late promoter used in the gene 10 fusions is only expressed if the T7 RNA polymerase is expressed in the cells. This is generally done by transferring the clone to a host that has the T7 RNA polymerase gene under the control of the *lacZ* promoter. The T7 RNA polymerase gene is then induced by IPTG as above. Although this seems redundant, the advantage is that the T7 gene 10 promoter is completely silent in a cell lacking the T7 RNA polymerase. If the fusion is toxic, there will be no expression in such a host, so the clone will survive. The other advantage is that when IPTG induces the T7 RNA polymerase, each molecule will then transcribe the target gene repeatedly, resulting in very high levels of expression.

Expression of genes in eukaryotic hosts requires a specific promoter for the host cell being used. Generally this is done using plasmid constructs, perhaps after screening for the gene using a λ library, and using automatic subcloning to generate a plasmid.

3.3. Fusion Proteins

An important consideration in expression vectors is that the gene will be expressed as a fusion with β-galactosidase or T7 gene 10. In λgt11 and its derivatives, the insertion site is very near the 3' end of the *lacZ* gene. This produces a very large fusion protein with approx 100 kDa of β-galactosidase fused to the foreign protein. An advantage to this is that the fusion can usually be purified on a β-galactosidase affinity column, and these fusions tend to be stable in *E. coli*, but the foreign protein expressed as part of such a fusion will only rarely have its native activity, and immunization with such a fusion will result in many anti-β-galactosidase antibodies being produced. It is also more difficult to accurately determine the sizes of the foreign portion of the fusion proteins on Western blots, since the differences in size are proportionately small.

In the case of λZAP, the insertion site is near the 5' end of the *lacZ* α fragment. This results in a fusion of 20–50 amino acids to the N-terminus of the foreign protein (depending on which insertion site in the polylinker is used). Fusion proteins with this small fragment of *lacZ* may have the native activity, and are likely to be useful antigens. Their stability may be lower, however, and they can not be purified by β-galactosidase affinity chromatography.

The T7 promoter vectors λEXLX and λMOSElox produce fusions of 260 amino acids from T7 gene 10 with the foreign gene. The same general comments apply to these as to the λZAP fusions.

Another vector, λfoo, has been designed to express foreign proteins as fusions to a λ tail protein *(11)*. It has been suggested that this vector may be useful in expressing large multimeric proteins. There is also a protease sensitive linker incorporated into the cloning site that may be used to cleave the foreign protein from the phage tail protein.

4. Recommendations

Many libraries have been made in λgt11 for historical reasons. It was the first general purpose λ expression vector, and is very useful for screening with antibodies. However for construction of a new library, the improvements in vector design have resulted in greater efficiency of expressing a fusion protein and greater ease in further manipulations and subcloning. The most important consideration is the ability to do directional cloning using an asymmetric cloning site. This cuts the number of phages that must be screened in half, and allows the investigator to subclone the insert in a directional way in another vector. The newer vectors also have reduced the number of recognition sites for commonly used restriction sites in the λ arms, and incorporated new unique restriction sites into the polylinker flanking the insert. This makes subcloning and restriction mapping significantly easier. There are some derivatives of λgt11 that have these features.

The automatic subcloning vectors provide another significant advantage. The λZAP clones can be readily converted to Bluescript clones, which means they can be used for isolating either double- or single-stranded DNA, giving this vector great versatility. The Cre/lox vectors λEXLX and λMOSElox provide an easier way to convert the phage into a plasmid, and also use the T7 late promoter system for expression. If the fusion gene is likely to be toxic, the regulation provided by the T7 system is tighter and may be advantageous.

Many cDNA libraries in these new vectors are available both from individual investigators and commercially. If the primary objective is to screen a library with an antibody or ligand, and then sequence the clone, λZAP has the advantage. If production of the protein in *E. coli* is the objective, then the T7 vectors may give better control of expression. Finally, if a clone is already in hand, but the protein's size or properties make expression difficult, λfoo may provide an alternative method for expressing the fusion protein.

References

1. Murrray, N. E. (1983) Lambda vectors, in *Lambda II* (Hendrix, R. W., Roberts, J. W., Stahl, F. W., and Weisberg, R. A., eds.), Cold Spring Harbor Laboratory, Cold Spring Harbor, NY, pp. 677–684.
2. Chauthaiwale, V. M., Therwath, A., and Deshpande, V. V. (1992) Bacteriophage Lambda as a cloning vector. *Microbiol. Rev.* **56,** 577–591.
3. Sambrook, J., Fritsch, E. F., and Maniatis, T. (1989) *Molecular Cloning: A Laboratory Manual*, Cold Spring Harbor Laboratory Press, Cold Spring Harbor, NY.
4. Young, R. A. and Davis, R. W. (1983) Efficient isolation of genes by using antibody probes. *Proc. Natl. Acad. Sci. USA* **80,** 1194–1198.
5. Short, J. M., Fernandez, J. M., Sorge, J. A., and Huse, W. D. (1988) λZAP: a bacteriophage λ expression vector with *in vivo* excision properties. *Nucleic Acids Res.* **16,** 7583–7600.
6. Palazzolo, M. J., Hamilton, B. A., Ding, D., Martin, C. H., Mead, D. A., Mierendorf, R. C., Raghavan, K. V., Meyerowitz, E. M., and Lipshitz, H. D. (1990) Phage lambda cDNA cloning vectors for subtractive hybridization, fusion-protein synthesis and Cre-*loxP* automatic plasmid subcloning. *Gene* **88,** 25–36.
7. Han, J. H. and Rutter, W. J. (1987) Lambda gt22, an improved lambda vector for directional cloning of full length cDNA. *Nucleic Acids Res.* **15,** 6304.
8. Langley, K. E., Villarejo, M. R., Fowler, A. V., Zamenhof, P. J., and Zabin, I. (1975) Molecular basis of beta-galactosidase alpha-complementation. *Proc. Natl Acad. Sci. USA* **72,** 1254–1257.
9. Huynh, T. V., Young, R. A., and Davis, R. W. (1985) Construction and screening of cDNA libraries in lambda gt10 and lambda gt11, in *DNA Cloning: A Practical Approach* (Golver, D. M., ed.), IRL, Oxford, pp. 49–78.
10. Murphy, A. J. M. and Schimke, R. T. (1991) pλZd39: a new type of cDNA expression vector for low background, high efficiency directional cloning. *Nucleic Acids Res.* **19,** 3403–3408.
11. Maruyama, I. N., Maruyama, H. I., and Brenner, S. (1994) λfoo: A λ phage vector for the expression of foreign proteins. *Proc. Natl. Acad. Sci. USA* **91,** 8273–8277.

5

Strategies for Gene Fusions

Stefan Ståhl, Per-Åke Nygren, and Mathias Uhlén

1. Introduction

Recombinant DNA technology enabling genes or gene fragments to be spliced to form fusions has become an important objective in many fields of research. The fusion proteins, expressed from such operatively linked genes, often display the combined properties of the parent proteins and are used extensively in microbiology, molecular biology, immunology, and biotechnology *(1–3)*.

When splicing genes or gene fragments together to create bi- or multifunctional fusion proteins it is advantageous if the three dimensional structures are known for the different protein entities. Fusions of gene fragments encoding discrete structural domains are more likely to fold independently and to retain the properties of the parent molecules. Knowledge of structural domain borders in order to perform "exon-shuffling-like" gene fusions is thus a rule of thumb, but not a guarantee for the successful generation of fusion proteins displaying the functions and properties of two or more proteins.

Underlying methodology to create gene fusions include a few basic techniques that have become standard laboratory procedures, such as: (a) enzymatic digestion of DNA with restriction endonucleases; (b) joining of complementary DNA ends using DNA ligase; (c) gene amplification by the polymerase chain reaction (PCR); and (d) synthesis of oligonucleotides to be used as PCR primers or for the assembly of synthetic genes. These techniques are described in detail by commercial manufacturers and in different manuals and handbooks *(4,5)*, whereas we will instead focus this chapter on some different fields of applications in which the use of fusion proteins is important.

From: *Methods in Molecular Biology, vol. 62: Recombinant Gene Expression Protocols*
Edited by: R. Tuan Humana Press Inc., Totowa, NJ

2. Generation of Genes and Gene Fragments for Gene Fusions

Depending on whether it is a prokaryotic or eukaryotic gene that is to be cloned and used for a gene fusion, the starting material could be either chromosomal DNA or cDNA (generated from purified mRNA), respectively. The genes or gene fragments to be used in the fusion can be generated by different methods. If the nucleotide sequence for a certain gene is known, the entire gene or a fragment thereof can be subcloned using suitable restriction endonucleases. However, the gene has to be isolated and introduced into a cloning vehicle such as a plasmid or phage vector to make this feasible. The introduction of the PCR in the middle of the 1980s *(6)* dramatically improved the possibilities for rational design of gene fusions since a target sequence can be selectively amplified from minute amounts of template DNA, and restriction sites can be introduced at selected positions by the introduction of recognition sites in noncomplementary sequences in the 5'-end of the PCR primers used for the amplification. However, DNA sequencing is always required to verify a correct sequence when PCR is used for cloning of genes, due to the misincorporation frequency of the *Taq* polymerase.

Gene fusions can also be accomplished by using completely synthetic genes or gene fragments, which thus allow *de novo* design of the nucleotide sequence. There are several situations when *de novo* gene design and construction is preferable to conventional subcloning of a gene fragment from the natural source: (a) a peptide/protein with known amino acid sequence can be expressed without knowledge of the nucleotide sequence or access to the genetic source; (b) for a gene construct to be expressed in a heterologous host it is possible to optimize the codon usage; and (c) straight-forward cassette mutagenesis can be performed using sets of alternative or degenerated oligonucleotides encoding a region of particular interest. In addition, any new gene or gene fragment can be designed and assembled. In spite of the development of automated oligonucleotide synthesizers, it is still difficult to manufacture oligonucleotides longer than 100–150 nucleotides with high yield and low error frequency. A general method for *de novo* assembly of gene fragments on paramagnetic beads with covalently coupled streptavidin was recently described (Fig. 1) *(7–8)*. The gene assembly was initiated by immobilizing a 5'-biotinylated oligonucleotide to the paramagnetic beads and with the subsequent step-wise addition of complementary overlapping oligonucleotides extended duplex DNA sequences were constructed *(7)*. To increase the yield, the assembled gene fragments could be amplified by PCR before subcloning *(8)*. Using this technique, the cloning step creating the gene fusion can also be accomplished without the use of restriction enzymes or ligation reactions, since single stranded DNA eluted after completed assembly can be annealed to a single stranded plasmid vector.

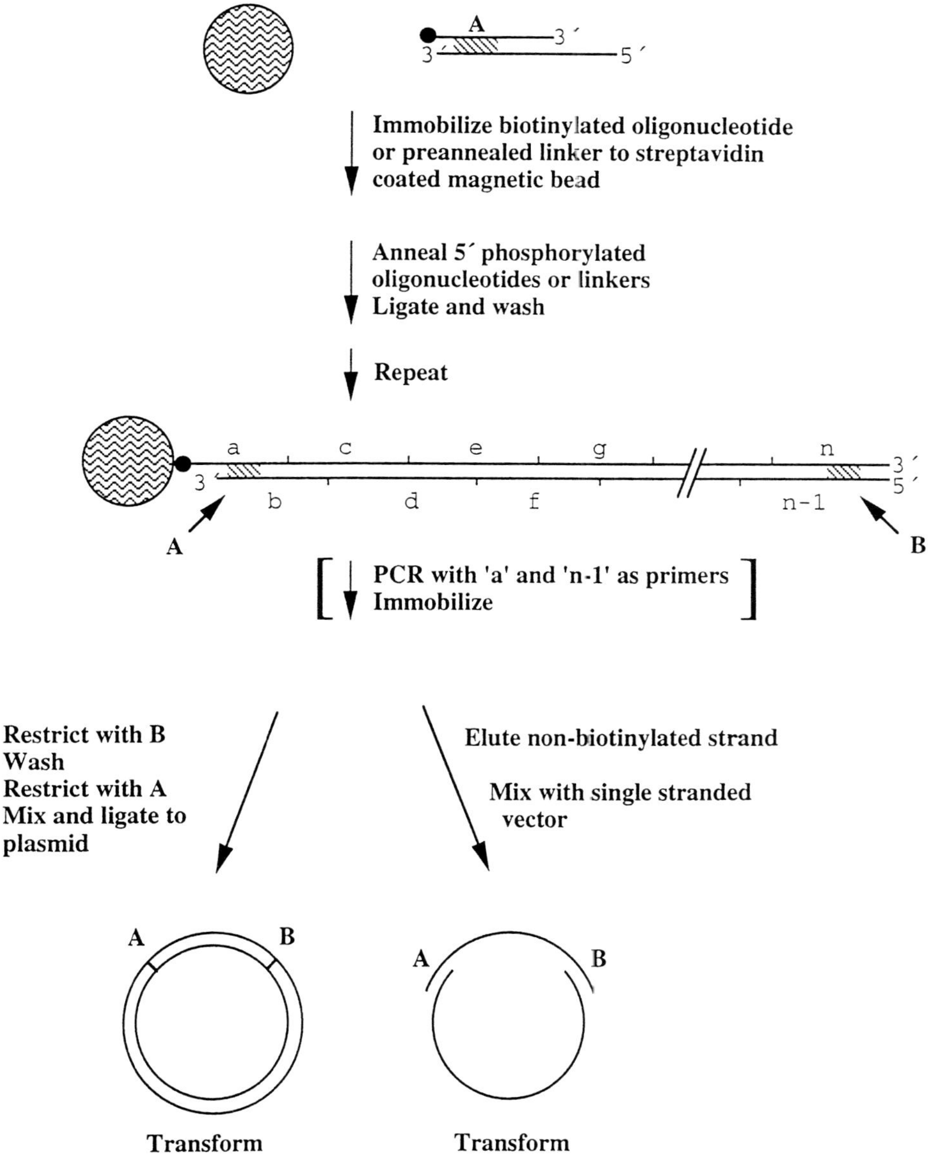

Fig. 1. Solid-phase gene assembly on paramagnetic beads with covalently coupled streptavidin. The gene construction is initiated from a 5'-biotinylated single-stranded oligonucleotide or a preannealed double-stranded linker fragment after immobilization to the magnetic beads. A PCR amplification step can be included if necessary to generate more material. The assembled gene constructs can be subcloned using conventional restriction and ligation, or alternatively by formation of a gap-duplex plasmid by elution of single-stranded construct and annealing to a previously prepared single-stranded vector.

The resulting duplex vector can thereafter be used directly to transform bacteria (Fig. 1) *(7,9)*. This strategy might not be significantly advantageous for the cloning of a single gene fragment, but could become of great importance in large-scale or routine cloning situations, e.g., for cDNA analysis in the ongoing genome projects.

2.1. Materials and Methods for Solid Phase Gene Assembly

1. A neodymium-iron-boron permanent magnet (Dynal AS, Oslo, Norway) is used to sediment the paramagnetic beads with covalently coupled streptavidin, Dynabeads M280-Streptavidin (Dynal AS, Oslo, Norway), during the washing procedures.
2. The gene assembly is initiated by washing 300 μg of beads in 60 μL washing/binding buffer (1*M* NaCl, 10 m*M* Tris-HCl, pH 7.5, 1 m*M* EDTA) before the addition of 0.2 pmol of biotinylated oligonucleotide or preannealed linker (Fig. 1) that is allowed to bind for 15 min at room temperature.
3. The beads with the immobilized oligonucleotide/linker are washed and 0.2 pmol of a 5'-phosphorylated oligonucleotide was added in 60 μL annealing/ligation buffer (50 m*M* Tris-HCl, pH 7.6, 10 m*M* $MgCl_2$, 1 m*M* ATP, 1 m*M* 1,4-dithiothreitol [DTT], 5% polyethylene glycol [PEG] 8000) to the dry beads.
4. The annealing mixture is incubated at 70°C for 5 min and allowed to cool to 37°C before adding 3 U T4 DNA ligase, followed by 15 min incubation at 37°C.
5. The reaction mixture is washed before adding 0.2 pmol of the subsequent oligonucleotide (5' phosphorylated). The annealing/ligation procedure can be repeated a selected number of times, adding fresh, 5' phosphorylated, complementary oligonucleotides each time.
6. After finished gene assembly of the DNA segment, the extreme 5' oligonucleotides in the assembled segments can be used as primers for PCR amplifications to generate more material for subcloning (Fig. 1), but this step is not always needed. The PCR-amplified fragments could be analyzed by agarose gel electrophoresis for a fragment of expected size.
7. After washing twice with 60 mL TE (10 m*M* Tris-HCl, pH 7.5, 1 m*M* EDTA) the paramagnetic beads carrying the assembled gene construct, the constructs should be digested with endonuclease B (in appropriate buffer). After washing twice with 60 mL TE the gene constructs should be restricted from the beads by incubation at 37°C for 30 min with enzyme A (in appropriate buffer). The gene constructs can be ethanol-precipitated or purified by any commercial DNA-fragment purification kit and ligated to a plasmid restricted with enzymes A and B.

   Alternatively, the nonbiotinylated strand of the completed gene constructs could be eluted with 40 mL 0.1*M* NaOH and neutralized by the addition of 5 mL 10X TE, pH 7.5 and 4 mL 1*M* HAc. The eluted and neutralized single-stranded gene fragment should carry at least 20 nucleotides, both upstream and downstream of the assembled region (Fig. 1), complementary to regions in the cloning plasmid, thus making it possible to form a transformable gap-duplex plasmid (Fig. 1) when mixed with single-stranded vector. The single-stranded vector could be provided by selective incorporation of biotin into one of the plasmid vector strands by enzymatic restriction, followed by a fill-in reaction using biotin-dUTP (Boehringer Mannheim, Mannheim, Germany) and Klenow DNA polymerase (Pharmacia, Stockholm, Sweden) as described earlier *(7,9)*. The double-stranded plasmid can be bound to streptavidin-coated magnetic beads and the non-biotinylated strand can be eluted and neutralized as described above and then mixed with the assembled single-stranded gene fragment. The mixture should be incubated at 70°C for 5 min and allowed to cool to room temperature to form a gap-duplex plasmid that thereafter can be used to transform *E. coli* cells.

3. Gene Fusions for Improved Heterologous Expression

The fusion of one gene to another is thought to be one of the ways in which the evolution has taken place. Recombinant DNA technology has allowed in vitro fusions of genes or gene fragments in a simple and predictable manner. A commonly used arrangement is where the gene encoding the protein of interest is fused to the 3' end of a gene fragment, which contains translation initiation signals for an efficiently expressed protein. This configuration is attractive because the signals that direct transcription, translation, and possibly secretion are located at the 5' end of the gene and will function in the presence of a variety of different fusions to the 3' end of the gene. There are several reasons for using gene fusions for the expression of recombinant proteins in heterologous hosts:

1. To localize the gene product to a certain cell compartment;
2. To stabilize the expressed fusion proteins;
3. To alter the solubility of a gene product;
4. To facilitate purification; and
5. To create multifunctional proteins *(1–3,10)*.

4. Localization of the Gene Fusion Product

A major consideration when setting up a production scheme for a recombinant protein, is whether the product should be expressed intracellularly or if a secretion system could be used to direct the protein out from the host cell. There is no generally applicable strategy that guarantees efficient expression. Instead, the inherent properties of the target protein determine which production route will be most successful. In addition, for certain applications it might be advantageous to target a heterologous protein to, and anchor it at, the outer cell surface of the bacteria by using different expression strategies.

4.1. Secretion

Fusion to a leader (signal) peptide can be used to direct the proteins out from the host cells for certain gene products. For *E. coli,* several systems have been described for the direction of the gene product to the culture medium *(11–16)* or the periplasm *(17–20)*. There are a number of advantages connected with the secretion strategy. First, the gene product will be less exposed to cytoplasmic proteases which might enable production of labile proteins *(21)*. Second, disulfide bond formation is enhanced in the oxidative environment outside the cytoplasm *(13,19)*. Third, the recovery of the recombinant protein is very much simplified since a large degree of the purification has been achieved through the secretion *(10,22)*.

4.2. Intracellular Production

There is no guarantee that a certain protein can be secreted even if an expression system with secretion signals is chosen for the production. Proteins with a strong tendency to precipitate intracellularly and proteins containing hydrophobic transmembrane regions have been demonstrated to be extremely difficult to secrete *(23)*. Intracellular expression of such proteins could be a more attractive alternative due to recent advances for in vitro renaturation of recombinant proteins from intracellular precipitates *(24)*. The mechanisms for inclusion body formation is not fully elucidated, but fusions to certain proteins, such as TrpE and to a lesser extent β-galactosidase, may result in inclusion body formation *(25)*. Expression from strong promoters such as the T7 promoter *(26)*, leading to very high expression levels, seems to increase the tendency for intracellular aggregation of the produced recombinant protein *(24)*. Production by the inclusion body strategy has the advantages that the recombinant product normally is protected from proteolysis and that it can be produced in large quantities. Levels of up to 50% of total cell protein content have been reported *(24)*. Gene fusion technology has recently been demonstrated to also greatly facilitate the recovery also of proteins with low solubility *(27; see below)*.

4.3. Surface Display on Bacteria

Surface display of heterologous proteins on bacterial cells is an important objective for many applications in microbiology, molecular biology and immunology. For gram-negative bacteria, different types of surface proteins have been exploited for this purpose. These include:

1. Outer membrane proteins, LamB *(28)*, PhoE *(29)* and OmpA *(30)*;
2. Lipoproteins, TraT *(31)*, the peptidoglycan-associated lipoprotein (PAL) *(32)* and pullulanase *(33)*;
3. Fimbriae protein, fimbrillin *(34)* and
4. Flagellae protein, flagellin *(35)*.

The different surface display systems have been used extensively to express heterologous antigenic determinants on the bacterial cells for the purpose of developing live bacterial vaccine vehicles. Enteric bacteria, such as *E. coli* and

Salmonella typhimurium, have been studied in this context and the cell surface presentation has been proven advantageous to induce an antibody response to the antigens using live cells for immunization *(36,37)*.

The expression of functional single-chain antibodies on the surface of *E. coli (32,38)* has opened the discussion whether this strategy would be an alternative to the rapidly developing phage display technology for the selection of peptides or recombinant antibody fragments from large libraries *(39,40)*. An interesting application for bacterial cells exposing heterologous proteins might be the development of whole-cell adsorbents by the surface expression of suitable protein ligands. Immobilized recombinant bacteria cannot be considered for separation processes in pharmaceutical industries, but might by their low production cost constitute competitive adsorbents for certain separations *(36,41)*. Genetically engineered bacteria as novel biocatalysts have also been envisioned, since enzymes with retained activity have been surface-displayed on *E. coli (42,43)*.

5. Stabilization of Gene Products During Production and for Therapeutic Applications

When expressing mammalian proteins in a bacterial host such as *E. coli*, proteolytic degradation might be a problem. This obviously reduces product yields, and may also cause problems in the separation of intact protein from partially degraded products. To overcome this, Hammarberg and coworkers *(44)* formulated a dual affinity fusion approach where the protein of interest was fused between two different affinity handle proteins. The concept was applied to the production of proteolytically-sensitive human insulin-like growth factor II (IGF-II) in *E. coli*. Besides the possibility of using two successive affinity chromatography steps, employing N- and C-terminal affinity handles, respectively, a dramatic stabilization was observed as compared with more conventional expression routes. Recently, the dual affinity fusion approach was also found to have a stabilizing effect on three other small and labile recombinant proteins, namely human proinsulin, a domain from rat protein disulphide isomerase, and a fragment from a human T-cell receptor *(21)* when expressed in *E. coli*.

Gene fusion technology might also be useful in prolonging the in vivo half-life of pharmaceutical proteins. Capon and coworkers *(45)* demonstrated that by fusing a soluble portion of CD4 (the target receptor for the human immunodeficiency virus particle) to the Fc part of IgG, the serum half-life of CD4 in rabbits increased 200-fold. Nygren and coworkers *(46)* described an alternative strategy where in vivo stabilization was achieved by fusion of the unstable protein to a bacterial receptor capable of binding to serum albumins. The serum half-lives of different fusion proteins containing the albumin-binding receptor were tested in mice. In these studies, the half-life of the CD4 fusion in vivo was shown to be comparable to the CD4-IgG molecule described above. Interestingly, the albumin-binding receptor itself was found in macaque monkeys to have a serum half-life similar to that of serum albumin. These examples suggest that gene fusion techniques might be useful ways to improve the in vivo stability of pharmaceutical protein drugs. It may also be possible to combine these approaches with new methods for drug targeting and/or slow drug release.

6. Solubilization to Improve In Vitro Refolding

The recovery of biologically active or native proteins by in vitro refolding from insoluble inclusion bodies is often hampered by the aggregation of the product during the procedure, leading to low overall yields. To make the procedure more efficient, several improved protocols have been developed, including the addition of different "folding enhancers" *(3)*. Alternatively, the protein itself can be engineered to facilitate the refolding. The presence of hydrophilic peptide extensions during the refolding can dramatically improve the folding yield, probably by conferring a higher overall solubility of the protein *(24)*. Samuelsson and coworkers *(47)* showed that the reshuffling of misfolded disulphides in recombinant insulin-like growth factor I (IGF-I) was greatly facilitated by fusion to the highly soluble ZZ fusion partner, derived from staphylococcal protein A *(48)*. Compared to unfused IGF-I, the fusion ZZ-IGF-I could be successfully refolded at a 100-fold higher concentration (1-2 mg/mL), without the formation of precipitates *(49)*.

7. Facilitated Purification by Fusions to Affinity Tags

More than a decade has passed since the first example of the use of gene fusions for affinity purification was reported *(50)*. Since then, a large number of affinity-fusion systems to facilitate the downstream protein purification have been developed which offer several alternative purification protocols *(1,3,51)*. However, several factors such as proteolytic stability, solubility, secretion competence, folding, and purification conditions have to be taken into consideration when choosing a suitable expression strategy for a specific protein. In particular, proteins produced as inclusion bodies must be refolded or at least solubilized prior to purification employing the tail features. Several of these tails have been commercialized and the corresponding expression vectors are manufactured in kits together with the affinity resin *(3)*. A selection of some of the more commonly used systems are shown in Table 1, and some of these will be thoroughly described later in this volume, in the section "Detection and Isolation of Expressed Proteins."

One of the most well characterized systems, staphylococcal protein A (SpA) and its derivative ZZ, has been used extensively in *E. coli (1,10,52)*, but has

Table 1
Examples of Commercially Available Affinity Purification Systems

Affinity tag	Size	Ligand	Elution conditions
ZZ	14 kDa	IgG	Low pH
His tail	6–10 amino acids	Ni^{2+}	Imidazole
Strep-tag	10 amino acids	Streptavidin	Iminobiotin
PinPoint	13 kDa[a]	Streptavidin[b]	Biotin
MBP	40 kDa	Amylose	Maltose
GST	25 kDa	Amylose	Maltose
Flag peptide	8 amino acids	Specific MAb	Low calcium

[a]Encodes a domain which is in vivo-biotinylated in *E. coli*.
[b]Monomeric.
Abbreviations: GST, glutathione-S-transferase; MAb, monoclonal antibody; MBP, maltose binding protein.

also been employed successfully in several different hosts such as gram-positive bacteria, yeast, CHO cells, and insect cells *(3)*. The low pH (pH 3) routinely used for elution, can be circumvented by competitive elution strategies based on engineered competitor proteins, which can be efficiently removed from the eluate mix *(53)*. The human polyclonal IgG used as ligand can be replaced by recombinant Fc fragments, avoiding the use of a human serum protein in the purification protocol. For detailed description of the use of protein A as affinity tag, *see* Chapter 10 in vol. 63.

Recently it was demonstrated that affinity purification can also be useful for the recovery of gene products with a strong tendency to precipitate. Murby and coworkers *(27)* used a fusion strategy for the *E. coli* production of various fragments of the fusion glycoprotein (F) from the human respiratory syncytial virus (RSV). Efficient recovery from inclusion bodies was achieved by affinity chromatography in the presence of a chaotropic agent (guanidine hydrochloride) throughout the purification process, which yielded substantial amounts of full-length proteins *(27)*.

Generally, a trend to design or select short tails can be observed, in particular since the affinity tag then can be genetically fused to the target gene by PCR techniques (Fig. 2). This trend might be further enhanced by modern phage display *(see below)* and other powerful selection techniques, which have the potential to provide a rich source of novel short tails suitable for purification and detection of recombinant proteins. For industrial production, economical aspects must be considered for systems based on proteinaceous ligands, due to reasons such as demands on possibilities for "cleaning-in-place" (CIP), in order to sanitize columns for repetitive use.

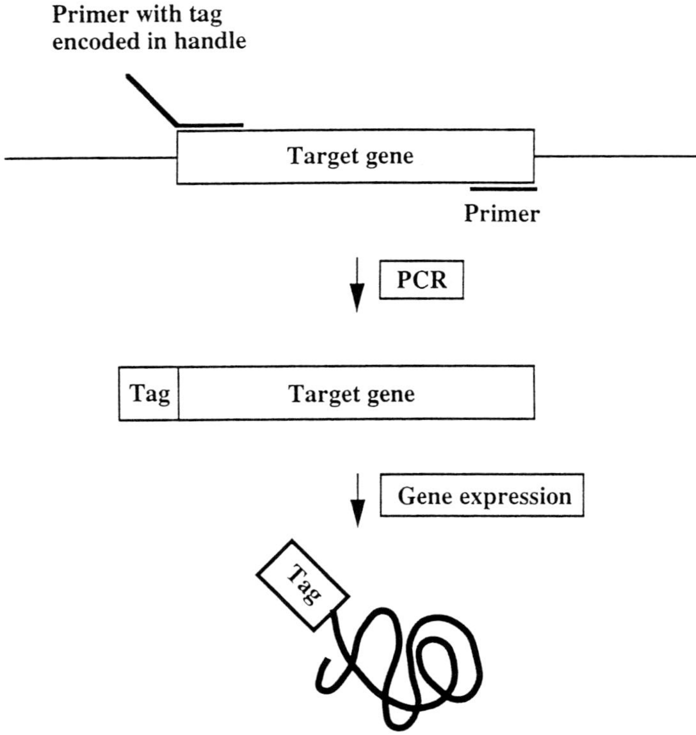

Fig. 2. An affinity tag can be fused to the target protein using PCR. The tag sequence is introduced via a noncomplementary handle sequence in one of the PCR primers used for the amplification. This PCR-based strategy for subcloning of a gene fragment to an expression vector makes it possible to recover the target protein upon expression using affinity chromatography.

8. Integration of Unit Operations in Downstream Processing

An important part of modern biotechnology is to develop simplified schemes for fermentation and downstream processing of recombinant proteins by the integration of unit operations. Genetic product design can be used in several ways to obtain such schemes, i.e., by influencing the yield and localization of the product as well as to adapt the gene product by fusion technology to specific unit operations suitable for large-scale downstream processing.

The isoelectric point (pI) of a protein can be altered by addition of charged amino acid residues, enabling the use of easily sanitized media for ion exchange chromatography in the separation process *(54)*. Recently, an expanded bed adsorption procedure was utilized for efficient recovery of a secreted recombinant fusion protein, directly from a crude fermentor broth without prior cell removal. The fusion protein was designed to have a relatively low pI to allow anionic exchange adsorption at pH 5.5 whereas most *E. coli* host proteins are not adsorbed. This strategy allowed integration of the cell separation step with ion exchange adsorption of the gene product with simultaneous volume reduction. An overall yield exceeding 90% was obtained including a subsequent affinity chromatography polishing step *(22)*.

An alternative strategy to simplify downstream processing was investigated by Köhler and coworkers *(55)*, where recombinant technology was applied to alter the partitioning properties of a gene fusion product which thereby could be efficiently recovered by aqueous two-phase extraction as the primary purification step. An attractive approach to simplify recovery of products, which may be produced with modified primary sequences, e.g., industrial enzymes, would be to alter the partitioning properties by substitution of one or more surface amino acid residues that are not involved in the biological activity, for tryptophans. This type of protein engineering will most probably be used in the future to reduce production costs for bulk proteins.

9. Fusion Proteins Used in Immunodetection Techniques

Various fusion proteins have been produced in *E. coli* to be used in different ELISA-like sandwich techniques for detection and/or purification of DNA fragments, protein antigens, or whole cells *(2)*. Taking advantage of the *E. coli* lac repressor (LacI), which binds to the *lac* operator sequence, a LacI-staphylococcal protein A (SpA) fusion protein was produced in *E. coli* for simplified analysis of PCR-amplified DNA fragments generated for diagnosis of pathogens *(56)*. In addition, this LacI-SpA fusion protein has been used for mild reversible recovery of protein antigens or whole cells (e.g. B-lymphocytes) using magnetic beads with a coupled DNA fragment containing the *lacO* recognition sequence *(57)*.

A competitive method was developed for quantification of PCR-amplified DNA *(58)*, involving the simultaneous amplification of a target and "competitor" DNA fragment to determine the initial amount of target DNA. The method, which takes advantage of an *E. coli* produced LacI-β-galactosidase fusion protein for colorimetric monitoring, has found use for quantification of different pathogens, such as malaria parasites *(58)* and hepatitis C virus *(59)*.

Recently, an antigen detection system, termed immuno-PCR was described *(60)*, combining in a microtiter format, the principles of immunodetection with the sensitivity of PCR. An *E. coli* produced streptavidin-SpA chimera was used to attach a biotinylated DNA fragment to antigen-IgG complexes immobilized on microtiter plate wells. Although the method gives a sensitivity greater than any existing antigen/antibody detection system, the sensitivity of the PCR assay might cause false positives resulting from nonspecific binding of IgG, fusion protein, or DNA.

10. Subunit Immunogens for Vaccine Research

Immunogenic proteins and peptides that are produced as fusion proteins have been frequently used for immunization purposes and represent a strategy to construct vaccines against infectious diseases. An important aspect regarding production of antigens by recombinant DNA technology is the purification of the gene product from contaminating host components. The approach to express peptides or proteins as one part of a fusion protein where the other part constitutes an affinity-handle allows rapid and efficient affinity purification of the gene product *(22)*.

A dual expression system has been developed, combining two similar affinity-fusion expression systems for the parallel expression of immunogenic peptides and proteins *(61)*. After affinity purification, one fusion protein may be used for immunization and the corresponding second fusion protein is used to analyze the induced antibody response to the fused peptide. This strategy eliminates any background originating from antibodies reactive with the affinity tag of the fusion used for immunization. In addition, the second fusion protein can, after immobilization, be used as a ligand for affinity purification of peptide-specific antibodies *(61)*. The dual expression system has been used to produce immunogenic structures derived from different *Plasmodium falciparum* malaria blood-stage antigens and the expressed fusion proteins have been shown to be immunogenic in mice, rabbits, and monkeys and to induce antigen-specific antibody and T-cell responses *(62)*. The influence of the affinity fusion tag on the immune responses to antigens expressed as fusion proteins has to be considered. It might, however, be possible to select affinity fusion partners with inherent positive immunomodulating effects, such as carrier function or adjuvant capacity *(62)*.

11. Phage Display Technology

A beautiful example of gene fusions making new technology possible is the display of peptides or proteins on filamentous phage surfaces, i.e., phage display (Fig. 3). Foreign proteins can be targeted to the phage (M13 or Fd phage) surface by genetic fusion to genes encoding phage coat proteins, pIII, or pVIII *(39)*. Phage particles decorated in such manner, can be enriched in vitro from a background of irrelevant phages by panning techniques (Fig. 3), employing an immobilized ligand capable of binding to the displayed protein. Moreover, the captured phage carries inside the protein coat the genetic information of the displayed protein, which allows later identification by DNA sequencing. Thus, the technology provides a link between genotype and phenotype and allows the selection of peptides or proteins of desired functions, expressed on the surface of filamentous phage. This technology, as pio-

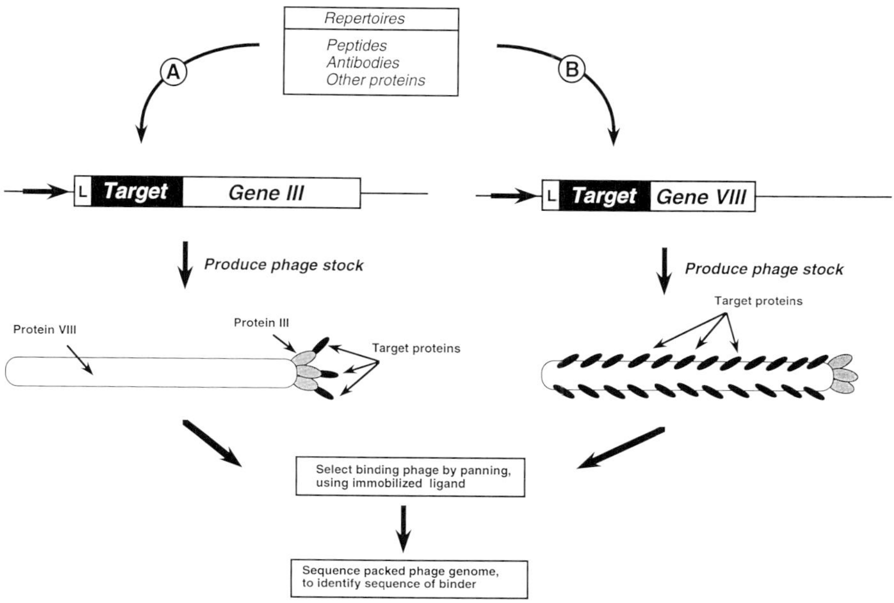

Fig. 3. Schematic description of the phage display technology. A mixture of DNA fragments encoding repertoires of, e.g., peptides, antibody fragments, or other proteins, are fused to genes encoding the phage coat protein III **(A)** or protein VIII **(B)**. Here, two different techniques can be utilized. If phage DNA is used as the cloning vector, all copies of the phage coat protein are produced as fusions to the target protein. Alternatively, if the repertoires are inserted into phagemid DNA, the phage stocks are prepared by the use of helper phage providing a source of the corresponding wild-type coat proteins. Thus, the resulting phages carry a mixture of wild-type and fusion proteins, resulting in a lower degree of decoration. Phage stocks prepared in either way are used in panning experiments, where phages displaying a variant of the repertoire capable of binding to an immobilized ligand are captured. After reinfection of cells, individual clones are formed. The amino acid sequence of a variant capable of binding to the ligand can be obtained by DNA sequencing of the packed phage genome. Thus, from large repertoires, rare clones can be selected and identified.

neered by Smith *(63)*, has emerged into a powerful technology to select, rather than screen, for rare phages capable of binding to a defined target molecule from libraries comprising up to 10^{10} phage species. Such phage libraries are obtained by the fusion to phage pIII or pVIII genes, cassettes encoding randomized peptides (usually 6–10-mers) *(64)*, antibody repertoires *(65–66)*, or multiple variants of a protein *(67–68)*. Thus, it is possible to rapidly and efficiently select high-affinity ligands or effector molecules that can be used for analytical or preparative applications.

12. Conclusions

Fusion proteins have become important tools in many fields of research, and several strategies for improved heterologous expression are based on gene fusion technology. In this chapter, we have focused on some current trends in certain fields of applications for fusion proteins. Fusion proteins will surely find extensive use in novel applications in medicine and biotechnology in the near future.

References

1. Uhlén, M., Forsberg, G., Moks, T., Hartmanis, M., and Nilsson, B. (1992) Fusion proteins in biotechnology. *Curr. Opin. Biotechnol.* **3,** 363–369.
2. Ståhl, S., Nygren, P.-Å., Sjölander, A., and Uhlén, M. (1993) Engineered bacterial receptors in immunology. *Curr. Opin. Immunol.* **5,** 272–277.
3. Nygren, P.-Å., Ståhl, S., and Uhlén, M. (1994) Engineering proteins to facilitate bioprocessing. *Trends Biotechnol.* **12,** 184–188.
4. Sambrook, J., Fritsch, E. F., and Maniatis, T. (1988) *Molecular Cloning: A Laboratory Manual, 2nd ed.*, Cold Spring Harbor Laboratory, Cold Spring Harbor, NY.
5. Wang, A. M. and Mark, D. (1990) *PCR Protocols: A Guide to Methods and Applications* (Innis, M. A., Gelfand, D. H., Sninsky, J. J., and White, T. J., eds.), Academic, San Diego, CA.
6. Mullis, K. and Faloona, F. (1987) Specific synthesis of DNA *in vitro* via a polymerase-catalyzed chain reaction. *Methods Enzymol.* **155,** 335–350.
7. Ståhl, S., Hansson, M., Ahlborg, N., Nguyen, T. N., Liljeqvist, S., Lundeberg, J., and Uhlén, M. (1993) Solid-phase gene assembly of constructs derived from the *Plasmodium falciparum* malaria blood-stage antigen Ag332. *BioTechniques* **14,** 424–434.
8. Nguyen, T. N., Uhlén, M., and Ståhl, S. (1994) *De novo* gene assembly using a paramagnetic solid support, in *Advances in Biomagnetic Separation* (Uhlén, M., Hornes, E., and Olsvik, Ø., eds.), Eaton, Natick, MA, pp. 73–78.
9. Hornes, E., Hultman, T., Moks, T., and Uhlén, M. (1990) Direct cloning of the human genomic apolipoprotein E gene using magnetic separation of single-stranded DNA. *BioTechniques* **9,** 730–737.
10. Uhlén, M. and Moks, T. (1990) Gene fusions for purpose of expression: an introduction. *Methods Enzymol.* **185,** 129–143.
11. Abrahmsén, L., Moks, T., Nilsson, B., and Uhlén, M. (1986) Secretion of heterologous gene products to the culture medium of *Escherichia coli. Nucleic Acids Res.* **14,** 7487–7500.
12. Suominen, I., Karp, M., Lähde, M., Kopio, A., Glumoff, T., Meyer, P., and Mäntsälä, P. (1987) Extracellular production of cloned a-amylase by *Escherichia coli. Gene* **61,** 165–176.
13. Moks, T., Abrahmsén, L., Österlöf, B., Josephson, S., Östling, M., Enfors, S.-O., Persson, I., Nilsson, B., and Uhlén, M. (1987) Large-scale affinity purification of human insulin-like growth factor I from culture medium of *Escherichia coli. Bio/Technol.* **5,** 379–382.

14. Kitai, K., Kudo, T., Nakamura, S., Masegi, T., Ichikawa, Y., and Horikoshi, K. (1988) Extracellular production of human immunoglobulin G Fc region (hIgG-Fc) by *Escherichia coli*. *Appl. Microbiol. Biotechnol.* **28,** 52–56.
15. Obukowicz, M. G., Turner, M. A., Wong, E. Y., and Tacon, W. C. (1988) Secretion and export of IGF-1 in *Escherichia coli* strain JM101. *Mol. Gen. Genet.* **215,** 19–25.
16. Thomas, W. D., Jr., Wagner, S. P., and Welch, R. A. (1992) A heterologous membrane protein domain fused to the C-terminal domain of HlyB can export *Escherichia coli* hemolysin. *J. Bacteriol.* **174,** 771–779.
17. Nilsson, B., Abrahmsén, L., and Uhlén, M. (1985) Immobilization and purification of enzymes with staphylococcal protein A gene fusion vectors. *EMBO J.* **4,** 1075–1080.
18. Ward, E. S., Güssow, D., Griffiths, A. D., Jones, P. T., and Winter, G. (1989) Binding activities of a repertoire of single immunoglobulin variable domains secreted from *Escherichia coli*. *Nature* **341,** 544–546.
19. Carter, P., Kelley, R. F., Rodrigues, M. L., Snedecor, B., Covarrubias, M., Velligan, M. D., Wong, W. L. T., Rowland, A. M., Kotts, C. E., Carver, M. E., Yang, M., Bourell, J.H., Shepard, H. M., and Henner, D. (1992) High level *Escherichia coli* expression and production of a bivalent humanized antibody fragment. *Bio/Technol.* **10,** 163–167.
20. Martineau, P., Charbit, A., Leclerc, C., Werts, C., O'Callaghan, D., and Hofnung, M. (1991) A genetic system to elicit and monitor antipeptide antibodies without peptide synthesis. *Bio/Technol.* **9,** 170–172.
21. Murby, M., Cedergren, L., Nilsson, J., Nygren, P.-Å., Hammarberg, B., Nilsson, B., Enfors, S.-O., and Uhlén, M. (1991) Stabilization of recombinant proteins from proteolytic degradation in *Escherichia coli* using a dual affinity fusion strategy. *Biotechnol. Appl. Biochem.* **14,** 336–346.
22. Hansson, M., Ståhl, S., Hjorth, R., Uhlén, M., and Moks, T. (1994) Single-step recovery of a secreted recombinant protein by expanded bed adsorption. *Bio/Technol.* **12,** 285–288.
23. Sjölander, A., Ståhl, S., Lövgren, K., Hansson, M., Cavelier, L., Walles, A., Helmby, H., Wåhlin, B., Morein, B., Uhlén, M., Berzins, K., Perlmann, P., and Wahlgren, M. (1993) *Plasmodium falciparum:* The immune response in rabbits to the clustered asparagine-rich protein (CARP) after immunization in Freund's adjuvant or immunostimulating complexes (ISCOMs). *Exp. Parasitol.* **76,** 134–145.
24. Rudolph, R. (1995) Successful protein folding on an industrial scale, in *Principles and Practice of Protein Folding* (Cleland,J. L. and Craik, C. S., eds.), Wiley, New York, pp. 283–298.
25. Ford, C. F., Souminen, I., and Glatz, C. E. (1991) Fusion tails for the recovery and purification of recombinant proteins. *Protein Expr. Purif.* **2,** 95–107.
26. Studier, F. W., Rosenberg, A. H., Dunn, J. J., and Dubendorff, J. W. (1990) Use of T7 RNA polymerase to direct expression of cloned genes. *Methods Enzymol.* **185,** 60–89.
27. Murby, M., Nguyen, T. N., Binz, H., Uhlén, M., and Ståhl, S. (1994) Production and recovery of recombinant proteins of low solubility, in *Separations for Biotechnology 3* (Pyle, D. L., ed.), Bookcraft, Bath, UK, pp. 336–344.

56. Lundeberg, J., Wahlberg, J., and Uhlén, M. (1990) Affinity purification of specific DNA fragments using a lac repressor fusion protein. *Gene Anal. Techn.* **7,** 47–53.
57. Ljungquist, C., Lundeberg, J., Rasmussen, A.-M., Hornes, E., and Uhlén, M. (1993) Immobilization and recovery of fusion proteins and B-lymphocyte cells using magnetic separation. *DNA Cell Biol.* **12,** 191–197.
58. Lundeberg, J., Wahlberg, J., and Uhlén, M. (1991) Rapid colorimetric quantification of PCR-amplified DNA. *BioTechniques* **10,** 68–75.
59. Yun, Z., Lundeberg, J., Johansson, B., Hedrum, A., Weiland, O., Uhlén, M., and Sönnerborg, A. (1994) Colorimetric detection of competitive PCR products for quantification of hepatitis C viremia. *J. Virol. Meth.* **47,** 1–14.
60. Sano, T., Smith, C., and Cantor, C. R. (1992) Immuno-PCR: very sensitive antigen detection by means of specific antibody-DNA conjugates. *Science* **258,** 120–122.
61. Ståhl, S., Sjölander, A., Nygren, P.-Å., Berzins, K., Perlmann, P., and Uhlén, M. (1989) A dual expression system for the generation, analysis and purification of antibodies to a repeated sequence of the *Plasmodium falciparum* malaria antigen Pf155/RESA. *J. Immunol. Meth.* **124,** 43–52.
62. Sjölander, A., Ståhl, S., and Perlmann, P. (1993) Bacterial expression systems based on protein A and protein G designed for the production of immunogens: applications to *Plasmodium falciparum* malaria antigens. *Immunomethods* **2,** 79–92.
63. Smith, G. P. (1985) Filamentous fusion phage: novel expression vectors that display cloned antigens on the virion surface. *Science* **228,** 1315–1317.
64. Scott, J. K. and Smith, G.P. (1990) Searching for peptide ligands with an epitope library. *Science* **249,** 386–390.
65. Marks, J. D., Hoogenboom, H. R., Bonnert, T. P., McCafferty, J., Griffiths, A. D., and Winter, G. (1991) By-passing immunization: Human antibodies from V-gene libraries displayed on phage. *J. Mol. Biol.* **222,** 581–597.
66. Nissim, A., Hoogenboom, H. R., Tomlinson, I. M., Flynn, G., Midgley, C., Lane, D., and Winter, G. (1994) Antibody fragments from a 'single pot' phage display library as immunochemical reagents. *EMBO J.* **13,** 692–698.
67. Lowman, H. B. and Wells, J. A. (1993) Affinity maturation of human growth hormone by monovalent phage display. *J. Mol. Biol.* **234,** 564–578.
68. Martin, F., Tonatti, C., Salvati, A. L., Venturini, S., Ciliberto, G., Cortese, R., and Solazzo, M. (1994) The affinity-selection of a minibody polypeptide inhibitor of human interleukin-6. *EMBO J.* **13,** 5303–5309.

6

Application of the *E. coli trp* Promoter

Daniel G. Yansura and Steven H. Bass

1. Introduction

The promoter of the *E. coli* tryptophan *(trp)* operon has proven to be a workhorse for the production of hundreds of proteins; from small scale to pharmaceutical production levels *(1)*. The *trp* promoter is strong, easily regulated, and well characterized. Transcription of a cloned gene from a *trp* promoter on a plasmid increases about 50-fold upon induction and the gene product can amount to 30% of the total cell protein *(2–4)*.

Transcription from the *trp* promoter is controlled by the level of free tryptophan in the cell. When grown in rich media, the dimeric Trp aporepressor is bound to two molecules of L-tryptophan forming an active repressor complex that competes with RNA polymerase for binding to the *trp* promoter/operator. The inducer 3-β-indoleacrylic acid (IAA) binds 30-fold more tightly than L-tryptophan to the Trp aporepressor, and forms an inactive Trp pseudorepressor that does not bind the *trp* operator *(5,6)*. This allows RNA polymerase to bind and transcribe the gene of interest.

The method described here includes the design of a *trp* promoter expression vector, the promoter induction in culture, and assessment of the expression level of one's protein of interest by means of SDS-polyacrylamide gel electrophoresis (SDS-PAGE). The design details of the expression vector can be critical to the levels of protein production achieved. Slight variations in the codons used at the amino terminal end of the coding sequence or large stretches of heterologous 3'-untranslated sequences, for example, can dramatically alter these levels. Therefore, it is best to incorporate as many of these details as early as possible to avoid excessive work in terms of plasmid construction. The actual steps involved in plasmid constructions are unique to the gene of interest, so they are not included in this chapter. Induction of the *trp* promoter in

From: *Methods in Molecular Biology, vol. 62: Recombinant Gene Expression Protocols*
Edited by: R. Tuan Humana Press Inc., Totowa, NJ

E. coli cultures with IAA is straightforward and a protocol for analyzing the results by SDS-PAGE is also provided. Normally the expression levels are high enough to see the protein of interest after staining with Coomassie blue. Other detection methods can be used such as those using antibodies against the protein of interest. Detection of enzymatic activity can also be used, although the possibility exists that one's protein may be present in insoluble inclusion bodies and require refolding *(7–9)*.

2. Materials

1. M9-casamino acids media for *trp* promoter inductions (per liter): 6 g Na_2HPO_4, 3 g KH_2PO_4, 0.5 g NaCl, 1 g NH_4Cl, adjust the pH to 7.4, autoclave, cool, then add 2 mL 1*M* $MgSO_4$, 10 mL 20% (w/v) glucose, and 36 mL casamino acids (135 g/L).
2. 3-β-indoleacrylic acid can be dissolved in ethanol at a convenient concentration of 25 mg/mL, and stored at 4°C.
3. TE: 10 m*M* Tris-HCl, pH 7.6, 1 m*M* EDTA.
4. The reducing agents dithiothreitol (DTT) and β-mercaptoethanol (BME) can be made into 1*M* solutions, and can be stored at 4°C for a few days. However, they will oxidize and become ineffective in reducing disulfide bonds. Freezing small aliquots or making fresh solutions is advisable.
5. For commercial SDS polyacrylamide gels, there is usually a sample buffer available. Alternatively, for lab cast gels, the following sample buffer may be used: 60 m*M* Tris-HCl, pH 6.8, 2% (w/v) SDS, 10% (v/v) glycerol, 0.025% (w/v) xylene cyanole, and 0.025% (w/v) bromophenol blue.
6. SDS polyacrylamide gels can be purchased from a number of commercial sources such as Novex (San Diego, CA) or prepared in the lab as described *(10)*.
7. Coomassie blue staining solution contains 25% (v/v) ethanol and 2.5 g Coomassie stain per liter. Acetic acid should be added to the staining solution prior to use (7% v/v). Destain contains 5% (v/v) acetic acid and 16.5% (v/v) methanol.
8. L-tryptophan is conveniently made into a 5 mg/mL solution and should be filter sterilized and stored at 4°C.
9. 10% (w/v) sodium dodecyl sulfate (SDS) is stored at room temperature.

3. Methods

3.1. Vector Design

The basic vector design involves placing the coding region of interest immediately downstream of the *E. coli trp* operon promoter and Shine-Dalgarno (SD) sequence in a pBR322-based plasmid. This is often, but not always sufficient to obtain high levels of protein. Some optimization may be required. The features and rationale for obtaining and maximizing expression are outlined below.

1. Plasmid pBR322 *(11)* is our preferred starting vector. It is more stable structurally and segregationally than higher copy number pUC-like plasmids.

CTGTTGACAATTAATCATCGA**ACTAGTTAACTAGTACGCAAGT**
-35 -10 +1
TCACGTAAAAAGGGTATCGACA**ATG**AAAGCAATTTTCGTACTG
SD M K A I F V L

Fig. 1. DNA sequence of the *E. coli trp* operon promoter region. The -35 and -10 regions are underlined and the Trp repressor binding site is in bold type. Bases complementary to the 3' end of the 16S rRNA (SD) are also underlined and the first seven amino acids of the Trp leader is shown.

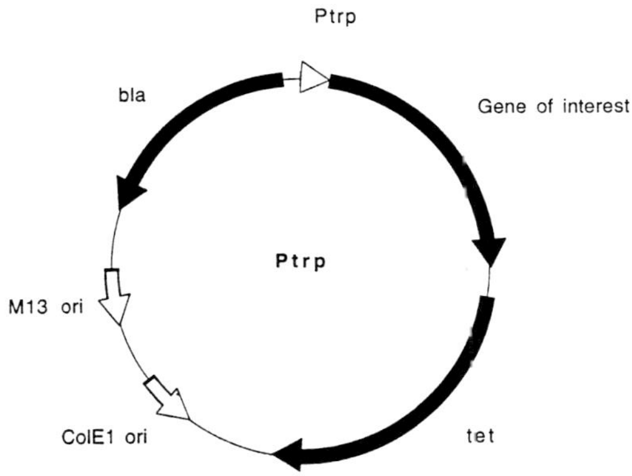

Fig. 2. Schematic view of a pBR322-based *trp* promoter expression plasmid.

2. The sequence of the *trp* promoter *(12)* is shown in Fig. 1. Vectors containing the trp promoter are available from many sources *(13–17)*, or can be easily cloned by PCR from the *E. coli* chromosome with convenient restriction sites added to match your gene of interest. We usually place the *trp* promoter at the EcoRI site of pBR322 and direct it toward the tetracycline resistance *(tet)* gene (Fig. 2).
3. The coding sequence should begin with an ATG and be placed 5–12 bp from the SD. Variations in this distance and its composition can have big effects on translation initiation rates *(18)*. The spacing in the natural *trp* operon is 9 bp, and this sequence and spacing is a good starting point. Avoid multiple restriction sites in this region.
4. The mRNA translation initiation region (TIR) must be devoid of secondary structures that obscure the SD or ATG to allow efficient formation of the initiation complex. Even weak structures can have big effects on translation *(19)* and are hard to predict, so again avoid multiple restriction sites in this area. The TIR includes the SD and extends ~15 bases into the coding region *(20,21)*, so its structure will vary with the gene of interest. The formation of an unfavorable secondary structure in the mRNA TIR is probably the main reason why some genes fail to be expressed even when placed behind a strong promoter and SD *(22)*. Strategies to overcome this potential problem will be discussed in the next section.

5. The 3' untranslated region of the mRNA can also affect protein production, although to a lesser extent than the TIR. We usually remove heterologous sequences 3' of the stop codon and anchor the 3' end in the *tet* gene. Different placements of the 3' end in the plasmid can affect production levels, so several may be tried. A transcription terminator can be included after the coding region to prevent transcription of vector sequences, but is not essential.
6. Addition of a filamentous phage (M13, f1) origin of replication to the vector is helpful to allow isolation of single-stranded plasmid DNA *(23)*. This will simplify DNA sequencing and mutagenesis, which is especially useful if structure/function studies are to be performed.

3.2. Optimizing Translation Initiation

Sometimes a protein fails to accumulate, even when these basic steps are followed. The protein of interest is either not being produced or it is made, then degraded by proteases. A pulse-chase experiment using labeled amino acids can differentiate between the two. The use of protease deficient strains can increase accumulation of unstable proteins (*see* Section 3.3.), but often the problem is due to inefficient translation initiation. As mentioned in the previous section, the TIR includes at least 15 bases of the gene of interest, so formation of an unfavorable secondary structure in this region may prevent efficient translation initiation. Changing the sequence of the TIR by either of the two general strategies described below can usually overcome this problem.

1. One method to overcome inefficient translation initiation is to fuse not only the SD, but the complete TIR of a highly translated *E. coli* gene such as *trpL* (Fig. 1; *24*) to the gene of interest. This will result in the production of a fusion protein where the N-terminal 6 or more amino acids are derived from the TIR. Many fusion partners have been described and most also include a means of purification. The fusion protein can be assayed directly or a cleavage site can be incorporated to separate the mature protein from the fusion partner. A variety of chemical and enzymatic methods have been used successfully for this purpose. The use of fusion proteins is the subject of several recent reviews and will not be discussed further in this chapter *(25–27)*.
2. The second method to improve the translation initiation rate of a cloned gene involves changing the initial DNA coding sequence. The correct protein sequence can be maintained by utilizing the degeneracy of the amino acid code. A library of plasmids is created using synthetic DNA with all possible codons for the first six amino acids, then induced cultures of single clones are screened by SDS-PAGE for production of the protein of interest. A method for direct selection of highly translated sequences from such a library is described by Yansura and Simmons *(28)*. Briefly, in this method, translation of the *tet* gene and the resulting tetracycline resistant phenotype is dependent upon translation of the gene of interest. Members of the codon library that grow with high levels of tetracycline are selected on plates, then induced cultures of these are screened for protein production by SDS-PAGE.

3.3. trp *Promoter Induction*

The expression plasmid should be freshly transformed into an appropriate host immediately before a *trp* induction is contemplated using any number of protocols (*10*; *see* Note 1). In theory, any *trpR*$^+$ strain is suitable, although it is a good idea to compare a number of strains, since expression levels can vary considerably. A number of protease deficient strains are available, which can improve protein yields. For intracellular protein production, strains deficient in the two major cytoplasmic proteases *lon* and *clpP*, as well as *rpoH* strains deficient in the heat shock response may be useful *(29)*. The outer membrane protease OmpT (protease VII) is known to cleave some proteins during the initial purification steps after cell lysis *(30–32)*. However, if one achieves high production levels, the protein of interest will generally form inclusion bodies and be resistant to further proteolysis.

The following procedure is designed for small scale inductions (2 mL) and provides enough material for several loadings of SDS gels. This size scale of induction is used primarily to determine the resulting expression level of a new protein before a scale up is undertaken.

1. Inoculate 3–5 mL of LB *(10)* containing the appropriate antibiotic with a freshly transformed culture of your expression plasmid (*see* Note 2). It is advisable to run a parallel negative control at this point, which is usually the parent plasmid pBR322 in the same strain.
2. Grow the LB culture at 37°C with shaking for a few hours until the optical density at 600 nm (OD_{600}) reaches approx 1–3.
3. Dilute 100 µL of the LB culture 20-fold into 2 mL of M9 media containing casamino acids and the appropriate antibiotic, and grow for an additional hour at 37°C.
4. Add 3-β-indoleacrylic acid to the M9 culture to a final concentration of 50 µg/mL and continue growing at 37°C for an additional 4 h (*see* Notes 3 and 4).
5. Measure the OD_{600} of the culture and spin down the equivalent of 1 mL of culture at an OD_{600} of one (1 OD_{600}-mL) in a microcentrifuge (16,000*g*). The cell pellets can be stored frozen at this point if desired.

3.4. Protein Detection by SDS-PAGE

In order to gauge the expression level of the protein of interest, it is usually necessary to analyze total cell lysates from induced cells by SDS-PAGE. If one runs a negative control cell lysate on the same gel, it usually becomes obvious as to which protein band is the expressed protein of interest. Overproduction of some proteins can induce a stress response in *E. coli*, so confirmation by other means such as an immunoblot or N-terminal sequencing is advisable. The procedure described here will enable one to obtain samples suitable for loading onto an SDS-PAGE starting with cell pellets obtained from the previous section.

1. Resuspend the 1 OD_{600}-mL cell pellets from the *trp* induction in 100 µL of TE.
2. Add 20 µL of 10% SDS and 10 µL 1*M* DTT or BME, and heat samples for 5 min at 90°C (*see* Note 5).
3. Cool slightly and add 1 mL of acetone. Vortex and let sit at room temperature for 15 min.
4. Spin in a microcentrifuge (16,000*g*) for 5 min, aspirate off the acetone, and dry the remaining pellet.
5. Resuspend the resulting pellet in 100 µL of SDS sample buffer. Generally, the sample buffers vary slightly depending on the SDS gel being used. However, one should include a reducing agent at this point such as 100 m*M* DTT or BME. Failure to do this may result in aggregation of the protein of interest, and poor detection on the SDS gel.
6. Heat the sample buffers to 90–100°C for 5 min, vortex well, and then spin in a microcentrifuge for 5 min. Remove the soluble fraction and transfer to a new tube.
7. The sample is now ready to load onto an SDS-PAGE. For most analytical gels, 5–10 µL is sufficient. Usually staining the gel with Coomassie blue will reveal the expression level of the protein of interest. Alternatively, one can transfer the protein bands to nitrocellulose and probe with an appropriate antibody.

4. Notes

1. Hosts transformed with the expression vector should always be kept on media containing sufficient tryptophan (such as LB) to keep the promoter turned off, except when doing an induction experiment. If it is necessary to grow cultures in a minimal media such as M9 for plasmid preps, one should supplement the media with tryptophan (50 µg/mL).
2. The host strain should be freshly transformed with the expression plasmid prior to an induction experiment and not stored on plates or passaged for any length of time.
3. Although 5 h growth in M9-casamino acid-IAA media is usually sufficient for most *trp* inductions, higher yields may be attained with a longer induction time (15–20 h). Analyzing a number of time points during the induction may be useful. Additionally, production may be improved by growth at 30°C rather than 37°C.
4. The inducer IAA works best when the cells are under mild tryptophan starvation. In a rich media with abundant tryptophan, full induction of the *trp* promoter will not occur *(13)*.
5. One can eliminate the reducing agent from the SDS sample buffer preparation if one is trying to preserve any intramolecular disulfide bonds that may have formed in the cell. In general, proteins expressed in the cytosol do not form disulfide bonds due to the reducing environment. Upon cell lysis these proteins often form highly polymerized aggregates. Proteins secreted into the periplasm however can and often do form at least some intramolecular disulfides, although they also can form highly polymerized aggregates depending on the protein of interest.

References

1. Yansura, D. G. and Henner, D. J. (1990) Use of *Escherichia coli trp* promoter for direct expression of proteins. *Methods Enzymol.* **185,** 54–60.
2. Olsen, M. K., Rockenbach, S. K., Curry, K. A., and Tomich, C. C. (1989) Enhancement of heterologous polypeptide expression by alterations in the ribosome-binding site sequence. *J. Biotechnol.* **9,** 179–190.
3. Hallewell, R. A. and Emtage, S. (1980) Plasmid vectors containing the tryptophan operon promoter suitable for efficient regulated expression of foreign genes. *Gene* **9,** 27–47.
4. Bogosian, G. and Somerville, R. L. (1984) Analysis in vivo of factors affecting the control of transcription initiation at promoters containing target sites for Trp repressor. *Mol. Gen. Genet.* **193,** 110–118.
5. Marmorstein, R. Q., Joachimiak, A., Sprinzl, M., and Sigler, P. B. (1987) The structural basis for the interaction between L-tryptophan and the *Escherichia coli* Trp aporepressor. *J. Biol. Chem.* **262,** 4922–4927.
6. Doolittle, W. F. and Yanofsky, C. (1968) Mutants of *Escherichia coli* with an altered tryptophanyl-transfer ribonucleic acid synthetase. *J. Bacteriol.* **95,** 1283–1294.
7. Williams, D. C., Van Frank, R. M., Muth, W. L., and Burnett, J. P. (1982) Cytoplasmic inclusion bodies in *Escherichia coli* producing biosynthetic human insulin proteins. *Science* **215,** 687–688.
8. Marston, A. O. (1986) The purification of eukaryotic polypeptides synthesized in *Escherichia coli*. *Biochem. J.* **240,** 1–12.
9. Kohno, T., Carmichael, D. F., Sommer, A., and Thompson, R. C. (1990) Refolding of recombinant proteins. *Methods Enzymol.* **185**, 187–195.
10. Sambrook, J., Fritsch, E. F., and Maniatis, T. (1989) *Molecular Cloning: A Laboratory Manual.* Cold Spring Harbor Laboratory, Cold Spring Harbor, NY.
11. Bolivar, F., Rodriguez, R. L., Green, P. J., Betlach, M. C., Heyneker, H. L., Boyer, H. W., Crosa, J. H., and Falkow, S. (1977) Construction and characterization of new cloning vehicles. II. A multi-purpose cloning system. *Gene* **2,** 95–113.
12. Yanofsky, C., Platt, T., Crawford, I. P., Nichols, B. P., Christie, G. E., Horowitz, H., VanCleemput, M., and Wu, A. M. (1981) The complete nucleotide sequence of the tryptophan operon of *Escherichia coli*. *Nucleic Acids Res.* **9,** 6647–6668.
13. Nichols, B. P. and Yanofsky, C. (1983) Plasmids containing the *trp* promoters of *Escherichia coli* and *Serratia marcescens* and their use in expressing cloned genes. *Methods Enzymol.* **101,** 155–164.
14. De Boer, H. A., Comstock, L. J., Yansura, D. G., and Heyneker, H. L. (1982) Construction of a tandem *trp-lac* promoter and a hybrid *trp-lac* promoter for efficient and controlled expression of the human growth hormone gene in *Escherichia coli*, in *Promoters: Structure and Function* (Rodrigues, R. L. and Chamberlin, M. J., eds.), Praeger, New York, pp. 462–481.
15. Windass, J. D., Newton, C. R., De Maeyer-Guignard, J., Moore, V. E., Markham, A. F., and Edge, M. D. (1982) The construction of a synthetic *Escherichia coli trp* promoter and its use in the expression of a synthetic interferon gene. *Nucleic Acids Res.* **10,** 6639–6657.

16. Tacon, W. C. A., Bonass, W. A., Jenkins, B., and Emtage, J. S. (1983) Expression plasmid vectors containing *Escherichia coli* tryptophan promoter transcriptional units lacking the attenuator. *Gene* **23,** 255–265.
17. Nishi, T., Sato, M., Saito, A., Itoh, S., Takaoka, C., and Taniguchi, T. (1983) Construction and application of a novel plasmid "ATG vector" for direct expression of foreign genes in *Escherichia coli. DNA* **2,** 265–273.
18. Ringquist, S., Shinedling, S., Barrick, D., Green, L., Binkley, J., Stormo, G. D., and Gold, L. (1992) Translation initiation in *Escherichia coli*: sequences within the ribosome-binding site. *Molecular Microbiol.* **6,** 1219–1229.
19. Gold. L. (1988) Post-transcriptional regulatory mechanisms in *E. coli. Ann. Rev. Biochem.* **57,** 199–233.
20. Hartz, D., McPheeters, D. S., and Gold, L. (1991) Influence of mRNA determinants on translation initiation in *Escherichia coli. J. Mol. Biol.* **218,** 83–97.
21. Steitz, J. A. (1969) Polypeptide chain initiation: nucleotide sequences of the three ribosomal binding sites in bacteriophage R17 RNA. *Nature* **224,** 957–964.
22. Gold, L. and Stormo, G. (1987) Translation initiation, in *Escherichia coli and Salmonella typhimurium vol. 2* (Neidhardt, F. C., Ingraham, J. L., Low, K. B., Magasanik, B., Schaechter, M., and Umbarger, H. E., eds.), American Society for Microbiology, Washington, DC, pp. 1302–1307.
23. Zagursky, R. J. and Berman, M. L. (1984) Cloning vectors that yield high levels of single-stranded DNA for rapid sequencing. *Gene* **27,** 183–191.
24. Yansura, D. G. (1990) Expression as *trpE* fusion. *Methods Enzymol.* **185,** 161–166.
25. Ford, C. F., Suominen, I., and Glatz, C. E. (1991) Fusion tails for the recovery and purification of recombinant proteins. *Protein expression and purification* **2,** 95–107.
26. Carter, P. (1990) Site-specific proteolysis of fusion proteins, in *Protein Purification: From Molecular Mechanisms to Large-Scale Processes* (Ladisch, M. R., Willson, R. C., Painton, C. C., and Builder, S. E., eds.), American Chemical Society, Washington, D.C., pp. 181–193.
27. Uhlén, M. and Moks, T. (1990) Gene fusions for the purpose of expression: an introduction. *Methods Enzymol.* **185,** 129–143.
28. Yansura, D. G., and Simmons, L. C. (1992) Nucleotide sequence selection for increased expression of heterologous genes in *Escherichia coli. Methods: A Companion to Methods Enzymology* **4,** 151–158.
29. Gottesman, S. (1990) Minimizing proteolysis in *Escherichia coli*: genetic solutions. *Methods Enzymol.* **185,** 119–129.
30. Grodberg, J., and Dunn, J. J. (1988) *ompT* encodes the *Escherichia coli* outer membrane protease that cleaves T7 RNA Polymerase during purification. *J. Bacteriol.* **170,** 1245–1253.
31. Henderson, T. A., Dombrosky, P. M., and Young, K. D. (1994) Artifactual processing of penicillin-binding proteins 7 and 1b by the OmpT protease of *Escherichia coli. J. Bacteriol.* **176,** 256–259.
32. Weber, C., Lee, V. D., Chazin, W. J., and Huang, B. (1994) High level expression in *Escherichia coli* and characterization of the EF-hand calcium-binding protein caltractin. *J. Biol. Chem.* **269,** 15,795–15,802.

7

Expression of Proteins in *E. coli* Utilizing a Dual Promoter-Based Vector: pLACT7

George A. Garcia and Shaorong Chong

1. Introduction

Most commonly used prokaryotic expression vectors fall into one of the following three categories:

1. They have a low copy number and a low yield of single-stranded DNA;
2. They have a high copy number and the ability to produce a high yield of single-stranded DNA, but require a suitable host to achieve an inducible protein expression; or
3. They lack the ability to produce single-stranded DNA.

The vector pLACT7 reported here has all the features necessary for inducible protein expression in virtually any *E. coli* host, and for site-directed mutagenesis, cloning, and DNA sequencing.

The vector pTGT1 overexpresses the enzyme tRNA-guanine transglycosylase (TGT) in the *Escherichia coli* strain BL21(DE3) pLysS without the induction of the T7 RNA polymerase gene located on the *E. coli* chromosome *(1)*. Previous mRNA primer extension analysis indicated that the *tgt* gene is actually transcribed from an upstream *lac* promoter (P_L) *(1)*. Though a *lac* operator *(Lac O)* is located immediately downstream from the *lac* promoter, the *tgt* gene is still constitutively expressed from pTGT1, presumably due to the inability of the cells to produce enough *lac* repressor to effectively suppress the *lac* promoter. This may be caused by the high copy number of pTGT1 (a derivative of pUC therefore ca. 300–500 per cell). In order to obtain a controllable expression system for TGT, we chose to subclone the *tgt* gene into pET21b(+) (Novagen). By cloning the *tgt* gene into pET21b, we were also able to transfer the *lac* I gene from pET21b to pTGT1 in the correct orientation in

From: *Methods in Molecular Biology, vol. 62: Recombinant Gene Expression Protocols*
Edited by: R. Tuan Humana Press Inc., Totowa, NJ

two simple and efficient steps. In the subsequently constructed plasmid pTGT5, protein expression by the *lac* promoter is effectively suppressed under noninducing conditions. The many positive features of pTGT5 led us to replace the *tgt* gene in pTGT5 with the multiple cloning site from pET21b *(2)*, generating the vector pLACT7 (Fig. 1).

Although transcription of the *tgt* gene from pTGT5 in *E. coli* BL21/pLysS is directed from the *lac* promoter rather than the stronger T7 promoter, as in the case of the pET21b construct in *E. coli* BL21(DE3)/pLysS, TGT is overexpressed in approximately equal amounts from both vectors. In addition, pTGT5 has a minimal basal level of TGT expression under noninducing conditions. It would seem that in order to effectively suppress the *lac* promoter on the high-copy-numbered pLACT7, one must incorporate a functional *lac* I gene on the expression vector itself. The TGT activity assays indicate that the amount of TGT expressed without IPTG induction is approx 2% of the amount expressed under IPTG induction (data not shown). Thus, a relatively toxic gene may be tolerated in the pLACT7 system under noninducing conditions. In order to demonstrate the generality of pLACT7, we have subcloned and expressed the *queA* gene from *E. coli (3)*. Whereas the expression level of the *queA* protein is not as high as that for TGT, we are obtaining essentially the same yield of *queA* as from a *queA* construct in the vector pET21b *(2)*.

The fact that pLACT7 can use *E. coli* RNA polymerase to direct the transcription of cloned genes renders this vector suitable for enzyme expression in a wide variety of *E. coli* hosts. Most importantly for us, this vector can be used to express TGT and TGT mutants with no background contamination of host wild type TGT in *E. coli* K12 (Δ*tgt*) (a *tgt-* strain that we have used previously to generate tRNA lacking queuosine *[4]*).

Since pLACT7 also contains the T7 promoter immediately upstream of the ribosome binding site, it can potentially be used in a host (e.g., *E. coli* BL21[DE3]) which uses the T7 RNA polymerase to direct transcription of the cloned gene. In this manner, an even wider variety of host cell lines (involving either *E. coli* or T7 RNA polymerase-directed transcription) could be screened for optimal protein expression for a given cloned gene. The presence of the T7 promoter should also allow pLACT7 to be used for in vitro transcription and translation.

The pLACT7 is a pUC-derived phagemid which contains the origin of replication and morphogenetic signal of phage f1 intergenic region. Upon infection with helper phage (e.g., M13KO7), cells (*E. coli* DH5αF') harboring pLACT7 produce a much higher yield of single-stranded DNA than those harboring the pET21b-derived plasmid. The large amount of single-stranded DNA obtained from pTGT5 has facilitated subsequent oligonucleotide-directed, site-specific mutagenesis and DNA sequencing *(5)*. The pUC-derived plasmids have higher copy numbers (ca. 300–500 per cell) than pET plasmids (ca. 30

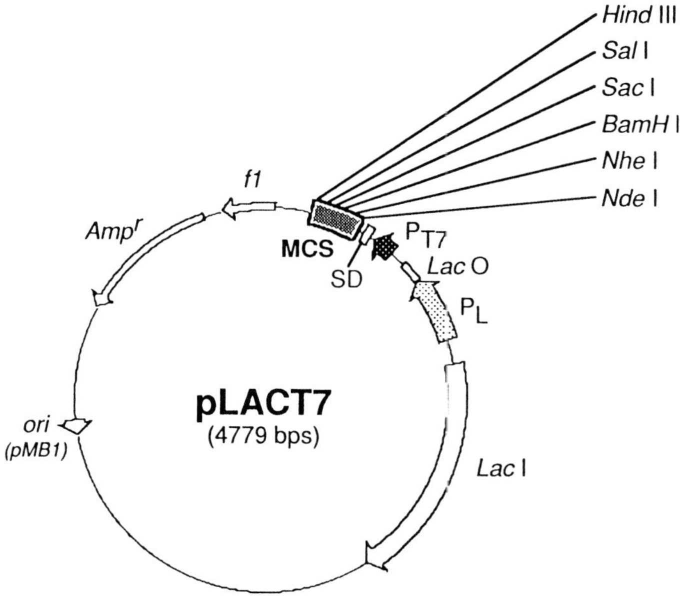

Fig. 1. Map of pLACT7; pertinent restriction sites and genes are noted.

per cell). Using UV spectrophotometry to estimate the concentrations of purified plasmids, we conclude that pLACT7 has a copy number at least four times higher than that of pET21b. Agarose gel electrophoresis of equivalent volumes of dsDNA obtained from equivalent cultures of pTGT5 and the pET21b construct is consistent with pLACT7 having a higher copy number. This makes pLACT7 a convenient vector for cloning and double-stranded DNA sequencing.

2. Materials

Standard sterile techniques should be followed. All buffers and solutions are made up in deionized/distilled water (e.g., MilliQ, Millipore Bedford, MA).

1. pLACT7: The vector is presently available from the author and has been submitted to the ATCC (American Type Culture Collection, #87063). The nucleotide sequence has been deposited in the GENbank database (accession number U16722).
2. Restriction enzymes: *Nde* I, *Nhe* I, *BamH* I, *Sac* I, *Sal* I, and *Hind* III are available from United States Biochemical, New England Biolabs, Boehringer Mannheim, and others.
4. Low melting agarose: SeaPlaque GTG agarose is from FMC BioProducts, Rockland, ME.
5. GELase: GELase is available from Epicentre Technologies, Madison, WI.
6. T4 DNA ligase: T4 DNA ligase is available from GIBCO BRL Life Technologies, Gaithersburg, MD.

7. QIAprep-spin Plasmid Kit: The QIAprep-spin Plasmid Kit is available from QIAGEN Inc., Chatsworth, CA.
8. *E. coli* strains: BL21/pLysS and BL21 (DE3)/pLysS are available from Novagen, Madison, WI.
9. Sequenase kit: Sequenase (Version 2.0) DNA Sequencing Kit is available from U. S. Biochemical, Cleveland, OH.
10. 2X TY media: 16 g Bactotryptone, 10 g yeast extract, and 5 g NaCl per 1 L water. Sterilize by autoclaving. If antibiotics are needed, then ampicillin (100 µg/mL of media), kanamycin (75 µg/mL), and/or chloramphenicol (30 µg/mL) is added after sterilization.
11. TBG media: 11.8 g Bactotryptone, 23.6 g yeast extract, 9.4 g K_2HPO_4, and 2.2 g KH_2PO_4 are brought to 1 L with water. The solution is sterilized by autoclaving, then 20 mL of 1*M* glucose (sterile filtered) is added after cooling to room temperature. If antibiotics are needed, then ampicillin (100 µg/mL of media), kanamycin (75 µg/mL), and/or chloramphenicol (30 µg/mL) is added after sterilization.
12. SDS-PAGE Sample Loading Buffer: 5% SDS, 10% β-mercaptoethanol, and 0.02% bromphenol blue.
13. HEPES buffer: 20 m*M* HEPES, pH 7.5.
14. PEG/NaCl: 20% PEG 8000 and 2.5*M* NaCl.
15. TE8: 10 m*M* Tris-HCl, 1 m*M* EDTA, pH 8.0.
16. 3*M* NaAc: Sodium acetate (24.6 g anhydrous) is first dissolved in 60 mL water and the pH is adjusted to 5.5 by the addition of glacial acetic acid. After the pH is adjusted, the final volume is brought to 100 mL and the solution is sterilized by autoclaving (*see* Note 1).
17. 5*M* KAc: Potassium acetate (17.6 g) is added to 28.5 mL water and 11.5 mL glacial acetic acid. The final volume is 60 mL and the solution is sterilized by autoclaving. (This yields a solution that is 3*M* in potassium and 5*M* in acetic acid/acetate.)

3. Methods

3.1. Cloning into pLACT7

pLACT7 has been designed to facilitate the subcloning of genes of interest into an *Nde* I site which contains the ATG start codon. In this manner, the open reading frame is correctly aligned with the ribosome binding site and spacer in the vector for protein translation (*see* Note 2). If the gene of interest does not contain an *Nde* I site at the beginning of the gene and/or one of the unique sites in the multiple cloning site of pLACT7, then these can be introduced by amplifying the gene via PCR utilizing primers that introduce the desired restriction sites. Care must be taken to avoid the usage of restriction sites that may be contained within the gene (*see* Note 3). Assuming that one has an open reading frame with suitable restriction sites, the following procedure (demonstrated for cloning into the *Nde* I/*Hind* III sites of pLACT7) may be followed for subcloning into pLACT7.

The pLACT7 (ca. 0.5 μg) and the gene of interest (ca. 3:1 molar ratio of insert to vector) are each double digested with 1 U each of *Nde* I and *Hind* III in 1X *Nde* I buffer at 37°C for 1 h. and separated on 1% SeaPlaque GTG agarose gel (containing 10 mg/mL ethidium bromide). The gene fragment and pLACT7 fragments are excised from the gel using a UV light transilluminator (*see* Note 4) to visualize the bands and a sterile scalpel. The gel slices are mixed, melted at 68°C for 10 min, cooled to 45°C, and the agarose is digested with 1 U of GELase for 1 h. To the digest (~50 μL), 0.5 U of T4 ligase and 15 μL 5X T4 ligase buffer are added. The ligation is performed at room temperature (ca. 23°C) for 1 h. A portion of the ligation product (15 μL) is used to transform 150 μL of competent *E. coli* DH5αF' cells (*see* Note 5).

3.2. Single-Stranded DNA Purification

The cell culture and phage infection are performed according to Karger and Jessee *(6)*. The procedure for the purification of the single-stranded DNA from the phage pellets is according to Garber *(7)*.

1. A well-isolated single colony (or 10 μL of a liquid N_2 glycerol stock) of DH5aF' cells harboring the plasmid DNA is inoculated into 10 mL of TBG medium (with ampicillin and kanamycin, *see* Note 6) containing 150 μL M13KO7 (5×10^8 to 10^9 pfu/mL). The culture is grown at 37°C with vigorous shaking (350 rpm) for 24 h. The cells are pelleted in a microcentrifuge (ca. 5000 rpm) and the phage-containing supernatant is filtered through a 0.2 mm filter.
2. To 1.2 mLs of phage supernatant, 250 μL of PEG/NaCl is added. After the mixture is incubated on ice for 30 min, the phage are pelleted by centrifugation for 15 min.
3. Phage pellets from up to 10 mL of culture can be combined and dissolved in 400 μL of TE8 buffer. To this, 50 μL of 10% (w/v) SDS and 30 μL of 0.1 mg/mL proteinase K are added. The mixture is vortexed briefly and incubated at 65 °C for 15 min. While the solution is still hot, 50 μL of 5*M* KAc is added. The mixture is again vortexed briefly and after incubation on ice for 10 min, it is centrifuged at room temperature for 15 min. The supernatant (containing the single-stranded DNA [ssDNA]) is removed to a new microcentrifuge tube, taking care not to disturb the pellet.
4. The ssDNA is precipitated by adding 1 mL of ethanol (100%) and incubating at –20°C for more than 1 h. After centrifugation for 20 min., the ssDNA pellet is resuspended in 40 μL TE8 and stored at –20°C.

3.3. Double-Stranded DNA Purification

All plasmid DNAs are purified using QIAprep-spin Plasmid Kit. From 3 mL cultures, 100 μL of the purified plasmid DNA is obtained for each vector (*see* Note 7).

3.4. DNA Sequencing

Both the single-stranded (ss) and double-stranded (ds) plasmid DNA sequencing are conducted using the Sequenase kit (USB). Once the sequence of the insert has been verified (*see* Note 8), the clone can be transformed into an appropriate cell line (or lines) for overexpression determination.

1. For single-stranded sequencing, 3 μL of ssDNA, prepared as described above, is used. The Sequenase reaction buffer (2 μL) and the primer (3 μL, 0.5 pmol/μL) are then added. The sample is heated at 85°C for 3 min and allowed to slowly cool to room temperature. The rest of the sequencing procedures are according to the vendor's protocols.
2. For double-stranded sequencing, 15 μL of dsDNA, prepared as described above, is denatured in a 20 μL alkaline-denaturation solution containing 200 m*M* NaOH, 0.2 m*M* EDTA at 65°C for 5 min. The mixture is neutralized by adding 2 μL of 3*M* NaAc so that the pH of the solution is ca. 6. The DNA is precipitated with 4 volumes of ethanol at –20°C for at least 1 h. After washing the pelleted DNA with 70% ethanol, it is redissolved in 5 μL of distilled water. The Sequenase reaction buffer (2 μL) and the primer (3 μL, 0.5 pmol/μL) are then added. The sample is heated at 85°C for 3 min and then allowed to slowly cool to room temperature. The rest of the sequencing procedures are according to the vendor's protocols.

3.5. Protein Overexpression from the lac Promoter

The resultant vector is transformed into the *E. coli* strain BL21/pLysS. One colony (more may be chosen to screen for optimal expression, *see* Note 9) harboring the plasmid is inoculated into 100 mL 2X TY media (containing ampicillin and chloramphenicol, see Note 10). After approx 5 h of shaking (300 rpm, 37°C, OD_{600} ca. 0.6–0.8), 0.5 m*M* final concentration of isopropyl-1-thio-β-D-galactoside (IPTG) is added to the media. After 2 h of IPTG induction, 60 μL of the cells from the culture are taken out for the determination of the total protein on SDS-PAGE, and the rest of the cells are spun down and stored at –20°C.

3.6. Protein Overexpression from the T7 Promoter

The resultant vector is transformed into the *E. coli* strain BL21(DE3)/pLysS. One colony (again more may be screened) harboring the plasmid is inoculated into 100 mL 2X TY media (containing 100 μg/mL ampicillin and 30 μg/mL chloramphenicol). After approximately 5 h of shaking (300 rpm, 37°C, OD_{600} ca. 0.6–0.8), 1 m*M* final concentration of IPTG is added to the media. After 2 h of IPTG induction, 60 μL of the cells from the culture are taken out for the determination of the total protein on SDS-PAGE, and the rest of the cells are spun down and stored at –20°C.

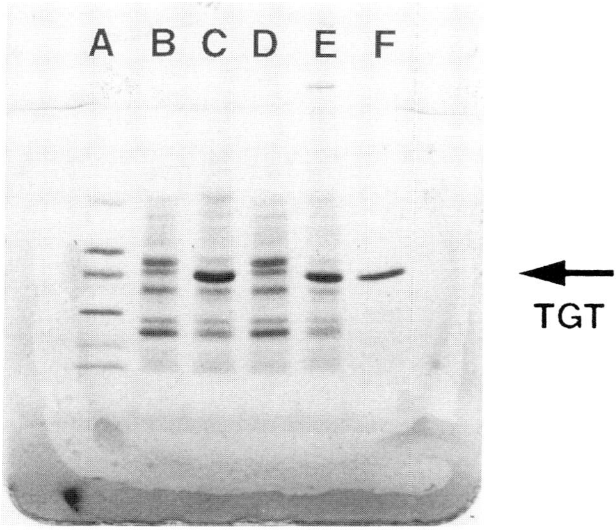

Fig. 2. SDS-PAGE of total protein preparations from pTGT5. Lane A: MW standards; lane B: BL21/pLysS/pTGT5 (uninduced); lane C: BL21/pLysS/pTGT5 (induced); lane D: BL21 (DE3)/pLysS/pTGT5 (uninduced); lane E: BL21 (DE3)/pLysS/pTGT5 (induced); lane F: purified TGT.

3.7. Total Protein by SDS-PAGE

For SDS-PAGE (denaturing), samples are mixed (1:1) with SDS-PAGE sample loading buffer and are heated in a boiling water bath for 10 min. The samples are then applied, electrophoresed, and stained/destained following standard protocols for the apparatus used (*see* Note 11).

1. Take 60 µL from a 3 mL overnight cell culture *(see above)* and spin down the cells in a 1.5 mL microcentrifuge tube (10,000 rpm, 5 min).
2. Discard the supernatant (*see* Note 12) and resuspend the cell pellet in 25 µL HEPES buffer (20 m*M*, pH 7.5) and add 25 µL SDS-PAGE sample loading buffer.
3. Heat the sample in a boiling water bath for 10 min. After cooling the samples to room temperature, load and run the sample on the gel electrophoresis apparatus. Stain and destain the gels after electrophoresis following your standard protocols.

Figure 2 shows an SDS-PAGE of pTGT5 expressed in BL21/pLysS and BL21 (DE3)/pLysS under noninducing and inducing conditions.

4. Notes

1. This yields a solution that is 3*M* in Na. We find that this is optimal for ethanol precipitation of nucleic acids, especially dsDNA during sequencing.

2. We find an excellent overexpression of TGT in this vector (Fig. 2). This overexpression may be due in part to the design of the spacer sequence between the ribosome binding site and the ATG start codon *(8)*. Additionally, DeBoer and Hui reported that highly efficient protein expression is correlated with proteins that start with a Met-Lys sequence, as is the case for TGT *(8)*.
3. In the subcloning of the *queA* gene, we found that the gene contained an *Nde* I site. This required us to insert the *queA* gene into pLACT7 in two fragments. This was an extremely tedious and inefficient process. It is recommended that if your gene contains an *Nde* I site, you remove the site by mutagenesis prior to subcloning into pLACT7. This may not always be practical if the *Nde* I site is in frame, as ATG is the only codon for Met and CAT is one of only two (the other is CAC) for His.
4. It is suggested that a 302 nm wavelength lamp be used in the transilluminator to minimize UV light-induced damage to the DNA *(9)*. We use a TM-15 unit from UVP, San Gabriel, CA.
5. We use the Hanahan procedure for generating competent cells as described in *Molecular Cloning: A Laboratory Manual, 2nd Ed.*, pp. 1.76–1.81, by Sambrook et al. *(10)*. Alternatively, one could use an electroporation technique to transform the cells.
6. The DH5αF' cells are resistant to low levels of kanamycin and, therefore, do not require an initial growth period for the induction of kanamycin resistance prior to the addition of kanamycin *(6)*.
7. We have found that optimal preparations of dsDNA are obtained when the culture time is limited to 8–10 h. If the culture is allowed to grow for longer times, we find contaminating DNA in the preparation. Relatively clean dsDNA is required for efficient dsDNA sequencing. The remainder of the dsDNA preparation is performed following the vendor's protocols.
8. Verification of the insert sequence and the flanking regions of the vector is necessary to ensure proper insertion of the open reading frame. This is especially necessary if DNA amplification techniques are used to ensure that no unwanted mutations were introduced during the amplification process.
9. Owing to the ease of qualitatively determining the level of protein overexpression via SDS-PAGE of total protein extracts (*see* Section 3.7.), it is suggested that a number of colonies (ca. 10–12) be screened. In this manner, a clone giving an optimal level of overexpression in the desired cell line can be identified.
10. Chloramphenicol is necessary to maintain the pLysS plasmid which contains the CAT gene imparting resistance to chloramphenicol.
11. We run our proteins on the Pharmacia Phastsystem using the vendor's precast gels and separation and development protocols. However, any gel electrophoresis apparatus may be used.
12. Standard safety procedures should be followed in disposal of culture supernatants, etc.

Acknowledgments

This work was supported by the National Institutes of Health (GM45968) and the University of Michigan, Program in Protein Structure and Design.

References

1. Garcia, G. A., Koch, K. A., and Chong, S. (1993) tRNA-guanine transglycosylase from *Escherichia coli*: overexpression, purification, and quaternary structure. *J. Mol. Biol.* **231,** 489–497.
2. Chong, S. and Garcia, G. A. (1994) A versatile and general prokaryotic expression vector, pLACT7. *BioTechniques* **17,** 686–691.
3. Reuter, K., Slany, R., Ullrich, F., and Kersten, H. (1991) Structure and organization of *Escherichia coli* genes involved in biosynthesis of the deazaguanine derivative queuine, a nutrient factor for eukaryotes. *J. Bacteriol.* **173,** 2256–2264.
4. Curnow, A. W., Kung, F. L., Koch, K. A., and Garcia, G. A. (1993) tRNA-guanine transglycosylase from *Escherichia coli*: Gross tRNA structural requirements for recognition. *Biochemistry* **32,** 5239–5246.
5. Chong, S. and Garcia, G. A. (1994) An oligonucleotide-directed, *in vitro* mutagenesis method using ssDNA and preferential DNA amplification of the mutated strand. *BioTechniques* **17,** 719–725.
6. Karger, B. and Jessee, J. (1990) Preparation of single-stranded DNA from phagemids. *Gibco-BRL Focus* **12,** 28,29.
7. Garber, A. (1993) Simplified purification of single-stranded DNA for dideoxynucleotide chain termination DNA sequencing. *BioTechniques* **14,** 540–542.
8. DeBoer, H. A. and Hui, A. S. (1990) Sequences within the ribosome binding site affecting messenger RNA translatability and method to direct ribosomes to single messenger RNA species. *Methods Enzymol.* **185,** 103–114.
9. Brunk, C. F. and Simpson, L. (1977) Comparison of various ultraviolet sources for fluorescent detection of ethidium bromide-DNA complexes in polyacrylamide gels. *Anal. Biochem.* **82,** 455–462.
10. Sambrook, J., Fritsch, E., and Maniatis, T. (1989) *Molecular Cloning: A Laboratory Manual.* Cold Spring Harbor Laboratory, Cold Spring Harbor, NY.

8

Expression and Secretion of Proteins in *E. coli*

Ophry Pines and Masayori Inouye

1. Introduction

In the past decade we have witnessed rapid advances in the development of systems for expression of recombinant proteins. A wide range of expression systems, host organisms, and processing procedures have been described. Among these, *E. coli* remains an important organism for production of recombinant proteins in both the laboratory and industry. This is owing to the fact that only a few organisms can satisfy all the criteria such as high growth rate, relatively simple nutritional demands, genetic stability, ease of product purification, and appropriate posttranslational modification. In addition to the advantage of genetic manipulation being a straightforward process in *E. coli*, this organism has the ability to accumulate many proteins to at least 20% of the total cellular protein, and to translocate them from the cytoplasm to the periplasm. This makes *E. coli* an attractive host for large scale production of recombinant proteins. Moreover, production of recombinant proteins in this organism presents advantages for many bioprocessing strategies.

To maximize expression of a cloned gene, it must be transcribed and translated efficiently and the mRNA and protein products must be stabilized. Since there is no universal solution applicable to all systems, an investigator must deal with each of these problems individually depending on each particular situation. A good example is the number of strong promoters that have been used to develop extremely efficient expression systems in *E. coli*. These include the T7, λ-pL, tryptophan, lactose, and lipoprotein promoters (*see* Chapters 2, 3, 5, and 6). It is not the intention of this chapter to review the different expression systems, but rather to describe in detail, the use of one system in which we can demonstrate high level expression of cloned genes, the targeting of their products to specific subcellular locations in the *E. coli* cell and the ability to overproduce and secrete eukaryotic proteins into the bacterial periplasm.

From: *Methods in Molecular Biology, vol. 62: Recombinant Gene Expression Protocols*
Edited by: R. Tuan Humana Press Inc., Totowa, NJ

2. Materials

The gram-negative bacterial cell is divided into four major subcellular compartments: the cytoplasm, the cytoplasmic (inner) membrane, the periplasmic space, and the outer membrane. Most proteins translocated across a bacterial membrane are synthesized as a protein precursor, containing an approx 20 amino acid extension at the amino terminus called the signal peptide *(1)*. This peptide is involved in the initial interaction between the membrane secretory apparatus and the translocated protein. The amino-terminus of the signal peptide contains several positively charged amino acids, which is followed by a 14–20 amino acid hydrophobic domain *(2)*. Correct processing by the signal peptidase requires certain carboxy-terminal residues such as alanine or glycine. Translocation of proteins into the periplasm may require only the signal peptide, whereas the targeting to a specific membrane requires additional sequence information *(2)*. Accordingly, we have constructed a series of cloning vectors designed to affect the localization of cloned gene products. For example, hybrid proteins containing a signal sequence for periplasmic localization, or a signal sequence plus part of the major lipoprotein mature protein sequence for outer membrane localization.

2.1. pIN III: Multipurpose Expression Vehicle in E. coli

We have used the pIN III high expression vector system as a basis for constructing several types of secretion cloning vectors *(3)*. This expression system possesses several characteristics of an ideal cloning vehicle. The vector is based on pBR322, a plasmid maintained at roughly 30 copies per bacterial cell *(4)*. The pIN III vector utilizes a high efficiency lipoprotein *(lpp)* promoter, one of the strongest promoters in *E. coli* to efficiently initiate transcription of the cloned gene *(5)*. These two factors ensure production of large amounts of protein from the cloned gene. In addition, using oligonucleotide-directed site-specific mutagenesis, we found that by changing the -35 region of the *lpp* promoter from TTCTCA to TTGACA, and by changing the -10 region from AATACT to TATACT, expression could be increased an additional fourfold *(6)*. We therefore incorporated this altered *lpp* promoter, called *lpp*^{p}-5, into certain pIN III cloning vectors. Transcription via this *lpp* promoter is controlled by the *lac* UV5 promoter-operator inserted downstream from the *lpp* promoter. Therefore, the expression of the cloned gene is repressed by the *lac* repressor (*lacI* gene product) in the absence of a *lac* inducer. This allows controlled induction of the cloned gene under desired growth conditions (e.g., when amplification of the cloned gene products is least detrimental to the host cell metabolism). The *lacI* gene is inserted into the plasmid, so that expression of the cloned gene is regulated by a *lac* repressor produced by the vector itself. A series of expression vectors based on pIN III have been developed allowing a protein to be

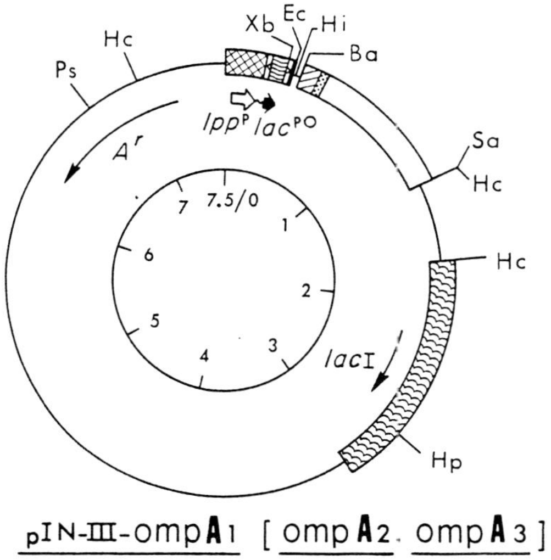

Fig. 1. Schematic representation of secretion vectors pINIII-*ompA*1, -*ompA*2, and -*ompA*3 (From Lunn et al., ref. *6*). Abbreviations: *Ba*, *Bam*HI; Ec, *Eco*RI; *Hc*, *Hinc*II; *Hi*, *Hind*III; *Hp*, *Hpa*I; *Ps*, *Pst*I; *Sa*, *Sal*I; *Xb*, *Xbal*I, *lpp*^{p}, *lpp* promoter; and *lac*^{po}, *lac* promoter-operator. Numbers indicate size in kilobase pairs.

targeted to any of the four compartments of the bacterial cell: pIN III-A for cytoplasmic localization, pIN III-B for localization in the cytoplasmic membrane, pIN III-*ompA* for localization in the periplasm, and pIN III-C for localization within the outer membrane. Details of the construction of the pIN III vector system is presented elsewhere *(7,8)*. The vectors of the pIN III system (pIN III-A, pIN III-B, pIN III-C, and pIN III-*ompA*) are designed to contain identical multirestriction sites. This permits exchange of cloned DNA fragments between vectors, from a unique upstream *XbaI* site through the multirestriction sites. This *XbaI* site also provides a convenient site for sequencing the junction between the cloning vehicle and the cloned gene. Here we have limited ourselves to the description of the secretory vectors, pIN III-*ompA* and pIN III-C and their applications.

2.2. Secretion Cloning Vectors for Periplasmic Localization (pIN III-ompA)

To direct hydrophilic proteins into the periplasmic space, we constructed a set of vectors that fuse the *ompA* signal sequence directly to the cloned gene *(6,9)*. A schematic representation of this secretion vehicle is presented in Fig 1. A 27-bp fragment of pIN III-A3, between *XbaI* and *Eco*RI restriction sites, was replaced with a DNA sequence including the *ompA* signal peptide sequence.

This signal peptide directs the translocation of OmpA protein across the cytoplasmic membrane *(10)*. This protein extension is known to be cleaved off from the *ompA* precursor protein by signal peptidase I, a nonspecific signal peptidase that is considered to process all secreted proteins, except for the lipoproteins *(10)*. The *ompA* signal peptide sequence was prepared from a 65 bp *HhaI* fragment from the cloned *ompA* gene (P. J. Green and M. Inouye, unpublished results) plus two synthetic oligonucleotides coding for the Shine-Dalgarno ribosome binding site *(11)*, and the *ompA* signal peptide cleavage site. This vector (pIN III-*ompA*3) has three unique restriction sites, *Eco*RI, *Hind*III and *Bam*HI; immediately after the sequence of the *ompA* signal peptide. In order to express any foreign DNA fragments with any reading frame, two other vectors (pIN III-*ompA*1 and pIN III-*ompA*2) have been constructed. These two vectors are identical to pIN III-*ompA*3, but have the multiple cloning sites in two different reading frames. The resulting junctional sequences for each vector, containing the unique *Eco*RI, *Hind*III, and *Bam*HI cloning sites, are shown in Fig. 2A. Because of these junctional sequences, each cloned gene product contains several extra amino acids at its amino-terminal end. However, the sequences coding for these extra amino acids can be easily removed by oligonucleotide-directed site-specific mutagenesis (Fig. 2B).

2.3. Secretion Cloning Vector for Outer Membrane Localization (pIN III-C)

Localization of a protein within the outer membrane requires more structural information than that contained within the signal sequence *(5)*. Our studies of the major lipoprotein suggested that the signal sequence, plus a short peptide (8–9 amino acids) from the amino terminus of the lipoprotein are sufficient to translocate β-lactamase to the outer membrane *(12)*. Based on this result, pIN III-C cloning vectors were constructed. These vectors contain three unique restriction sites (*Eco*RI, *Hind*III, and *Bam*HI) for cloning a foreign DNA fragment located immediately after the *lpp* signal sequence, plus eight amino acids from the *lpp* structural gene. Construction of the pIN III-C vectors has been described elsewhere *(13)*. As with the pIN III-*ompA* vectors described above, the unique *Eco*RI, *Hind*III, and *Bam*HI sites for cloning are constructed in three different reading frames (pIN III-C1, pIN III-C2, and pIN III-C3). The sequences of these vectors at the multirestriction sites are shown in Fig. 3.

3. Methods

3.1. Targeting Prokaryotic Proteins to the Periplasmic Space

3.1.1. RTEM-β-Lactamase

The structural gene for β-lactamase (minus its own signal sequence) was inserted into the *Eco*RI site of pIN III-*ompA*3 to yield pJG105 *(9)*. This cloning

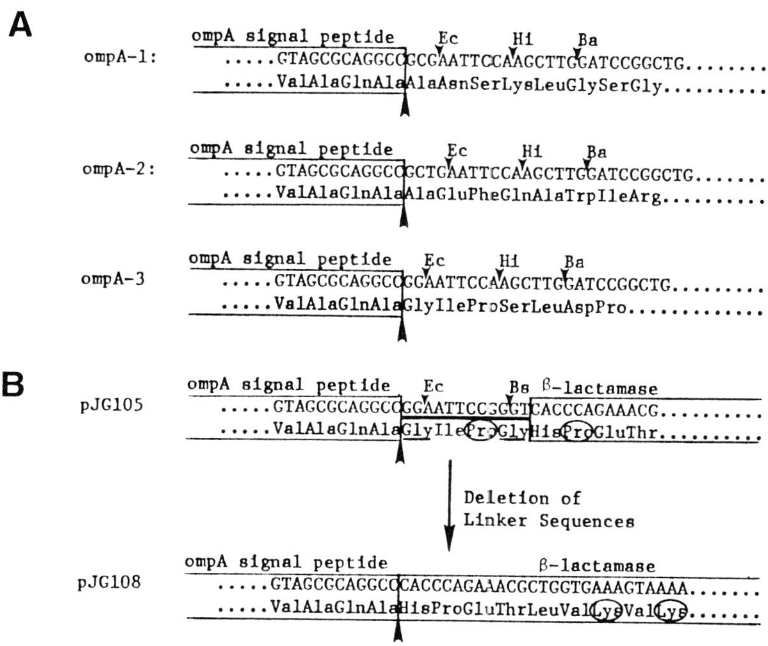

Fig. 2. Partial DNA sequences of linker polynucleotides (From Ghrayeb et al., ref. *9*). **(A)** pINIII-*ompA*1, pIN III-*ompA*2, pIN III-*ompA*3, **(B)** pJG105 and pJG108. Abbreviations are the same as in Fig. 1. Arrows indicate the cleavage site of the *ompA* signal peptide. The linker sequence of pJG105 is underlined.

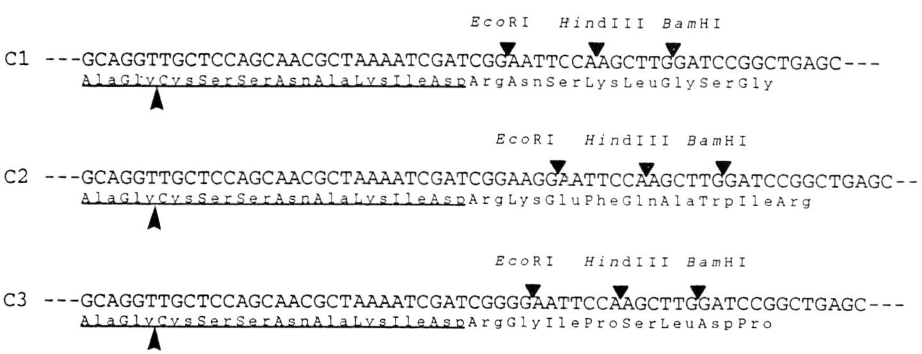

Fig. 3. Partial DNA sequences of linker polynucleotides of pIN III-C1, pIN III-C2, and pIN III-C3 (From Masui et al., ref. *8*). Arrows below the amino acid sequences indicate the cleavage site of the *lpp* signal peptide. The sequences underlined represent the amino-terminal region of lipoprotein.

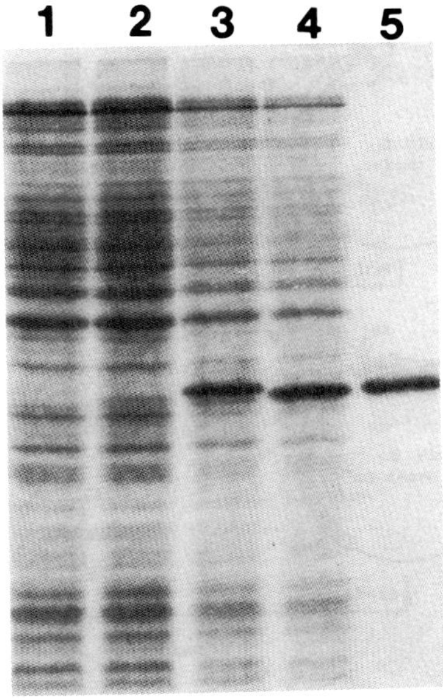

Fig. 4. SDS-polyacrylamide gel electrophoresis of total cell fractions of *E. coli* JA221 *lpp*/F'*lac*I^q/pJG105, and pJG108 (From Ghrayeb et al., ref. *9*). Cells were grown at 37°C in M9 medium to a Klett reading of 30. IPTG (2 m*M*) was then added to half of the culture. After 3 h, the cells were collected by centrifugation, washed twice with 10 m*M* Tris-HCl, pH 7.0, lysed by sonication, and aliquots were submitted to SDS-polyacrylamide gel electrophoresis. Lanes 1 and 3: 0.1 mL culture of *E. coli* JA221 *lpp*/F' *lacI*^{q}/pJG105 grown with (lane 1) and without (lane 3) IPTG. Lanes 2 and 4: 0.1 ml culture of *E. coli* JA221 *lpp*/F'lacIq/pJG108 grown with (lane 2) and without (lane 4) IPTG. Lane 5: purified β-lactamase.

strategy predicts the presence of 12 extra base pairs between the *ompA* signal peptide sequence and the structural gene for β-lactamase (*see* Fig 2B). Thus, four extra amino acids (Gly-Ile-Pro-Gly) were presumed to be attached to the amino terminus of the amplified gene product. Figure 4 shows that pJG105 produced a single product, with an apparent molecular weight slightly higher than authentic β-lactamase, after induction with 2 m*M* isopropyl-β-D-galactopyranoside (compare lane 3 to lane 5). No product was observed in the absence of inducer (lane 1). The protein yield reached approx 20% of total cellular protein after induction for 3 h. The new product exhibited β-lactamase activity, and was immunoprecipitated with anti-β-lactamase serum.

We next probed the amino acid sequence of the amino terminus of the pJG105 gene product. Following induction, the cloned gene product was labeled with [^{3}H]proline, purified, and then subjected to sequential Edman degradation. Significant radioactivity was observed after cycle 3 and cycle 6. This indicates that the *ompA* signal peptide was cleaved at its normal cleavage site, yielding β-lactamase gene product with four extra amino acids (Gly-Ile-Pro-Gly), as predicted from the construction (*see* Fig. 2). The conclusion was further supported by Edman degradation of the pJG105 gene product labeled with [^{3}H]glycine (radioactivity appeared at cycle 1 and cycle 4). These results indicate that no cleavage occurred at histidine, the normal amino terminus of β-lactamase.

The four extra amino acids at the amino terminus of the pJG105 gene product were removed using oligonucleotide-directed site-specific mutagenesis. For this purpose, a 24-mer oligonucleotide (GTAGCGCAGGCCCACCCAGAAACG) was synthesized, and used to remove the linker sequence. The DNA sequence of one resulting transformant, pJG108, is shown in Fig. 2B. As with pJG105, the induction of pJG108 yielded a single product (Fig. 4, lanes 2 and 4). The product migrated to the same position as authentic β-lactamase (lane 5). This protein was also produced to 20% of the total cellular protein. Sequential Edman degradation of the gene product demonstrates that the *ompA* signal peptide was correctly removed at a cleavage site immediately preceding the amino-terminal histidine (Fig 2B).

In *E. coli*, the β-lactamase encoded by pBR322 is localized within the periplasmic space *(14)*. Because the cloned β-lactamase gene produced a product identical to the natural gene product, we expected the pJG108 gene product would be contained within the periplasmic space. However, neither the pJG108 nor the pJG105 gene products were released from IPTG-induced cells following osmotic shock or digestion of the outer cell envelope with lysozyme-EDTA. Both proteins were recovered in a low speed pellet fraction following sonication. The pellet contained mainly outer membrane proteins. Pure (greater than 96% pure) β-lactamase was extracted from the low speed pellet fraction with 0.3% sodium lauryl sarcosinate. We concluded that overproduction of β-lactamase leads to its aggregation, presumably within the periplasmic space. Protein aggregates associated with the peptidoglycan layer of the outer membrane were also observed upon overproduction of a hybrid lipoprotein *(15)*. In our case, aggregation was not the result of intermolecular disulfide bonds, as no high-mol wt forms of the pJG105 gene product were observed upon SDS-polyacrylamide gel electrophoresis of samples prepared without reducing agent. Because correct disulfide bonds are formed as β-lactamase is translocated into the periplasmic space *(16)*, this result supports periplasmic localization of the cloned protein.

3.1.2. Staphylococcal Nuclease A

The structural gene for staphylococcal nuclease A was inserted into the *Bam*HI site of pIN III-*ompA*3 *(17)*. The gene was included within a 518-bp

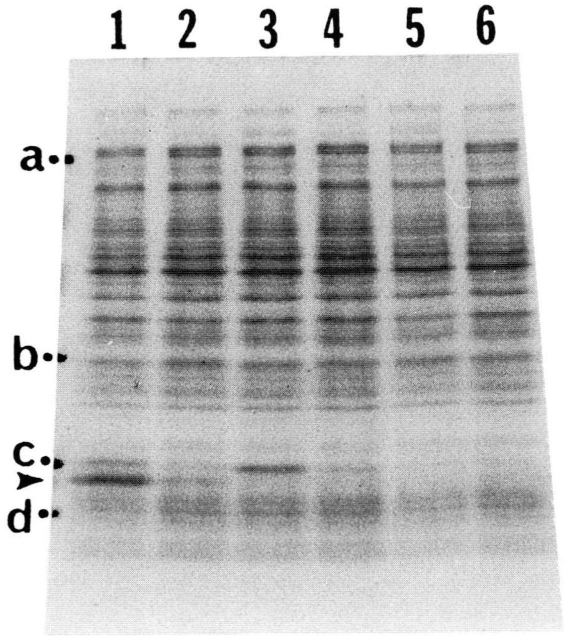

Fig. 5. SDS-polyacrylamide gel electrophoresis of total cellular proteins (From Takahara et al., ref. *17*). *E. coli* JA221 *lpp* harboring pONF1 (lanes 1 and 2), pIN III-*ompA*-98 (lanes 3 and 4), or pIN III-*ompA*3 (lanes 5 and 6) was grown at 37°C in L-broth containing 50 μg/mL ampicillin. At approx 1×10^8 cells/ml IPTG was added to a final concentration of 2 m*M*, and the cultures were further incubated for 110 min. The gel was then stained with Coomassie brilliant blue. Lanes 1, 3, and 5 show total cellular proteins produced in the presence of IPTG; lanes 2, 4, and 6 show proteins produced in the absence of IPTG. The letters represent migration positions of molecular weight standards: (a) phosphorylase *b* (92,500); (b) carbonate dehydrase (31,000); (c) soybean trypsin inhibitor (21,500); and (d) egg white lysozyme (14,400). The arrow indicates the migration position of commercially obtained staphylococcal nuclease A.

Sau3A1 fragment containing the nuclease structural gene, a 6-residue amino-terminal extension, and a 53-bp 3'-nontranslated region. This produced a plasmid, pIN III-*ompA*3-#98, with 33 extra base pairs between the *ompA* signal peptide sequence and the structural gene for nuclease A. The extra bases derive from the linker sequence (15 bp) and from the amino-terminal extension of nuclease A (18 bp) present within the cloned Sau3AI fragment. The protein produced by this plasmid was characterized. Figure 5 shows that pIN III-*ompA*3-98 produced a single product comprising 3% of total cellular protein after induction for 70 min with 2 mM IPTG. Little of this protein was present without induction. The product exhibited nuclease activity, and was precipitated with antiserum raised against nuclease A. Edman degradation demonstrated that the *ompA* signal peptide was removed at its normal cleavage site, yielding a protein composed of nuclease A with an 11 amino acid extension. This protein was localized within the periplasmic space.

The 33-bp linker sequence between the *ompA* signal peptide sequence and the nuclease A structural gene was removed by oligonucleotide-directed site-specific mutagenesis. The plasmid resulting from this manipulation (pONF1) produced one peptide comprising 10% of total cellular protein (mature nuclease A) and one peptide comprising 2% of total cellular protein, with a higher molecular weight (unprocessed nuclease A). The processed nuclease was localized within the periplasmic space. The reason for the fourfold increase in protein production after removal of the linker sequence is not known. Furthermore, by using the altered lpp^{p}-5 promoter in conjunction with pIN III-*ompA*-nuclease, much more nuclease was produced *(6)*. Quantitation of a Coomassie-stained gel by densitometry showed that the pIN III-(lpp^{p}-5)-*ompA*-nuclease gene product corresponds to roughly 50% of the total cellular protein. This represents approx 1.5 g nuclease produced/liter culture in which 40% of the nuclease produced by pIN III(lpp^{p}-5)-*ompA*-nuclease was secreted and processed by the signal peptidase.

3.2. Targeting to the Bacterial Outer Membrane

The same staphylococcal nuclease Sau3A1 fragment used above was also inserted into the *Bam*HI site of pIN III-C3 *(3)*. The localization of the gene product from this plasmid (pIN III-C3 nuclease A) was then compared to that of the pONF1 plasmid. Cells harboring these two plasmids were induced with 2 mM IPTG for 3 h, then collected by centrifugation, lysed by sonication, and fractionated into a soluble and membrane fraction. Each fraction was then analyzed by SDS-polyacrylamide gel electrophoresis. Figure 6 shows that the pIN III-C3-nuclease A gene product is found exclusively in the membrane fraction, migrating with an apparent molecular weight of 18,800. The pONF1 gene product, on the other hand, resides exclusively in the soluble (including periplasmic) fraction at an apparent molecular weight of 17,000. The increased size of nuclease A cloned in pIN III-C3 reflects addition to the nuclease gene of a 24 bp portion of the *lpp* structural gene. Thus, the staphylococcal nuclease protein with nine amino acids from the lipoprotein is processed as is the major lipoprotein. Further characterization of the pIN III-C3 nuclease gene product showed that the protein was not modified by a glyceride group, and its processing was sensitive to globomycin. The hybrid protein was localized mainly in the outer membrane, but also in the inner cytoplasmic membrane fraction.

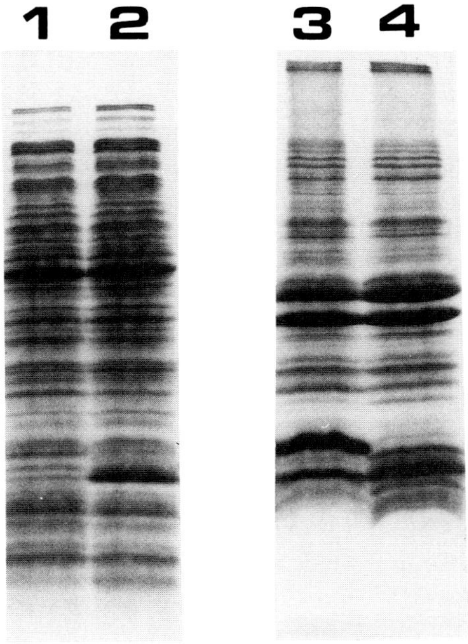

Fig. 6. Localization of staphylococcal nuclease A gene product after cloning into pIN III-C or pIN III-*ompA* (From Lunn et al., ref. *3*). Staphylococcal nuclease A was cloned into pIN- III-C (yielding pIN III-C3 nuclease A) or pIN III-*ompA* (yielding pONF1). These plasmids were used to transform *E. coli* JA221 *lpp*/F' *lacI*^{q}. Selected transformants were then grown at 37°C in M9 medium. At a Klett of 50, IPTG (2 m*M*) was added, and cells incubated for an additional 3 h. Cells were then collected by centrifugation, separated into a soluble and a membrane fraction, and aliquots analyzed by electrophoresis on an SDS-polyacrylamide gel using a Tris buffering system. Lanes 1 and 2: 0.15 mL culture of soluble fraction from cells harboring pIN III-C3-nuclease A (lane 1) or pONFI (lane 2). Lanes 3 and 4: 0.45 mL culture of membrane fraction from cells harboring pIN III-C3-nuclease A (lane 3) or pONF1 (lane 4). Arrows indicate positions of mature staphylococcal nuclease A (a) and the membrane associated form of nuclease A (b).

3.3. Secretion of Eukaryotic Proteins Using pIN III-OmpA

The pIN III-OmpA vector has been successfully used to secrete human proteins such as human growth hormone and human interleukin-6 *(18–20)*. These proteins are secretory proteins, designed to be secreted across membranes, and folded into the active configuration of the protein after completing translocation. Nevertheless, the pIN III-*ompA* secretion vector can also direct the secretion of a eukaryotic cytoplasmic enzyme into the bacterial periplasm. This we have shown with the secretion of human superoxide dismutase *(21)*. A series of DNA manipulations were used to create a pIN III-OmpA-hSOD expression vector. Oligonucleotide-directed mutagenesis was performed for the conversion of the G of codon Cys^{111} (TGC) to a C (TCC:Ser). This amino acid substitution is considered to stabilize the gene product, since Cys^{111} of hSOD has been reported to be responsible for the unstability of the human enzyme *(22)*. One of the plasmids obtained from these constructions was designated pIN-III(*lpp*-5)-OmpA-SOD14 *(21)*.

The protein patterns of the total cell extract of cultures expressing this hybrid were analyzed by SDS-PAGE, and a new protein band was detected which was absent from the vector control (not shown). The total amount of the hybrid protein was approx 7% (6 h induction) of total protein as determined by densitometric scanning of the gel. A longer induction (about 19 h) gave a much higher value (about 20%). The SDS-polyacrylamide gel autoradiogram of subcellular fractions from induced cells (6 h) is shown in Fig. 7. An additional 1*M* Tris-HCl (pH 7.6) treatment *(17)* was added to the conventional osmotic shock procedure. The induced band (shown by a dot) appears as the major one (64% purity by densitometric scanning of the lane) in the conventional periplasm fraction (lane 3). Washing the cells with 1*M* Tris-HCl after the osmotic shock yielded some more hSOD (lane 4). The remainder was found in the cytoplasmic fraction (lane 2) and is considered to be due to the periplasmic contamination. A small amount of the induced protein (less that 10%) was recovered in the membrane fraction (lane 5).

The total cell extract and the subcellular fractions of cells harboring pIN-III(*lpp*-5)-OmpA-SOD14 induced by IPTG were electrophoresed on an agarose gel film, and were examined for the position of the SOD activity by the NBT (nitro blue tetrazolium) method *(23)*. A new hSOD activity in the lane of the pIN-III(*lpp*-5)-OmpA-SOD14 total cell extract, was indistinguishable from that of purified nonsecreted SOD (21). This new SOD activity was distributed between the cytoplasmic, the periplasmic, and 1*M* Tris-HCl wash fractions, but hardly any was detectable in the membrane fraction. No such band was observed in the total cell extract of control cells (data not shown). The protein band from the periplasmic fraction and the SOD activity band were detected at identical positions (data not shown). The activity in the periplasmic fraction was found to be dramatically increased by the addition of Cu^{2+} ion (final 1 m*M* $CuSO_4$, 50 m*M* Tris-HCl, pH 7.5). The specific activity of the secreted SOD was compared with that of the cytoplasmic SOD produced by a nonsecretion vector and was found to exhibit a specific activity approx half that of the cytoplasmic SOD.

These results show that a cytoplasmic protein can be secreted across the membrane to form an enzymatically active protein by fusing it with the OmpA

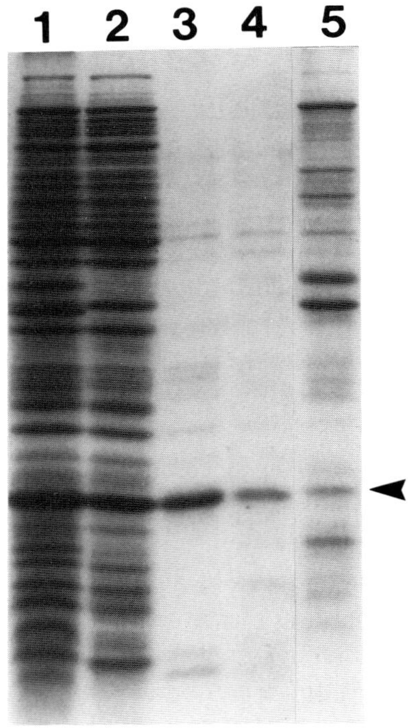

Fig. 7. SDS-PAGE of various cellular fractions of *E. coli* DH1 harboring pIN-III (*lpp*ᵖ-5)-OmpA-SOD14 (From Takahara et al., ref. *21*). Cells were grown and induced by IPTG and cellular fractionation was carried out as described previously *(14,17)*. The gel was stained with Coomasie brilliant blue. Lane 1, total cell extract; Lane 2, cytoplasmic fraction; Lane 3, periplasmic fraction; Lane 4, 1M Tris-HCl fraction; and Lane 5, membrane fraction. The arrow indicates the position of hSOD.

signal peptide. The secreted superoxide dismutase was indistinguishable from the enzyme produced in the cytoplasm in terms of mobility in SDS-polyacrylamide gel electrophoresis. The amino-terminal sequence of the purified secreted enzyme was Ala-Thr-Lys-Ala, indicating that the OmpA signal was cleaved at the expected cleavage site to form human superoxide dismutase having an identical amino terminal structure as the authentic enzyme except for the acylation at the amino-terminal group.

3.4. Concluding Remarks

We have described efficient cloning and expression systems that direct a cloned gene product out of the *E. coli* cytoplasm. The system is applicable for the production of both prokaryotic and eukaryotic proteins that can be targeted to the different compartments of the bacterial cell. Of particular importance is the ability to accumulate a specific gene product in the periplasm thus facilitating its purification. Furthermore, this allows one to separate the gene product from proteases and modification enzymes in the cytosol, to avoid potentially toxic effects of the product in the cytosol (e.g., nucleases and proteases), and to yield efficient disulfide bond formation in the oxidizing environment of the periplasm. Note that disulfide bonds are formed only after the protein is translocated across the bacterial inner membrane *(25)*. In addition, production of proteins in the cytosol requires an initiator methionine at the polypeptides amino terminus, which may not exist in the mature form of the original product. This problem is overcome by taking advantage of signal peptide removal upon secretion into the periplasm, thereby exposing different amino-terminal amino acids.

As mentioned in the introduction many factors can affect the expression and secretion of proteins in *E. coli*. One must therefore bear in mind that depending on the specific situation, additional approaches should be taken in order to improve product recovery. A good example is provided by Perez-Perez et al. *(20)* who have recently shown that they can increase the efficiency of OmpA-human-interleukin-6 export (from a pIN-III vector) by over-expressing components of the bacterial secretory apparatus (secY and secE). The recovery of secreted proteins in the periplasm can also be enhanced by using strains that are deficient in periplasmic proteases such as DegP, ProteaseIII, Tsp (Prc), and OmpT *(24)*. Other factors that can affect expression rather than secretion directly, such as mRNA stability, codon usage, and cytosolic proteases, should also be taken into account. Thus, the full potential of protein secretion and recovery may be attained by using not only the efficient expression and secretion system described in this chapter, but also additional factors considered above.

4. Notes

1. The experimental approaches used for exploiting the cloning and secretion vectors described in this chapter fall into the category of standard and routine techniques in molecular biology. The specific techniques such as DNA cloning, IPTG *lac* promoter induction, polyacrylamide gel electrophoresis of proteins, and subcellular fractionation have been described in numerous laboratory manuals, experimental reviews, and books.
2. The procedures used for the experimental examples described in this chapter are specified in the figure legends.
3. When transforming *E. coli* cells with derivatives of the pIN III vectors, M9 media should be used instead of L-broth (which contains a small amount of a *lac* inducer) in order to avoid any toxic effects of the cloned gene expression.

References

1. Gennity, J. M. and Inouye, M. (1991) Protein secretion in bacteria. *Curr. Opin. Biotechnol.* **266,** 661–667.
2. Gennity, J., Goldstein, J., and Inouye, M. (1990) Signal peptide mutants of *Escherichia coli. J. Bioenerg. Biomembr.* **22,** 233–269.
3. Lunn, C. A., Takahara, M., and Inouye, M. (1986) Use of secretion cloning vectors for guiding the localization of proteins in *Escherichia coli. Methods Enzymol.* **125,** 138–149.
4. Stueber, D. and Bujard, H. (1982) Transcription from efficient promoters can interfere with plasmid replication and diminish expression of plasmid specified genes. *EMBO J.* **1,** 1399–1404.
5. Inouye, M. and Halegoua, S. (1980) Secretion and membrane localization of proteins in *Escherichia coli. CRC Crit. Rev. Biochem.* **7,** 339–371.
6. Lunn, C. A., Takahara, M., and Inouye, M. (1986) Secretion cloning vectors for guiding the localization of proteins *in vivo. Curr. Top. Microbiol. Immunol.* **125,** 59–74.
7. Masui, Y., Mizuno, T., and Inouye, M. (1984) Novel high level expression cloning vehicles: 10^4-fold amplification of *E. coli* minor proteins. *Bio/Technology* **2,** 81–85.
8. Masui, Y., Coleman, J., and Inouye, M. (1983) Multipurpose expression cloning vehicles in *E. coli*, in *Experimental Manipulation of Gene Expression* (Inouye, M., ed.), Academic, New York, pp. 15–32.
9. Ghrayeb, J., Kimura, H., Takahara, M., Hsiung, H., Masui, Y., and Inouye, M. (1984) Secretion cloning vectors in *Escherichia coli. EMBO J.* **3,** 2437–2442.
10. Wickner, W. (1983) M13 coat protein as a model of membrane assembly. *TIBS* **8,** 90–94.
11. Shine, J. and Dalgarno, L. (1975) Determination of cistron specificity in bacterial ribosomes. *Nature* **254,** 34–38.
12. Ghrayeb, J. and Inouye, M. (1984) Nine amino acid residues at the amino terminal of lipoprotein are sufficient for its modification, processing and localization in the outer membrane of *Escherichia coli. J. Biol. Chem.* **259,** 463–467.
13. Inouye, M., Nakamura, K., Inouye, S., and Masui, Y. (1983) Versatile expression cloning vehicles using the lipoprotein gene of the *Escherichia coli* and their application, in *The Future in Nucleic Acid Research* (Watanabe, I., ed.), Academic, Japan, Tokyo, pp. 419–436
14. Koshland, D. and Botstein, D. (1980) Secretion of β-lactamase requires the carboxy end of the protein. *Cell* **20,** 749–760.
15. Yu, F., Furukawa, H., Nakamura, K., and Mizushima, S. (1984) Mechanism of localization of major outer membrane lipoprotein in *E. coli. J. Biol. Chem.* **259,** 6013–6018.
16. Pollitt, S., and Zalkin, H. (1983) Role of primary structure and disulfide bond formation in β-lactamase secretion. *J. Bacteriol.* **153,** 27–32.
17. Takahara, M., Hibler, D. W., Barr, P. J., Gerlt, J. A., and Inouye, M. (1985) OmpA signal peptide directed secretion of Staphylococcal nuclease A by *Escherichia coli. J. Biol. Chem.* **260,** 2670–2674.

18. Hsuing, H. M., Mayne, N. G., and Becker, G. W. (1986) High level expression, efficient secretion and folding of human growth hormone in *E. coli*. *Bio/Technology* **4,** 991–995.
19. Barthelemy, I., Gonzalez de Buitrago, G., Carreiro, C., Roncal, F., Perez-Aranda, A., Marquez, G., and Barbero, J. L. (1993) Production and secretion of human interleukin-6 into the periplasm of *E. coli*: Efficient processing of N-terminal variants of hIL-6 by the *E. coli* signal peptidase. *J. Biotechnol.* **27,** 307–316.
20. Perez-Perez, J., Marquez, G., Barbero, J. L., and Gutierrez, J. (1994) Increasing the efficiency of protein export in *E. coli*. *Bio/Technology* **12,** 178–180.
21. Takahara, M., Sagai, H., Inouye, S., and Inouye, M. (1988) Secretion of human superoxide dismutase in *E. coli*. *Bio/Technology* **6,** 195–198.
22. Jabush, J. R., Farb, P. F., Kerschensteiner, D. A., and Deutsh, H. F. (1980) Some sulfhydryl properties and primary structure of human erythrocyte superoxide dismutase. *Biochemistry* **19,** 2310–2316.
23. Beauchamp, C. and Fridovich, I. (1971) Superoxide dismutase: Improved assays and an assay applicable to polyacrylamide gels. *Annal. Biochem.* **44,** 276–287.
24. Meerman, H. J. and Georgiou, G. (1994) Construction and characterization of a set of *E. coli* strains deficient in all known loci affecting the proteolytic stability of secreted recombinant proteins. *Bio/Technology* **12,** 1107–1110.
25. Bardwell, J. C. A. and Beckwith, J. (1993) The bonds that tie: Catalyzed disulfide bond formation. *Cell* **74,** 769–771.

9

Synthetic Two-Cistron Expression System

Brigitte E. Schoner

1. Introduction

Expression systems for the production of heterologous proteins are now available in many different host organisms such as *Escherichia coli*, yeast, baculovirus and mammalian cells. The choice of an appropriate host organism is dictated by the particular protein to be expressed, and by requirements for posttranslational modifications, solubility and yield. The primary advantages that *E. coli* offers as a host organism include the ease and speed with which recombinant strains can be constructed and expectations that the protein yield will be high when compared with other organisms. This is owing to extensive knowledge about mechanisms that control gene expression in *E. coli*, availability of well-characterized plasmids and phages that can serve as cloning and expression vehicles, and relatively inexpensive and efficient large-scale fermentation properties. Many proteins have been expressed at very high levels in *E. coli*, representing in some cases up to 50% of total cell protein *(1)*. Typically, to achieve such high levels of expression, multicopy plasmids, which carry strong promoters to optimize transcription and strong ribosome binding sites to optimize translation, have been used. However, there are instances where little or no protein expression is achieved even though all the required elements appear to have been optimized and incorporated into the expression plasmids. The most accepted explanation for this observation is inefficient translation of mRNA containing eukaryotic coding sequences *(2–6)*. It has been speculated that the cause for inefficient translation of mRNA is the potential for secondary structure formation involving the 5' ends of mRNA, which encompass the ribosome binding site, the translational start codon, and the 5' end of the coding sequence. Presumably, such secondary structures can lower translation initiation whenever the Shine-Dalgarno (SD) sequence and/or the

From: *Methods in Molecular Biology, vol. 62: Recombinant Gene Expression Protocols*
Edited by: R. Tuan Humana Press Inc., Totowa, NJ

translation initiation codon are sequestered in a double-stranded structure that prevents the binding of the 30S ribosomal subunit to the mRNA. This explanation is based largely on the empirical observation that changes within the 5' end of mRNA have led to improved protein production *(6)*. However, this approach is not very efficient because the bases at the 5' ends of coding sequences are different for each gene, and it is difficult if not impossible to predict which bases to change in order to minimize the propensity of mRNA to form secondary structures and thus achieve optimal expression. Further, any base changes within the 5' ends of coding sequences are limited by the need to retain the proper amino acid sequence at the amino terminus of the desired protein product.

In this chapter, a general method for optimizing translation initiation, which does not require any alterations at the 5' end of the sequences encoding proteins of interest, is described. This method is based on the finding that if the sequence within the 5' end of mRNA can be optimized and kept constant for the expression of different genes, the probability of obtaining high level expression can be greatly improved. To this end, a short coding sequence (first cistron) is placed in front of the initiation codon for the gene of interest (second cistron) in such a way that this initiation codon is no longer part of the ribosome binding site. Since the polypeptide encoded by the first cistron has no apparent function, its sequence can be manipulated for optimal translation initiation without regard for altering the amino acid sequence it encodes.

2. Materials

1. Bacterial strains. Competent *E. coli* cloning hosts can be purchased from GibcoBRL (Gaithersburg, MD) or from Novogen (Madison, WI). These cells need to be stored at –70°C.
2. Expression plasmids. There are many commercial sources for expression plasmids including New England BioLabs (Beverly, MA), Promega (Madison, WI), and Stratagene (La Jolla, CA). Plasmid maps and sequence information are usually listed in the product catalogs.
3. Growth media for culturing *E. coli* cells can be purchased from Difco (Detroit, MI). Antibiotics can be obtained from Sigma (St. Louis, MO) or from Calbiochem (La Jolla, CA). Antibiotics dissolved in aqueous solution are filter sterilized prior to being added to the growth medium. Antibiotics such as tetracycline, that are dissolved in organic solvents do not need to be sterilized.
4. Suppliers for restriction enzymes, ligases, and respective buffers include New England BioLabs (Beverly, MA), Boehringer Mannheim (Mannheim, Germany), GibcoBRL (Gaithersburg, MD), and several others. To prolong their shelf life, enzymes should be stored at –20°C in a nonself-defrosting freezer.

5. DNA purification systems can be purchased from Promega (Madison, WI) or from Qiagen (Chatsworth, CA).
6. Synthetic DNA sequences can be prepared on standard DNA synthesizers (from-Applied Biosystems, Foster City, CA) or they can be purchased from a number of different companies, e.g., Genosys Biotechnologies (The Woodlands, TX).
7. Required laboratory equipment includes gel boxes and power supplies for agarose gel electrophoresis and polyacrylamide gel electrophoresis, a microfuge, waterbaths, temperature-controlled incubators, and an autoclave. A phase contrast microscope is useful, but optional.

3. Methods

3.1. Design of a Two-Cistron Expression System

The basic components of a two-cistron expression system are:

1. A multicopy plasmid containing a strong, regulatable promoter and a ribosome binding site;
2. A first cistron sequence; and
3. A second cistron sequence that constitutes the coding sequence for the protein of interest.

In this section, considerations for choosing the individual components that are part of a two-cistron expression system are described. The most commonly used multicopy plasmids for *E. coli* contain the ColE1 origin of replication. Many versions of these plasmids are commercially available and they differ primarily with respect to the antibiotic resistance marker, the promoter, and other elements that make these plasmids suitable for a particular application. Commercially available plasmids such as these can serve as starting materials for the subsequent insertion of a first and second cistron sequence.

Several issues need to be considered for the design of a first cistron sequence. One is its length, which should be kept to a minimum to avoid unnecessary expenditure of energy and precursors for the synthesis of the peptide it encodes. As shown in Fig. 1, a minimal first-cistron sequence must begin with an ATG start codon, followed by an in-frame stop codon. An SD sequence (SD2) must be placed between the ATG start codon (start 1) and the TAA stop codon (stop 1) to ensure proper and efficient read-through translation into the second cistron *(4,7)*. It is also important that SD2 be placed at a proper distance (i.e., between 5 and 9 bases) 5' from the start codon (start 2) for the second cistron. The stop codon which terminates translation of the first cistron (stop 1), must be 3' to SD2. The precise distance and framing of stop 1 and start 2 are relatively less important. To minimize the potential for secondary structure formation, the first-cistron sequence is enriched for A and T bases. Inclusion of a unique restriction site (e.g., *Cla*I) will facilitate subsequent sequence changes

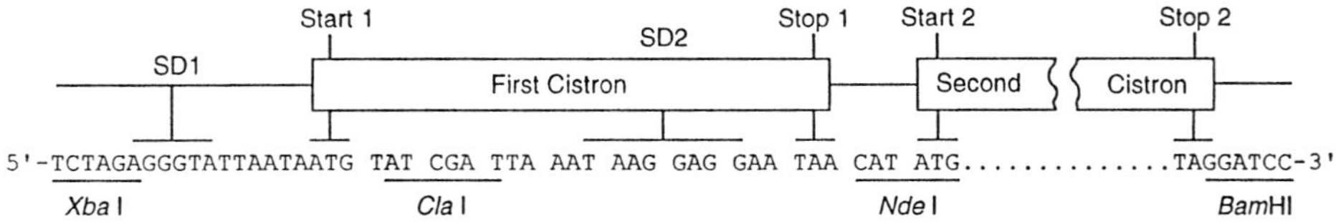

Fig. 1. Schematic of a synthetic two-cistron expression system.

should they become necessary. A second restriction site for *Nde*I (CATATG) is placed at the end of the first cistron in such a way that the ATG within the recognition sequence for *Nde*I can serve as the translational start codon for the second cistron. Downstream from the *Nde*I site is an additional restriction site for the insertion of a second cistron. It is important for cloning purposes that these restriction sites are different and that they are unique within the expression plasmid. In the example shown in Fig. 1 and Fig. 2, *Bam*HI was chosen as a second restriction site; however, other restriction sites are equally acceptable as long as they are unique within the plasmid.

3.2. Construction of Two-Cistron Expression Plasmids

The construction of two-cistron expression plasmids requires standard molecular biology techniques, including DNA isolation, restriction enzyme digestion, agarose gel electrophoresis, PCR amplification, ligation, and transformation of bacterial cells. Details for these methods are described in various molecular biology manuals *(8–11)* and by the manufacturers of the reagents.

A typical starting point for employing a two-cistron expression system is the observation that the protein yield obtained with commercially available expression plasmids is lower than expected. Since most promoters present on commonly used expression plasmids are strong promoters, such as those from bacteriophage T7 or from the lactose operon, the lack of high protein yield can usually be attributed to inefficient translation of the mRNA. When this is the case, the insertion of a first cistron can be expected to improve expression levels. As discussed in the previous section, a first-cistron sequence containing features shown in Fig. 1, can be chemically synthesized and, depending on the availability of compatible restriction sites, it can be cloned directly in front of the gene encoding the protein of interest. If compatible restriction sites do not exist, they can be readily introduced by PCR. Since the presence of a first-cistron sequence does not usually lower the expression levels of genes which express well without a first cistron sequence, a two-cistron expression system

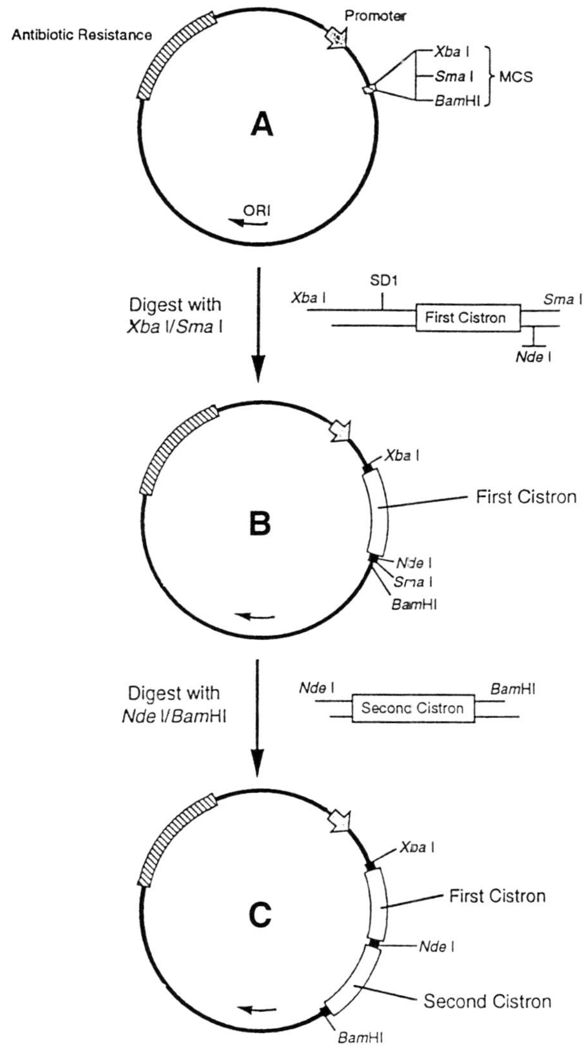

Fig. 2. Cloning scheme for the construction of a synthetic two-cistron expression system.

can be used routinely in all experiments designed to obtain high-level protein production *(12)*. Figure 2 gives a general outline of steps involved in constructing a two-cistron expression plasmid (plasmid C) using commercially available starting materials . Plasmid A represents a commercially available plasmid containing an origin of replication, an antibiotic resistance marker, an inducible promoter, and a multiple cloning site. A chemically synthesized first cistron sequence with *Xba*I and *Sma*I compatible ends can be inserted between the *Xba*I and *Sma*I sites within the multiple cloning site, giving rise to plasmid B. Because plasmid B has unique *Nde*I and *Bam*HI sites, these two sites can be used for the insertion of any coding sequence wherein the ATG start codon is part of the *Nde*I site. Although these sites may not normally be associated with the particular coding sequence of interest, especially when the source of the coding sequence is a cDNA, they can be introduced using PCR and primers containing the *Nde*I and *Bam*HI recognition sequences. If PCR amplification is used, sequence verification after cloning is strongly recommended.

3.3. Induction of Gene Expression and Analysis of Protein Yield

The construction of the two-cistron expression plasmids should be carried out in an *E. coli* host that does not allow for the recombinant protein to be expressed, particularly in those cases where the recombinant protein might be detrimental or lethal to the *E. coli* host. In this respect, the use of the T7 promoter is particularly convenient because this promoter is only utilized in cells expressing the T7 polymerase *(13)*. In the *E. coli* host BL21(DE3) (Novogen, Madison, WI), the T7 polymerase is encoded by the T7 polymerase gene, which resides on a λ prophage inserted into the host genome *(13)*. The expression of the T7 polymerase gene is under the control of the *lac*UV5 promoter and can be induced with the nonhydrolyzable lactose analog, IPTG. Upon addition of IPTG, usually to late log phase cultures, the T7 polymerase is synthesized, which will transcribe genes associated with the T7 promoter. After 2–6 h of continued incubation, the appearance of the recombinant protein can be expected. Most recombinant proteins expressed at high levels form intracellular protein granules (also referred to as inclusion bodies or refractile bodies) that are visible under the phase-contrast microscope *(19)*. Cells containing such granules tend to elongate and fail to undergo normal cell division. However, these cells are perfectly competent in carrying out protein synthesis as was demonstrated by the incorporation of radiolabeled amino acid precursors. A standard way of analyzing the production of recombinant proteins is by SDS-polyacrylamide gel electrophoresis (PAGE). Induced cultures (i.e., through the addition of IPTG) compared with uninduced cultures, differ primarily with respect to the presence of the expressed recombinant protein, which typically represents one of the major bands on a stained SDS-PA gel. Recombinant proteins present in granules can be readily isolated by simple differential centrifugation *(14)*. However, these proteins tend to be biologically inactive and require solubilization under high salt conditions and subsequent refolding to restore biological activity. Production of soluble protein can be encouraged by lowering the growth temperature to decrease the rate of protein synthesis allowing the protein to fold in vivo.

4. Notes

1. Choice of plasmid origin of replication. For some experiments, plasmids containing the ColE1 origin of replication may not be suitable, in which case plasmids containing the pl5A origin of replication (e.g., pACYC177 or pACYC184) can be used *(15)*. The copy number of the pl5A-based plasmids tends to be lower than those containing the ColE1 origin, however they are compatible with ColE1-based plasmids and therefore can coexist in the same *E. coli* cell, which is useful when the presence of two recombinant plasmids is required.
2. Choice of antibiotic resistance marker. The most commonly used antibiotics are ampicillin, tetracycline, kanamycin, neomycin, and to a lesser extent, chloramphenicol and spectinomycin. These antibiotics are relatively inexpensive and are soluble in typical culture media. An additional advantage of tetracycline is that the antibiotic is stable and, unlike ampicillin, is not degraded during culturing.
3. Choice of a promoter. Promoters that are commonly present on commercially available expression plasmids include the *lac* promoter and various derivatives thereof, the pL promoter from bacteriophage λ, and promoters from bacteriophages T7 and SP6. The T7 promoters are very strong promoters that require the T7 polymerase for transcription. The *E. coli* polymerase will not recognize these promoters and, for this reason, upon induction (usually of expression of the T7 polymerase, which has been inserted into the *E. coli* host genome), the only genes that are transcribed are those driven by the T7 promoter, leading to high level of expression. The pL promoter from bacteriophage λ *(16)* requires the coexpression of the temperature-sensitive λ repressor (cI857), which can be inactivated by a temperature shift to 40–42°C. The *lac* promoter and variants thereof can be induced with IPTG.
4. Choice of a first-cistron sequence. The sequence shown in Fig. 1 is only one of several possibilities for a first-cistron sequence. Other sequences may work equally well provided that the essential features are present. If addition of a particular first-cistron sequence does not significantly elevate the expression levels, additional bases can be added at the *Cla*I restriction site, which may disrupt any existing secondary structures. Since it is difficult if not impossible to predict which bases to add, linkers with random sequences can be cloned into the *Cla*I site, and the resulting clones can be screened for improved protein expression.
5. Choice of expression plasmids. Most commercially available expression plasmids contain multiple, unique restriction sites within the multiple cloning sites that are provided for cloning. A careful inspection of these sites prior to the selection of an expression plasmid and the design of a cloning scheme is recommended. Although *Nde*I is a very useful restriction site for the insertion of a second cistron (i.e., the gene of interest); in cases where a *Nde*I site is present within the coding region of the gene of interest, another restriction site containing an ATG initiation codon must be selected. *Nco*I (CCATGG), *Sph*I (GCATGC),

*Bsp*HI (TCATGA), and *Nsi*I (ATGCAT) are possibilities; however they are less useful because they determine the first base or the sequence of the second codon.

6. Choice of a host organism. Although *E. coli* is the most widely used host organism for the production of recombinant proteins, two-cistron expression systems have been used for the overexpression of proteins in other prokaryotic hosts such as *Bacillus (17)* and *Streptomyces (18)*. Because the translation of two-cistron expression systems resemble mechanistically the translation of polycistronic messages *(7)*, the use of a two-cistron expression system for enhancing gene expression in eukaryotic hosts is not appropriate.

Acknowledgments

I thank Barbara Fogleman for typing the manuscript and Steven Kahl for preparing the figures.

References

1. Gold, L. (1990) Expression of heterologous proteins in *Escherichia coli. Methods Enzymol.* **185,** 11–14.
2. Gold, L. and Stormo, G. D. (1990) High level translation initiation. *Methods Enzymol.* **185,** 899–993.
3. Schoner, B. E., Hsiung, H. M., Belagaje, R. M., Mayne, N. G., and Schoner, R. G. (1984) Role of mRNA translational efficiency in bovine growth hormone expression in *Escherichia coli. Proc. Natl. Acad. Sci. USA* **81,** 5403–5407.
4. Schoner, B., Belagaje, R., and Schoner, R. G. (1986) Translation of a synthetic two-cistron mRNA in *Escherichia coli. Proc. Natl. Acad. Sci. USA* **83,** 8506–8510.
5. Makolf, A. J., and Smallwood, A. E. (1990) The use of two-cistron constructions in improving the expression of a heterologous gene in *E. coli. Nucleic Acids Res.* **18,** 1711–1718.
6. Stanssens, P., Remaut, E., and Fiers, W. (1985) Alterations upstream of the Shine-Dalgarno region and their effect on bacterial gene expression. *Gene* **36,** 211–223.
7. Das, A. and Yanofsky, C. (1984) A ribosome binding site is necessary for efficient expression of the distal gene of a translationally-coupled gene pair. *Nucleic Acids Res.* **12,** 4757–4768.
8. Sambrook, J., Fritsch, E. F., and Maniatis, T. (1989) *Molecular Cloning. A Laboratory Manual*, 2nd ed. Cold Spring Harbor Laboratory, Cold Spring Harbor, NY.
9. Ausubel, F. M., Brent, R., Kingston, R. E., Moore, D. D., Seidman, J. G., Smith, J. A., and Struhl, K. (eds.) (1993) *Current Protocols in Molecular Biology.* Greene, Wiley, New York.
10. Perbal, B. (1988) *A Practical Guide to Molecular Cloning.* Wiley, New York.
11. Promega (1991) *Protocols and Applications Guide*, 2nd ed.
12. Schoner, B., Belagaje, R. M., and Schoner, R. (1990) Enhanced translational efficiency with a two-cistron expression system. *Methods Enzymol.* **185,** 94–103.

13. Studier, F. W., Rosenberg A. H., Dunn, J., and Dubendorff, J. W. (1990) Use of T7 RNA polymerase in direct expression of cloned genes. *Methods Enzymol.* 183, 60–89.
14. Schoner, R., Ellis, L. R., and Schoner, B. (1985) Isolation and purification of protein granules from *Escherichia coli* cells overproducing bovine growth hormone. *Bio/Technology* **3,** 151–154.
15. Chang, A. C. Y. and Cohen, S. N. (1978) Construction and characterization of amplifyable, multicopy DNA cloning vehicles derived from the pISA cryptic mini plasmid. *J. Bacteriol.* **134,** 1141–1156.
16. Simons, G., Remaut, E., Allet, B., Devos, R., and Fiers, W. (1984) High-level expression of human interferon gamma in *Escherichia coli* under control of the P_L promoter of bacteriophage lambda. *Gene* **28,** 55–64.
17. Parenti, D., Deferra, F., Galli, G., and Grandi, G. (1991) Prochymosin expression in Bacillus subtilis. *FEMS Microbiol. Lett.* **77,** 243–249.
18. Kuhstoss, S. and Rao, R. N. (1991a) A thiostreptoninducible expression vector for use in Streptomyces spp. *Gene* **103,** 97–99.

10

Expression of cDNA-Encoded Proteins by Cell-Free Transcription and Translation

Chistopher Coleclough

1. Introduction

Expression of proteins from cDNA clones by transcription and translation in vitro can be an attractive option for analytical applications that do not require significant quantities of product. Our laboratory has been interested in this technique because the translation of nucleic acid into protein sequence allows use of the unsurpassed resolving power of 2D protein electrophoresis to analyze complex mixtures of gene products. Thus, typically we might compare two cell types that differ in some phenotypic trait, initially by translation of mRNA purified from both sources and 2D electrophoresis of the translation products. If a 2D gel spot varies in an interesting manner between the cell types, we convert the population of mRNA molecules expressed in the positive cell type into a population of cDNA clones, then search among the clones for one that produces the spot of interest after in vitro transcription and translation. The 2D gel spot is used only as the diagnostic signature of an mRNA species, and does not need to be physically handled itself to permit the recovery of cognate cDNA clones; in this way, this approach differs radically from strategies based on microscale sequencing. However, other investigators may wish to use cell–free expression to monitor mutagenesis, epitope mapping or the biological activity of already isolated clones encoding known proteins: The methods described here can be applied equally well to single clones and to complex mixtures of clones.

We have constructed a family of cDNA-cloning vectors based on λ phage, specifically designed for cell-free expression of cDNA information; however, many other phage-cloning vectors exist with promoters for bacteriophage T7, T3, or SP6 RNA polymerase immediately adjacent to the site of foreign DNA

From: *Methods in Molecular Biology, vol. 62: Recombinant Gene Expression Protocols*
Edited by: R. Tuan Humana Press Inc., Totowa, NJ

integration, and the techniques described here can be used just as well with these other vectors—provided that the orientation of cDNA inserts is confidently known. This chapter presents methods for recombinant phage growth, DNA preparation, transcription, and translation; the construction of cDNA clones has been thoroughly described elsewhere *(1)*. Though the focus is on expression of cDNA cloned in λ phage vectors, the sections concerning transcription and translation are also applicable to plasmid clones.

The family of vectors that we use currently includes four members: λjac, λecc, λjacII, and λeccII. λjac and λecc have been described before *(1)*; λjacII and λeccII are illustrated in Fig. 1. All four vectors share a common system for the imposition of strong genetic selection for true recombinants: The oligo(dT)-containing primer used in cDNA synthesis also includes the 8-nucleotide *chi* motif, and preparation of the vector to receive cDNA inserts deletes the parental *gam* gene. The parental vectors, being gam^+, are spi^+ and will not form plaques on *E. coli* strains lysogenic for bacteriophage P2: only true, correctly oriented recombinants, that have lost the *gam* gene and incorporated a *chi* site will grow well on such strains (*[1,2]* aspects of λ phage biology relevant to the cloning of foreign DNA are covered, e.g., in the handbook by Sambrook, Fritsch, and Maniatis *[3]*, and might well be reviewed if the last few sentences are opaque). The new vectors, λjacII and λeccII, both incorporate a novel feature: Promoters for T7 and SP6 RNA polymerases are positioned in tandem upstream of the cDNA insertion site, in the same orientation, separated by most of the 5' untranslated region (UT) of encephalomyocarditis (EMC) virus. As shown in Fig. 1, transcription of recombinant genomes with T7 RNA polymerase generates RNA molecules, all of which contain the EMC 5' UT at their 5' termini, while use of SP6 RNA polymerase produces transcripts that lack EMC sequences, and closely resemble the original natural mRNA in structure. The 5' UT of EMC virus contains an internal ribosome binding site that confers cap-independence on the translation of any coding block immediately downstream, and can greatly enhance the translation of cellular mRNA sequences that are poorly translated in their natural context *(4)*.

2. Materials

1. Water of qualities appropriate for media and for biochemical manipulations (*see* Note 1).
2. *E.coli* strains LE392(P2) and TAP90.
3. L–broth (LB): 1.0% bacto-tryptone, 0.5% yeast extract, 0.5% NaCl, sterilized by autoclaving.
4. 20% maltose, sterilized by filtration.
5. 1*M* $MgSO_4$, sterilized by filtration.

6. 1.2% bacto-agar in LB/10 m*M* $MgSO_4$; sterilize by autoclaving and store in a screw-capped glass bottle. Melt in microwave oven and use as required.
7. Dilute top agar: 0.3% bacto-agar in LB/10 m*M* $MgSO_4$. Prepare from sterile 1.2% agar and LB as required.
8. 50 mL sterile, disposable polypropylene centrifuge tubes, e.g, Falcon #2098 (Becton Dickinson, Lincoln Park, NJ).
9. Sterile plastic Petri dishes, e.g., Falcon #1029.
10. Sterile 6-mL polypropylene tubes, e.g., Falcon # 2063.
11. Sterile 14 mL polypropylene tubes, e.g., Falcon # 2059.
12. Screw cap, sterile cryovials.
13. 1.5-mL microfuge tubes.
14. Glycerol.
15. $CHCl_3$.
16. Polyethyleneglycol (PEG) 6000.
17. Phenylmethylsulfonylflouride (PMSF).
18. Dimethylsulfoxide (DMSO).
19. Isopropanol.
20. 5*M* NaCl.
21. 1*M* Tris-HCl, pH 8.0.
22. 0.5*M* EDTA.
23. 20% SDS.
24. 1*M* Na acetate/acetic acid, pH 4.0.
25. 3*M* $MgCl_2$.
26. 2*M* Dithiothreitol (DTT); store at –20°C.
27. 1*M* spermidine; store at –20°C.
28. 1*M* HEPES/KOH, pH 7.8; store at –20°C.
29. 100-m*M* solutions each of ATP, CTP, GTP, and UTP; store at –20°C (available from, e.g., Pharmacia, Piscataway, NJ).
30. m^7G(5')ppp(5')G (cap analogue, e.g., NE BioLabs #1404, New England Biolabs, Beverly, MA).
31. 50:50:0.1 phenol/$CHCl_3$/8-hydroxyquinoline, saturated with 0.1*M*NaCl, 10 m*M* Na acetate, 1 m*M* EDTA.
32. 90:10 phenol/H_2O.
33. DNase I, 10 mg/mL in 10 m*M* Tris-HCl, pH 7.5, 1 m*M* $MgCl_2$; store at –20°C.
34. 10 mg/mL RNase A in H_2O. Hold in boiling water bath for 20 min and store at –20°C.
35. 10 mg/mL proteinase K in H_2O. Store at –20°C.
36. RNase inhibitor (RNasin), e.g., Promega #N2511 (Madison, WI).
37. T7 RNA polymerase.
38. SP6 RNA polymerase.
39. Rabbit reticulocyte lysate, micrococcal nuclease treated, e.g., Promega #L4960 (*see* Note 24).
40. ^{35}S-labeled methionine and cysteine, e.g., NEN #NEG009A and #NEG022T (DuPont NEN, Boston, MA) (*see* Note 25).
41. Unlabeled amino acid set, e.g., Sigma (St. Louis, MO) #LAA-21.

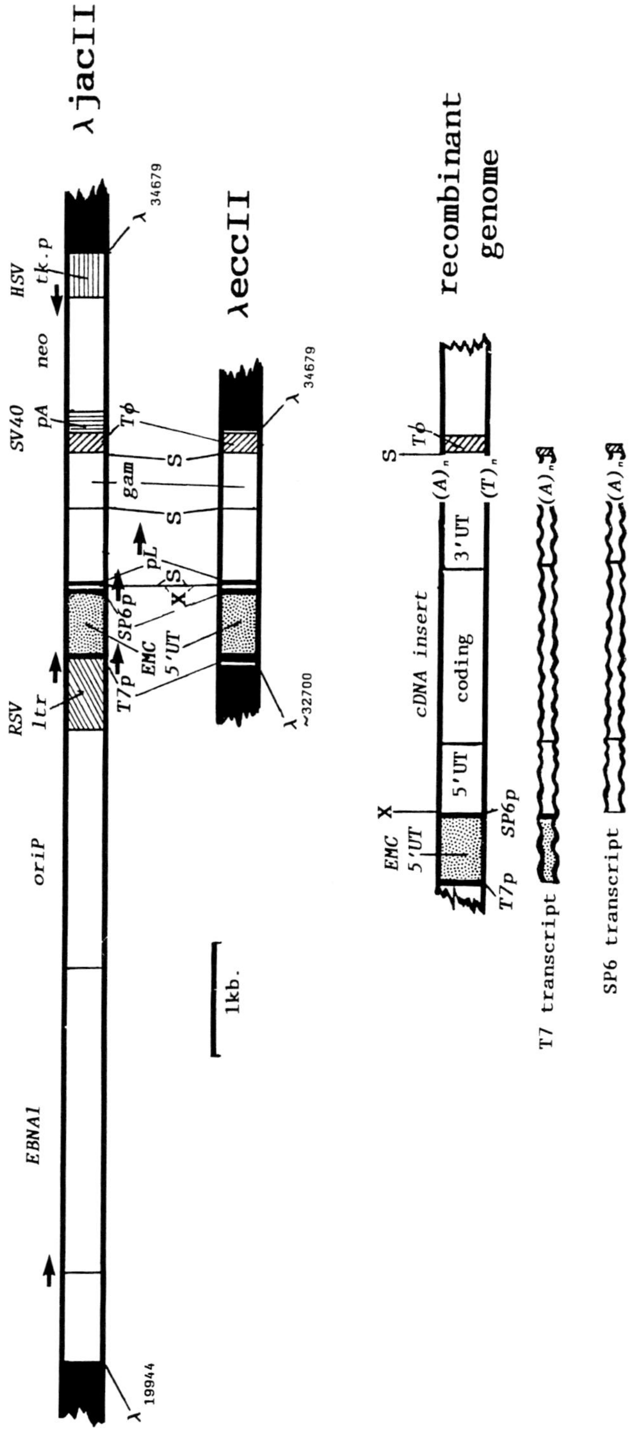
λjacII
λeccII
recombinant genome
HSV tk.p
neo
SV40 pA
Tφ
gam
S
pL
SP6p
X
EMC 5'UT
T7p
RSV ltr
oriP
EBNA1
λ 34679
λ 19944
λ ~32700
1kb.
cDNA insert
5'UT
coding
3'UT
(A)n
(T)n
T7 transcript
SP6 transcript

Fig. 1. Structure and transcription of λjacII and λeccII; λjacII and λeccII are based on λgtWES.λB *(9)*, as were λjac and λecc *(1)*, and maintain all of its relevant genetic features. The structure of the features around the site of cDNA insertion is illustrated for both new vectors as indicated, showing positions of junction with left and right λ phage arms shaded black; note that these numbers refer to coordinates of the standard λ map, and not to physical distances in these vectors.

λjac II incorporates the *EBNA1* gene and *oriP* replication origin from Epstein-Barr virus, as well as the Rous sarcoma virus long terminal repeat, positioned upstream of the cDNA insertion site. λjacII also contains a *neo* gene expression cassette, which includes the thymidylate kinase gene promoter from herpes simplex virus and the bidirectional polyadenylation signal from SV40. Sites for restriction enzymes XhoI (X) and SalI (S) are indicated; there are no sites for these enzymes in either vector other than those indicated. Digestion with XhoI plus SalI releases the λ phage *gam* gene, driven by a copy of the *pL* promoter, as DNA fragments, allowing insertion of cDNA downstream of T7 and SP6 promoters and upstream of *Tϕ*, the major transcriptional termination site from bacteriophage T7, which is recognized by both T7 and SP6 RNA polymerase. Arrows indicate the direction of transcription from relevant promoters.

λeccII omits all of the features relevant to expression in eukaryotic cells, but is identical to λjacII in the regions immediately flanking the cDNA insertion sites that contain the motifs used for cell-free expression.

Here is indicated the structure of a typical cDNA-recombinant genome of either λjacII or λeccII, as well the products of its transcription with T7 or with SP6 RNA polymerase. SP6 polymerase initiates immediately upstream of the cDNA insert, producing a transcript generally similar in structure to the natural mRNA molecule from which the cDNA was originally copies, including a 3' polyadenylate trace, and terminating at *To*. T7 polymerase initiates further upstream and results in the incorporation of the encephalomyocarditis virus 5' untranslated region (EMC 5' UT) at the 5' end of all transcripts. The genetic and biochemical effects that permit cDNA insertion only in the indicated orientation are described in ref. *1*.

3. Methods

3.1. Preparation of High-Titer Stocks of Recombinant Phage

1. Using colonies less than 2 wk old, innoculate 25 mL of LB/0.4% maltose/10 m*M* $MgSO_4$ with LE392(P2) (*see* Notes 1 and 2).
2. Grow with vigorous shaking to late log phase: 4–6 h at 38°C, or overnight at room temperature (*see* Note 3).
3. Pellet bacteria by centrifugation at 1500*g* for 10 min.
4. Resuspend bacteria in 15 mL sterile 10 m*M* $MgSO_4$; store at 4°C (*see* Note 4).
5. Pour LB/1.2% agar/10 m*M* $MgSO_4$ Petri plates (*see* Note 5).
6. Mix >10^5 PFU of recombinant phage with 0.15 mL resuspended LE392(P2) in a sterile 6-mL polypropylene tube; hold at 37°C for 15 min.
7. Add 3 mL 0.3% top agar; pour onto Petri plate.
8. Incubate right side up at 38°C for 5–7 h, when clearing should be evident (*see* Note 6).
9. Prepare 14 mL polypropylene tubes each containing 3 mL $CHCl_3$ and 0.5 mL glycerol.
10. Using a flamed, flat-end spatula, scrape the top agar from one plate into a tube containing $CHCl_3$ and glycerol; cap and shake well (*see* Note 7).
11. Spin tubes at 5000*g* for 10 min (*see* Note 8).
12. Transfer clear upper liquid to cryovial; store at –70°C.

3.2. Preparative Growth of Recombinant Phage and Purification of DNA

1. Grow and resuspend *E.coli* strain TAP90 exactly as described in Section 3.1. for strain LE392(P2) (*see* Note 9).
2. Using a glass Pasteur pipet, scrape off a small amount of frozen high-titer stock and transfer to a microfuge tube, without letting the stock vial thaw.
3. Transfer 1 µL of the phage stock to a sterile 50-mL polypropylene tube (*see* Note 10).
4. Add 0.3 mL resuspended TAP90; hold at 38°C for 15 min.
5. Add approx 22 mL of LB/10 m*M* $MgSO_4$.
6. Cap tightly and shake vigorously in a horizontal position at 38°C for 5–6 h, when lysis should be evident (*see* Note 11).
7. Add 4 µL 5 mg/mL DNaseI and 1 drop of $CHCl_3$. Shake at 38°C for 15 min.
8. Add 1.2 g solid NaCl; shake to dissolve (*see* Note 12).
9. Spin at 1500*g* for 10 min.
10. Transfer supernatant to a fresh 50-mL polypropylene tube, to which has been added 2.3 g PEG 6000; shake to dissolve.
11. Hold at 0°C for at least 1 h.
12. Spin at 1500*g* for 10 min at 4°C; discard supernatant.
13. Drain pellet, then resuspend in 0.8 mL 10 m*M* Tris-HCl, pH 8.0, 10 m*M* EDTA, 1% SDS; transfer to a 1.5-mL microfuge tube.
14. Add 0.5 mL phenol/$CHCl_3$, shake by hand, spin in microfuge for 3 min.
15. Transfer upper aqueous phase to a fresh microfuge tube.
16. Add 25 µL 5*M* NaCl, followed by 0.5 mL isopropanol; shake.

17. Spin for 5 min in microfuge at 4°C; carefully remove all supernatant.
18. Resuspend pellet in 0.5 mL 10 m*M* Tris-HCl, pH 8.0, 2 m*M* EDTA, 10 μg/mL RNase A; incubate at 50°C for at least 20 min (*see* Note 13).
19. Add 50 mL 5% SDS containing 500 μg/mL freshly diluted proteinase K; swirl to ensure adequate rinsing of tube and continue incubation at 50°C for at least 1 h (*see* Note 14).
20. Add 5 μL fresh 100 m*M* PMSF in DMSO; swirl to ensure adequate rinsing of tube (*see* Note 15).
21. Add 0.5 mL phenol/$CHCl_3$, shake by hand, spin in microfuge for 3 min.
22. Transfer upper aqueous phase to a fresh microfuge tube.
23. Add 15 μL 5*M* NaCl, followed by 0.27 mL isopropanol; shake (*see* Note 16).
24. Hold for 5 min at 0°C.
25. Spin for 10 min in microfuge at 4°C; carefully remove all supernatant.
26. Rinse pellet with 0.5 mL 50% isopropanol, remove all liquid, and allow to dry in air for 10 min (*see* Note 17).
27. Resuspend in 40 μL H_2O (*see* Note 18).

3.3. Transcription

Two variations are described; one generates 5'-capped RNA, the second produces triphosphate terminated RNA and should be used only for transcripts containing the EMC 5' UT.

1. By dilution of concentrated stocks, make 150 m*M* $MgCl_2$, 100 m*M* DTT, 20 m*M* spermidine.
2. Transfer 1 μL of DNA into a microfuge tube.
3. Make a reagent cocktail: for 10 reactions, **either** (a) mix in order, at room temperature, 38 μL H_2O, 20 μL 15 m*M* each of ATP, CTP, UTP and GTP, 10 μL $MgCl_2$/DTT/spermidine, 8 μL 1*M* HEPES/KOH pH 7.8, 6 μL RNasin, and 8 μL T7 or SP6 RNA polymerase (50 U/μL); **or** (b) keeping the other components the same, substitute for the A/U/C/GTP mix, 20 μL of a solution 15 m*M* in each of ATP, CTP, UTP and $m^7G(5')ppp(5')G$, and 4.5 m*M* in GTP (*see* Note 19).
4. Rapidly mix 9 μL aliquots of reagent cocktail with the 1 μL DNA samples (*see* Note 20).
5. Hold in an air incubator for 2 h at 38°C (*see* Notes 21 and 22).

(Transcripts containing the EMC 5' UT can now be translated immediately, or purified as follows.)

6. Add 100 μL 0.15*M* Na acetate, pH 4.0, followed by 70 μL 90% phenol/H_2O.
7. Vortex; spin in microfuge for 3 min.
8. Carefully remove upper, aqueous phase, and transfer to a fresh microfuge tube.
9. Add 45 μL isopropanol, vortex, place in ice for 10 min (*see* Note 23).
10. Spin for 10 min in microfuge at 4°C, noting the orientation of each tube.
11. From the side of the microfuge tube opposite to the (frequently invisible) pellet, carefully remove all liquid with a pipet.

12. Add 200 μL 50% aqueous isopropanol; spin in 4°C microfuge for 3 min.
13. Carefully pipet off all liquid; allow pellet to dry in air for 10 min.
14. Dissolve pellet in 5–10 μL H_2O; if desired, store at –70°C (*see* Note 24).

3.4. Translation

The method described here produces a small mass of ^{35}S labeled protein, and is suitable for gel electrophoresis or other experiments in which expressed proteins can be traced by radioactive emission. If the aim is to assess the activity of an expressed protein on some other substrate, it may be preferable to omit the radiolabel. Again, two variants are described; one for 5'-capped RNA, the other for RNA containing the EMC 5' UT.

1. For simplicity, purchase micrococcal nuclease-treated rabbit reticulocyte lysate that is dispensed in small aliquots, e.g., Promega #L4960. Store in liq. N_2 (*see* Note 25).
2. Purchase ^{35}S methionine and ^{35}S cysteine, e.g., NEN #NEG 009A and #NEG 022T, store at –70°C, and when first thawing, working in a fume hood, dispense into 10-L aliquots and store them at –70°C (*see* Note 26).
3. Prepare a stock that is 5 m*M* in each of the 18 standard L-amino acids except for cysteine and methionine; store at –20°C.
4. Prepare a reaction cocktail: For 10 reactions, mix **either** (a) 200 μL reticulocyte lysate, 10 μL ^{35}S methionine, 10 μL ^{35}S cysteine, 7.5 μL H_2O and 5 μL amino acid stock; **or** (b) keeping all other components the same, substitute 7.5 μL 2*M* KCl for 7.5 μL H_2O. Use the first recipe for 5' capped transcripts, the second only for transcripts containing the EMC 5' UT (*see* Note 27).
5. Dispense 23-μL aliquots of reaction cocktail into microfuge tubes.
6. Add 2 μL of RNA, transcribed as in Section 3.3.
7. Incubate for 2 h at 30°C.
8. Add 0.5 μL 10 mg/mL RNase A to each reaction, and continue incubation for 20 min (*see* Note 28).
9. Store reactions at –70°C until analysis (*see* Notes 29 and 30).

4. Notes

1. Contamination of water used to prepare media can result in DNA preparations that are inactive with some enzymes, presumably because trace microbial acidic polysaccharide contaminants are concentrated along with bacteriophage DNA. If DNA preparations cannot be transcribed, or cut with certain restriction enzymes, suspect the water. HPLC purity-grade water can be purchased, e.g., from Fisher, and used to dissolve purified nucleic acids and in biochemical procedures
2. The λjac and λecc vectors that we use all have nonsense mutations that require both *sup E* and *sup F* for effective suppression, which LE392 supplies. Other vectors are less demanding, but LE392 is a good general host. Note that the P2 lysogenic derivative of LE392, denoted LE392(P2), will only allow growth of *spi*⁻ phage, and is the selective host for true recombinants in the jac/ecc system. Parental λjac or λecc vectors will not form plaques on LE392(P2), and neither

will any phage, recombinant or not, that contains a functional *gam* gene; LE392 should be substituted for LE392(P2) for the growth of such phages.

3. Sterile, disposable plastic tubes can be conveniently used both as culture vessels and for centrifugation.
4. Bacteria can be used for 10 d when stored in this way.
5. Conditions of phage amplification can be varied considerably for convenience. For growth of large numbers of individual clones, we use 24-well tissue culture dishes, the wells each containing 1 mL of sterile LB/10 m*M* $MgSO^4$/1% agarose. 10^3–10^5 PFU of phage plus 10^6 plating bacteria are mixed in a 200-µL LB/10 m*M* $MgSO_4$ overlay in each well. Lysis is evident in 8–10 h, and the overlay can be removed with a pipet.
6. As with any virulent phage, plaque size can be controlled by environmental factors: moist, fresh plates, dilute top agar, and decreased input of bacteria, all favor large plaque formation and thus the generation of completely-lysed plates when the input of phage PFU is low or unknown.
7. When processing many plates, crosscontamination can be conveniently avoided by immersing the used spatula tip in a small beaker of ethanol, flaming it until the ethanol has burnt off, then cooling it in a beaker of 10 m*M* $MgSO_4$ before scraping the next plate.
8. 14-mL polypropylene tubes like Falcon #2069, without caps but closed with Saran Wrap, fit directly in the Sorvall S*M* 24 rotor.
9. TAP90 is a *recD*$^-$ *supEsupF* strain derived specifically for the growth of *gam*$^-$ phage (5). Yields of DNA are higher with this host than with LE392, but it is emphasized that *chi* will have no effect on the growth of phage in TAP90. Therefore, TAP90 cannot be used as a host to select true recombinants in the λjac/ecc system, and should only be used for recombinant phage that have been selected by two rounds of growth on LE392(P2).
10. With many simultaneous cultures, we find it convenient to shake Falcon #2098 tubes horizontally in the styrofoam rack and plastic sleeve in which they are supplied.
11. With TAP90, fully lysed cultures may still appear somewhat turbid. Cultures that show no evidence of lysis should be aborted, as the final phage DNA product is likely to be heavily contaminated with bacterial DNA.
12. Though addition of NaCl prior to the removal of bacterial debris is preferable, it may be found more convenient to add both NaCl and PEG simultaneously to cleared lysates, when processing many cultures.
13. To be effective, the DNase and RNase treatments cannot be simultaneous: The integrity of the phage particle, which confers DNase resistance on DNA, requires Mg^{2+}, which stabilizes ribosomes and protects ribosomal RNA from RNase. It would clearly be advantageous to avoid the use of RNase for a DNA preparation that is ultimately to be used to generate RNA. However, we have found this to be the simplest (and least bacterial RNA, which can otherwise inhibit the RNA polymerases; but it is critical that the subsequent proteinase treatment be effective.
14. 50°C is ideal, but the RNase and proteinase digestions can be performed at 37°C, in which case the incubation times should be doubled.

15. Avoid skin contact with PMSF. As PMSF binds rapidly and covalently to proteinase K and is unstable in aqueous solution, its concentration is relatively unimportant. We prepare a PMSF solution of approx 100 m*M* by shaking a few crystals into a microfuge tube and adding a few hundred liters of DMSO.
16. It is important that a minimal amount of isopropanol be used to precipitate the DNA, if severe contamination with RNA fragments is to be avoided. Aim to add 0.55 solution volumes of isopropanol.
17. Nucleic acids precipitated with minimal isopropanol form a pellet that is qualitatively distinct from the more familiar ethanol precipitate, in being more compact and translucent.
18. This procedure typically yields around 20 µg of DNA.
19. Commercial SP6 and T7 RNA polymerase is frequently supplied at about 50 U/µL; compensate for more or less concentrated enzyme by appropriately altering input volumes of enzyme and H_2O. The methylated cap dinucleotide is preferentially used for initiation by these polymerases, so its inclusion results in a large proportion of transcripts being 5' capped. Cap dinucleotide should be included for all transcripts that are to be translated, except where they are to contain the EMC 5' UT, unless there is definitive proof that 5' caps are unimportant for the translation of mRNA species of interest in rabbit reticulocyte lysate.
20. Use of ice-cold reaction cocktail or slow mixing can result in precipitation of the long linear phage DNA by spermidine.
21. Use of an air incubator ensures uniform temperature distribution and so avoids the possibility of loss of volume by condensation, which would be disastrous with these tiny reaction volumes.
22. The transcription reaction conditions are based on those described by Gurevich et al. *(6)*, and result in remarkably more efficient transcription than the former Tris-based buffer. Particularly when the T7 polymerase is used without the cap dinucleotide, so much RNA may be produced that it precipitates—this is not a problem, but should be monitored especially if, as may be the case if the EMC 5' UT is included, RNA is to be added directly to reticulocyte lysate without purification.
23. Phenol extraction under acid conditions removes both DNA and protein, and precipitation with minimal isopropanol leaves in solution the nucleotides, in particular the cap dinucleotide, which must be eliminated to avoid its competition for cap binding protein. Extraction with 90% phenol will typically result in about a 30% reduction in the volume of the aqueous phase. The amount of isopropanol used should be about 0.6 vol of the aqueous phase.
24. It is not critically important to know accurately the yield of RNA, which can vary considerably. However, the relative yield should be roughly judged from the bulk of the isopropanol pellet, and the volume of water used to dissolve it varied appropriately: use 5 µL for invisible pellets.
25. We have used wheat germ extract successfully, but the nuclease-treated rabbit reticulocyte lysate is our standard. Variation between lysate from different sup-

pliers and from batch to batch is not a trivial issue: we used to use Amersham #N90 with routine success, until a batch completely failed to translate any RNA made in vitro, though it was highly active on natural mRNA. We subsequently discovered that lysate from Ambion and from Promega could translate the in vitro transcripts that were inactive with the Amersham lysate. The ideal solution to this problem, of course, is to prepare and characterize one's own lysate, as the yield from a single rabbit is sufficient for years of experiments on the small scale described here. This is not difficult, and is described in many publications (e.g., ref. *7*); if this is undertaken, process 5–6 rabbits individually, as there is considerable animal-to-animal variation. Select only the lysate that is most active, treat with nuclease and supplement as described *(7)*, and store in small aliquots in liq. N_2. Remember that activity on natural mRNA may not predict activity on in vitro transcripts.

26. Comparison of 2D gels of translation products of one RNA preparation labeled only in methionine, and only in cysteine, reveals surprising differences in relative spot intensity. Therefore, unless it is known that the protein of interest is well labeled using only one of the ^{35}S amino acids, use of both is recommended.
27. Addition of KCl to about 60 m*M* over the existing level in the lysate as supplied considerably enhances the translation of RNA molecules that contain the EMC 5' UT, and suppresses the translation of those that do not (*see* ref. *8*).
28. Addition of RNase eliminates ^{35}S aminoacyl-tRNAs that can confuse electrophoretic analysis of translation products.
29. In the simplest case, translation products may be analyzed by SDS polyacrylamide gel electrophoresis. It must be borne in mind that reticulocyte lysate typically contains about 60 mg/mL hemoglobin, and it is crucial to add sufficient SDS to saturate this protein mass if reasonable gel profiles are to be obtained: We routinely dilute translation products with 5 vol of loading buffer that contains 3% SDS. With conventional 2D gel electrophoresis, the main technical difficulty is that all of the hemoglobin focuses as a single band, which tends to make the IEF gel brittle at that point, limiting the amount that can be loaded to about 15–20 μL of translation reaction, diluted with at least 3 vol of standard urea solubilization buffer. For assays based on the biological activity of translated protein, remember that reticulocyte lysate contains a very complex mixture of proteins in addition to hemoglobin, so preliminary controls must ensure detectability above the lysate background.
30. We have very recently established that transcripts produced from a single collection of cDNA clones in λjacII or λeccII with T7 polymerase (thus containing the EMC 5' UT), generate a very different set of translation products from those produced by SP6 polymerase transcripts of the same collection (that lack the EMC 5' UT), as assayed by 2D gel electrophoresis. We are currently assessing the biological significance of this observation, but, technically, it strongly suggests that users should initially compare transcripts of their species of interest containing and lacking the EMC 5' UT, to determine which mode of transcription is more appropriate for their use.

Acknowledgments

These methods were developed in the context of a collaboration with Drs. Ivan Lefkovits, Basel Institute for Immunology, Switzerland, and Charles Auffray, Genethon, France. St. Jude Children's Research Hospital is supported by Grant CA 21765 from the National Cancer Institute and by the American Lebanese Syrian Association (ALSAC).

References

1. Coleclough, C. (1993) Cell-free expression of large collections of cDNA clones using positive-selection λ phage vectors. *Methods Enzymol.* **217,** 152–170.
2. Lefkovits, I., Kettman, J., and Coleclough, C. (1990) A strategy for forming a global lymphocyte proteinpaedia and gene catalogue. *Immunol. Today* **11,** 157–162.
3. Sambrook, J., Fritsch, E. F., and Maniatis, T. (1989) *Molecular Cloning, a Laboratory Manual*, 2nd ed. Cold Spring Harbor Laboratory, Cold Spring Harbor, NY.
4. Parks, G. D., Duke, G. M., and Palmenberg, A. C. (1986) Encephalomyocarditis virus 3C protease: efficient cell–free expression from clones which link viral 5' noncoding sequences to the P3 region. *J. Virol.* **60,** 376–384.
5. Patterson, T. A. and Dean, M. (1987) Preparation of high titer lambda phage lysates. *Nucleic Acids Res.* **15,** 6298.
6. Gurevich, V. V., Pokrovskaya, I. D., Obukhova, T. A., and Zozulya, A. (1991) Preparative *in vitro* mRNA synthesis using SP6 and T7 RNA polymerases. *Anal. Biochem.* **195,** 207–213.
7. Clemens, M. J. (1984) Translation of eukaryotic messenger RNA in cell-free extracts, in *Transcription and Translation: A Practical Approach* (Hames, B. D. and Higgins, S. J., eds.), IRL, Oxford, pp. 232–270.
8. Jackson, R. J. (1991) Potassium salts infuence the fidelity of mRNA translation in rabbit reticulocyte lysates: unique features of encephalomyocarditis virus RNA translation. *Biochim. Biophys. Acta* **1088,** 345–358.

III

Expression in Yeast

11

Yeast Plasmids

Keshav K. Singh and Jack A. Heinemann

1. Introduction

The yeast *Saccharomyces cerevisiae* provides an excellent system to study genes of eukaryotes because it has been extensively characterized genetically and because the molecular mechanisms governing many cellular processes in yeasts are conserved in other organisms. For example, yeasts provide a powerful system for the study of mammalian proteins. However, to study the function of a cDNA encoding a heterologous protein in yeast, the cDNA needs to be cloned in an appropriate vector that permits expression, correct localization, and the posttranslational modification of the product. In this chapter we describe the yeast vectors available for analysis of a new gene and its product and provide two recommended transformation protocols.

1.1. Yeast Cloning Vectors

A generic structure of yeast transformation vectors is shown in Fig. 1. As depicted, a typical yeast plasmid vector contains:

1. Two types of genes, one of which confers a selectable phenotype in *Escherichia coli*, and the other in *S. cerevisiae*. Sometimes one gene, such as *LEU2* or *URA3*, can be used to select either organism harboring the plasmid because several yeast genes involved in different biosynthetic pathways complement analogous mutations in *E. coli*. Nevertheless, a drug resistance gene is generally used to select the plasmid in *E. coli*, thus circumventing the need to use particular auxotrophic strains and minimal medium.

   Many laboratory yeast strains are auxotrophic owing to a recessive mutation in at least one gene, making it easy to design a selection for plasmids that encode a complementing allele. For example, *LEU2, HIS3, TRP1*, and *URA3*, singly or in combination, and other dominant markers, such as HSV-1 *TK* (thymidilate kinase), are used routinely in auxotrophic strains with corresponding recessive

From: *Methods in Molecular Biology, vol. 62: Recombinant Gene Expression Protocols*
Edited by: R. Tuan Humana Press Inc., Totowa, NJ

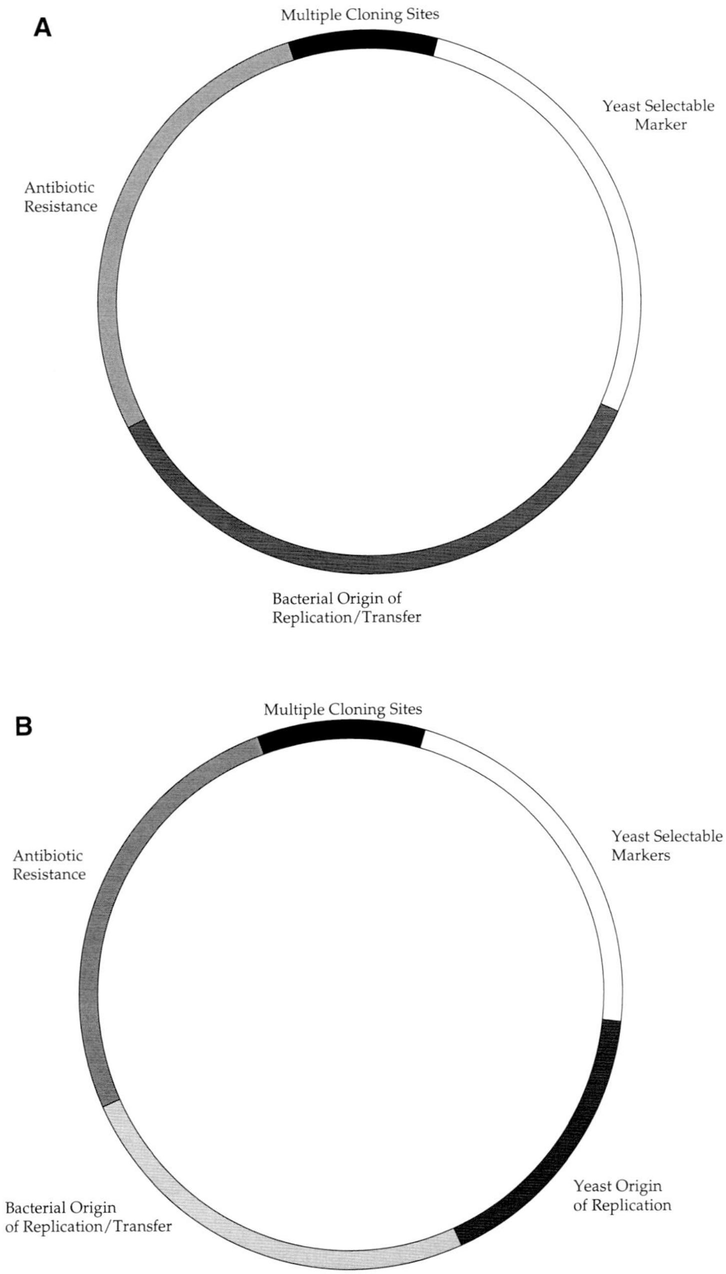

Fig. 1. Essential features of yeast transformation vectors are shown. **(A)** Yeast integrating vectors usually encode an antibiotic resistance gene, yeast selectable marker, and a multiple cloning site. **(B)** Yeast extrachromosomal vectors encode a yeast origin of replication in addition to possessing the essential features of integrating vectors.

mutations *(1,2)*. Other selectable markers are available for use in wild-type strains. These include genes that confer resistance to copper *(3)*, methotrexate *(4)*, hygromycin *(5)*, tunicamycin *(6)*, phleomycin *(7)* and chloramphenicol *(8)*.
2. Unique restriction sites to facilitate cloning.
3. An *E. coli* origin of replication. Depending on the origin, vector DNA can be amplified considerably providing large quantities of DNA for yeast transformation.

Yeast transformation vectors can be classified into two general categories: integrative vectors and extrachromosomal vectors. Extrachromosomal plasmid vectors contain an origin of DNA replication, whereas integrative vectors do not (Fig. 1A,B). Sikorski and Hieter have constructed a uniform set of yeast vectors specifically for cloning *(9)*. Based on the pBluescript backbone, they offer the many advantages of using vectors designed for cloning in *E. coli*, such as color screens to detect inserts, multiple unique restriction sites, the ability to generate ssDNA, and unidirectional deletions, and high DNA yields in bacteria. The set consists of four integrating and four autonomous plasmids, each cognate pair harboring a different marker for selection in yeast. A complementary set of yeast auxotrophs with nonrevertible alleles was also constructed for use with these plasmids.

1.1.1. Yeast Integrating Plasmid Vectors

Yeast integrating plasmids (YIp) cannot replicate efficiently in yeast because they lack autonomous replication functions. The YI plasmids are useful for generating mutations in yeast by homologous recombination. The frequency of transformation for integrating plasmids is very low (1–10 transformants per μg of DNA). The transformation frequency can, however, be increased 10–1000-fold following restriction of the plasmid. The double strand break introduced by the enzyme must occur within a region of the plasmid that has sequences in common with some region of the yeast genome to direct the site of integration (to either a chromosome or another plasmid).

Integrated plasmids are usually very stable, but can be lost during culture in a medium that does not select for the integrated marker. The YI plasmids are lost at a frequency of approx 0.1–2% per generation. Weak replication of plasmids in transformants, rather than integration, may also produce microcolonies *(10)*.

1.1.2. Extrachromosomal Plasmid Vectors

1.1.2.1. Yeast Episomal Plasmids (YEp)

The YE plasmids replicate autonomously using the origin of 2 μm circle, a multicopy endogenous yeast episome. Plasmids with 2 μm sequences transform yeast efficiently (10^4–10^5 transformants per μg of DNA). YE plasmids replicate to 20–50 copies per cell, depending on their selectable marker *(6)*.

However, YE plasmids are unstable in neutral environments, being lost at a frequency of 0.2–2% per generation. This instability has been put to creative use in certain genetic screens (*11*; discussed below).

1.1.2.2. Yeast Replicating Plasmids (YRp)

The YR plasmids replicate autonomously using yeast chromosomal sequences. These autonomous replicating sequences (ARS) act as origins of replication and transform yeast at high frequency (10^4–10^5 transformants per μg of DNA) but are highly unstable (10–20% of the cells lose plasmids per generation). Instability is attributed to a strong bias in segregation that favors the mother cell. Although these plasmids can be present in extremely high numbers (up to 100 copies per cell), their utility is limited by their instability.

1.1.2.3. Yeast Centromeric Plasmids (YCp)

When centromeric sequences are incorporated in YRps, the plasmids become stable both mitotically and meiotically and, unlike YR plasmids, they show no segregation bias. The YCps are maintained at a low copy number (one per cell) and have a loss rate of 1% per generation.

Segregational instability is not always a deleterious property. Several ingenious schemes have used plasmid instability to indirectly identify synthetic lethal, multicopy suppressee, and redundant genes *(11–13)*. Not all genes confer a phenotype that can be selected or screened for several different reasons. First, the function of two distinct genes may be redundant, overlapping sufficiently to compensate for the loss of either gene alone. Therefore, two or more mutations could be required for a scorable phenotype. It may also be desirable to create an artificial redundancy in yeast using a heterologous gene from another organism in order to characterize the function of the heterolog (and perhaps identify a new gene in yeast!). Of course, this approach requires isolating a mutation in the yeast analog, a laborious or impossible task if the heterolog's function is unknown. Second, not all genes are essential and thus designing a means of detecting their loss or complementation may be beyond the intuitive insights of the researcher. Finally, it may just be impossible to raise conditional mutations in some genes.

One means of circumventing these problems makes use of yeast plasmid instability and a color screen. Yeast with an *ade2 ADE3* genotype, for example, accumulates a red pigment which is easily visualized, whereas *ade2 ade3* strains are white. When plasmids carrying the *ADE3* gene are maintained in an *ade2 ade3* background, plasmid instability can be monitored by colony sectoring: that is, areas of the colony devoid of pigment due to the growth of clones that have lost *ADE3*. The YE and YC plasmids are sufficiently instable to produce sectoring in most colonies when the plasmid itself is not required for survival. Sectored colonies are rare or extremely infrequent when the plasmid carries another gene essential for survival. Therefore, to isolate mutants that require a complementing allele of one gene present on a plasmid (e.g., a gene whose function is normally made redundant), one can screen for clones that give rise to wholly red colonies after random mutagenesis (to mutate the redundant gene). A particularly good discussion of the advantages and pitfalls of indicator genes used for colony sectoring assays is given in Bender and Pringle *(11)*. Occasionally, loss of sectoring may be owing to dependence on plasmid sequences other than the gene of interest, a condition that can usually be determined by a few simple controls. Alternatively, introduced sequences may increase plasmid stability by other mechanisms, such as fortuitous ARS activity. Cvrckova and Nasmyth *(12)* overcame this problem by introducing a destabilizing mutation in their YC vector.

1.2. Yeast Expression Vectors

Expression plasmids contain two types of transcription cassettes: inducible and constitutive. The inducible cassettes may include the promoters of *GAL4 (14)*, *ADH2 (15)*, and Cu metallothionin protein *(16)*. The constitutive cassettes include *PGK* (17) and *G3PDH* promoters *(18)*. Both inducible and constitutive promoters respond to changes in growth conditions. The detailed structure of vectors and the protocol for expression of fusion proteins are described in Chapters 12–14 of this book.

An alternative expression system utilizes the inducible mammalian glucocorticoid response element. Schena and Yamamoto demonstrated that the mammalian steroid receptor functions as a conditional transcriptional regulator in yeast *(19)*. Based on these observations Picard et al. developed a glucocorticoid-inducible expression system *(20)*. The expression system requires two yeast vectors (Fig. 2). One vector (p2UG) contains *URA3* for selection and three 26 bp glucocorticoid response elements (GRE) upstream of the *CYC1* promoter region in a YE plasmid. The other vector (pG-N795) contains *TRP1* for selection and the glucocorticoid receptor cDNA. When glucocorticoid is added to the culture medium, the receptor binds GRE resulting in increased transcription of the gene fused to the *CYC1* promoter. Detailed structures and the important features of the glucocorticoid expression vectors are described by Schena et al. *(21)*.

1.3. Plasmid Vectors for Detecting Protein–Protein Interaction

Protein-protein interactions in a multiprotein complex play critical roles in essential cellular functions. Two elegant systems, the two-hybrid system and interaction trap *(22–24)*, were recently developed to detect protein–protein interactions (Fig. 3). Both systems involve a set of two vectors containing

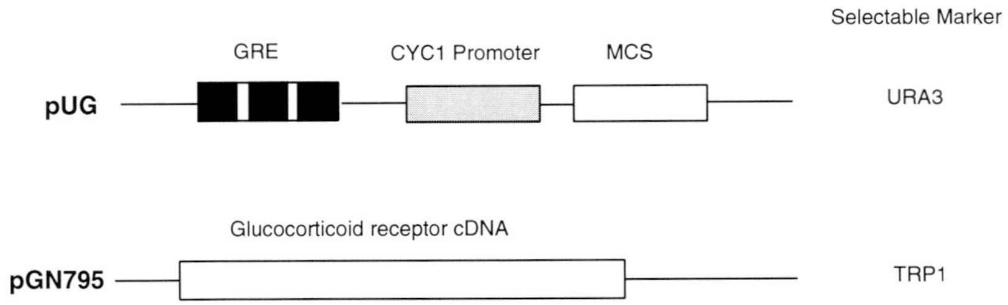

Fig. 2. Essential features of hormone-inducible expression vectors are shown. The plasmid p2UG contains three tandem copies of the 26 bp glucocorticoid response elements (GRE) fused upstream of the *CYC1* promoter. A downstream multiple cloning site permits insertion of cloned genes. The plasmid pGN795 encodes the glucocorticoid receptor cDNA. Glucocorticoid added to the culture medium induces transcription by stimulating receptor binding to GRE.

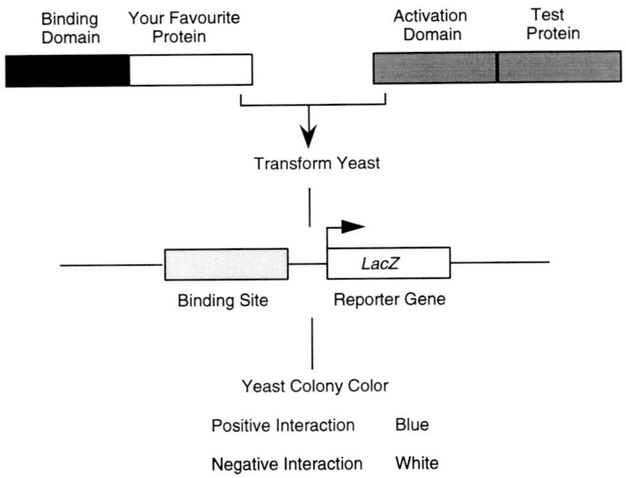

Fig. 3. Essential features of yeast vectors used in two hybrid and interaction trap systems are shown. A set of two vectors are used. One vector contains a sequence that encodes the DNA-binding domain of a transcription factor and the protein of your interest. The second vector encodes the activation domain of the transcription factor (in the first vector) and a test protein or library of cDNA sequences. Successful expression and localization of the two fusion proteins may result in transcriptional activation of a *lacZ* reporter gene, which creates blue colonies on plates seeded with X-gal (5-Bromo-4-chloro-3-indolyl-β-D-galactoside).

hybrid genes. One vector contains a sequence that encodes the DNA-binding domain of a transcription factor in frame with "your favorite protein" (YFP, the gene of the protein to be tested for interaction with another protein). The second vector contains a gene fusion composed of the cognate transcriptional activation domain of the transcription factor from the first vector and the test protein. If YFP interacts with the test protein, the two domains of the transcription factor may form an active complex. In such cases, the interaction of YFP and test proteins induces transcription of a reporter gene (e.g., *lacZ* yielding blue colonies) in transformants expressing both YFP and the test protein fusions. The two-hybrid system utilizes the yeast *GAL4* transcriptional factor and the interaction trap system depends on bacterial *lexA* to assay protein-protein interactions in vivo in yeast.

These vector sets are useful for multiple analyses. Interactions can be quantitated by measurement of β-galactosidase levels. In addition, these plasmid systems can be used to develop a screen for proteins that interact. A cDNA library made in the complementary vector can be screened for gene products that interact with YFP. However, negative results obtained with these systems must be interpreted cautiously. For example, a positive signal requires that the YFP and test proteins fold in such a way as to assemble an active transcriptional activator and localize to the nucleus. Therefore, it should be noted that an absence of signal does not necessarily mean that the proteins do not interact. When yeasts are used to evaluate the potential for interaction of heterologous proteins, it is also possible that absence of a positive signal is because the proteins of interest do not interact in yeast. Further details of these methods can be found in refs. *22* and *25*.

1.4. Other Yeast Vectors

1.4.1. YAC Vectors

Yeast artificial chromosome (YAC) vectors permit the cloning and genetic modification of extremely large DNA fragments (up to thousands of kilobases in size), that cannot be cloned by conventional techniques *(26,27)*. Large fragments of DNA are cloned by ligation in vitro to two separate vector arms. One arm contains the yeast centromere, a selectable marker, and a telomere. The other arm contains a second yeast selectable marker and telomere. Both markers and both telomeres are required for successful transformation, reducing the likelihood of false positives. The YAC cloning system has been extremely useful for physical mapping of the nematode genome *(28)*, cloning human genes *(29,30)*, and cloning centromeric fragments from other systems *(31)*. YAC technology has also provided a powerful new technique to study regulation and function of human genes *(32)*.

The plasmid family YAC3–55 are widely used vectors *(27)*. Family members primarily differ in their repertoire of restriction sites useful for cloning. These plasmids replicate as a circular molecule in *E. coli* and as either a circular or a linear molecule, depending on the preparation, in *S. cerevisiae*. Proper insertion of DNA fragments into a unique restriction site in the *SUP4* (ochre-suppressor) gene produces red transformants in an *ade2-1* background. A second digestion of pYAC with *Bam*HI liberates the *HIS3* gene linking the two telomeres in the circular molecule and produces a linear molecule that functions as a chromosome in yeast (Fig. 4).

The YAC technology has been limited by vector stability. The YACs can be unstable as a result of segregational loss and recombination. Many existing YAC libraries contain chimeric DNA sequences (i.e., sequences from different regions of the genome cloned into a single YAC) *(33)*. Such clones are vulnerable to deletions and rearrangements *(34)*.

1.4.2. Plasmids for Conjugal Transfer of DNA

Naturally-occurring bacterial plasmids can have large host ranges and certain plasmids also encode the means to transmit themselves to a large number of bacterial species. It nonetheless came as a bit of a surprise that plasmids which largely exist within the cytoplasm of prokaryotes had the means to transfer themselves from bacteria to eukaryotes such as yeast. Previously, such a transmission process was thought to be the exclusive practice of the phytopathogen, *Agrobacterium tumefaciens*, which causes tumorigenesis following transmission of plasmid DNA to plants *(35)*. That DNA transfer process appears to be genetically and physically the same as conjugation between bacteria *(35,36)*, suggesting that the plasmid-encoded mechanism of DNA transfer is generically suited to move DNA across phylogenic, even kingdom, boundaries. Since the first announcement that plasmids could be introduced into *S. cerevisiae* by conjugation with *E. coli* *(37)*, *E. coli* has hosted the transfer of plasmids to six different species of yeast in five genera *(10,38,39)*.

The YEp, YCp, YRp and YIp vectors have been successfully transmitted to yeast by conjugation. The general requirements of the conjugative transfer of plasmids from bacteria to yeast have been described in earlier reviews *(35,40)*. At least three high frequency systems have been put into practice *(10,30,37)*. The systems differ in the origin of plasmid-encoded functions required for DNA transmission but all use a two-plasmid approach. The first plasmid, or shuttle vector, contains the functions which permit its replication and selection in both bacteria and yeast and a *cis*-acting sequence called *oriT*, for the origin of conjugal transfer. The second plasmid, or helper, is much larger and usually only replicates in bacteria, but encodes most or all the factors required *in trans* for DNA transfer. The three systems offer some flexibility for cloning and for pairing of plasmids. With increasing numbers of different plasmids in a single cell, comes increasing complexity in finding plasmid combinations that are replication compatible and carry unique selectable markers.

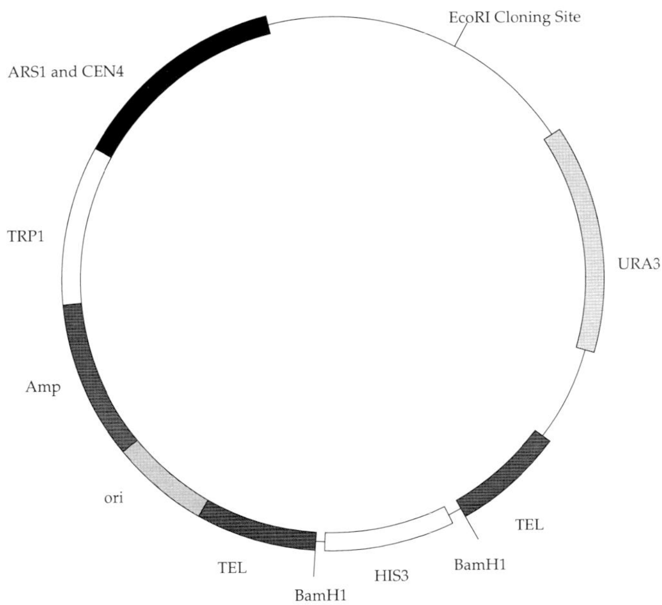

Fig. 4. Essential features of YAC vectors are shown. The plasmid pYAC4 is illustrated. pYAC4 contains the ARS1/CEN4 origin of yeast replication and two telomeres (TEL), the pMB1 (pBR322) origin of replication in bacteria, the yeast *TRP1* and *URA3* genes (the *HIS3* gene, which is deleted upon linearization of pYAC4), and encodes amp^r in bacteria.

Nevertheless, the two-plasmid system has important advantages. First, the frequency of plasmid transmission has been inversely correlated with the size of the transferred molecule *(37)* and since 30–40 different *trans*-acting genes may be required for conjugation, separating the helper and shuttle optimizes transmission frequencies. Second, smaller vectors are easier to manipulate in vitro and conjugative plasmids tend to be large (on the order of 100 kb) low copy number replicons, making them poor choices for cloning.

Current protocols optimize the frequency of yeast transconjugant formation by YE plasmids, with reports of transmission frequencies as high as 0.1 transconjugants per bacterium *(35,37)*. Lower frequencies could be a result of many difficult to determine factors from DNA mobilization of the shuttle by the helper (within the bacterial cytoplasm), to DNA transfer (especially between different combinations of bacterial and yeast strains), to plasmid establishment, replication, and phenotype expression (after transfer to the eukaryote). We have noticed, for example, that two YEp vectors, YEp13 and YEp24, that differ by one gene, *LEU2* and *URA3*, respectively, can have significantly different transmission frequencies. The most likely explanation for this observation is differences in the potential for phenotype conversion of yeast exconjugants starved for leucine or uracil *(40)*. Stability of the plasmid in yeast transconjugants may also affect the apparent frequency of transfer *(10)*. It is advisable to check the frequency of transmission of each component of the two-plasmid system by first mating to another strain of bacteria before attempting to mate the bacteria with yeast *(36)*. Low frequencies between bacteria have, in our experience, generally predicted a lack of success in ultimately accomplishing the transmission of the shuttle to yeast.

Since the last review of bacteria with yeast mating procedures, a protocol has been developed for the relatively efficient transmission of YI plasmids. YIp transfer is difficult to detect using the protocols developed for replicating plasmids. We detected transmission only when a plasmid with significant homology to the YIp vector was already present in the yeast recipient, providing a target for integration. This may have been due to the copy number of the target plasmid (~30/cell), increasing the number of homologous sequences available for recombination (increasing target size), or the presence of a particular sequence within the target plasmid (site-specificity). Since yeast homologous recombination is directed by the sequences near the ends of broken molecules, Heinemann and Sprague previously speculated that the target plasmid was the preferred recombination substrate because it contained sequences homologous to *oriT*, which potentially biased subsequent recombinational events *(40)*.

Preliminary evidence suggests that conjugatively transferred molecules prefer to recombine with molecules of bacterial origin (i.e., that retain *oriT*) over those of yeast origin even when the target size should favor recombination between yeast sequences. Yeast recipients, with pBR322 integrated into one chromosomal site, mated with bacterial donors of pBR322 based YIp vectors, carrying both the *URA3* and *LEU2* genes, produced independent yeast transconjugants with the YIp markers integrated at the "pBR322 locus" about twice as frequently as at the *URA3* and *LEU2* loci combined (Heinemann, unpublished observation). The pBR322 locus was favored despite the fact that *URA3* and *LEU2* comprised two-thirds of the sequences of the YIp vector used in this study.

However, yeast recipients with homology to the *oriT* of the shuttle plasmid are not required for transmission of YIps. We routinely transmit YIps to yeast using the replica plate procedure described previously *(40)*. Although this method cannot be quantitated on a per donor or per recipient basis, it can be quantitated on a per colony basis. As many as 90% of the bacterial colonies imprinted on a fresh lawn of yeast recipients have produced one or more yeast transconjugants. The replica plate method also ensures the independence of the transconjugants.

1.4.3. Vectors for Promoter Analysis

These vectors contain the basic structure of an extrachromosomal plasmid plus the *lacZ* gene (encoding β-galactosidase) of *E. coli* to which an appropriate DNA fragment of the gene of interest is fused in frame to study promoter structure and function. β-galactosidase hydrolizes a variety of β-D-galactosides to give a colored product. The assay for β-galactosidase activity is simple and very sensitive. Colorimetric assays can be performed on plates or cell extracts. The pRS series of vectors *(9)* contain the *lacZ* gene and can be used to construct *lacZ*-fusions. A detailed protocol for construction of fusions and measurement of β-galactosidase in yeast is provided in *(41)*.

1.4.4. Multispecies Gene Expression Vectors

A series of plasmid vectors that express cloned genes in both yeast and mammalian cells have been constructed *(42)*. The early SV40 promoter is used to drive transcription. A set of these vectors also contains the *lacZ* gene and can be used to monitor the efficiency of transfection of mammalian cells. Camonis et al. showed a linear relationship between levels of β-galactosidase and number of transfected mouse fibroblast cells *(42)*. These studies also demonstrated that the quantity of protein expressed by the gene can be modulated by culture conditions.

1.4.5. Epitope Tagged Vectors

Epitope tagged vectors are useful for analyzing the function and localization of cloned gene products. A protein is "tagged" by translational fusion to a sequence of amino acid residues that fold into an epitope recognized by a MAb. The nonapeptide (HA) of the influenza virus hemagglutin is a convenient tag *(43)* because the MAb to HA is well characterized and commercially available. Tagging to HA offers an excellent alternative to raising antibodies against multiple proteins.

Tagging can be used to determine a protein's size, subcellular localization, abundance and to indicate whether it has been posttranslationally modified *(44)*. Subcellular localization of the tagged protein can be monitored in vivo by microscopic analysis using immunofluorescence or immunogold labelling *(45,46)*. For the technique to be successful, the protein must be accessible to the antibody. Moreover, the epitope:protein fusion cannot be toxic to the cell. However, toxicity can be diagnosed quickly by monitoring the growth characteristics of transformants expressing the fusion and confirming expression of the fusion by Western blot analysis.

A tag's DNA sequence is usually incorporated into the coding sequence of a protein by site-directed mutagenesis or through the polymerase chain reaction (PCR) with oligonucleotides encoding the epitope. However, Cullin et al. have streamlined the process *(47)*. Their vectors allow a PCR-amplified gene to be transcriptionally fused to the coding sequence of the epitope, bypassing several steps of site directed mutagenesis and sequencing.

Two types of vectors are available, one for fusing the tag to the N- and the other to the C-terminus of the protein. Transcription of the fusion is controlled by the GAL10-CYC1 hybrid promoter in both vectors. The pYeF1 vector was designed to fuse the tag to the N-terminus of the protein by placing the epitope-coding sequences 5' to the insertion site of the PCR amplified gene *(47)*. In contrast, pYeF2 has the insertion site 5' to the epitope coding sequences *(47)*. Both vectors are available with *URA3, HIS3*, or *TRP1* markers.

2. Materials

1. Yeast strain to be transformed.
2. YPD medium *(53)*.
3. 10X TE: 100 m*M* Tris-HCl, pH 7.5, 10 m*M* EDTA; filter sterilized and stored at 4°C.
4. 10X lithium acetate (LiAc): 1.0*M* lithium acetate, adjust pH to 7.5 with dilute acetic acid. Filter sterilize.
5. DNA:
   a. Transforming DNA (store at –20°C).
   b. High molecular ss carrier DNA (Sigma D 1626). The carrier (salmon sperm) DNA is prepared as described by Schiestl et al. *(48)*. Store at –20°C.
6. PEG: 3350 (Sigma P 3640) 50% (w/v). Filter sterilized and stored at 4°C in a tightly capped bottle to prevent evaporation.
7. Selection medium: appropriate drop-out medium *(53)*. Store at 4°C.
8. Sterilized distilled water (dH_2O).
9. Incubator shaker at 30°C.
10. Sorvall GSA and SS-34 rotor or equivalent.
11. Water bath at 42°C.
12. Incubator at 30°C.
13. A source of electric current, such as the Bio-Rad gene pulser (for electroporation).

3. Method

Two popular and easy protocols are presented for transforming yeast with plasmid DNA. For transforming yeast with YACs containing large inserts, *see* the procedure recommended by *(27)*. Notes follow presentation of both transformation procedures.

3.1. Lithium Acetate Procedure

Yeast transformation by lithium acetate is quick, inexpensive, and does not require special equipment or reagents. Transformation efficiency is comparable (as high as 10^6 transformants per μg of transforming DNA) to the electroporation and spheroplast *(48–50)* procedures. Lithium acetate treatment increases cell wall and membrane permeability *(51)*. After brief treatment with lithium acetate, cells are incubated with plasmid DNA, high-mol-wt single-stranded (ss), and carrier DNA, and polyethylene glycol (PEG). The PEG appears to deposit the plasmid vectors and carrier DNA onto the cells *(48)*. Cells are then heat shocked and plated on the selection medium. A heat shock treatment increases recovery of transformants by several orders of magnitude *(52)*.

The yeast transformation protocol used in our lab is essentially the same as described by Schiestl et al. *(48)*. Enough competent cells are made to perform up to 10 transformations. We routinely carry out transformations in duplicate. In addition we include two controls: a positive (vector DNA only) and a negative control (no DNA). The positive control provides a measure of transformation efficiency and the negative control helps to confirm the identity of the transformants.

1. Prepare a 5 mL overnight culture of yeast in YPD medium (*see* Note 1). Incubate shaking at 30°C. Yeast usually grow to a density of about 1–2 × 10^8 cells/mL.
2. Prewarm 50 mL YPD medium to 30°C in a 250 mL flask.
3. Inoculate prewarmed YPD medium with 2.5 mL of overnight liquid culture and adjust the optical density (OD_{600}) of the mixture to 0.1–0.2 with YPD.
4. Incubate the culture at 30°C and agitate at 200 rpm to a final OD_{600} of 0.5–1.0 (usually 3–4 h).
5. Harvest cells by centrifugation for 10 min at 3000*g* in a 50 mL sterilized tube.
6. Decant supernatant and rinse cells with 25 mL sterilized dH_2O. Repeat as in step 5.
7. Resuspend cell pellet again with 1.0 mL of sterilized dH_2O. Transfer to 1.5 mL microfuge tube. Spin at maximum speed for 20 s. Remove dH_2O.
8. Resuspend cells in 500 µL of 100 m*M* LiAc (50 µL 1.0*M* stock, 450 µL of dH_2O) (*see* Note 2). Incubate cell suspension at 30°C for 15 min. Mix by gently vortexing at conclusion of incubation.

Meanwhile:

9. Boil the carrier DNA (1.0 mg/mL salmon sperm) for 10 min and chill on ice (*see* Note 3).
10. Prepare the transformation mixture by adding the solutions shown in Table 1 to a microfuge tube and vortex mixing after each addition.
11. Transfer 345 µL of transformation mixture to fresh microfuge tubes and add 5 µg of transforming DNA (*see* Notes 4 and 5). Mix by pipetting.
12. Aliquot 50 µL of LiAc treated cells to each transformation mix. Mix thoroughly by pipetting.
13. Incubate 30 min at 30°C.
14. Incubate 20 min at 42°C (heat shock).
15. Pellet cells in microfuge (20 s) at top speed. Remove supernatant.
16. Rinse cells with 1.0 mL of sterile dH_2O. Pellet cells as in step 15 and resuspend in 500 µL sterile dH_2O. Plate a 1–10 and a 1–100 dilution of cells onto medium that selects transformants.

Table 1
Transformation Mixture

	For single reaction	For 10 reactions
PEG (50% w/v)	240 µL	2.4 mL
LiAc (1.0*M*)	36 µL	360 µL
carrier DNA (1.0 mg/mL)	50 µL	500 µL
sterilized dH_2O	19 µL	190 µL

3.2. Electroporation Procedure

Introduction of DNA into yeast by electroporation is also a commonly used means of transformation. Electrical fields increase the permeability of cell membranes *(54)*, obviating the need for chemical agents. Schiestl et al. *(48)* recently optimized the existing protocols for yeast transformation by this procedure. They demonstrated that as in the LiAc procedure, the efficiency of transformation by electroporation can also be increased by use of ssDNA and a brief heat shock treatment of the cells before electroporation.

1. Prepare a 100 mL overnight culture of yeast (cell density $1–2 \times 10^8$/ mL) in YPD medium (*see* Notes 1 and 6). Incubate shaking at 30°C.
2. Harvest cells by centrifugation for 10 min at 3000 *g*. Decant the supernatant and wash cells in an equal volume of sterilized dH_2O. Repeat once.
3. Resuspend pellet in 200 µL of sterilized dH_2O. The actual volume of resuspended cells should be about 1 mL. This should allow at least 10 transformations.
4. Add to a microfuge tube: 100-150 ng of transforming DNA, 20 g of ss carrier DNA (to a total volume of 50 µL) and 100 L of cells (*see* Notes 3–5). Mix thoroughly by pipetting.
5. Add 250 µL of 50% PEG 3350. Vortex mix thoroughly but briefly.
6. Incubate at 42°C for 15 min (heat shock).
7. Transfer the mixture to a 0.2 cm electroporation cuvette. Carry out electroporation at 1.5 kV, 25 µF, and 200Ω. When transferring the mixture take care not to create air bubbles. Air bubbles may result in arching and shorting the current that mediates DNA uptake.
8. Immediately add 600 µL of ice-cold YPD medium.
9. Plate a 1–10 and a 1–100 dilution of cells onto medium that selects transformants.
10. Incubate the plates at 30°C for 3–4 d.

4. Notes

1. It is extremely important that aseptic techniques are used throughout the procedure to avoid contamination.
2. Frozen competent cells (10^5 transformants per µg of DNA) can be stored for long periods when prepared as described by Dohmen et al. *(55)*.

3. Complete denaturation of high mol wt carrier (salmon sperm) DNA is critical for success. If transformation frequency is low, prepare a new batch of carrier DNA.
4. DNA quality can influence transformation efficiency. Miniprep DNA is usually adequate, but higher frequencies result with DNA purified in CsCl or other agents.
5. Up to 10 μg of DNA (if necessary) can be used for transformation by the LiAc procedure. However, this is not true for the electroporation procedure. Schiestl et al. demonstrated that the efficiency of transformation declined with greater than 200 ng of transforming DNA *(48)*.
6. Efficiency of transformation by electroporation is strain dependant. If electroporation does not give satisfactory results, we suggest use of the LiAc method.

Acknowledgments

We thank K. Keshav, J. Klena, and J. P. Healy for critical reading of this manuscript. This work was supported in part by Individual National Research Service Award from National Cancer Institute, National Institutes of Health (K. K. S).

References

1. Lundblad, V. (1991) Yeast vector. in *Protocols in Molecular Biology* (Ausubel, F. D., Brent, R. G., Kingston, R. E., Moore, D., Seidman, J. G., Smith, J. A., and Struhl, K., eds.), Wiley, New York, pp. 13. 4. 1.
2. Zealey, G. R., Goodey, A. R., Piggot, J. R., Watson, M. E., Cafferkey, R. C., Doel, S. M., Carter, B. L. A., and Wheals, A. E. (1988) Amplification of plasmid copy number by thymidine kinase expression in *Saccharomyces cerevisiae. Mol. Gen. Genet.* **211,** 155–159.
3. Henderson, R. C. A., Cox, B. S., and Tubb, R. (1985) Transformation of brewing yeasts with a plasmid containing the gene for copper resistance. *Curr. Genet.* **9,** 133–138.
4. Mortimer, R. K., Schild, C. R., Cantopolou, C. R., and Kans, J. A. (1989) Genetic map of *Saccharomyces cerevisiae . Edition 10. Yeast* **5,** 321–329.
5. Gritz, L. and Davies, J. (1983) *Plasmid*-encoded hygromycin B resistance: the sequence of hygromycin B phosphotransferase gene and its expression in *E. coli* and *Saccharomyces cerevisiae . Gene* **25,** 179–185
6. Rine, J., Hansen, W., Hardeman, E., and Davis, R. W. (1983) Targeted selection of recombinant clones through gene dosage effects. *Proc. Natl. Acad. Sci. USA* **80,** 6750–6754.
7. Brake, A. J. (1990) Alpha-factor leader-directed secretion of heterologous proteins from yeast. *Methods Enzymol.* **185,** 408–421.
8. Hadfield, C., Cashmore, A. M., and Meacock, P. A. (1986) An efficient chloramphenicol-resistance marker for *Saccharomyces cerevisiae. Gene* **45,** 149–155.
9. Sikorski, R. and Hieter, P. (1989) A system of shuttle vectors and host strains designed for efficient manipulation of DNA in *Saccharomyces cerevisiae. Genetics* **122,** 19–27.

10. Inomata, K., Nishikawa, M., and Yoshida, K. (1994) The Yeast *Saccharomyces kluyveri* as a recipient eukaryote in transkingdom conjugation: behavior of transmitted plasmids in transconjugants. *J. Bacteriol.* **176,** 4770–4773.
11. Bender, A. and Pringle, J. (1991) Use of a synthetic lethal and multicopy suppressee mutants to identify two new genes involved in morphogenesis in *Saccharomyces cerevisiae. Mol. Cell. Biol.* **11,** 1295–1305.
12. Cvrckova, F. and Naysmyt, K. (1993) Yeast G1 cyclins CLN1 and CLN2 and a GAP-like protein have a role in bud formation. *EMBO* **12,** 5277–5286.
13. Kranz, J. and Holm, C. (1990) Cloning by function: An alternative approach for identifying yeast homologs of genes from other organisms. *Proc. Natl. Acad. Sci. USA* **87,** 6629–6633.
14. Mylin, L. M., Hofmann, K. J., Schultz, L. D., and Hopper, J. E. (1990) Regulated GAL4 expression cassette providing controllable and high-level output from high-copy galactose promoters in yeast. *Methods Enzymol.* **185,** 297–308.
15. Price, V. L., Taylor, W. E., Clevenger, W., Worthington, M., and Young, E. T. (1990) Expression of heterologous proteins in *Saccharomyces cerevisiae* using the *ADH2* promoter. *Methods Enzymol.* **185,** 308–318.
16. Etcheverry, T. (1990) Induced expression using yeast copper metallothionein promoter. *Methods Enzymol.* **185,** 319–329.
17. Kingsman, S. M., Cousens, D., Stanway, C. A., Chambers, A., Wilson, M., and Kingsman, A. J. (1990) High efficiency yeast expression vectors based on the promoter of the phosphoglycerate kinase gene. *Methods Enzymol.* **185,** 329–341.
18. Rosenberg, S., Coit, D., and Tekamp-Olson, P. Glyceraldehyde-3-phosphate dehydrogenase-derived expression cassettes for constitutive synthesis of heterologous proteins. *Methods Enzymol.* **185,** 341–351.
19. Schena, M. and Yamamoto, K. R. (1988) Mammalian glucocorticoid receptor derivatives enhance transcription in yeast. Science **241,** 965–967.
20. Picard, D., Schena, M., and Yamamoto, K. R. (1990) An inducible expression vector for both fission and budding yeast. *Gene* **86,** 257–261.
21. Schena, M., Picard, D., and Yamamoto, K. R. (1991) Vectors for constitutive and inducible gene expression in yeast. *Methods Enzymol.* **194,** 389–398.
22. Chien, C., Bartel, P. L., Sternglanz, R., and Fields, S. (1991) The two hybrid system: a method to identify and clone genes for proteins that interact with a protein of interest. *Proc. Natl. Acad. Sci. USA* **88,** 9578–9582.
23. Touchette, N. (1991) New approach detects protein interactions *in vivo. J. NIH Res.* **3,** 44–45.
24. Fields, S. (1993) The two hybrid system to detect protein-protein interactions. *Methods* **5,** 116–124.
25. Finely, R. L. and Brent, R. G. Interaction trap cloning with yeast, in *DNA Cloning-Expression System: A Practical Approach* (Glover, D. and Hames, B. D., eds.), Oxford University Press, Oxford, England, in press.
26. Burke, D. T., Carle, G. F., and Olson, M. V. (1987) Cloning of a large segment of exogenous DNA in to yeast by means of artificial chromosones vectors. *Science* **236,** 806–812.

27. Burke, D. T. and Olson, M. V. (1991) Preparation of clone libraries in yeast artificial chromosomes. *Methods Enzymol.* **194,** 251–270.
28. Coulson, A., Kozono, Y., Lutterbach, B., Shownkeen, R., Sulston, J., and Waterston, R. (1991) YACs and the *C. elegans* genome. *Bioessays* **13,** 3–417.
29. Little, R. D., Porta, G., Carle, G. F., Schlessinger, D., and D'Urso, M. (1989) Yeast artificial chromosones with 200- to 800-kilobase inserts of human DNA containing HLA, V kappa, 5s and *X*q24-*X*q28 sequences. *Proc. Natl. Acad. Sci. USA* **85,** 98–1602.
30. Brownstein, B. H., Silverman, G. A., Little, R. D., Burke, D. T., Korsmeyer, S. J., Schlessinger, D., and Olson, M. V. (1989) Isolation of single-copy human genes from a library of yeast artificial chromosome clcnes. Science **244,** 348–1351.
31. Hahnenbeger, K. M., Baum, M. P., Polizzi, D. M., Carbon, J., and Clarke, L. (1989) Construction of functional artificial minichromosones in the fission yeast *Schizosaccharomyces pombe*. *Proc. Natl. Acad. Sci. USA* **86,** 577–581.
32. Jakobovits, A. (1994) Humanizing the mouse genome. *Curr. Biol.* **4,** 761–763.
33. Kouprina, N., Eldarov, M., Moyzis, R., Resnick, M., and Larionov, V. (1994) A model system to assess the integrity of mammalian YACs during transformation and propogation in yeast. *Genomics* **21,** 7–17.
34. Sleister, H. M., Mills, K. A., Blackwell, S. E., Killany, A. M., Murray, J. C., and Malone, R. E. (1992) Construction of a human chromosome 4 YAC pool and analysis of artificial chromosome stability. *Nucleic Acids Res.* **20,** 3419–3425.
35. Heinemann, J. A. (1991) Genetics of gene transfer between species. *Trend. Genet.* **7,** 181–185.
36. Lessl, M. and Lanka, E. (1994) Common mechanisms in bacterial conjugation and Ti-Mediated T-DNA transfer to plant cells. *Cell* **77,** 321–324.
37. Heinemann, J. A. and Sprague, G. F. Jr. (1989) Bacterial conjugative plasmids mobilize DNA transfer between bacteria and yeast. *Nature* **340,** 205–209.
38. Hayman, G. T. and Bolen, P. L. (1993) Movement of shuttle plasmids from *Escherichia coli* into yeasts other than *Saccharomyces cerevisiae* using trans-kingdom conjugation. *Plasmid* **30,** 251–257.
39. Nishikawa, M., Suzuki, K., and Yoshida, K. (1990) Structural and functional stability of IncP plasmids during stepwise transmission by trans-kingdom mating: Promiscuous conjugation of *Escherichia coli* and *Saccharomyces cerevisiae*. *Jpn. J. Genet.* **65,** 323–334.
40. Heinemann, J. A. and Sprague, G. F. Jr. (1991) Transmission of plasmid DNA to yeast by conjugation with bacteria. *Methods Enzymol.* **194,** 187–195.
41. Gurante, L. (1983) Yeast promoters and lacZ fusions designed to study expression of cloned genes in yeast. *Methods Enzymol.* **101,** 181–191.
42. Camonis, J. H., Cassan, M., and Roussel, J.-P. (1990) Of mice and yeast: versatile vectors which permit gene expression in both budding yeast and higher eukaryotic cells. *Gene* **86,** 263–268.
43. Park, E. C., Finely, D., and Szostak, J. W. (1992) A strategy for the generation of conditional mutations by protein destabilization. *Proc. Natl. Acad. Sci. USA* **89,** 1249–1252.

44. Kolodziej, P. A., and Young, R. A. (1991) Epitope tagging and protein surveillance. *Methods Enzymol.* **194,** 508–519.
45. Pringle, J. R., Adams, A. E. M., Drubin, D. G., and Haarer, B. K. (1991) Immunofluorescence methods for yeast. *Methods Enzymol.* **194,** 565–602.
46. Clark, M. (1991) Immunogold labeling of yeast ultrathin sections. *Methods Enzymol.* **194,** 608–626.
47. Cullin, C. and Minvielle-Sebastia, L. (1994) Multipurpose vectors designed for the fast generation of N- or C- terminal epitope-tagged proteins. *Yeast* **10,** 105–112.
48. Schiestl, R. H., Manivasakam, P., Woods, R. A., and Gietz, R. D. (1993) Introducing DNA into yeast by transformation. *Methods* **5,** 79–85.
49. Manivasakam, P., and Schiestl, R. H. (1993) High efficiency transformation of *Saccharomyces cerevisiae* by electroporation. *Nucleic Acids Res.* **21,** 4414–4415.
50. Rose, M. D., Winston, F., and Hieter, P. (1990) *Methods in Yeast Genetics—A Laboratory Manual.* Cold Spring Harbor Laboratory, Cold Spring Harbor, NY.
51. Brzobohaty, B. and Kovac, L. (1986) Factors enhancing genetic transformation of intact yeast cells modify cell walls porosity. *J. Gen. Microbiol.* **132,** 3089–3093.
52. Gietz, R. D., Weinberg, O., and Woods, R. A. (1992) Ultra high efficiency yeast transformation using the LiAc/ssDNA/PEG method. *Yeast* **8,** S259.
53. Sherman, F. (1991). Getting started with yeast. *Methods Enzymol.* **194,** 3–21.
54. Neumann, E., Schaefer-Ridder, M., Wang, Y., and Hofschneider, P. H. (1982) Gene transfer into mouse loyoma cells by electroporation in a high electrical field. *EMBO* **1,** 841–845.
55. Dohmen, R. J., Srasser, A. W. M., Honer, C. B., and Hollenberg, C. P. (1991) An efficient transformation procedure enabling long term storage of competent cells of various genera. *Yeast* **7,** 691–692.

12

Inducible Expression Cassettes in Yeast: GAL4

Lawrence M. Mylin and James E. Hopper

1. Introduction

The yeast *Saccharomyces cerevisiae* can be grown in the laboratory with ease and at relatively low expense, and can be propagated in large scale fermentation cultures when preparation of larger amounts of a recombinant protein is desired. The *S. cerevisiae* genome can be manipulated to reduce proteolysis or prevent unwanted posttranslational processing of recombinant proteins *(1,2)*. Genes that control protein degradation and glycosylation have been identified, and mutants are available for constructing genetic backgrounds appropriate for overcoming problems specific to expression of the recombinant protein of choice *(3–5)*. The availability of such mutations combined with the ability to introduce and express foreign genes, isolate temperature sensitive mutants, or alter nonessential genes by targeted disruption makes the yeast *Saccharomyces cerevisiae* an attractive host for the production of heterologous recombinant proteins.

Constitutive high level production of heterologous recombinant proteins in *S. cerevisiae* is not always desirable, as some proteins may be toxic to the host. For such proteins, it is necessary to use inducible expression systems. Several yeast transcriptional regulators have been exploited for the controlled expression of heterologous proteins in yeast (*see* Chapters 11, 13, and 14). The yeast *GAL1*, *GAL10*, and *GAL7* promoters (*GAL* genes or *GAL* promoters) have been widely used for regulated and high-level expression of heterologous proteins in yeast. These promoters are tightly controlled by three regulators encoded by the *GAL4*, *GAL80*, and *GAL3* genes *(6,7)*. The GAL4 protein (Gal4p) is a potent transcriptional activator that activates transcription when bound to UAS_{gal} sequences that are found in the 5' region of *GAL* promoters *(8–14)*. Activation by Gal4p is controlled by the GAL80 (Gal80p) and GAL3

From: *Methods in Molecular Biology, vol. 62: Recombinant Gene Expression Protocols*
Edited by: R. Tuan Humana Press Inc., Totowa, NJ

(Gal3p) proteins *(6,7,15)*. Gal80p binds to Gal4p and prevents all but basal levels of *GAL* gene transcription when *S. cerevisiae* cells are cultured in media containing noninducing carbon sources such as glycerol, lactate, ethanol, or acetate *(16–19)*. Upon addition of galactose, Gal3p rapidly relieves Gal80p inhibition, allowing for high level Gal4p-dependent transcription from *GAL* promoters *(20,21)*. Expression of the *GAL* promoters is also tightly controlled by carbon catabolite, or glucose repression *(6,22–25)*. Transcription of *GAL* promoters is severely repressed following growth in glucose-containing medium *(16,26)*. Therefore, rapid induction of high level transcription from *GAL* promoters requires that yeast cells be cultured in glucose-free (nonrepressing) media prior to galactose induction *(see below)*.

Despite the ease of high level and inducible expression afforded by *GAL4*-dependent control, the abundance of the Gal4p itself appears to be limiting for optimal expression of heterologous recombinant proteins from *GAL* promoters in *S. cerevisiae (19,27)*. The Gal4p is found at relatively low abundance in wild-type yeast cells. Maximal expression of even the wild-type complement of *GAL4*-dependent genes in *S. cerevisiae* is limited by the abundance of Gal4p *(19)*. Expression of heterologous recombinant proteins from high copy expression vectors bearing *GAL4*-inducible promoters can, therefore, be limited by the availability of Gal4p. It is not advisable, however, to constitutively increase the abundance of Gal4p in yeast cells in which overproduction of recombinant proteins under *GAL* promoters is intended. Increasing the abundance of Gal4p upsets the delicate balance between activator (Gal4p) and repressor (Gal80p), and leads to constitutive activation of *GAL* promoters *(19,28)*. To overcome the limitations imposed by low abundance of Gal4p and the need to maintain appropriate levels of *GAL* regulators, we have used a fusion construct in which the *GAL10* promoter controls transcription of the *GAL4* coding sequence *(GAL10p-GAL4) (29–31)*. The *GAL10p-GAL4* expression cassette has been tailored to allow for integration into the yeast genome at the *HIS3* locus. This allows for subsequent introduction into the same cell of a high copy yeast plasmid vector bearing another fusion construct in which coding sequences for the recombinant protein of choice *(Gene X)* are also fused to a *GAL* promoter *(GALp-Gene X)*. By this strategy, increased expression of Gal4p is induced only when induced expression the recombinant protein is desired. Combining galactose-inducible Gal4p overproduction from a *GAL10p-GAL4* construct integrated into the yeast genome with the use of high copy *GAL* promoter vectors has been shown to increase production of a recombinant protein five- to 10-fold above the level realized by the use of high copy *GAL* promoter vectors alone *(29)*. Here we present information regarding the use of the inducible *GAL10p-GAL4* cassette and galactose inducible promoters for high-level production of heterologous proteins in *S. cerevisiae*. Instructions are also given for metabolic labeling of yeast proteins with ^{35}S-methionine and ^{32}P-orthophosphate under conditions of galactose induction.

2. Materials

2.1. Yeast Culture Media—General

1. YEP media. Both broth and solid (agar-containing) media supplemented with appropriate carbon sources to maintain yeast strains that do not harbor plasmids containing selectable markers *(32)*. (*See* Note 1.)
2. 1X synthetic media (liquid and solid) lacking the appropriate nutrient(s) for maintenance of yeast strains harboring plasmids *(16,32)*. (*See* Note 2).
3. Concentrated solutions of carbon sources, including 20% (w/v) glucose, 20% (w/v) galactose, 60% (v/v) glycerin, and 40% (w/v) potassium lactate (pH 5.7). (*See* Notes 3 and 4.)
4. Phase-contrast microscope, hemacytometer, and/or spectrophotometer.
5. Orbital water bath or air shaker incubator, and incubator suitable for maintaining Petri dish cultures at 30°C (CO_2 not required). Roller drum rotator or other device suitable for agitating culture tubes. Best aeration is achieved when the tubes can be tilted to at least a 30° angle.
6. Glass or plastic (autoclavable) Erlenmeyer flasks of various sizes (usually from 100 mL to 2 L) with autoclavable closures. For increased aeration, flasks that have indentations molded into the bottom edge can be used. Closures should be easily removed and replaced so that solutions can be added to or removed from the culture using pipets. Inverted disposable polypropylene beakers work well as closures for the larger flasks, and can be held in place with a piece of tape.
7. Glass culture tubes (18 × 150 m*M*) with autoclavable polypropylene or metal closures are useful for small scale starter cultures.
8. Standard microbiological laboratory apparatus like an inoculating loop (or sterile toothpicks), sterile pipets, a sterilizing flame, and cloth squares and an appropriate holder for replica plating of yeast colonies or patches on Petri dishes. Standard procedures and equipment for manipulation of *S. cerevisiae* have been described elsewhere *(33)*.

2.2. Yeast Strains and Plasmids

1. Construction of the plasmid pKHint-C, which contains the integrable 5'-*his3-GAL10p-GAL4-URA3-his3-3'* expression cassette that directs galactose-inducible production of the GAL4 protein has been described previously *(29,31)*. The plasmid pKHint-C is illustrated in Fig. 1 and is available upon request.
2. Yeast strain Sc340 is a derivative of the Gal$^+$ strain SJ21R *(ura3-52 leu2-3 leu2-112 ade1 MEL1)* and was generated by targeted replacement of chromosomal *HIS3* sequences in SJ21R with the 5'-*his3-GAL10p-GAL4-URA3-his3-3'* expression cassette *(29,31)*. The strain Sc340 overexpresses GAL4 protein more than 60-fold over wild type levels following 4 h of galactose induction *(31)*. The strain Sc340 is *his3*$^-$ and *leu2*$^-$ and, therefore, can be used for expression of recombi-

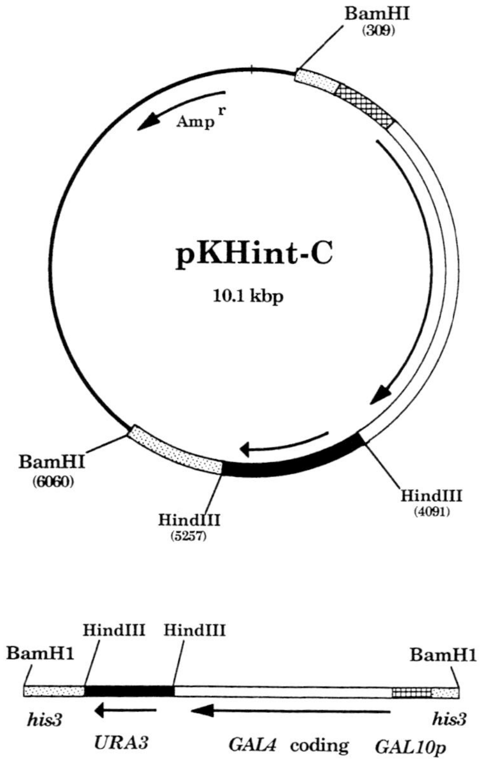

Fig. 1. Plasmid pKHint-C which contains a *GAL10p-GAL4* cassette tailored for integration into the *S. cerevisiae HIS3* locus. Coordinates of relevant restriction endonuclease cleavage sites are given in the figure (upper) in parenthesis, and are numbered relative to an *Eco*R1 site, which occurs within pBR322-derived *Tet* sequences. Arrows indicate direction of transcription. The 5'-*his3-GAL10p-GAL4-URA3-his3*-3' cassette can be liberated from pKHint-C as a 5.8-kb *Bam*H1 fragment. Segments used to construct the cassette are illustrated as hatched boxes which are labeled in the lower portion of the figure. Detailed nucleotide sequence information is available upon request.

nant proteins from yeast plasmid vectors that carry these selectable markers. These yeast strains are available from the author upon request.

3. If necessary, the plasmid pKHint-C (or other related derivatives) can be used to insert the *GAL10p-GAL4-URA3* cassette into other yeast strains by targeted replacement of chromosomal sequences at the *HIS3* locus *(34)*. If the cassette is to be used as designed, the recipient must be ura$^-$ (*ura3*$^-$) and preferably *HIS3*. Targeted insertion of the cassette to the *HIS3* locus is guided by *his3* sequences that flank the *GAL10p-GAL4-URA3* construct and are liberated following digestion of pKHint-C with *Bam*H1 (*see* Fig. 1). If desired, the *Hin*dIII restriction sites that bracket the *URA3* gene in the plasmid pKHint-C (*see* Fig. 1) can be used to

replace the *URA3* gene with a different selectable marker. Inserts used to replace the *URA3* gene must not contain a *Bam*H1 restriction site, as digestion of the plasmid pKHint-C with *Bam*H1 is used to prepare the plasmid for targeted disruption/replacement of the *HIS3* locus. Alternately, other derivatives of the *GAL10p-GAL4* cassette can be tailored for integration into a genomic locus other than *HIS3 (31)*.

4. Yeast vectors that contain *GAL* promoter cassettes have been described *(35–38)*. Options described include promoter cassettes that contain a variety of restriction sites for cloning of inserts; insertion sites that do or do not supply a translation start codon; availability of *LEU2*, *HIS3*, *TRP1* or *URA3* selectable markers; and multicopy vs low copy number options. Several vectors include coding sequences for an epitope tag (influenza virus hemagglutinin) or glutatione S-transferase that are useful for subsequent immunoaffinity purification of the recombinant fusion protein *(37,38)*. Alternatively, new expression constructs can be designed. The upstream activating sequence (UAS_{gal}) located in the *GAL1-10* intragenic region contains four closely spaced GAL4p binding sites and can be subcloned as a 365 bp *Sau*3a-*Dde*I fragment *(8,35)* or amplified by the polymerase chain reaction *(39)* with primers chosen so as to incorporate appropriate restriction sites on either end. Since this fragment contains no TATA box sequences or mRNA initiation sites, it can be inserted in front of another promoter to allow for galactose regulated transcription. Inserts representing genes from higher eukaryotes should be inserted in the form of cDNA. The intron excision machinery of yeast may not be able to properly process complicated transcripts, since most yeast genes do not contain introns.

2.3. Solutions for Introduction of Plasmids into S. cerevisiae by Transformation with Alkali Cations

1. YEP liquid medium containing 2% (w/v) glucose (*see* Notes 1 and 3).
2. TE: 10 m*M* Tris-HCl, pH 7.5, 1 m*M* EDTA. Autoclave or filter sterilize (0.45 µm).
3. LiAc solution: 10 m*M* Tris-HCl, pH 7.5, 1 m*M* EDTA, 100 m*M* lithium acetate. Filter sterilize.
4. 50% PEG solution: 50% (w/v) polyethylene glycol (3350) in glass distilled water (GDW). Filter sterilize.
5. Intact or linearized plasmid DNA (*see* Note 5).

2.4. 5X Synthetic Liquid Media for GAL4-Induced Protein Overexpression

We have routinely used buffered 5X synthetic media for galactose induction experiments *(29–31)*. This media differs from standard 1X synthetic media in that it contains five times increased concentrations of many of the nutrients, and is buffered. Although generation time in glycerol- and lactic acid-containing media is normally rather slow (>4 h per generation compared to <2 h in glucose-containing synthetic media), the 5X synthetic media promotes better growth than is obtained with the standard 1X formula (*see* Note 2) under noninducing/nonrepressing conditions.

1. Leu, Met, Ura drop out amino acid premix base is prepared by thoroughly mixing the following components, and can be stored at room temperature dry and protected from light: 8.0 g Tryptophan, 8.0 g Histidine-HCl, 8.0 g Arginine, 12.0 g Tyrosine, 12.0 g Lysine, 20.0 g Phenylalanine, 80.0 g Threonine, 12.0 g Isoleucine, 8.0 g Adenine.
2. 5X synthetic complete media containing 3% (v/v) glycerol and 2 % (v/v) lactic acid (1000 mL final volume):
   a. Nutrient solution: Suspend 7.15 g yeast N base without amino acids and ammonium sulfate (Difco 0335-15), 6.5 g ammonium sulfate (Sigma A-5132), 3.5 g Leu, Met, Ura drop out amino acid premix (above), 300 mg leucine, 100 mg methionine, and 100 mg uracil in 450 mL GDW. (Omit amino acids when appropriate for maintenance of plasmids.) Autoclave and allow to cool to room temperature.
   b. Buffer solution: Dissolve 10 g succinic acid and 6 g sodium hydroxide in 450 mL of GDW. Autoclave and allow to cool. (The volume of the buffer solution can be reduced if higher volumes of other reagents are to be added to make the final media.)
   c. In a sterile flask, combine 450 mL nutrient solution and 450 mL buffer solution. Add 50 mL 60% (v/v) glycerol and 50 mL 40 % (v/v) lactic acid pH 5.7 solutions (*see* Notes 3 and 4).

2.5. Solutions and Other Items Used to Prepare Yeast Cell Extracts for Western Blotting

1. Buffer A: 50 m*M* sodium phosphate pH 7.2 (pH achieved by combining concentrated stocks of mono and dibasic sodium phosphate), 5 m*M* EDTA, 1 m*M* DTT, 1 m*M* PMSF (added from a 17.5 mg/ml stock freshly prepared in ethanol), 20 μg/mL Aprotinin, 2 μ*M* Pepstatin, 0.6 μ*M* Leupeptin, and other inhibitors as required.
2. Buffer B: Buffer A, which additionally contains 2% (w/v) SDS.
3. 5X electrophoresis sample buffer (5X goo): 500 m*M* Tris-HCl, pH 8.0, 5 m*M* EDTA, 50% (w/v) sucrose, 5% (w/v) SDS, 160 m*M* DTT, 0.4 mg/ml Pyronin Y. (Store in aliquots at –70°C.)
4. Glass beads (0.45 mm diameter, VWR).
5. 13 × 100 mm disposable glass tubes with inside diameter such that standard 1.5 mL microfuge tubes will slide down into the glass tube until the rim of the microfuge tube rests on the rim of the glass tube (for "piggy-back" centrifugation).
6. 18-gage needles, 1.5 in. long.

2.6. Phosphate-Limiting Media for Growth of Yeast Strains Prior to Labeling with ^{32}P-Orthophosphate

These recipes were adapted from conditions routinely used to prepare phosphate-limiting media for the induction of phosphate repressed transcription *(40,41)*.

1. Premix stock A (Salts and Nitrogen): 103.5 g $(NH_4)_2SO_4$, 25.9 g $MgSO_4 \cdot 7H_2O$, 103.5 g Asparagine, 12.9 g $CaCl_2$ (anhydrous). Mix well, store at room temperature protected from light.
2. Premix stock B (vitamins and minerals): 704 mg KI, 403 mg Boric Acid, 215 mg $MnSO_4.H_2O$, 2.16 mg $ZnSO_4 \cdot 7H_2O$, 178 mg $CuSO_4 \cdot 5\ H_2O$, 130 mg $(NH_4)_6Mo_7O_{24} \cdot 4H_2O$, 1.25 g $FeCl_2 \cdot 4H_2O$, 1.76 g Thiamine-HCl, 1.69 g Pyridoxine-HCl, 1.41 g Niacin, 3.06 g Calcium pantothenate, 704 mg Riboflavin, 7.04 g Inositol, 1.41 mg Biotin. Mix, and store dry at –20°C protected from light.
3. 4X minimal liquid stock: Dissolve 19.32 g Premix Stock A and 26 mg Premix Stock B in 1 L GDW. Sterilize by passing through a 0.45-μ*M* filter. Store 4°C.
4. 2X Pi-depleted YEP (pH 4.7): Dissolve 20 g yeast extract (Difco) and 40 g bactopeptone (Difco) in 960 mL GDW. Add 20 mL of 1*M* $MgSO_4$ and 20 mL of concentrated NH_4OH. Stir 30 min at room temperature. Carefully filter twice through Whatman #1 filter paper. Adjust pH to 4.7 with concentrated HCl. Autoclave and store at room temperature.
5. Complete amino acid liquid concentrate (omit nutrients as needed for plasmid maintenance): Dissolve in 500 mL heated GDW 5.25 g Leu, Met, Ura drop out amino acid premix (*see* Section 2.4.1.), 1.88 g valine, 1.25 g aspartic acid, 750 mg leucine, 250 mg methionine, and 250 mg uracil. Allow to cool, filter sterilize (0.45 μm), and store at 4°C.
6. 3/5 hi Pi (900 μg/mL) citrate buffered 5X synthetic complete media containing 2% (w/v) glucose is made by aseptically combining the following sterile components (100 mL total volume): 25 mL 4X min liquid stock, 10 mL 20% (w/v) glucose, 20 mL complete (or drop out) amino acid liq. conc., 10 mL 0.25*M* citric acid, pH 4.75 (with NaOH), 0.6 mL 150 mg/mL KH_2PO_4, 0.4 mL 1.1 M KCl, 34 mL sterile GDW.
7. Low Pi (10 μg/mL) citrate buffered 5X synthetic complete media containing 3% (v/v) glycerol and 2% (v/v) ethanol is made by aseptically combining the following sterile components (500 mL total volume): 125 mL 4X min liquid stock, 100 mL complete (or drop out) amino acid liq. conc., 50 mL 0.25*M* citric acid, pH 4.75 (with NaOH), 0.333 mL 15 mg/mL KH_2PO_4, 5 mL 1.1 M KCl, 163 mL sterile GDW, 7.1 mL 2X Pi-depleted YEP, 25 mL 40 % (v/v) ethanol, 25 mL 60 % (v/v) glycerol.

3. Methods

3.1. Introduction of GAL Promoter Vectors into Gal$^+$ Yeast Strains

Plasmids can be readily introduced into Sc340 (SJ21R-related strains) by transformation procedures that involve use of alkali cations *(42–45)* or enzymes that degrade the yeast cell wall *(46)*. Electroporation represents another option *(47,48)*. Transformants are initially selected by plating the transformation suspensions on 1X synthetic media containing 2% (w/v) glucose and lacking the nutrient appropriate to select for growth of only yeast cells that take up and express plasmid DNA (sorbitol must be added for additional osmotic stability following electroporation). Selected candidates are grown and screened for production of the recombinant gene product *(see below)*. A procedure using lithium acetate (LiAc), which works well for introduction of plasmids into strains related to Sc340, is described below *(42)*:

1. Inoculate yeast colony into 5 mL YEP 2% (w/v) glucose (YEPD) in an 18 × 150 mm culture tube. Grow this starter culture overnight at 30°C (roller drum or air shaker) to saturation.
2. Late in the day, inoculate a fresh culture of 100 mL YEPD medium (in 500 mL Erlenmeyer flask) with an appropriately diluted aliquot of the starter culture (75 µL of 1/10 diluted aliquot should work well) such that the cell density will be approx 2 × 10^7 cells/mL (A_{600} of about 0.2) the next morning (*see* Notes 6 and 11). This allows the culture to be in the exponential growth phase, which is important for efficient transformation (*see* Note 7).
3. Pellet yeast cells by centrifugation (3000*g* for 5 min).
4. Suspend yeast cell pellet in 10 mL TE. Transfer cells to a smaller screw capped tube and centrifuge as above.
5. Resuspend cell pellet in 5.0 mL of LiAc solution. Incubate at 30°C for 1 h with gentle agitation.
6. Centrifuge as above, and resuspend the cell pellet in 2.0 mL LiAc solution.
7. Add 200 µL of the cell suspension to 10 µL of DNA solution (10 µg) in a 1.5 mL Eppendorf tube and mix by vortexing briefly at low speed.
8. Incubate at room temperature for 10 min.
9. Add 1.0 mL 50% PEG solution and mix by inverting the tube five times.
10. Incubate at 30°C for 1 h.
11. Heat shock the cells by placing in a 42°C water bath for 5 min. Return to room temperature.
12. Centrifuge in an eppendorf centrifuge with a fixed angle rotor at full speed (12,000*g*) for 5 s (sufficient only to pellet cells on the side of the tube).
13. Carefully decant supernatant. Rinse pellet by gently adding 1.0 mL GDW, close cap, invert tube several times, and centrifuge as above.
14. Remove majority of supernatant. Suspend cells in 200 mL GDW, and spread on 1X synthetic media that contains 2% glucose and lacks the appropriate nutrient to select for growth of cells that have taken up the plasmid.
15. Allow the cell suspension liquid to be absorbed by the media, invert, and incubate at 30°C. Colonies should appear after 2–3 d.

3.2. Introduction of the GAL10p-GAL4-URA3 Cassette into Yeast Strains

1. If it is necessary to introduce the *GAL10p-GAL4-URA3* cassette into a different genetic background to allow for expression of a particular recombinant protein, several precautions should be noted. Many laboratory strains of *S. cerevisiae* are *gal2* mutants. *GAL2* encodes a galactose permease that is necessary for rapid and

efficient *GAL* gene induction *(49,50)*. The strain to be used should grow well on YEP media containing 2% galactose. We do find that respiratory competent gal$^-$ strains can grow slowly on galactose-containing plates. A simple way to confirm the Gal$^+$ phenotype is to supplement YEP galactose or 1X synthetic galactose plates with ethidium bromide *(51)*, which prevents "background" growth of gal$^-$ strains (*see* Note 8).

2. Linearize the plasmid pKHint-C by digestion with *Bam*H1 (*see* Fig. 1). Introduce the plasmid into yeast cells by LiAc transformation (*see* Section 3.1.). Select for Ura$^+$ colonies by plating on 1X synthetic 2% glucose media lacking uracil ("ura drop out glu"). Ura$^+$ colonies should appear in 2–3 d. Select several colonies, and streak for single colonies on "ura drop out glu" plates. Using sterile toothpicks or an inoculating loop, patch selected isolates onto duplicate ura drop out glucose master plates, and allow to regrow at 30°C overnight. Confirm the his$^-$ phenotype by replica plating one of the master plates to a synthetic glucose plate lacking histidine (and a control plate lacking uracil). Additionally, we find that Gal$^+$ strains properly transformed with the *GAL10p-GAL4-URA3* cassette become phenotypically gal$^-$, most likely owing to toxicity, which results from overexpression of the GAL4 protein when cells are replica plated onto media containing galactose. Replacement of the *HIS3* locus by the cassette can be confirmed by Southern blotting *(52)*.

3.3. Growth and Induction of GAL4-Dependent Overexpression

1. Maintain yeast strains bearing plasmids in liquid or on solid 1X synthetic media containing 2% glucose. Yeast transformants may be stored for several weeks at 4°C when patched onto synthetic media. Wrap plates with plastic wrap to reduce desiccation. Aliquots of freshly grown cultures can be preserved by storage at –70°C following freezing in medium supplemented with glycerol (*see* Note 9).
2. Start experiments from a fresh "overnight" culture. That is, revive transformants by overnight growth to saturation (30°C) in 5 mL of 1X synthetic media containing 2% glucose and lacking the nutrient appropriate for maintaining selection of the plasmid. Use of older "starter cultures" can result in a prolonged lag phase at the next step.
3. Prepare for galactose induction by diluting an aliquot of the starter culture 1/200 or 1/100 into a larger culture (in an Erlenmeyer flask; *see* Note 6) of 5X synthetic noninducing/nonrepressing media that contains 3% (v/v) glycerol and 2% (v/v) lactic acid as carbon sources. Incubate overnight (usually about 24 h) with vigorous shaking in an orbital air shaker or water bath. During this period, the yeast cells should go through five to six doublings, and residual glucose repression is relieved (*see* Note 10).
4. When the cell density has reached 2–4 × 10^6 cells/mL (*see* Note 11), induce *GAL* promoters by adding galactose to the culture to a final concentration of 2% (w/v). Continue vigorous shaking. Galactose induction is allowed to proceed until the protein of interest has accumulated to sufficient levels, usually 4–8 h after galactose has been added to the culture (*see* Note 12 and refs. *29,31*).

5. Harvest the cells by centrifugation. (The culture may be chilled on ice prior to pelleting in a refrigerated centrifuge.) Wash the cell pellet at least once with ice-cold GDW, freeze on dry ice, and store in liquid nitrogen or at –70°C, or proceed directly to the extraction step. The cell pellet should be frozen in a tube that can be used for homogenization (*see* Sections 3.4. and 3.5.).

3.4. Preparation of Yeast Cell Extracts for Detection of Overexpressed Recombinant Proteins (or Gal4p) by Western Blotting

The yeast cell wall must be disrupted to allow for efficient extraction of nonsecreted proteins. This can be accomplished by the use of enzyme preparations which digest the cell wall *(53)*, or by simply disrupting the cell wall by vortexing the yeast cells in the presence of small glass beads *(29,31)*. For the analysis of GAL4 protein (overproduced from the *GAL10p-GAL4* cassette), or for monitoring relative levels of soluble enzymes encoded by the *GAL* genes, we routinely use the latter method. The composition of the buffers used for the lysis, fractionation of the homogenate, and specific preparation of extracts will be determined by the characteristics of your recombinant protein and the condition of the desired protein. For example, the GAL4 protein is not released into the soluble fraction of homogenates prepared by disrupting yeast cells with glass beads in buffers of low ionic strength. In this case, discarding soluble proteins with the supernatant fraction obtained following a brief centrifugation of the homogenate resulted in a threefold enrichment of GAL4p in the pellet fraction (Mylin and Hopper, unpublished observations). A simple denaturing extraction procedure that yields protein extracts that represent whole cell lysates and are useful for western blotting analysis is described below:

1. Allow cell pellet in 1.5 mL microfuge tube to thaw briefly on ice. (For analytical scale experiments, pellets of approx 5×10^8 cells, which represent 25 mL taken from a culture at cell density of 2×10^7 cells per mL, work well.)
2. Add 0.3 mL ice-cold buffer A. Suspend cell pellet by vortexing.
3. Add an equal volume of ice-cold 0.45 mm glass beads with a scoop prepared by cutting a microfuge tube to the appropriate size (*see* Note 13).
4. Lyse the cells by six or more 20-s pulses on a vortex mixer interrupted by equal periods of cooling on ice.
5. To remove extracts from the glass beads, place the microfuge tubes "piggyback" on top of 13 × 100 mm glass tubes. Poke a hole through both the lid and the bottom of the microfuge tube using an 1.5 in. 18-gage needle. Briefly centrifuge piggyback tubes in a refrigerated low speed centrifuge. Remove microfuge tubes and discard.
6. Resuspend pelleted material by a brief pulse on a vortex mixer.
7. Add 0.3 mL buffer B to tubes individually, vortex immediately, and place in a boiling water bath for 7 min. Vortex samples briefly at least once during the heating period.

8. Transfer extracts to a 1.5 mL Eppendorf tubes. Centrifuge 15 min in a room temperature microcentrifuge at maximum speed.
9. Transfer 0.5 mL of the supernatants to microfuge tubes containing 125 µL 5X electrophoresis sample buffer (5X goo). Mix well, freeze on dry ice, or proceed directly to analysis of protein concentration and/or Western blotting. Reheat in a boiling water bath prior to use for SDS-PAGE. Because of high concentrations of DTT and SDS in the extracts, protein concentrations should be determined using 10 µL aliquots of the extracts following precipitation with trichloroacetic acid according to Peterson *(54)*.

3.5. Metabolic Labeling of Recombinant Proteins with ^{35}S-Methionine During Galactose Induced Overproduction

It may be of interest to generate ^{35}S-labeled recombinant proteins for use in other in vitro assays. The following conditions are useful for generating labeled protein suitable for immunoprecipitation analyses. Isolation of recombinant protein under less denaturing conditions will require a different extraction procedure. The induction procedure summarized below differs from the standard galactose induction procedure in that methionine must be omitted from the 5X synthetic medium used for growth and labeling.

1. Grow yeast strain/transformant to saturation in 5X synthetic buffered medium containing 2% (w/v) glucose but lacking methionine (and appropriate nutrient required to maintain plasmid.)
2. Dilute 1/100 or 1/200 into Erlenmeyer flask containing 5X synthetic buffered medium containing 3% (v/v) glycerol and 2% (v/v) lactic acid. Incubate (overnight) with vigorous agitation at 30°C in an orbital shaker (*see* Notes 6 and 10).
3. When the cell density has reached approx 2–4 × 10^7 cells/mL (*see* Notes 10 and 11), add galactose to a final concentration of 2% (w/v). Continue vigorous shaking at 30°C.
4. After sufficient time has been allowed for induction of synthesis of the recombinant protein (*see* Note 12), add 10 µCi/mL ^{35}S-methionine (specific activity 1500 Ci/mmol; Amersham) to the culture. Continue shaking at 30°C for 1 h.
5. Add unlabeled L-methionine to the culture to a final concentration of 1 m*M* (from a freshly prepared 0.1*M* stock prepared in GDW). Shake at 30°C for an additional 5–10 min.
6. Chill on ice, and harvest by centrifugation.
7. Preparation of extracts for immunoprecipitation and SDS-PAGE analysis follows the outline described above (*see* Section 3.4.) excepting that special precautions are taken to avoid radioactive aerosols during extract preparation. Differences include (refer to steps in Section 3.4.):
   a. (Steps 5–8) Buffer B is added directly to glass beads vortexed homogenates in 1.5 mL tubes (or 12 × 75 mm polycarbonate snap cap tubes) in the presence of the glass beads, mixed briefly, and heated with lids sealed. Gas expansion vents are provided by piercing the caps with 1 cc tuberculin syringes plugged with glass wool.

b. Extracts are recovered from glass beads using an 18 gage needle fitted to a 1-cc tuberculin syringe, placed in a fresh tube, centrifuged, and the supernatants recovered and placed into a fresh tube. To increase recovery of extract, rinse beads with a second aliquot of solution prepared by mixing equal portions of buffers A and B, and combine with the initial extract.
c. Volumes of extract to be used for immunoprecipitation are determined by assaying trichloroacetic acid precipitable radioactivity in 10 μL aliquots of the supernatant.
d. Extracts can be stored frozen (dry ice, or –70°C) until used for immunoprecipitation and SDS-PAGE.

3.6. Metabolic Labeling of Recombinant Proteins with ^{32}P-Orthophosphate During Galactose Induced Overproduction

It may become necessary to verify the phosphorylation state of a recombinant protein expressed under galactose inducing conditions. If phosphorylation of the protein leads to altered migration during SDS-PAGE, it should be possible to demonstrate a mobility shift for the ^{35}S-labeled, immunoprecipitated protein following incubation in the presence of calf intestinal alkaline phosphatase *(30,55)*. Alternatively, metabolic labeling can be performed during galactose induction in the presence of ^{32}P-orthophosphate. To metabolically label proteins expressed in *S. cerevisiae* cells with inorganic ^{32}P-orthophosphate, it is first necessary to deplete the yeast cells of intracellular stores of phosphate. This is accomplished by growth of the cells in phosphate-limiting media prior to galactose induction. The noninducing carbon sources have been altered because we have found that derepression of phosphate-controlled gene expression (acid phosphatase) occurs more readily when cells are grown in synthetic media containing glycerol and ethanol than in synthetic media containing glycerol and lactic acid (unpublished results).

1. Grow yeast strain/transformant to saturation in 3/5 hi Pi synthetic media containing 2% (w/v) glucose and lacking any nutrient appropriate for maintenance of a plasmid.
2. Inoculate larger scale cultures of low Pi media using GDW washed cells. Wash cells twice by pelleting an aliquot of the 3/5 hi Pi grown stationary culture, resuspending in GDW, and repelleting. Resuspend the cell pellet in a volume of GDW corresponding to the original culture aliquot. Inoculate into low Pi synthetic media containing 3% (v/v) glycerol and 2% (v/v) ethanol as carbon sources to achieve a final dilution of between 1/300 to 1/1000 (dilution is strain dependent; *see* Note 6.) It is advisable to grow a larger volume than will be required for the labeling step.
3. Incubate with vigorous shaking at 30°C (for 24–48 h) until the culture reaches an A_{600} of 0.3–0.4 (*see* Note 14).
4. Place 20 mL culture samples into a 125 mL flask. Add 1 mCi/mL ^{32}P-orthophosphate (carrier free, 285 Ci/mg; ICN, #64041). Continue vigorous agitation at 30°C

for 2 h. Use remote handling tools and appropriate shielding, as each flask will represent at least 20 mCi ^{32}P.

5. Induce *GAL* promoters by adding 2 mL 20% (w/v) galactose solution to flasks. Continue agitation for two or more hours.
6. Chill cultures on ice and harvest cells by centrifugation. To minimize the chance of rotor contamination, remove supernatants using a disposable pipet.
7. Wash cells at least three times using ice-cold GDW. Freeze cell pellet on dry ice in tube appropriate for preparing extracts. Precautions should be taken to minimize aerosol formation during extract preparation (*see* Section 3.5.).

4. Notes

1. YEP media contains per L of deionized or glass distilled water (GDW) 5 g yeast extract (Difco, Detroit, MI), 10 g Bacto Peptone (Difco), and 25 mg adenine. Components are dissolved in GDW representing 90% of the final volume, and autoclaved in aliquots. For solid media, add 20 g agar (Sigma) and a magnetic stir bar prior to autoclaving. After autoclaving, allow to cool (until flask is hot to the touch, but can be touched without great discomfort), add appropriate amounts of concentrated sterile sugar solutions *(see below)*, and dispense 25 mL aliquots into Petri dishes. Allow plates to sit at room temperature 2 d to dry. Store plates inverted in plastic bags (bags in which plates were supplied work well) at 4°C.
2. Synthetic complete (1X) media contains per L 6.7 g yeast nitrogen base without amino acids (Difco), 300 mg of threonine, 30 mg each of leucine and lysine, 25 mg of tryptophan, 20 mg each of histidine-HCl, arginine, methionine, tyrosine, isoleucine, phenylalanine and adenine, and 10 mg of uracil. The amino acids and bases are routinely combined into a powdered premix in advance using a mortar and pestle, or a blender. Omit nutrients from the premix when maintenance of a plasmid with the corresponding selectable marker is desired. Dissolve components (excluding carbon sources) in 900 mL GDW, dispense in aliquots, autoclave, and store at room temperature. The balance of the media (100 mL or 10% of final volume) is made up with sugar solutions and/or water as needed at the time of use. To prepare solid media, combine nutrients, agar (15 g/L Bacto-agar, Difco) and a stir bar, autoclave, allow to cool, add carbon source solutions, mix, and dispense into Petri dishes (25 mL/plate). Protect solid media from light when drying and during storage at 4°C.
3. Dextrose (glucose) and glucose-free galactose can be obtained from Pfahnsteil (Waukegan, IL). Although substitutions should be possible, it is especially important that galactose not be contaminated with glucose, as this could prevent optimal induction. Prepare 20% (w/v) solutions in GDW (galactose solution requires heating), autoclave, and store at room temperature. Add to media at 10% of the final volume for a 2% (w/v) final concentration. Glycerin can be obtained commercially as a 99% solution. Dilute with GDW to arrive at a stock solution of 60% (v/v), autoclave, and store at room temperature. Add at 5% of final media volume for 3% (v/v) final concentration.
4. Following is a recipe for preparation of 500 mL of 40% (v/v) potassium lactate solution. If possible, prepare in a fume hood. A pH meter must be available at

the same location. Place a 1 L beaker in an ice bath on top of a magnetic stir plate. Add 235 mL of 85% lactic acid syrup (J. T. Baker) to the beaker. Slowly add KOH pellets while stirring until the solution has reached pH 5.7. More than 100 g of KOH pellets will be required. Increase the volume to near 500 mL with GDW and recheck the pH. Adjust as necessary by adding KOH pellets or lactic acid syrup. Adjust final volume to 500 mL with GDW. Autoclave in aliquots and store at room temperature. Add to media at 5% of the final volume for a final concentration of 2% (v/v).

5. Plasmid DNA used for the LiAc transformation procedure should be of a quality comparable to that obtained by alkaline lysis method, or better *(15)*. Plasmid DNA digested with restriction enzymes should be extracted with phenol and chloroform, precipitated with ethanol, and dissolved in TE prior to use in yeast transformation *(56)*.
6. When preparing cultures in Erlenmeyer flasks, the culture volume should not exceed 40% of the flask volume (i.e., maximum of 40 mL in a 100 mL flask). Less is better.
7. Not all strains of *S. cerevisiae* are efficiently transformed following growth to similar culture densities. Some strains require growth to near stationary phase (A600 ≥ 1.2) before the cells can be made competent by the LiAc method described.
8. Prepare a 1.5 mg/mL solution of ethidium bromide in GDW and sterilize by passing through a 0.45 μ*M* filter. Use a sterile glass pipet or glass rod to spread 0.4 mL of this solution on a YEP galactose plate and allow to dry overnight at room temperature. Replica plate yeast patches from glucose media to YEP gal + EtBr and appropriate control plates. Incubate at 30°C overnight. Score for growth. **Caution: Ethidium bromide is toxic.** Protect solutions and plates containing ethidium bromide from light.
9. Grow transformants to saturation in 5 mL of 1X synthetic glucose containing medium (overnight). Dilute an aliquot 1/20 into YEP media containing 2% glucose and allow to grow to saturation. Mix 0.5 mL sterile 30% glycerin solution with 0.5 mL culture in a suitable cryovial, and place at –70°C.
10. For some strains it may be necessary to include 0.05% (w/v) dextrose in the 5X noninducing (glycerol and lactic acid or ethanol containing) media to expedite this preinduction growth phase *(55)*. When used at this concentration, glucose repression does not noticeably reduce *GAL* induction, but growth rate slows with time in culture, and is harder to predict doubling times during the final few generations. Once the cultures have approached A_{600} = 0.2, the cells can be concentrated by centrifugation, and resuspended in half the volume of fresh media (lacking 0.05% glucose) to achieve a higher cell density during the induction and labeling periods *(55)*.
11. Cell density can be measured by counting with a hemacytometer, or estimated by measuring the optical density at 600 nm (A_{600}). It is necessary to initially determine a calibration curve for A_{600} vs cell number for yeast cells grown under the desired conditions as strain type, culture phase, and media conditions will affect the relationship between cell concentration and A_{600}. Under the conditions routinely used in our laboratory, an A_{600} of 0.2 corresponds approximately to a density of 1–2 × 10^7 cells per mL for strains related to Sc340.

12. The length of the optimal galactose induction period should be determined empirically for each protein. A small scale pilot experiment in which culture samples are removed (for Western blotting analysis) at 1–2 h intervals following galactose induction should allow the optimal induction period to be determined. Samples representing 20 mL aliquots of a culture having a cell density of 2×10^7 cells/mL usually provide adequate amounts of protein for accurate quantitation and several Western blotting determinations. If antibodies are not available for your recombinant/fusion protein, another functional or enzymatic assay will have to be used. Antisera raised against the DNA binding and transcriptional activation domains of the GAL4 protein are commercially available (Upstate Biotechnology, Lake Placid, NY; Santa Cruz Biotechnology, Santa Cruz, CA) and should provide a positive control for galactose induction of *GAL10p-GAL4* bearing cells.
13. Volumes of buffer and beads should be increased proportionately during scale up. Be careful to avoid foaming during the homogenization (vortexing) step. Consult Jaswinski *(57)* for additional guidance.
14. Once the cultures have reached a density of $A_{600} \geq 0.2$, the cells can be concentrated by centrifugation, and resuspended at a higher density in a smaller portion of the Pi-depleted culture media from which the cells were pelleted. It is important *not* to use fresh low Pi media for resuspension at this step since it contains added Pi (*see* Section 2.6.7.).

Acknowledgments

We wish to thank T. Eric Blank and Michael Woods for helpful comments during the preparation of this manuscript. JEH is supported by National Institutes of Health grant GM 27925. LMM is supported by a Postdoctoral Fellowship from The Cancer Research Institute of New York.

References

1. Gardell, S. J., Hare, T. R., Han, H., Markus, H. Z., Keech, B. J., Carty, C. E., Ellis, R. W., and Schultz, L. D. (1990) Purification and characterization of human plasminogen activator inhibitor type I expressed in *Saccharomyces cerevisiae. Arch. Biochem. Biophys.* **278,** 467–474.
2. Schultz, L. D., Markus, H. Z., Hoffman, K. H., Montgomery, D. L., Dunwiddie, C. T., Kniskern, P. J., Freedman, R. B., Ellis, R. W., and Tuite, M. F. (1994) Using molecular genetics to improve the production of recombinant proteins by the yeast *Saccharomyces cerevisiae. Ann. New York Acad. Sci.* **721,** 148–157.
3. Ballou, C. E. (1990) Isolation, characterization, and properties of *Saccharomyces cerevisiae mnn* mutants with conditional protein glycosylation defects. *Methods Enzymol.* **185,** 440–470.
4. Jones, E. W. (1991) Three proteolytic systems in the yeast *Saccharomyces cerevisiae. J. Biol. Chem.* **266,** 7963–7966.
5. Jones, E. W. (1991) Tackling the protease problem in *Saccharomyces cerevisiae. Methods Enzymol.* **194,** 428–453.

6. Johnston, M. and Carlson, M., Regulation of carbon and phosphate utilization, in *The Molecular and Cellular Biology of the Yeast Saccharomyces. Gene Expression* (1992) (Jones, E. W., Pringle, J. R., and Broach, J. R., eds.), Cold Spring Harbor Laboratory, Cold Spring Harbor, NY, pp. 193–282.
7. Johnston, M. (1987) A model fungal gene regulatory mechanism: the *GAL* genes of *Saccharomyces cerevisiae. Microbiol. Rev.* **51,** 458–476.
8. West, R. W., Yocum, R. R., and Ptashne, M. (1984) *Saccharomyces cerevisiae GAL1-GAL10* divergent region: location and function of the upstream activating sequence UASg. *Mol. Cell. Biol.* **4,** 2467–2478.
9. Bram, R. J. and Kornberg, R. D. (1985) Specific binding to far upstream activating sequences in polymerase II promoters. *Proc. Natl. Acad. Sci. USA* **82,** 43–47.
10. Giniger, E., Varnum, S. M., and Ptashne, M. (1985) Specific DNA binding of GAL4, a positive regulatory protein of yeast. *Cell* **40,** 767–774.
11. Lohr, D. and Hopper, J. E. (1985) The relationship of regulatory proteins and DNase I hypersensitive sites in the yeast *GAL1-10* genes. *Nucleic Acids Res.* **13,** 8409–8423.
12. Bram, R. J., Lue, N. F. , and Kornberg, R. D. (1986) A family of upstream activating sequences in yeast: roles both in induction and repression of transcription. *EMBO J.* **5,** 603–608.
13. Selleck, S. B. and Majors, J. E. (1987) In vivo DNA-binding properties of a yeast transcription activator protein. *Mol. Cell. Biol.* **7,** 3260–3267.
14. Bajwa, W., Torchia, T. E., and Hopper, J. E. (1988) Yeast regulatory gene *GAL3*, carbon regulation; UASgal elements in common with *GAL1*, *GAL2*, *GAL7*, *GAL10*, *GAL80*, and *MEL1*; encoded protein strikingly similar to yeast and *Escherichia coli* galactokinases. *Mol. Cell. Biol.* **8,** 3439–3447.
15. Birnboim, H. C. and Doly, J. (1979) A rapid alkaline extraction procedure for screening recombinant plasmid DNA. *Nucleic Acids Res.* **7,** 1513–1523.
16. Torchia, T. E., Hamilton, R. W., Cano, C. L., and Hopper, J. E. (1984) Disruption of the regulatory gene *GAL80* in Saccharomyces cerevisiae: effects on carbon-controlled regulation of the galactose/melibiose pathway genes. *Mol. Cell. Biol.* **4,** 1521–1527.
17. Ma, J. and Ptashne, M. (1987) The carboxyl 30 amino acids of GAL4 are recognized by GAL80. *Cell* **50,** 113–119.
18. Lue, N. F., Chasman, D. I., Buchman, A. R., and Kornberg, R. D. (1987) Interaction of *GAL4* and *GAL80* gene regulatory proteins in vitro. *Mol. Cell. Biol.* **7,** 3446–3451.
19. Johnston, S. A., and Hopper, J. E. (1982) Isolation of the yeast regulatory gene *GAL4* and analysis of its dosage effects on the galactose/melibiose regulon. *Proc. Natl. Acad. Sci. USA* **79,** 6971–6975.
20. Winge, O. and Roberts, C. (1948) Inheritance of enzymatic characters in yeast and the phenomenon of long-term adaptation. *C. R. Trav. Lab. Carsberg. Ser. Physiol.* **24,** 264–315.
21. Torchia, T. E., and Hopper, J. E. (1986) Genetic and molecular analysis of the *GAL3* gene in the expression of the galactose/melibiose regulon of *Saccharomyces cerevisiae. Genetics* **113,** 229–246.

22. Carlson, M. (1987) Regulation of sugar utilization in *Saccharomyces* species. *J. Bacteriol.* **169,** 4873–4877.
23. Gancedo, J. M. (1992) Carbon catabolite repression in yeast. *Eur. J. Biochem.* **206,** 297–313.
24. Entian, K.-D., and Barnett, J. A. (1992) Regulation of sugar utilization by *Saccharomyces cerevisiae. Trends. Biochem. Sci.* **17,** 506–510.
25. Trumbly, R. J. (1992) Glucose repression in the yeast *Saccharomyces cerevisiae. Mol. Microbiol.* **6,** 15–21.
26. Adams, B. (1972) Induction of galactokinase in *Saccharomyces cerevisiae*: kinetics of induction and glucose effects. *J. Bacteriol.* **111,** 308–315.
27. Baker, S. M., Johnston, S. A., Hopper, J. E., and Jaehning, J. A. (1987) Transcription of multiple copies of the yeast *GAL7* gene is limited by specific factors in addition to *GAL4. Mol. Gen. Genet.* **208,** 127–134.
28. Johnston, S. A., Zavortnik, M. I., Debouck, C., and Hopper, J. E. (1986) Functional domains of the yeast regulatory protein GAL4. *Proc. Natl. Acad. Sci. USA* **83,** 6553–6557.
29. Schultz, L. D., Hoffman, K. J., Mylin, L. M., Montgomery, D. L., Ellis, R. W., and Hopper, J. E. (1987) Regulated overproducton of the *GAL4* gene product greatly increases expression from galactose-inducible promoters on multicopy expression vectors in yeast. *Gene* **61,** 123–133.
30. Mylin, L. M., Bhat, J. P., and Hopper, J. E. (1989) Regulated phosphorylation and dephosphorylation of GAL4, a transcriptional activator. *Genes Dev.* **3,** 1157–1165.
31. Mylin, L. M., Hoffman, K. H., Schultz, L. D., and Hopper, J. E. (1990) A regulated GAL4 expression cassette providing controllable and high output from high copy galactose promoters in yeast. *Methods Enzymol.* **185,** 297–308.
32. Hopper, J. E., Broach, J. R., and Rowe, L. B. (1978) Regulation of the galactose pathway in *Saccharomyces cerevisiae*: induction of uridyl transferrase mRNA and dependency on *GAL4* gene function. *Proc. Natl. Acad. Sci. USA* **75,** 2878–2882.
33. Guthrie, C. and Fink, G. (1991) Guide to yeast genetics and molecular biology, in *Methods in Enzymology*, vol. 194 (Abelson, J. N. and Simon, M., eds.), Academic, San Diego, CA.
34. Rothstein, R. (1983) One step gene disruption in yeast. *Methods Enzymol.* **101,** 202–211.
35. Schneider, J. C. and Guarente, L. (1991) Vectors for expression of cloned genes in yeast: regulation, overproduction, and underproduction. *Methods Enzymol.* **194,** 373–388.
36. Romanos, M. A., Scorer, C. A., and Clare, J. J. (1992) Foriegn gene expression in yeast: a review. *Yeast* **8,** 423–488.
37. Mithcell, D. A., Marshall, T. K., and Deschenes, R. J. (1993) Vectors for the inducible overexpression of glutathione S-transferrase fusion proteins in yeast. *Yeast* **9,** 715–723.
38. Foreman, P. K. and Davis, R. W. (1994) Cloning vectors for the synthesis of epitope-tagged, truncated and chimeric proteins in *Saccharomyces cerevisiae. Gene* **144,** 63–68.

39. Saiki, R. K., Gefland, D. H., Stoffel, S., Scharf, S. J., Higuchi, R., Horn, G. T., Mullis, K. B., and Erlich, H. A. (1988) Primer-directed enzymatic amplification of DNA with a thermostable DNA polymerase. *Science* **239,** 487–491.
40. Toh-E, A., Ueda, Y., Kakimoto, S., and Oshima, Y. (1973) Isolation and characterization of acid phosphatase mutants in *Saccharomyces cerevisiae. J. Bacteriol.* **113,** 277–287.
41. Bostian, K. A., Lemire, J. M., Cannon, L. E., and Halverson, H. O. (1980) *In vitro* synthesis of repressible yeast acid phosphatase: identification of multiple mRNAs and products. *Proc. Natl. Acad. Sci. USA* **77,** 4504–4508.
42. Ito, H., Fukudua, Y., Murata, K., and Kimura, A. (1983) Transformation of intact yeast cells treated with alkali cations. *J. Bacteriol.* **153,** 163–168.
43. Schiestl, R. H. and Gietz, R. D. (1989) High efficiency transformation of intact yeast cells using single stranded nucleic acids as a carrier. *Curr. Genet.* **16,** 339–346.
44. Gietz, D., St. Jean, A., Woods, R. A., and Schiestl, R. H. (1992) Improved method for high efficiency transformation of intact yeast cells. *Nucleic Acids Res.* **20,** 1425.
45. Chen, D. C., Yang, B. C., and Kuo, T. T. (1992) One-step transformation of yeast in stationary phase. *Curr. Genet.* **21,** 83–84.
46. Hinnen, A., Hicks, J. B., and Fink, G. R. (1978) Transformation of yeast. *Proc. Natl. Acad. Sci. USA* **75,** 1929–1933.
47. Becker, D. M. and Guarente, L. (1991) High-efficiency transformation of yeast by electroporation. *Methods Enzymol.* **194,** 182–187.
48. Manivasakam, P., and Schiestl, R. H. (1993) High efficiency transformation of *Saccharomyces cerevisiae* by electroporation. *Nucleic Acids Res.* **21,** 4414–4415.
49. Tshopp, J. F., Emr, S. D., Field, C., and Scheckman, R. (1986) *GAL2* codes for a membrane-bound subunit of the galactose permease in *Saccharomyces cerevisiae. J. Bacteriol.* **166,** 313–318.
50. Szkutnicka, K., Tschopp, J. F., Andrews, L., and Cirillo, V. P. (1989) Sequence and structure of the yeast galactose transporter. *J. Bacteriol.* **171,** 4486–4493.
51. Long, R. M., Mylin, L. M., and Hopper, J. E. (1991) GAL11 (SPT13), a transcriptional regulator of diverse yeast genes, affects the phosphorylation state of GAL4, a highly specific transcriptional activator. *Mol. Cell. Biol.* **11,** 2311–2314.
52. Southern, E. M. (1975) Detection of specific sequences among DNA fragments separated by gel electrophoresis. *J. Mol. Biol.* **98,** 503–517.
53. Aris, J. P. and Blobel, G. (1991) Isolation of yeast nuclei. *Methods Enzymol.* **194,** 735–749.
54. Peterson, G. L. (1977) A simplification of the protein assay method of Lowry et al. which is more generally applicable. *Anal. Biochem.* **83,** 346–356.
55. Mylin, L. M., Johnston, M., and Hopper, J. E. (1990) Phosphorylated forms of GAL4 are correlated with ability to activate transcription. *Mol. Cell. Biol.* **10,** 4623–4629.
56. Maniatis, T., Fritsch, E. F., and Sambrook, J. (1982) *Molecular Cloning: A Laboratory Manual.* Cold Spring Harbor Laboratory, Cold Spring Harbor, NY.
57. Jaswinski, S. M. (1990) Preparation of extracts from yeast. *Methods Enzymol.* **182,** 154–174.

13

Inducible Expression Cassettes in Yeast: ADH2

Virginia L. Price

1. Introduction

The ADH2 promoter belongs to a class of promoters subject to catabolite repression. Many of these promoters have strong transcriptional start signals, which are repressed by the presence of glucose. The strength of the ADH2 promoter coupled with the fact that the promoter is tightly regulated make it highly amenable for use in the expression of heterologous genes.

Synthesis of ADH2 mRNA is repressed in cells grown in the presence of glucose and is derepressed more than 200-fold when glucose is absent *(1,2)*. Genetic and biochemical analysis of ADH2 expression has revealed several genes that encode trans-acting factors that are required for full derepression of the gene *(3–6)*. These include factors such as SNF1 and CCR1, which are common to other glucose-repressed genes *(4)*. The transcription factor ADR1 has a central role in ADH2 regulation and is necessary for the full derepression of ADH2 transcription *(2,7)*.

Deletion analysis of the ADH2 promoter has identified DNA sequences that are necessary for transcription, known as upstream activation sequences (UAS) *(6,8,9)*. Analogies have been made between the yeast UASs and enhancer sequences present in higher eukaryotes *(6)*, although the yeast UASs do not function 3' to the TATA box as do mammalian enhancers. Two such UASs have been identified that are necessary for maximal derepression of the promoter. One of these, UAS1, is characterized by a 22 base pair (bp) inverted repeat that is the binding site for the transcription factor ADR1 *(9)*. The second, UAS2, consists of a 5 bp element composed of the consensus sequence GGAGA *(6)*.

The wealth of information available regarding this promoter and its regulation suggests numerous changes can potentially be made to increase transcription. Several altered forms of the promoter that increase transcriptional activity

From: *Methods in Molecular Biology, vol. 62: Recombinant Gene Expression Protocols*
Edited by: R. Tuan Humana Press Inc., Totowa, NJ

have been identified. For example, two ADH2 promoter mutants described by Russell et al. *(10)* were characterized by an expansion of the poly (A)·poly (T) tract 222 bp upstream of the ADH2 gene from the usual 20–54 or 55. These mutations were originally called ADR3-4^{c} and 5^{c}, respectively, and are now referred to as ADH2-4^{c} and 5^{c}. Yeast containing this expanded A tract are constitutive for expression of ADH2 because they express approx 15-fold more ADH2 enzyme under repressing conditions (glucose in the medium) than yeast with a wild-type (WT) ADH2 promoter *(10)*. Under derepressing conditions, the ADH2-4^{c} and 5^{c} mutations increased the levels of ADH2p by up to five times that of the WT promoter *(11)*.

Another modification of the ADH2 promoter that has been described in the literature is the duplication of the UAS1 to create two ADR1 binding sites (each site is known to bind two ADR1 monomers) *(12)*. Yu et al. fused the UAS1 region of the ADH2 promoter upstream of a CYC1 promoter which lacked its own UAS. This hybrid promoter was then fused to the lacZ gene *(6)*. β-galactosidase expression was monitored under repressed and derepressed conditions. Yu et al. demonstrated that the ADH2 UAS1 could confer glucose regulation on a heterologous promoter; and that two copies of UAS1 acted cooperatively to increase derepression of the promoter 10-fold better than a single copy in this system. They also demonstrated the requirement of the UAS2 (GGAGA) sequence for maximal derepression: Deletion of a 20 bp fragment from the yeast genome corresponding to 244–264 in the promoter reduced ADH2 activity under derepressed conditions from 1500 mU/mg to less than 9 mU/mg. The transcriptional activities described above result from the hybrid ADH2 UAS/CYC1 promoter (with the exception of the UAS2 deletion from the chromosomal locus). It is not known whether increasing the UAS upstream of the ADH2 promoter itself would affect transcription in exactly the same way, although one would predict some increase to occur especially in combination with excess ADR1p *(see below)*.

The transcription factor ADR1p has been identified as the factor that binds to the UAS1 in the ADH2 promoter and serves to increase transcription of the gene under derepressing conditions *(2,11)*. The gene was cloned and characterized in 1983 *(1)*, and was subsequently found to be a Zn finger protein that bound the UAS1 dyad as two monomers *(7,12)*.

As early as 1979, Ciriacy *(11)* described mutants in *trans*-acting elements that allowed constitutive expression of ADH2 on glucose and derepressed the gene to a greater extent than WT in the absence of glucose (ADR1c mutants): Repressed levels of ADH2 enzyme activity were nearly 100-fold higher than in a WT strain and derepressed levels were three- to fourfold higher. One of these mutants, ADR1-5^{c}, was further analyzed by Cook and Denis *(13)* and Denis and Gallo *(14)* and found to contain an arginine to lysine mutation at position 228 in the ADR1 protein. This mutation (and others in the amino acid 227–239 region of the ADR1 protein) is believed to disrupt the binding of an as yet unidentified repressor of the ADR1p *(15,16)*. Denis et al. *(15)* also obtained a constitutive phenotype for ADH2 expression similar to the ADR1-5^c mutation by deleting the region between amino acids 220–262 genomically.

Alternatively, more ADR1 activity could be obtained by simply overexpressing the gene. Denis *(17)* compared the effects of extra copies of both a WT ADR1 gene and the ADR1-5^c mutant gene on ADH2 expression. He obtained as many as 75 integrated copies of the ADR1 gene and eight copies of the ADR1-5^c gene. It has been observed that overexpression of ADR1 has shown some toxicity as demonstrated by increased petit formation and increased doubling time *(18,19)*. Derepressed levels of ADH2 (the single genomic copy) in both cases were about three- to fourfold higher than WT, similar to a single ADR1-5^c gene. However, when multiple copies of the ADH2 gene were introduced into these strains, as would be the case from a multicopy plasmid, the derepressed levels of ADH2 enzyme activity increased about eightfold (21,000 mU/mg as compared to 2500 mU/mg from WT cells).

The WT ADH2 promoter has been used to drive expression of heterologous proteins from yeast in both research and commercial settings (20–23) from shake flask cultures to the 1600 L scale.

2. Materials

1. Promoter Fragment: The fragment we currently use extends from -1 to -1200, relative to the ATG initiation codon for the ADH2 structural gene. The sequence has been published *(24)*, and is accessible through the Genbank database under accession number J01314. Based on published information characterizing the ADH2 promoter *(6,8,9)*, as little as 300 bp should suffice for transcription initiation, because the region between –1 and –300 appears to contain all the requisite UAS, TATA box, and transcription start sites for the promoter. (We have not yet, however, tried this smaller fragment.)
2. Vectors: The promoter fragment can be inserted into numerous vectors, including the commercially available pYES vector from Invitrogen (San Diego, CA) or the YEpFLAG expression vector available from Eastman Kodak Co./Scientific Imaging Systems (New Haven, CT; cat. no. IB 14303), the vectors described by Hill et al. (YEp351 and YEp352; ref. *25*), or the integrating vector YIp5 for the chromosomal integration of the ADH2 promoter upstream of a desired gene. The sequence of a vector described by Brunelli and Pall *(26)* is available from the Genbank database, accession number L11060 (which includes the ADH2 promoter). Also, plasmids containing the ADH2 promoter are available from the Washington Research Foundation (4225 Roosevelt Way NE, Seattle, WA, 98105).
3. Yeast Strains. A number of strains are available from the Yeast Genetics Stock Center (Berkeley, CA). We have successfully used strains X2181 and YNN281.

The protease-deficient strains BJ2168 and BT3505 are also available from this source. Many of the strains have auxotrophies in TRP, URA, and HIS and can be used with several of the available vectors. It is often worth testing several different strains for expression of a heterologous protein.

4. Media and Growth Conditions. Media are essentially as described in Price et al. *(20)* and Bishop et al. *(23)*. Selective medium for yeast consists of: 6.7 g/L yeast nitrogen base, 20 g/L bacto agar (for plate medium), 1.4 g/L drop-out powder (below), and 2% glucose (glucose is prepared as a 10% solution, autoclaved separately, and added to the medium before use).

Drop-out powder for yeast selective medium (omit the amino acid used in selection): 1.0 g Adenine (Ade), 1.0 g Arginine (Arg), 5.0 g Aspartic Acid (Asp), 5.0 g Glutamic Acid (Glu), 1.0 g Histidine (His), 4.0 g Isoleucine (Iso), 3.0 g Leucine (Leu), 3.0 g Lysine (Lys), 1.0 g Methionine (Met), 2.5 g Phenylalamine (Phe), 20.0 g Serine (Ser), 10.0 g Threonine (Thr), 3.0 g Tyrosine (Tyr), 1.0 g Tryptophan (Trp), 1.0 g Uracil (Ura), and 7.5 g Valine (Val). Mix thoroughly, then add 1.4 g/L to medium while still hot after autoclaving.

Rich medium for yeast consists of: 10 g/L yeast extract, 20 g/L peptone, 1% glucose (for derepressing cultures) or 2% Glucose for plates and growing strains. This medium is referred to as YPD.

We typically maintain cells on selective medium (plates) and inoculate small shake flask cultures of YPD. For cultures larger than 50 mL, a preculture is grown in selective medium and used to inoculate the larger volume of YPD. The inoculum into YPD should be fairly dense: an A_{600} of 0.2–0.4.

3. Method

3.1. Use of the ADH2 Promoter to Express Heterologous Proteins

A 1.2-kb fragment derived from the 5' region of the ADH2 structural gene (–1 to –1200 relative to the ATG initiation site) includes the necessary UAS for full derepression of the ADH2 promoter. This fragment has been used in previous studies describing the expression of heterologous genes from the ADH2 promoter *(20–22)*. With the ease and availability of the polymerase chain reaction (PCR; Perkin-Elmer Cetus, Emeryville, CA), the desired promoter fragment can be readily generated using yeast genomic DNA as a template. We routinely use the technique described by Sathe et al. *(27)*. They use the PCR reaction with whole yeast cells rather than purified DNA as template. This technique is reproducible and generates large amounts of the desired product. The PCR method of Barnes *(28)* that uses Klentaq (Antibody Peptides, St. Louis, MO) and Vent (New England Biolabs, Beverly, MA) polymerases works well with whole yeast cells as templates.

The promoter fragment is generated using primers based on the published sequence *(24)* including desired restriction sites at the 5' and 3' ends and then subcloned into the vector of choice. A typical set of PCR primers that could be used to generate an ADH2 promoter fragment containing the restriction sites *Eco*RI and *Xho*I at the 5' and 3' ends, respectively, would be: 5'-GAATTCGATCCTTCAATATGCGCACATACG-3' for the 5' primer and 5'-CTCGAGCTTTGTGTATTACGATATAGTTAATA-3' for the 3' primer. The reaction condition for PCR are virtually identical to those originally described by Perkin-Elmer Cetus.

We have fused this promoter fragment to a number of heterologous genes *(20,29,30)* both on 2 μ-based multicopy plasmids as well as singly integrated expression cassettes in the chromosome. This promoter has also been fused to a secretion signal and drives successful secretion of product *(20,29,30)* as well as intracellular expression of proteins. We have fused the ADH2 promoter to the yeast TyA open reading frame as described *(31)* to make virus-like particle fusions in yeast.

3.2. Derepression of the Promoter

Derepression of the promoter typically occurs through exhaustion of the glucose supply in the medium. Shake flask cultures are grown in YEP containing 1% glucose and incubated at 30°C for 18–24 h. Despite the fact that a YEP medium is not selective for auxotrophic markers on the plasmids, we find that in the five or six generations of cell growth (initial A_{600} = ca. 0.2–0.4 to a final A_{600} of ca. 10–12), the decrease in copy number of the 2 μ-based plasmid is negligible. We obtain several-fold more product when the cells are grown in this medium than in a synthetic-defined selective medium. The glucose in the medium is rapidly exhausted when the cells are inoculated at this cell density; growth-rate slows considerably, thus minimizing the number of doublings (Fig. 1). Cells are removed by centrifugation and the supernatant is filtered through a 0.2 μ Acrodisk filter.

Large-scale fermentation protocols have been described *(22,23)*. Both groups of investigators used yeast extract in their medium and a fed-batch protocol. (Media containing yeast extract without peptone are considered to be selective for tryptophan because the tryptophan, being acid labile, is destroyed in the yeast extract.)

3.3. Modified Forms of the Promoter that Increase Transcription

Based on the results of Russell et al. *(10)* described above, the ADH2-5^c promoter can be used to drive expression of a heterologous gene. We obtained 20–50% greater expression of a heterologous protein using this promoter relative to a WT promoter. However, it must be considered that in this case the

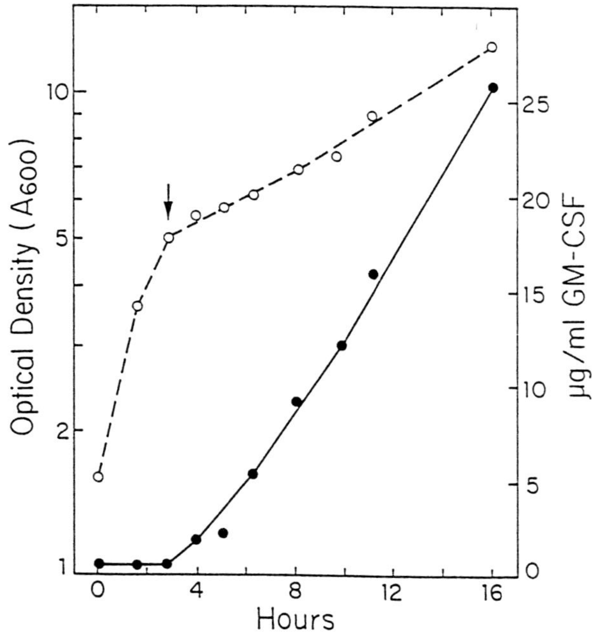

Fig. 1. Derepression of the ADH2 promoter following glucose exhaustion. The dashed line represents the growth of yeast transformed with a plasmid containing the ADH2 promoter regulating transcription of the murine granulocyte-macrophage colony stimulating factor (GM-CSF) gene. Cells were grown in YPD containing 1% glucose as described in Section 2. The arrow indicates the point at which glucose is no longer detectable in the culture medium (Diastix glucose test strips, Miles, Elkhart, IN). The solid line shows the amount of secreted GM-CSF (μg/mL) in the culture medium as determined by radioimmunoassay.

protein is being secreted rather than expressed intracellularly and other factors undoubtedly become limiting. This promoter is available through the Washington Research Foundation.

Bishop et al. *(23)* used the ADH2-4^c promoter to express human factor XIII intracellularly in yeast. They described high-yield fermentation using this promoter and obtained product accumulation to 2% of total soluble protein. They switched from glucose to ethanol during the fermentation and observed an increase in protein level for about 20 h followed by a plateau that was maintained for about 24 h.

4. Notes

1. When considering modifications that one might make to a promoter to increase its transcriptional activity, one must first ask the question, "Is transcription limit-

ing in the system?" A multicopy vector combined with a strong promoter used as a vehicle for expression of heterologous genes usually provides very high mRNA levels. Increasing the copy number or stability of the vector (as seen in the work by Chinery and Hinchliffe *[32]*) or increasing the transcription from the promoter are ways of maintaining or increasing mRNA levels, respectively. Correlation of mRNA levels with expression levels of the protein will allow one to determine if mRNA is limiting. Also, too much mRNA accumulation might be detrimental to growth if other factors involved in translation, for example, become limiting; however, we have no direct evidence for this.

2. Potential toxicity. The potential toxicity of a heterologous gene product must also be considered when choosing an expression strategy. The constitutive ADH2 promoter, ADH2-5^c, or a strain expressing hyperactive ADR1 (described above) may give increased expression from the ADH2 promoter under derepressed conditions, but the promoter is also active under repressed conditions as well. Thus, the WT ADH2 promoter, which is more tightly regulated by glucose, may be more desirable when toxicity of a heterologous product is suspected.
3. Strain differences: We and others have observed differences among strains in their ability to express foreign proteins. Similarly, we have observed differences among strains in their response to altering the ADH2 promoter. (For example, one observes more or less of an effect of the ADH2-5^c promoter in different strains.)
4. Other promoters: We have compared expression levels of the GM-CSF/IL-3 fusion protein *(30)* using the ADH2, ADH1, triose-phosphate isomerase (TPI), and GAL 10 promoters. In our hands, the ADH2 and GAL 10 promoters gave similar expression levels under derepressed conditions in a shake flask. The TPI and ADH1 promoters both gave expression levels approximately four- to five-fold lower than the ADH2 promoter. Because the product was secreted from the yeast cells in all cases, other factors may have influenced the level of secreted product obtained. Also, mRNA levels were not determined in comparing the promoters. The cytoplasmic expression of Factor XIII was reported by Bishop et al. *(23)* using both the ADH2-4^c promoter and the TPI promoter. It appeared from their Western data that the levels of expression were comparable although quantitation of the two systems was not described.

References

1. Denis, C. L. and Young, E. T. (1983) Isolation and characterization of the positive regulatory gene *ADR1* from *Saccharomyces cerevisiae*. *Mol. Cell. Biol.* **3,** 360–370.
2. Denis, C. L., Ciriacy, M., and Young, E. T. (1981) A positive regulatory gene is required for accumulation of the functional messenger RNA for the glucose-repressible alcohol dehydrogenase from Saccharomyces cerevisiae. *J. Mol. Biol.* **148,** 355–368.
3. Ciriacy, M. (1975) Genetics of alcohol dehydrogenase in *Saccharomyces cerevisiae*. II. Two loci controlling synthesis of the glucose-repressible ADHII. *Mol. Gen. Genet.* **138,** 157–164.

4. Ciriacy, M. (1977) Isolation and characterization of mutants defective in intermediary carbon metabolism and in carbon catabolite repression. *Mol. Gen. Genet.* **154,** 213–220.
5. Denis, C. L. (1984) Identification of new genes involved in the regulation of yeast alcohol dehydrogenase II. *Genetics* **108,** 833–844.
6. Yu, J., Donoviel, M. S., and Young, E. T. (1989) Adjacent upstream activation sequence elements synergistically regulate transcription of *ADH2* in *Saccharomyces cerevisiae. Mol. Cell. Biol.* **9,** 34–42.
7. Eisen, A., Taylor, W. E., Blumberg, H., and Young, E. T. (1988) The yeast regulatory protein ADR1 binds in a zinc-dependent manner to the upstream activating sequence of *ADH2. Mol. Cell. Biol.* **8,** 4552–4556.
8. Beier, D. R. and Young, E. T. (1982) Characterization of a regulatory region upstream of the *ADR2* locus of *S. cerevisiae. Nature* **300,** 724–728.
9. Beier, D. R., Sledziewski, A., and Young, E. T. (1985) Deletion analysis identifies a region, upstream of the *ADH2* gene of *Saccharomyces cerevisiae*, which is required for *ADR1*-mediated derepression. *Mol. Cell. Biol.* **5,** 1743–1749.
10. Russell, D. W., Smith, M., Cox, D., Williamson, V. M., and Young, E. T. (1983) DNA sequences of two yeast promoter-up mutants. *Nature* **304,** 652–654.
11. Ciriacy, M. (1979) Isolation and characterization of further cis- and trans-acting regulatory elements involved in the synthesis of glucose-repressible alcohol dehydrogenase (ADHII). *Mol. Gen. Genet.* **176,** 427–431.
12. Thukral, S. K., Eisen, A., and Young, E. T. (1991) Two monomers of yeast transcription factor ADR1 bind a palindromic sequence symmetrically to activate ADH2 expression. *Mol. Cell. Biol.* **11,** 1566–1577.
13. Cook, W. J. and Denis, C. L. (1993) Identification of three genes required for the glucose-dependent transcription of the yeast transcriptional activator *ADR1*. *Curr. Genet.* **23,** 192–200.
14. Denis, C. L. and Gallo, C. (1986) Constitutive RNA synthesis for the yeast activator *ADR1* and identification of the *ADR1-5*^{c} mutation: implications in posttranslational control of *ADR1*. *Mol. Cell. Biol.* **6,** 4026–4030.
15. Denis, C. L., Fontaine, S. C., Chase, D., Kemp, B. E., and Bemis, L. T. (1992) ADR1^c mutations enhance the ability of ADR1 to activate transcription by a mechanism that is independent of effects on cyclic AMP-dependent protein kinase phosphorylation of Ser-230. *Mol. Cell. Biol.* **12,** 1507–1514.
16. Cook, W. J., Chase, D., Audino, D. C., and Denis, C. L. (1994) Dissection of the ADR1 protein reveals multiple, functionally redundant activation domains interspersed with inhibitory regions: evidence for a repressor binding to the ADR1^c region. *Mol. Cell. Biol.* **14,** 629–640.
17. Denis, C. L. (1987) The effects of *ADR1* and *CCR1* gene dosage on the regulation of the glucose-repressible alcohol dehydrogenase from *Saccharomyces cerevisiae*. *Mol. Gen. Genet.* **208,** 101–106.
18. Cherry, J. R. and Denis, C. L. (1989) Overexpression of the yeast transcriptional activator ADR1 induces mutation of the mitochondrial genome. *Curr. Genet.* **15,** 311–317.

19. Price, V. L., Taylor, W. E., Clevenger, W., Worthington, M., and Young, E. T. (1990) Expression of heterologous proteins in *Saccharomyces cerevisiae* using the ADH2 promoter. *Methods Enzymol.* **185,** 308–318.
20. Price, V., Mochizuki, D., March, C. J., Cosman, D., Deeley, M. C., Klinke, R., Clevenger, W., Gillis, S., Baker, P., and Urdal, D. (1987) Expression, purification and characterization of recombinant murine granulocyte-macrophage colony stimulating factor and bovine IL-2 from yeast. *Gene* **55,** 287–293.
21. Kaslow, D. C., Hui, G., and Kumar, S. (1994) Expression and antigenicity of *Plasmodium falciparum* major merozoite surface protein (MSP_{19}) variants secreted from *Saccharomyces cerevisiae. Mol. Biochem. Parasitol.* **63,** 283–289.
22. Kaslow, D. C. and Shiloach, J. (1994) Production, purification and immunogenicity of a malaria transmission-blocking vaccine candidate: TBV25H expressed in yeast and purified using nickel-NTA agarose. *Biotechnology* **12,** 494–499.
23. Bishop, P. D., Teller, D. C., Smith, R. A., Lasser, G. W., Gilbert, T., and Seale, R. L. (1990) Expression, purification, and characterization of human factor XIII in *Saccharomyces cerevisiae*. Biochemistry, **29,** 1861–1869.
24. Russell, D. W. and Smith, M. (1983) Nucleotide sequence of the yeast alcohol dehydrogenase II gene. *J. Biol. Chem.* **258,** 2674–2682.
25. Hill, J. E., Myers, A. M., Koerner, T. J., and Tzagoloff, A. (1986) Yeast/*E. coli* shuttle vectors with multiple unique restriction sites. *Yeast* **2,** 163–167.
26. Brunelli, J. P. and Pall, M. L. (1993) A series of yeast/*Escherichia coli* l expression vectors designed for directional cloning of cDNAs and *cre/lox*-mediated plasmid excision. *Yeast* **9,** 1309–1318.
27. Sathe, G. M., O'Brien, S., McLaughlin, M. M., Watson, F., and Livi, G. P. (1991) Use of polymerase chain reaction for rapid detection of gene insertions in whole yeast cells. *Nucleic Acids Res.* **19,** 4775.
28. Barnes, W. M. (1994) PCR amplification of up to 35-kb DNA with high fidelity and high yield from λ bacteriophage templates. *Proc. Natl. Acad. Sci. USA* **91,** 2216–2220.
29. Gearing, D. P., Thut, C. J., VandenBos, T., Gimpel, S. D., Delaney, P. B., King, J., Price, V., Cosman, D., and Beckmann, M. P. (1991) Leukemia inhibitory factor receptor is structurally related to the IL-6 signal transducer, gp130. *EMBO J.* **10,** 2839–2848.
30. Curtis, B. M., Williams, D. E., Broxmeyer, H. E., Dunn, J., Farrah, T., Jeffery, E., Clevenger, W., deRoos, P., Martin, U., Friend, D., Craig, V., Gayle, R., Price, V., Cosman, D., March, C. J., and Park, L. S. (1991) Enhanced hematopoietic activity of a human granulocyte/macrophage colony-stimulating factor-interleukin 3 fusion protein. *Proc. Natl. Acad. Sci. USA* **88,** 5809–5813.
31. Adams, S. E., Dawson, K. M., Gull, K., Kingsman, S. M., and Kingsman, A. J. (1987) The expression of hybrid HIV:Ty virus-like particles in yeast. *Nature* **329,** 68–70.
32. Chinery, S. A. and Hinchliffe, E. (1989) A novel class of vector for yeast transformation. *Curr. Genet.* **16,** 21–25.

14

Constitutive Expression Vectors: *PGK*

Ian R. Graham and Alistair Chambers

1. Introduction

The promoter of the yeast gene encoding the glycolytic enzyme phosphoglycerate kinase (PGK) has been used to construct vectors for expression of heterologous proteins in budding yeast (*Saccharomyces cerevisiae*) *(1–4)*. This promoter is one of the most efficient yeast promoters and is used when a high level of constitutive gene expression is required. Most *PGK*-based vectors are constructed using high copy number plasmids, to maximize expression levels of heterologous sequences. When present on such a plasmid, the promoter can drive production of PGK protein up to 30–40% of total cell protein, although much lower levels are obtained when heterologous sequences are expressed *(5,6)*. When the human IFNα2 gene was expressed using the high copy number plasmids pMA230-1 and pMA301-1, interferon was only produced to the level of 1–3% of total cell protein *(6)*. Similar low yields have also been described for several other proteins expressed using *PGK* systems *(7,8)*. The reasons for these less than maximum levels of expression are not completely understood, although several suggestions have been made. These range through positive feedback by the PGK protein, RNA stability effects, and the presence of a transcriptional enhancer within the *PGK* coding region *(9,10)*. Despite this underperformance, *PGK* promoter systems are useful because they allow relatively high level gene expression under very simple conditions.

The *PGK* promoter has been intensively studied as a model high efficiency promoter system. The UAS consists of binding sites for the multifunctional transcription factors Rap1p and Abf1p, as well as the glycolytic transcription factor Gcr1p *(11–13)*. Transcriptional activation is mediated via a complex set of interactions between these factors and several other proteins that are brought to the promoter via protein:protein interactions. The basal promoter has also

From: *Methods in Molecular Biology, vol. 62: Recombinant Gene Expression Protocols*
Edited by: R. Tuan Humana Press Inc., Totowa, NJ

been well defined and the site of transcription initiation determined. There is one major transcription start site, a considerable advantage when designing expression vectors *(14)*.

The *PGK* promoter drives expression of a glycolytic protein required at a high level in yeast cells utilizing glucose by fermentation, precisely the conditions in which yeast cells are grown in many different applications. Although the *PGK* promoter is generally regarded as constitutive it can be regulated to some extent by carbon source *(2,11)*. Yeast cells utilize nonfermentable carbon sources such as pyruvate and acetate via the TCA cycle. Under these conditions, expression of the PGK enzyme is controlled via a transcriptional mechanism that results in approximately fivefold reduction of *PGK* mRNA *(11)*. Although this degree of regulation is not sufficiently tight for expression of toxic products, it can be utilized under circumstances where only a low level of protein product is required.

In this method we will describe the use of the basic expression vector, pMA91 (*see* Fig. 1.) but the same principles apply to most *PGK*-based vectors. pMA91 is a yeast/*E. coli* shuttle vector *(7)*. It contains a large (1.5 kb) *PGK* promoter fragment with a *Bgl*II linker inserted at position -2. There is no ATG codon. Downstream of the *Bgl*II linker is a 370 bp fragment from the 3' end of the *PGK* gene, often referred to as a terminator. This extends from a *Bgl*II site at position +1157 within the *PGK* coding region, to a downstream *Hind*III site. The plasmid also contains the *leu2d* gene, the 2 μm origin of replication and sequences from the *E.coli* plasmid pBR322. Sequences for expression are inserted at the unique *Bgl*II site.

2. Materials

2.1. Vectors

The first *PGK*-based vectors were produced in the early 1980s, shortly after the *PGK* gene was cloned *(1,2,7,15)*. These consist of 2 μm based, high copy number plasmids, usually containing a large *PGK* promoter fragment (1.5 kb) and a single unique restriction enzyme site for insertion of foreign DNA. Vectors of this type contain either *leu2d* or *TRP1* as selectable markers and sequences from pBR322 to provide bacterial plasmid functions. They can be used in any yeast strain that contains endogenous 2 μm plasmids and a suitable auxotrophic mutation. Plasmids that carry the *leu2d* gene, a poorly expressed version of the gene with a truncated promoter, replicate to a copy number of about 100. Vectors such as pMA91 and YEpIPT contain the *PGK* promoter but no ATG codon and are designed for expression of complete coding sequences *(7,15)*. Other vectors such as pMA230 contain both the promoter and N-terminus of the *PGK* gene and are designed for the expression of fusion proteins *(2)*. The more recent vector, pPE282, contains a promoter constructed from artificial oligonucleotides and short DNA fragments *(16)*. The basal promoter was derived by joining two oligonucleotides representing the TATA box and RNA initiation site of the *PGK* promoter (T'R). The *PGK* UAS was then inserted immediately upstream of this to make a highly active minipromoter. To prevent any interference by read-through transcription, a 0.75 kb transcriptional blocker fragment from the *TRP1* locus was inserted upstream of the UAS.

One disadvantage with early *PGK* vectors is that they contain only a single restriction enzyme site for insertion of foreign DNA (*Bgl*II, *Bam*HI, or *Eco*RI). More recently, vectors that have a polylinker replacing the single insertion site have been constructed. They also have the added advantage of being significantly smaller than earlier vectors. These pYPGE vectors are also 2 µm-based and contain the *TRP1* selectable marker *(17)*. The pYPGE2 vector was specifically designed for high level expression of cDNAs. It has a polylinker containing eight unique restriction enzyme sites, isolated from pBluescript KS (+) . A further derivative based on *PGK* is the plasmid pYEULS1 *(18)*. This is a vector designed for high level expression of secreted proteins in yeast. It contains a large *PGK* promoter fragment, a secretion signal from the α subunit of the *K. lactis* killer toxin and a *Sac*I site for insertion of foreign sequences. The secretion signal can be cleaved from the translation product by KEX2. Another recent development is a bifunctional yeast-*E.coli* promoter based on *PGK*. The bifunctional promoter was constructed by joining the *PGK* promoter (from -581) to a sequence from the 2 µm *REP2* gene *(19)*. This fused promoter allows the expression of heterologous sequences in both yeast and *E.coli*. Attempts have also been made to combine the high efficiency of the *PGK* promoter with the tight regulation of the *GAL* promoter. The vector pKV49 and derivatives contain the *PGK* promoter with the UAS replaced by the *GAL1-10* UAS (20). There is a single *Bgl*II site for insertion of foreign sequences and the 2 µm origin of replication. This vector has been used successfully to regulate the production of human serum albumin in yeast.

2.2. Yeast Strains and Media

1. *Saccharomyces cerevisiae* DBY745 (α, *ade1-100, leu2-3, leu2-112, ura3-52*) or any suitable yeast strain that is auxotrophic for leucine metabolism.
2. YEPD liquid medium: 2% Bacto-peptone (Difco), 1% yeast extract, 2% glucose. For YEPD agar, add 2% agar (Difco bacto-agar).
3. SC liquid medium: 0.67% yeast nitrogen base without amino acids (Difco), 2% glucose. For SC agar, add 2% agar. To reduce the strength of the *PGK* promoter by about fivefold, replace the glucose with 2% pyruvate.

   Autoclave and when cool add supplements as required.
4. Regeneration agar: 1*M* sorbitol, 0.67% yeast nitrogen base without amino acids, 1% glucose, 3% agar.

   Autoclave and when cool add supplements as required.

5. 100X supplements mix: 0.2% adenine, histidine, arginine, and methionine; 0.3% tyrosine, isoleucine, and lysine; 0.5% phenyl alanine; 1% glutamic acid and aspartic acid; 1.5% valine; 2% threonine; 4% serine.

   Autoclave once only and store in the dark at 4°C. Add 1 mL of mix per 100 mL of agar or liquid medium. All components of supplements mix can be purchased from Sigma.
6. Individual supplement stocks: 0.3% leucine, 0.2% tryptophan, and 0.2% uracil. Autoclave each and store in the dark at 4°C. These are the commonest selectable markers on *PGK* plasmids and should be added at 1 mL/100 mL agar or liquid medium as required. For plasmids carrying the *leu2d* gene, both tryptophan and uracil should be added to the medium.

2.3. Yeast Transformation

1. Sorbitol solution: 1*M* sorbitol pH 5.6. pH adjust water to 5.6 using a very dilute solution of HCl. Use this pH-adjusted water to make the 1*M* sorbitol solution.
2. Lyticase: Resuspend lyophilized lyticase powder (Sigma) in 50 m*M* sodium phosphate buffer pH 7, to an activity of 8000 U/mL. Store in 1 mL aliquots at –20°C. To make the 50 m*M* sodium phosphate buffer first make 0.2*M* sodium phosphate buffer, pH 7. Dissolve 8.19 g Na_2HPO_4 and 6.6 g NaH_2PO_4 in 500 mL H_2O. Autoclave, and store at room temperature. Alternatively, add 57.7 mL and 42.3 mL of 1*M* stocks of Na_2HPO_4 and NaH_2PO_4 , respectively, to 400 mL of H_2O, prior to autoclaving. Dilute 1 + 3 with H_2O to make 50 m*M* stock just before use.
3. Sorbitol/$CaCl_2$/Tris solution: 1*M* sorbitol, 10 m*M* $CaCl_2$, 10 m*M* Tris-HCl, pH 7.5.
4. PEG solution: 44% (w/v) polyethylene glycol 4000 made up in sorbitol/$CaCl_2$/ Tris solution.
5. OneStep transformation buffer: 0.2*M* lithium acetate, 40% (w/v) polyethylene glycol (PEG) 4000, 0.1*M* dithiothreitol (DTT). Make up in small quantities (typically 10–20 mL), autoclave, and store at 4°C for up to 2 mo. It should not be used after any discoloration has occurred.

2.4. Total RNA Extraction

1. LET buffer: 100 m*M* Tris-HCl, pH 7.4, 100 m*M* LiCl, 0.1 m*M* EDTA. Autoclave and store at room temperature.
2. TNE buffer (1X): 140 m*M* NaCl, 1 m*M* EDTA, 10 m*M* Tris-HCl, pH 7.4. Autoclave and store at room temperature.
3. TNES solution: 1% (w/v) SDS in 10X TNE buffer. This should be made fresh and stored for short periods at 37°C.
4. Phenol for RNA extraction: High quality phenol (Sigma Molecular Biology Grade) is melted by placing at 65°C for a short time. The desired amount (typically 10 mL) is placed in a 50 mL, conical-bottomed, polycarbonate centrifuge tube, and to it is added an equal volume of 1X TNE buffer. Shake well. This should be made as fresh as possible, and must be used at room temperature, but may be stored at 4°C for short periods.

Also required: chloroform, 3*M* sodium acetate pH 5, absolute ethanol, 80% ethanol.

2.5. Total Protein Extracts

1. Phosphate buffer: 0.2*M* sodium phosphate buffer pH 7.5. Mix 77.4 mL 1*M* Na_2HPO_4 with 22.6 mL 1*M* NaH_2PO_4, and make up to 500 mL with H_2O. Dilute 1 + 7 just before use to make 25 m*M* buffer for protein extract.
2. Phosphate buffer plus PMSF: 25 m*M* sodium phosphate buffer containing 1 m*M* PMSF. Make PMSF as a 100 m*M* solution in ethanol and store at –20°C.

   Also required: glycerol.

2.6. Measurement of Plasmid Copy Number

1. Sorbitol/EDTA/2-mercaptoethanol: 0.9*M* sorbitol, 0.02*M* EDTA, 14 m*M* 2-mercaptoethanol.
2. TE buffer: 10 m*M* Tris-HCl, pH 7.4, 1 m*M* EDTA.
3. RNase: DNase-free RNase (0.5 mg/mL; Boehringer).

   Also required: lyticase solution (*see* Section 2.3.), 0.5*M* EDTA pH 8.5, 1*M* Tris-HCl, pH 7.4, 10% (w/v) SDS, diethylpyrocarbonate (DEPC), 5*M* potassium acetate, 70% ethanol, absolute ethanol, 3*M* sodium acetate, isopropanol.

3. Methods

We describe methods for use of the basic *PGK* expression vector, pMA91 (*see* Fig. 1). The DNA sequence to be expressed should contain an initiating ATG codon and should be inserted into the unique *Bgl*II site. The resulting plasmid should be transformed into yeast, and gene expression analyzed in yeast cells growing in selective medium containing glucose.

3.1. Yeast Transformation (Spheroplast Method) (see Note 1)

1. Grow DBY745 overnight with shaking in 100 mL YEPD to a cell density of 1.5 × 10^7 cell/mL.
2. Harvest cells by centrifugation at 1200*g* for 10 min at room temperature.
3. Resuspend cell pellet in 20 mL sorbitol solution.
4. Repellet by centrifugation at 1200*g* for 10 min as previously.
5. Resuspend cell pellet in 10 mL sorbitol solution. Add 20 µL lyticase solution.
6. Incubate at 30°C with occasional shaking for 1–2 h. From this stage, treat the spheroplasts gently, do not vortex.
7. Check spheroplasting by adding 100 µL of cells to 1 mL aliquots of water and sorbitol solution. If the majority of cells have been converted to spheroplasts they will lyse in the water but remain intact in the sorbitol. This can be observed by light microscopy. If spheroplasts have not formed, add more lyticase and leave for a further 1–2 h. If spheroplasts do not form, it is indicative of the starting culture being overgrown.
8. Harvest spheroplasts by centrifugation at 1200*g* for 5 min.

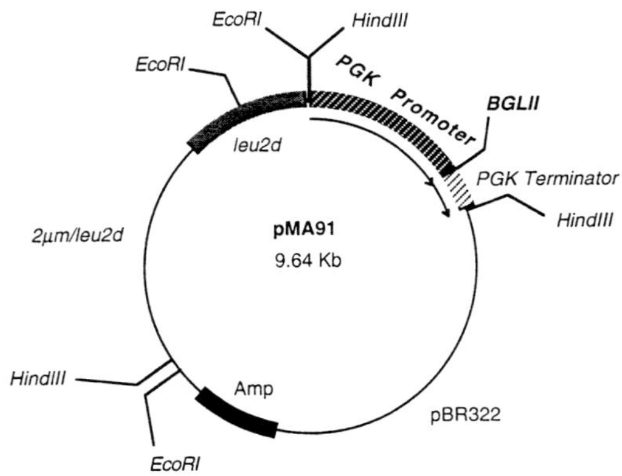

Fig. 1. The basic *PGK* expression vector pMA91 *(7)*. Sequences for expression from the *PGK* promoter are inserted at the unique *Bgl*II site. The plasmid is leucine selectable and replicates to high copy number in yeast.

9. Wash spheroplasts twice with 10 mL sorbitol solution.
10. Wash spheroplasts once with 10 mL sorbitol/$CaCl_2$/Tris solution.
11. Resuspend in 0.5 mL sorbitol/$CaCl_2$/Tris solution. Leave static at room temperature for 30 min.
12. Add 2–5 µg of plasmid DNA in a volume of about 20 µL to a 100-µL aliquot of yeast spheroplasts.
13. Incubate at room temperature for 20 min.
14. Add 1 mL of PEG solution.
15. Mix by inverting and incubate at room temperature for 10 min.
16. Collect cells by gentle centrifugation in a microfuge. Use short pulses for a total time of 40 s.
17. Resuspend the pellet in 1 mL of sorbitol solution, then add to 18 mL of regeneration agar (at 50°C) and pour into plates.
18. Incubate at 30°C for several days.
19. *PGK* vectors carrying the *leu2d* gene should give 10^3–10^5 transformants per microgram using this spheroplast method.

3.2. One Step Transformation of Yeast (21) (see Note 1)

1. Grow culture of yeast cells (DBY745 or similar) overnight in YEPD medium, at 30°C with shaking. The cell density is not critical for this method, but is better if the culture is approaching stationary phase (2.5×10^8 cells/mL).
2. Transfer 0.5–1 mL of culture to microfuge tubes. One tube is required per sample of transforming DNA. This should give approx 5×10^7 cells per experiment.
3. Spin the cells for 30 s at low speed in a microcentrifuge, then remove supernatant completely.

4. Add 100 µL of One Step buffer, and 2–5 µg of transforming DNA, then resuspend pellet by vortexing gently, using several short (2–5 s) pulses.
5. Incubate at 45°C for 30–60 min. Ensure cells remain in suspension by gently vortexing every 10 min.
6. Plate suspensions directly on to selective (SC) medium, and incubate at 30°C for up to 1 wk.

3.3. Selection of Transformants

1. Streak new transformed colonies onto selective SC plates. Isolate single colonies and restreak. Use at least two different transformants for analysis of gene expression.
2. Analyze gene expression by growing transformed cells in 100 mL aliquots of liquid SC selective medium to required cell density. Depending on available assay, either prepare RNA and measure gene expression directly, or prepare crude protein extracts and perform protein assays.
3. Transformed cells can be stored for about 1 mo on selective plates at 4°C. For longer storage, prepare glycerol stocks. Use 1 mL sterile 15% glycerol in a 2 mL freezing vial. Fresh cells can be transferred to the vials from the surface of SC plates using either a sterile loop or sterile toothpick. The cells should be well mixed with the glycerol and stored at lower than –60°C.

3.4. Total RNA Extraction (see Note 4)

1. Grow yeast cells in 100 mL of selective SC medium to required cell density, typically midlog phase (6×10^6 cells/mL).
2. Harvest cells by centrifugation at 1200*g* for 5 min at 4°C.
3. Wash cell pellet twice using 1 mL LET buffer for each wash.
4. After the second wash resuspend the cells in 0.2 mL LET buffer. Transfer cell suspension to microfuge tube. Add 100 µL phenol and glass beads (BDH, 40 mesh) to fill half the volume of the liquid.
5. Vortex hard for 30 s then add 20 µL of TNES solution.
6. Add 100 µL of chloroform and 100 µL of sterile distilled water. Vortex sample for ten seconds then shake for 1 min.
7. Centrifuge at top speed in a microfuge for 2 min and transfer supernatant to a fresh microfuge tube.
8. Re-extract supernatant three more times, each time using 100 µL phenol and 100 µL chloroform.
9. Add 0.1 vol of 3*M* sodium acetate pH 5, and 2.5 vol of absolute ethanol to precipitate the RNA. Place at –80°C for 30 min before pelleting RNA at high speed in a microfuge for 10 min. Remove supernatant, then add 0.5 mL of 80% (v/v) ethanol. Spin for 2 min and remove supernatant.
10. Resuspend dried pellet in about 50 µL sterile distilled water.
11. Heat RNA at 65°C for 5 min and cool rapidly on ice before removing samples to determine concentration by spectrophotometry and agarose gel electrophoresis.

3.5. Total Protein Extracts from Yeast Cells (see Note 5)

1. Grow 50–100 mL of yeast culture to required cell density (typically midlog phase; 4–6 × 10^6 cells/mL) and harvest cells by centrifugation at 1200*g* for 5 min at room temperature.
2. Wash cell pellet twice, each time using 1 mL of phosphate buffer and spinning at 1200*g* for 5 min.
3. After second wash resuspend cell pellet in 0.3 mL of ice-cold phosphate buffer containing PMSF. Transfer cell suspension to microfuge tube. Add baked glass beads (BDH, 40 mesh) to fill half of the liquid volume.
4. Vortex hard for three periods of 20 s, interspersed with periods of 20 s on ice.
5. Centrifuge for 10 s at full speed in a microfuge and transfer supernatant to a fresh tube.
6. Add 0.2 mL phosphate buffer containing PMSF, and invert several times to wash glass beads and cell debris. Spin at full speed for 10 s. Add the supernatant to the previous supernatant fraction.
7. Centrifuge supernatant at full speed in a microfuge for 15 min at 4°C.
8. Place supernatant in a fresh tube, and add glycerol to 10% for storage at –20°C.

3.6. Measurement of Plasmid Copy Number

1. Collect yeast cells by centrifugation at 1200*g* for 5 min. Typically, use about 50 mL of a midlog phase culture.
2. Resuspend the cells in 0.7 mL of sorbitol/EDTA/2-mercaptoethanol.
3. Transfer the cells to a 15 mL centrifuge tube (Corex or similar).
4. Add 10 µL lyticase solution and incubate at 37°C for one to 2 h with occasional gentle shaking.
5. Centrifuge at 1200*g* for 5 min at room temperature.
6. Resuspend spheroplasts in 0.7 mL TE buffer.
7. Add 70 µL 0.5*M* EDTA, pH 8.5, 30 µL 1*M* Tris-Hcl, pH 7.4, 100 µL 10% SDS, 2 µL diethylpyrocarbonate (DEPC).
8. Mix by gentle swirling and incubate uncovered in a fume hood at 65°C for 30 min.
9. Add 0.16 mL 5*M* potassium acetate and incubate on ice for 20–30 min.
10. Centrifuge at 8000*g* for 15 min at 4°C.
11. Transfer supernatant to a fresh 15 mL centrifuge tube and add two volumes of ethanol. Mix gently.
12. Centrifuge at 1200*g* for 5 min at room temperature.
13. Wash with 70% ethanol and resuspend in 0.6 mL TE buffer. Leave overnight at room temperature (*see* Note 6).
14. Add 5 µL RNase and incubate at 37°C for 30 min.
15. Add 60 µL 3*M* sodium acetate and 0.375 mL isopropanol. Mix gently.
16. Centrifuge at 1200*g* for 5 min at room temperature.
17. Wash pellet with isopropanol, allow to dry, and resuspend in 0.2 mL TE buffer. Leave overnight at room temperature.
18. Transfer to microcentrifuge tube. Extract twice with phenol/chloroform and once with chloroform.

19. Ethanol precipitate and wash twice with 70% ethanol.
20. Dry and resuspend in 100 µL of sterile distilled water. Leave for 4 h at room temperature.
21. Run 5 µL on mini-gel.
22. Digest 40–70 µL of the DNA sample with an appropriate restriction enzyme. For many *PGK* 2 µm vectors *EcoRI* can be used. Digest in a reaction volume of 200 µL and use 4 µL of enzyme. Leave digesting overnight.
23. After digestion, ethanol precipitate the DNA, and electrophorese on a 1% agarose TAE gel. Transfer the DNA to nitrocellulose or similar membrane by Southern blot.

4. Notes

1. When selecting the method to be used for transformation, it is worth bearing in mind that the only real advantage of using the spheroplast method is the increased efficiency of transformation (10^5 transformants per µg of DNA, versus 10^4 per µg with the One Step method). If this is not critical to your experiment, the One Step method is likely to be better–we now use it for most applications. It is extremely quick to perform, and requires only a 45°C water bath or heating block, and a microcentrifuge.
2. For most applications described in this chapter, it is ideal to assess the number of yeast cells in a culture accurately, using a hemacytometer. If one of these is not available, a good estimate of a midlog phase culture is an OD_{600} reading of 0.2–0.4. In studies of yeast gene expression, it is important to ensure that the cultures have grown for approximately six doubling times (15–20 h for most strains). This represents a 64-fold increase in the number of cells. Cultures are normally inoculated using a suspension of cells for which the OD_{600} has been established. As a guide, 0.3 mL of a suspension with an OD_{600} of 0.75 is used to inoculate a 100 mL culture.
3. Harvesting of yeast cells is performed by centrifugation in 50 mL, conical-bottomed tubes using a benchtop centrifuge. In general, it is better to use a swing-out rotor, particularly if cell numbers are low. A fixed angle rotor may be used, but extreme care must be taken not to dislodge the cell pellet when decanting the supernatant.
4. RNA preparation from yeast is a particularly tricky procedure—the cells are very rich in ribonucleases. It is essential to keep a separate set of solutions for RNA work. If it is not feasible to use an entirely separate set of pipetors, then they must be cleaned thoroughly with ethanol, both inside and out, before embarking on the RNA preparation. Try to work as quickly as possible once the cells have been lysed. The RNA used for your application should be as fresh as possible, but if it needs to be stored, –80°C is a must. If possible, keep a separate electrophoresis gel tank for use with RNA gels only, particularly if RNase is used on a regular basis in your lab.
5. To avoid degradation of total protein extracts, their preparation should be performed in a 4°C room. If this facility is not available, great care must be taken to ensure that samples are kept on ice.

6. In the copy number DNA preparation, resuspension of the DNA should be left overnight to ensure that both plasmid and chromosomal DNA is resuspended completely. Failure to do this may result in inaccurate determination of the plasmid DNA to chromosomal DNA ratio.

   Two probes should be used to screen the Southern blot in assessment of plasmid copy number. As a standard, we use a probe which hybridizes to the 18s rDNA region. The other probe should be specific to the plasmid, usually derived from the heterologous sequence to be expressed. Probes are labelled to high activity by nick translation or by random oligo-labeling.

References

1. Hitzeman, R. A., Hagie, F. E., Levine, H. L., Goeddel, D. V., Ammerer, G., and Hall, B. D. (1981) Expression of a human gene for interferon in yeast. *Nature* **24,** 717–722.
2. Tuite, M. F., Dobson, M. J., Roberts, N. A., King, R. M., Burke, D. C., Kingsman, S. M., and Kingsman, A. J. (1982) Regulated high efficiency expression of human interferon-alpha in Saccharomyces cerevisiae. *EMBO J.* **1,** 603–608.
3. Hitzeman, R. A., Chang, C. N., Matteucci, M., Perry, L. J., Kohr, W. J., Wulf, J. J., Swartz, J. R., Chen, C. Y., and Singh, A. (1986) Construction of expression vectors for secretion of human interferons by yeast. *Methods Enzymol.* **119,** 424–433.
4. Kingsman, S. M., Cousens, D., Stanway, C. A., Chambers, A., Wilson, M., and Kingsman, A. J. (1990) High efficiency yeast expression vectors based on the promoter of the phosphoglycerate kinase gene. *Methods Enzymol.* **185,** 329–341.
5. Chen, C. Y., Oppermann, H., and Hitzeman, R. A. (1984) Homolgous versus heterologous gene expression in the yeast Saccharomyces cerevisiae. *Nucleic Acids Res.* **12,** 8951–8970.
6. Mellor, J., Dobson, M. J., Roberts, N. A., Kingsman, A. J., and Kingsman, S. M. (1985) Factors affecting heterologous gene expression in Saccharomyces cerevisiae. *Gene* **33,** 215–226.
7. Mellor, J., Dobson, M. J., Roberts, N. A., Tuite, M. F., Emtage, J. S., White, S., Lowe, P. A., Patel, T., Kingsman, A. J., and Kingsman, S. M. (1983) Efficient synthesis of enzymatically active calf chymosin in *Saccharomyces cerevisiae.* *Gene* **24,** 1–14.
8. Hitzeman, R. A., Chen, C. Y., Hagie, F. E., Patzer, E. J., Liu, C. C., Estell, D. A., Miller, J. V, Yaffe, A., Kleid, D. G., Levinson, A. D., and Oppermann, H. A. (1983). Expression of hepatitis b surface antigen in yeast. *Nucleic Acids Res.* **11,** 2745–2763.
9. Chen, C. Y. and Hitzeman, R. A. (1987) Human, yeast and hybrid 3-phosphoglycerate kinase gene expression in yeast. *Nucleic Acids Res.* **15,** 643–660.
10. Mellor, J., Dobson, M. J., Kingsman, A. J., and Kingsman, S. M. (1987) A transcriptional activator is located in the coding region of the yeast *PGK* gene. *Nucleic Acids Res.* **15,** 6243–6259.

11. Chambers, A., Tsang, J. S. H., Stanway, C., Kingsman, A. J., and Kingsman, S. M. (1989) Transcriptional control of the Saccharomyces cerevisiae *PGK* gene by RAP1. *Mol. Cell. Biol.* **9,** 5516–5524.
12. Chambers, A., Stanway, C., Tsang, J. S. H, Henry, Y., Kingsman, A. J., and Kingsman, S. M. (1990) ARS binding factor 1 binds adjacent to RAP1 at the UASs of the yeast glycolytic genes *PGK* and *PYK1*. *Nucleic Acids Res.* **18,** 5393–5399.
13. Henry, Y. A. L., Lopez, M. C., Gibbs, J. M., Chambers, A., Kingsman, S. M., Baker, H. V., and Stanway, C. (1994) The yeast protein Gcr1p binds to the *PGK* UAS and contributes to the activation of transcription of the *PGK* gene. *Mol. Gen. Genetics* **245,** 506–511.
14. Rathjen, J., and Mellor, J. (1990) Characterisation of sequences required for RNA initiation from the PGK promoter of Saccharomyces cerevisiae. *Nucleic Acids Res.* **18,** 3219–3225.
15. Hitzeman, R. A., Leung, D. W., Perry, L. J., Kohr, W. J., Levine, H. L., and Goeddel, D. V. (1983) Secretion of human interferons by yeast. *Science* **219,** 620–625.
16. Stanway, C. A., Chambers, A., Kingsman, A. J., and Kingsman, S. M. (1989) Characterisation of the transcriptional potency of sub-elements of the UAS of the yeast *PGK* gene in a *PGK* mini-promoter. *Nucleic Acids Res.* **17,** 9205–9218.
17. Brunelli, J. P. and Pall, M. L. (1993) A series of yeast shuttle vectors for expression of cDNAs and other DNA sequences. *Yeast* **9,** 1299–1308.
18. Castelli, L. A., Mardon, C. J., Strike, P. M., Azad, A. A., and Macreadie, I. G. (1994) High level secretion of correctly processed beta-lactamase from Saccharomyces cerevisiae using a high copy number secretion vector. *Gene* **142,** 113–117.
19. Faulkner, J. D. B., Anson, J. G., Tuite, M. F., and Minton, N. P. (1994) High-level expression of the phenylalanine ammonia lyase-encoding gene from *Rhodosporidium toruloides* in *Saccharomyces cerevisiae* and *Esherichia coli* using a bifunctional expression system. *Gene* **143,** 13–20.
20. Cousens, D. J., Wilson, M. J., and Hinchliffe, E. (1990) Construction of a regulated *PGK* expression vector. *Nucleic Acids Res.* **18,** 1308.
21. Chen, D.-C., Yang, B.-C., and Kuo, T.-T. (1992). One-step transformation of yeast in stationary phase. *Curr. Genet.* **21,** 83–84.

IV

Viral Expression Systems

15

Design and Construction of Recombinant Vaccinia Viruses

Christopher C. Broder and Patricia L. Earl

1. Introduction

The structural and functional analyses of proteins have benefited enormously from the use of technologies of recombinant gene expression. The recombinant vaccinia virus system has been widely employed to express genes from eukaryotic, prokaryotic, and viral origins (for reviews, *see* refs. *1–7*) and several detailed protocols for the generation, identification, isolation, and characterization of recombinant vaccinia viruses have been published *(8–10)*.

Vaccinia virus, a large double-stranded DNA virus, is the prototypic and best characterized member of the poxvirus family. Replication and gene expression occur in the cytoplasm of the infected host cell *(11,12)*. The expression of vaccinia viral genes occur in succession through the regulated transcription of early, intermediate, and late classes of genes, as dictated by viral promoter structures (for review, *see* ref. *13*). Since the first description and use of the recombinant vaccinia virus expression system in the early 1980s *(14,15)*, it has been modified, improved, and extensively used. Foreign gene expression by recombinant vaccinia viruses offers several advantages:

1. Proteins are processed and modified correctly;
2. Proteins are properly transported and localized in the infected cell;
3. Uniform protein production is achieved within a target cell population using a high multiplicity of infection (moi);
4. The extremely broad host range of vaccinia virus allows a wide array of primary and transformed tissue culture cell lines to be utilized;
5. A variety of natural and synthetic vaccinia virus promoters, as well as hybrid systems using the bacteriophage T7 *(16–18)*, T3 *(19)*, and SP6 *(20)* promoters,

From: *Methods in Molecular Biology, vol. 62: Recombinant Gene Expression Protocols*
Edited by: R. Tuan Humana Press Inc., Totowa, NJ

and repression via the *Escherichia coli lac* repressor/operator *(21–23)* permit varying levels and control of gene expression;

6. The problems and limitations associated with expression in permanently transformed cell lines (e.g., production of cytotoxic proteins) are avoided owing to the transient nature of the vaccinia virus system; and
7. The cytoplasmic localization of transcription bypasses requirements for regulated export of unspliced mRNAs out of the nucleus (e.g., for structural proteins of primate lentiviruses *[24]*).

However, since messenger RNAs (mRNAs) are not spliced in the vaccinia virus system *(25)*, open reading frames must be continuous.

1.1. The Mechanism of Recombinant Vaccinia Virus Generation

The original and still most widely used method for the generation of recombinant vaccinia viruses relies on homologous recombination in vivo *(14,15)*. The general scheme for incorporation of foreign coding sequences into the virus genome by homologous recombination is diagrammed in Fig. 1. First, the gene of interest is cloned into a plasmid transfer vector that contains the following elements: (1) a vaccinia virus promoter; (2) a multiple cloning site adjacent to the promoter; (3) flanking sequences derived from a nonessential site within the vaccinia virus genome; and (4) the necessary elements for replication and selection in bacteria. In addition, screening and/or selection markers may be included to facilitate identification of recombinant virus. A list of commonly used transfer vectors is shown in Table 1. Second, tissue culture cells are infected with a parental strain of vaccinia virus, such as WR, and transfected with the transfer vector containing the gene of interest. Homologous recombination between the vaccinia virus DNA and the transfer vector results in incorporation of the foreign gene into the viral genome. This recombination process yields approx 1 recombinant virion in 1000 progeny. Replication of the recombinant genome continues and maturation of virions occurs. Third, the desired recombinant vaccinia virus is plaque purified by several rounds of selection and/or screening. Finally, high-titer recombinant virus stocks are prepared from infected cell lysates. Purification of vaccinia virus can be performed if the presence of host cell proteins is undesirable or if very high titers of virus are required ($1–5 \times 10^{10}$ plaque-forming-units (PFU)/mL).

In addition to generation of recombinant viruses by in vivo homologous recombination, protocols for in vitro ligation of genomic vaccinia virus DNA with foreign DNA have recently been described *(26,27)*. Incorporation of very large DNA segments (up to 26,000 bp) have been demonstrated using this method.

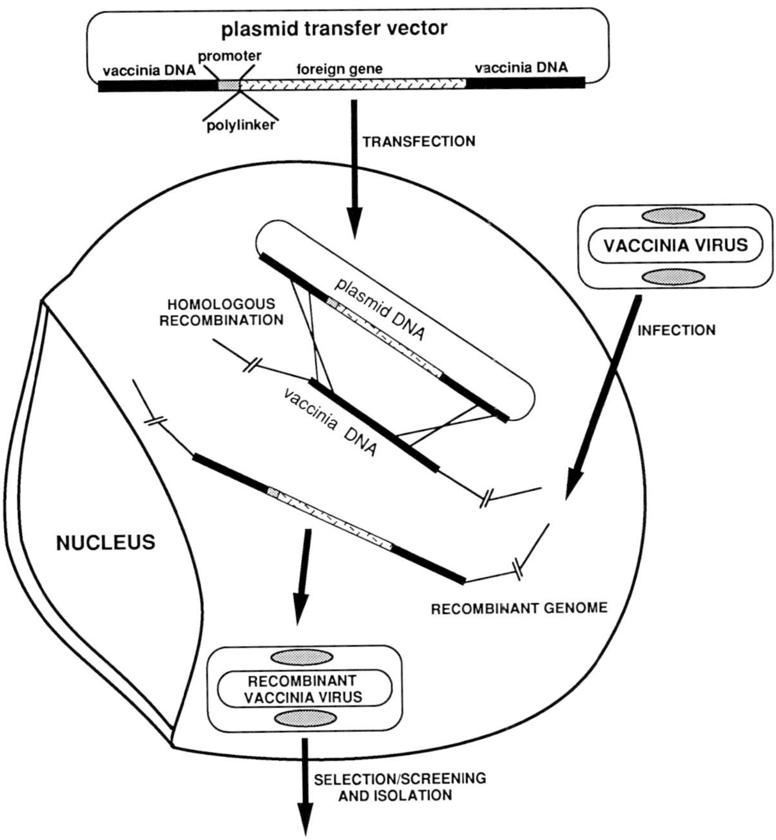

Fig. 1. The generation of recombinant vaccinia viruses by homologous recombination. Cells are infected with vaccinia virus and transfected with a plasmid transfer vector that contains a foreign gene driven by a viral promoter and flanked by vaccinia virus DNA segments. Homologous recombination between the vaccinia virus sequences in the transfected vector DNA and the viral genome occurs during the replication cycle of virus. The resulting DNA genome is packaged to form progeny recombinant vaccinia virus (diagram not drawn to scale).

1.2. Choice of Transfer Vector and Promoter Considerations

There are numerous combinations of promoter and selection systems. The type of promoter employed dictates both the level and time of expression. Quantitative analysis of the expression of vaccinia virus genes has revealed that early promoters express genes from 0.5 h to a peak at 1.5 h postinfection, intermediate promoters from approx 1.5 h to a peak at 2 h, and late promoters from approx 3 h onward (*see* ref. *13* for review). Constitutive, or compound,

Table 1
Vaccinia Virus Transfer Vectors[a]

Selection/ screening	Vector	Promoter[b]	Flanking[c] vaccinia	Ref.
TK	pGS20	p7.5 (E/L)	TK	*(51)*
	pSC59	Synthetic (E/L)	TK	Chakrabarti and Moss, unpublished
TK and β-gal	pMJ601	Synthetic (L)	TK	*(28)*
	pSC65	Synthetic (E/L)	TK	Chakrabarti and Moss, unpublished
	pSC11[d]	p7.5 (E/L)	TK	*(36)*
β-gal	pCF11	p7.5 (E/L)	*Hind*III C	*(52)*
Ecogpt and/or TK	pTKgptF1s[d]	p11 (L)	TK	*(39)*

[a]Representive plasmid transfer vectors utilizing the types of selection/screening protocols outlined in this chapter are shown. The table is not intended to be exhaustive. A more complete list can be found in ref. *8*.
[b]L, late; E/L, early and late.
[c]Vaccinia virus genome region used for directing homologous recombination.
[d]Represented in Fig. 2.

promoters are those that contain both early and late transcriptional elements. Factors that influence the choice of promoter system come from assessing the desired use of the recombinant vaccinia virus, or from the known properties of the gene product of interest. For example: for large scale protein production a strong vaccinia virus promoter such as the synthetic late *(28)* or early/late (Chakrabarti and Moss, unpublished) promoter or the hybrid vaccinia/T7 polymerase system *(16–18)* should be used *(29,30)*; for induction of class I restricted cytotoxic T-cell response in vivo, a natural early or tandem early/late promoter is recommended *(7)*; for production of a potentially cytotoxic protein, use of the *E. coli* lac repressor/operator system *(31)* or the hybrid vaccinia/T7 system allows for initiation of gene expression when appropriate. When early gene expression is important, the coding sequence should be scanned for the presence of the sequence TTTTTNT. This sequence signals early transcriptional termination in vaccinia virus *(32)* and should be changed without altering the amino acid sequence *(33)*. Finally, if a specialized cell type is to be used, such as primary cell cultures, it may prove useful to characterize that cell type for its ability to support vaccinia virus infection, replication, and gene expression by different classes of promoters *(34,35)*. This can be achieved in a straightforward manor by use of an available reporter gene (e.g., *E. coli lacZ*) linked to different promoters.

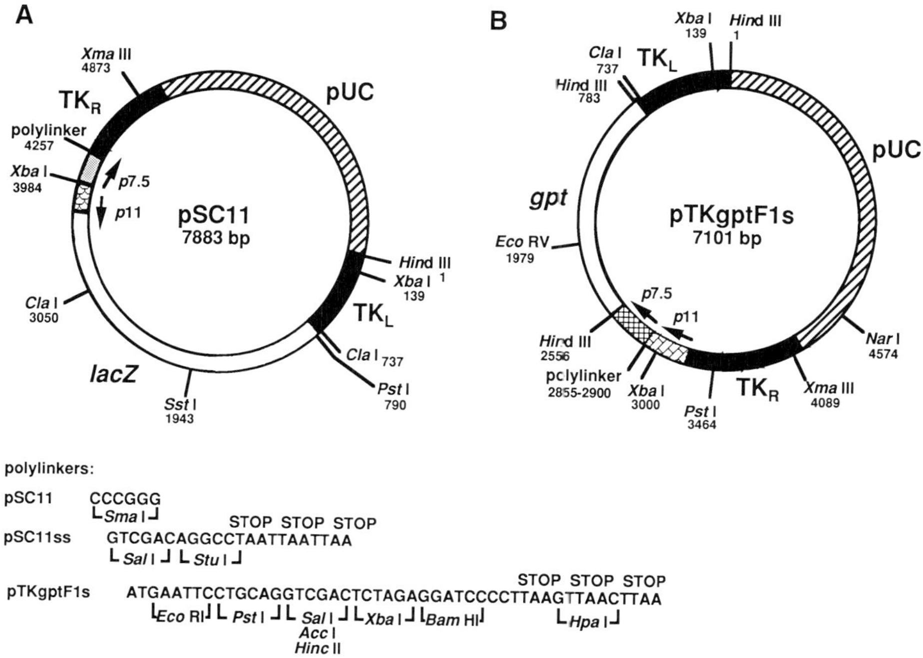

Fig. 2. Examples of plasmid transfer vectors. **(A)** pSC11 *(36)* and pSC11ss *(33)* provide for both tk selection and beta-galactosidase screening. The late vaccinia virus promoter *p*11 drives the *E. coli lacZ* gene. Genes can be inserted at the polylinker site and are controlled by the compound early/late vaccinia virus promoter *p*7.5. **(B)** pTKgptF1s *(39)* provides for *tk*⁻ and/or XGPRT selection. The compound early/late vaccinia virus promoter *p*7.5 drives the expression of the *Ecogpt* gene. The polylinker provides an ATG start and expression is driven by the late vaccinia virus promoter *p*11. Additional variants, pTKgptF2s and pTKgptF3s, contain one or two additional G residues, respectively, following the ATG to allow all three codon phasing possibilities. The expression cassette in both of these plasmids is flanked by segments of the vaccinia virus *tk* gene to direct the insertion into the vaccinia virus genome. Nucleotide numbers for both these plasmids have been estimated and may not be exact.

1.3. Selection and Screening of Recombinant Vaccinia Viruses

One of the most widely used type of transfer vector utilizes recombination into the nonessential thymidine kinase *(tk)* gene of vaccinia. An example of such a vector is pSC11 *(36)*, shown in Fig. 2A. Not only is the *tk* gene nonessential, but disruption of this function provides a means of selecting recombinant viruses with a *tk*⁻ phenotype by growth in the presence of the thymidine analogue 5-bromodeoxyuridine (BrdU) *(14)*. Using spontaneous *tk*⁻ vaccinia viruses, the first foreign gene to be introduced and expressed in vaccinia virus was the herpes simplex virus *tk* gene *(14,15)*. Incorporation of a functional *tk* gene into the transfer vector allows selection of recombinant vaccinia viruses with a *tk*$^+$ phenotype when using a *tk*$^-$ parental virus *(37)*. Another widely used selection mechanism employs the incorporation of the *E. coli* xanthine-guanine phosphoribosyl transferase (XGPRT) gene *(Ecogpt)* into the transfer vector *(38,39)*. An example is pTKgptF1s, shown in Fig. 2B. Mycophenolic acid (MPA), an inhibitor of purine metabolism, blocks replication of vaccinia virus. Expression of *Ecogpt* by vaccinia virus and inclusion of xanthine and hypoxanthine in the growth medium rescues the virus from this blockage. Thus, plasmid transfer vectors including the *Ecogpt* gene controlled by a vaccinia virus promoter in the recombination cassette will yield recombinant vaccinia viruses expressing both the *Ecogpt* gene and the gene of interest. The use of *Ecogpt* is advantageous for several reasons: selection of recombinant vaccinia viruses is not restricted to a *tk*$^-$ cell line; homologous recombination can be directed to any nonessential site in the vaccinia virus genome; spontaneous MPA resistant mutations do not occur so only *Ecogpt* expressing recombinant viruses will replicate and form a plaque; and MPA is nonmutagenic. In addition, once incorporated into the vaccinia virus genome, the *Ecogpt* gene can be removed using a reverse selection mechanism *(40)*. The drug 6-thioguanine (6-TG) is toxic to mammalian cells that express *Ecogpt* or hypoxanthine-guanine phosphoribosyl transferase (HGPRT), the mammalian homolog of *Ecogpt*. Thus, HGPRT negative cells are used to select recombinant vaccinia viruses that have undergone homologous recombination to remove the *Ecogpt* gene and replace it with another gene. The ability to select for or against *Ecogpt* expression provides a technique for introducing multiple genes into a recombinant virus through successive rounds of insertion and removal of an *Ecogpt* cassette. A potential disadvantage of this method is that viruses resulting from a single crossover event (containing both the foreign gene and *Ecogpt*) can be stable. If such a virus is isolated it may later undergo recombination to yield a mixed population containing parental and recombinant viruses. A modification of this method, known as transient dominant selection, provides a means of introducing foreign DNA followed by removal of the *Ecogpt* selection marker *(41)*. In this method the *Ecogpt* gene is located outside of the vaccinia virus DNA segments in the plasmid transfer vector. Thus, MPA resistant recombinant viruses acquire the *Ecogpt* gene through a single recombination event in which the entire plasmid is incorporated into the virus genome. This arrangement is unstable and the *Ecogpt* gene is readily lost when selection is removed. The transient dominant selection method simplifies the technique for introducing multiple genes in succession, and is advantageous when the presence of selectable marker in the recombinant vaccinia virus is undesirable, such as for vaccine purposes. Other mechanisms of selection include: antibiotic resistance through the use of neomycin *(42)* or hygromycin *(43)*, changes in plaque size *(44,45)* or red blood cell agglutination phenotypes *(46)*, and alterations of host range *(47)*.

Because homologous recombination occurs with a low frequency and most selection methods allow for growth of some parental virus, plaques must be screened to identify ones containing recombinant virus. By far, the easiest method involves use of the *E. coli lacZ* gene, which has been included in many transfer vectors. This provides a positive colorimetic assay for the identification of recombinant viruses through the production of β-galactosidase (β-gal) *(36)* (Fig. 2A). The *E. coli lacZ* gene can be used alone or in conjunction with one of the selection markers described in Section 1.3. However, in the absence of β-gal screening, plaques containing recombinant virus can be identified either by DNA or immunological analyses. The presence of DNA containing the foreign gene can be identified by DNA dot blot or polymerase chain reaction (PCR) analyses *(31)*. Alternatively, the gene product can be identified by Western blot, immunoprecipitation, or immunostaining if an antibody is available.

The purpose of this chapter is to provide a detailed outline of the processes involved in the generation of recombinant vaccinia viruses. All manipulations with live virus and virus-infected cells should be performed in a biological safety cabinet using sterile techniques. Waste should be decontaminated chemically or by autoclaving before disposal. For simplicity, the outlined protocols describe the procedures performed when utilizing vaccinia virus transfer vectors, which provide for selection via the *tk*⁻ phenotype or acquisition of the *Ecogpt* gene (XGPRT selection) in the context of the wild-type vaccinia virus strain WR. Also included are the steps performed to identify recombinant vaccinia viruses that express the *E. Coli lacZ* gene. Several other screening protocols are also included. These protocols can be used for identification of recombinant vaccinia viruses, as well as for characterization of the foreign gene product.

2. Materials

2.1. Cell Culture

1. Cell lines: HeLa (American Type Culture Collection (ATCC) (ATCC #CCL 2); HeLa S3 (ATCC #CCL 2.2); BS-C-1 (ATCC #CCL 26); CV-1 (ATCC #CCL 70); HuTK-143B (ATCC #CRL 8303).
2. Cell culture media: Eagle's minimal essential medium (MEM); Dulbecco modified Eagle's MEM (DMEM); MEM spinner medium (Quality Biologicals, Gaithersburg, MD).
3. Cell culture supplements: fetal bovine serum (FBS); horse serum (HS); 2 m*M* L-glutamine (100X); 50 mg/mL gentamicin sulfate in water (1000X; stable at room temperature); 5 mg/mL 5-bromodeoxyuridine (BrdU) in water (200X; filter sterilize, store in the dark at –20°C); Saline A (350 mg/L $NaHCO_3$, 400 mg/L KCl, 8 g/L NaCl) containing 0.1% dextrose, 0.002% phenol red, 0.25% trypsin, and 0.02% EDTA for passage of monolayer cell lines.

mg/mL trypsin stock) in a sterile tube and vortex vigorously (*see* Note 6). Incubate in a 37°C water bath for 30 min vortexing at 5–10-min intervals. Sonicate the mixture in a cup sonicator in ice-water for 30 s.

5. Dilute the trypsinized virus in MEM-2.5 to 2.5–7.5 × 10^7 PFU/mL.
6. Aspirate the medium from the flasks containing the HeLa S3 cells and overlay with 2 mL of the diluted, trypsinized virus suspension (the optimal moi is 1–3 PFU/cell). Incubate the flasks at 37°C in a CO_2 incubator for 2 h, rocking the flasks by hand at 15–30-min intervals to prevent drying of the monolayer.
7. Overlay the cells with 25 mL of MEM-2.5 and incubate for 3 d at 37°C in a CO_2 incubator.
8. Shake, thump, or scrape the flasks to loosen the cells, and pipet into sterile plastic screw-cap centrifuge tubes. Centrifuge for 10 min at 1800*g* at 4°C. Resuspend the cell pellets in MEM-2.5 (2 mL/5 × 10^7 cells). Disperse the cells by vortexing.
9. Lyse the cell suspension with three freeze-thaw cycles using a dry ice/ethanol bath and 37°C water bath. Disperse the cells by vortexing during each thaw. Sonicate the lysate in an ice-water filled cup sonicator for 30 s.
10. Aliquot and store the virus stock at –70°C. Volumes of 0.5 mL are convenient for later experiments. This virus stock can now be titered (*see* Section 3.3.).

3.2. Purification of Vaccinia Virus

1. Just prior to use, mix equal volumes of vaccinia virus stock and 0.25 mg/mL trypsin (prepared from the 2.5 mg/mL trypsin stock); vortex. Incubate 30 min at 37°C, vortexing at 10-min intervals.
2. Count the HeLa S3 spinner culture cells. Remove 5 × 10^8 cells for each liter to be infected (see Note 4). Centrifuge cells for 10 min at 1800*g* at room temperature. Resuspend cells in MEM-S-5 to a final density of 2 × 10^7 cells/mL. Transfer to a sterile Erlenmeyer flask (50–200 mL) containing a plastic stir bar and a cotton stopper.
3. Add the trypsinized virus to a moi of 5–8 PFU/cell. Stir gently for 30 min at 37°C. Transfer cells to a vented spinner flask containing MEM-S-5 equilibrated to 37°C (1 L/5 × 10^8 cells) and stir for 3 d at 37°C.
4. Harvest the cells by centrifugation for 10 min at 1800*g* at 4°C. Resuspend in 10 m*M* Tris-HCl, pH 9.0 (14 mL/5 × 10^8 cells). Keep cell suspension on ice for remainder of the protocol.
5. Homogenize the cell suspension with 30–40 strokes in a tight-fitting, glass Dounce homogenizer. Examine a sample of the lysed cells for breakage by light microscopy using the Trypan blue dye exclusion technique. Transfer the cell suspension to a sterile plastic screw-cap centrifuge tube or bottle. If necessary, the cell suspension may be stored at –70°C after a quick freeze in a dry ice-ethanol bath.
6. Centrifuge the lysed cells for 10 min at 300*g* at 4°C to remove nuclei. Save the supernatant (virus stock) on ice in a sterile 50-mL plastic screw-cap centrifuge tube. Resuspend the cell pellet in 10 m*M* Tris-HCl, pH 9.0 (3 mL/5 × 10^8 cells), and centrifuge for 10 min at 300*g* at 4°C. Combine with the previous supernatant and keep on ice.
7. Sonicate the virus stock using a probe sonicator as follows:
   a. Sterilize the probe by dipping it in 95% ethanol and passing it through a flame;

b. Let probe cool;
c. Remove cap from tube containing the virus stock and place probe into the virus stock;
d. Sonicate at full power for 15 s;
e. Wait 15 s and repeat sonication four times.

If a probe sonicator is unavailable, sonication can be performed using a cup (*see* Note 7).

8. Layer the sonicated virus stock onto a cushion of 17 mL of 36% sucrose (in 10 m*M* Tris-HCl, pH 9.0) in a sterile SW-27 centrifuge tube. Centrifuge for 80 min at 32,900*g* (13,500 rpm in SW-27 rotor) at 4°C. Aspirate to remove supernatant; virus is in the pellet.
9. Resuspend the viral pellet in 1 mL of 1 m*M* Tris-HCl, pH 9.0. Sonicate once for 15 s with a probe sonicator or 1 min in a cup sonicator (*see* step 7).
10. Prepare sterile 24–40% continuous sucrose gradients in sterile SW-27 centrifuge tubes the day before needed by carefully layering 6.8 mL of each sucrose solution (in 1 m*M* Tris-HCl, pH 9.0) in the following order: 40, 36, 32, 28, and 24%. Place the gradients at 4°C overnight.
11. Carefully overlay each sucrose gradient with 1 mL of the sonicated viral suspension from step 9. Centrifuge for 50 min at 26,000*g* (12,000 rpm in SW-27 rotor) at 4°C.
12. After centrifugation, the virus appears as a milky band in about the middle of the gradient. Carefully aspirate to remove the sucrose above the virus band; discard. Carefully collect the virus band (about 10 mL) with a sterile pipet and place in a sterile screw-cap plastic centrifuge tube on ice.
13. Aspirate the remaining sucrose from the tube and recover the pellet containing aggregated virus from the bottom of the tube. Resuspend in 1 mL of 1 m*M* Tris-HCl, pH 9.0 by pipeting; sonicate as in step 7.
14. Repeat the virus-banding procedures in steps 10–12 with the viral pellet from step 13. Combine this viral band (about 10 mL) with the previous one from step 12 and add 2 vol of 1 m*M* Tris-HCl, pH 9.0; vortex. The total volume should be about 60 mL. Transfer to sterile SW-27 centrifuge tubes and centrifuge for 60 min at 32,900*g* (13,500 rpm in SW-27 rotor).
15. Aspirate the supernatants and resuspend the virus pellets in 1 m*M* Tris-HCl pH 9.0 (0.5–1.0 mL/5 × 10^8 infected cells) (*see* Note 4). Sonicate in cup sonicator, divide into 0.25-mL aliquots, and store at –70°C. The purified virus stock can now be titered (*see* Section 3.3.).

3.3. Titration of Vaccinia Virus Stocks

1. Prepare 6-well (35 mm diameter) tissue culture plates of BS-C-1 cells by seeding 5 × 10^5 cells/well in a total volume of 2 mL of MEM-10. Do not swirl the plates, as this results in clumping of the cells in the middle of the well. Incubate overnight at 37°C in a 5% CO_2 atmosphere to reach confluence (*see* Note 8).
2. For titration of a vaccinia virus stock, trypsinize as described in step 4 of Section 3.1. For titration of a purified virus stock, trypsinization is not required; however, the purified stock should be sonicated using a cup sonicator.

3. Prepare eight tenfold serial dilutions (beginning with a 10^{-2} dilution) of the virus stock in MEM-2.5, using a fresh pipet for each dilution. This is most easily done by aliquoting 2.7 mL of MEM-2.5 into tubes 2–9 and 3 mL into tube 1. Remove 30 µL of medium from tube 1 and add 30 µL of the virus stock. Vortex to mix. The serial dilutions are then prepared by the sequential passing of 0 3 mL. (Note: For titration of purified virus stocks, prepare nine tenfold serial dilutions.)
4. Aspirate the medium from the 6-well cultures of BS-C-1 cells and infect the cell monolayers in duplicate with 1-mL aliquots of the 10^{-7},10^{-8}, and 10^{-9} dilutions. (Note: For titration of purified virus stocks, plate the 10^{-8}, 10^{-9}, and 10^{-10} dilutions.) Incubate 1–2 h at 37°C in a 5% CO_2 atmosphere, rocking the plates at 15-min intervals to prevent drying of the monolayer.
5. Overlay each well with 2 mL MEM-2.5 and incubate for 2 d at 37°C in a 5% CO_2 atmosphere.
6. Aspirate the medium and add 0.5 mL of 0.1% crystal violet solution to each well. Incubate 5 min at room temperature, and then aspirate. Keeping the lids of the plates off, rest the plates on their lids at an angle in the biological safety cabinet to air dry.
7. Determine the virus titer by counting plaques in both wells, dividing by 2, and multiplying by the dilution factor of those wells. Most accurate results are obtained from wells with 20–80 plaques. Remember to take into account the 1:1 dilution of the virus stock and trypsin.

3.4. Infection and Transfection

1. Prepare a 25-cm^2 flask of CV-1 cells by seeding 10^6 cells in 4 mL of MEM-10 and incubating overnight at 37°C in a 5% CO_2 atmosphere. This will usually result in a culture that is just reaching confluence the next day.
2. Prepare an aliquot of trypsinized virus (usually WR strain) as in step 4 of Section 3.1.
3. Dilute the trypsinized virus in MEM-2.5 to 1.5×10^5 PFU/mL. Aspirate the medium from the flask of CV-1 cells and infect with 1 mL of the diluted virus (this yields a moi of 0.05 PFU/cell). Incubate the cells for 2 h at 37°C in a 5% CO_2 atmosphere, rocking the flask by hand at 15-min intervals to prevent drying of the monolayer.
4. At 30 min prior to the end of the infection period, prepare the transfection mixture as follows: place 1 mL of transfection buffer (HBS) into a 12 × 75-mm polystyrene tube and add 5–10 µg of the recombinant transfer vector DNA (the volume of DNA should be no more than 50 µL), mix by gently tapping the tube 2–3 times, slowly add 50 µL of 2.5*M* $CaCl_2$ drop-wise to the DNA solution, and again mix by gently tapping the tube 2–3 times. Incubate the mixture for 20–30 min at room temperature; a fine milky precipitate should appear (*see* Note 9).
5. Aspirate the virus inoculum and overlay the cell monolayer with the transfection mixture from step 4. Incubate for 30 min at room temperature. Add 9 mL of MEM-10 and incubate for 3.5 h at 37°C in a 5% CO_2 atmosphere.
6. Aspirate the medium, add 10 mL of fresh MEM-10, and incubate for 2–3 days at 37°C in a 5% CO_2 atmosphere until the entire monolayer of cells is infected from the spreading virus.

7. Harvest the cells by scraping with a sterile disposable cell-scraper or rubber policeman and transfer to a sterile 15-mL plastic screw-cap centrifuge tube. Centrifuge for 10 min at 1800*g* at 4°C. Aspirate the supernatant, and resuspend the cell pellet in 1 mL of MEM-2.5.
8. Lyse the cell suspension with three freeze-thaw cycles as described in step 9 in Section 3.1. Store the lysate at –70°C if not used immediately.

3.5. Selection of Recombinant Vaccinia Viruses

1. Prepare 6-well (35-mm diameter) tissue culture plates of BS-C-1 cells (for XGPRT selection) or HuTK$^-$143B cells (for *tk*$^-$ selection) by seeding 5×10^5 cells/well in a total volume of 2 mL of MEM-10. Do not swirl the plates as this results in clumping of the cells in the middle of the well. Incubate the cultures overnight at 37°C in a 5% CO_2 atmosphere to reach confluence (*see* Note 8).
2. For XGPRT selection, preincubate the cell culture monolayers for 12–24 h in MEM-2.5 containing 25 µg/mL MPA, 250 µg/mL xanthine, and 15 µg/mL hypoxanthine at 37°C in a 5% CO_2 atmosphere.
3. Thaw and sonicate the transfected cell lysate (from step 8 in Section 3.4.) for 30 s in an ice-water filled cup sonicator.
4. Prepare four tenfold serial dilutions (10^{-1} to 10^{-4}) of the sonicated cell lysate in MEM-2.5. For XGPRT selection, MPA, xanthine, and hypoxanthine are included in the serial dilutions at the concentrations indicated in step 2.
5. Aspirate the medium from the 6-well cell cultures (step 1) and infect the cell monolayers in duplicate with 1-mL aliquots of the 10^{-2}, 10^{-3}, and 10^{-4} dilutions. Incubate for 1–2 h at 37°C in a 5% CO_2 atmosphere, rocking the plates at 15-min intervals.
6. Before the end of the infection period, melt a bottle of sterile 2% LMP agarose and place in a 42–45°C water bath to equilibrate. Equilibrate a bottle of 2X MEM-10 in the 42–45°C water bath.
7. Prepare 25 mL of the agarose overlay for each 6-well plate as follows: For XGPRT selection, mix 12.5 mL of 2X MEM-10 and 12.5 mL of melted 2% LMP agarose (both equilibrated to 42–45°C) in a tube and add MPA, xanthine, and hypoxanthine to the final concentrations noted in step 2. Mix by gently swirling or inverting the tube. For *tk*$^-$- selection, mix 12.5 mL of 2X MEM-10 and 12.5 mL of melted 2% LMP agarose to a tube and add 125 µL of 5 mg/mL BrdU (*see* Note 10) and mix by gently swirling or inverting the tube.
8. Remove the virus inoculum, overlay each well with 4 mL of the appropriate agarose overlay mixture, swirl the plates to mix, and allow to solidify at room temperature or briefly at 4°C. Incubate for 2 d at 37°C in a 5% CO_2 atmosphere.
9. After the 2-d incubation period, prepare a second agarose overlay by mixing equal volumes of melted 2% LMP agarose with 2X MEM-10 (both equilibrated to 42–45°C). Add neutral red to a final concentration of 100 µg/mL, mix by gently swirling or inverting the tube. If β-gal screening is used include 1/120 vol of 4% Xgal. Overlay each well with 2 mL of the second agarose preparation, allow to solidify, and incubate the plates at 37°C in a 5% CO_2 atmosphere until plaques

can be easily visualized (6 h to overnight). Plaques will appear as clear areas surrounded by a red background. Plaques containing β-gal producing virus will appear blue owing to hydrolysis of the Xgal substrate.

10. When virus plaques are readily detectable, either by the neutral red stain (which visualizes all plaques) or by the Xgal stain (which identifies β-gal producing plaques), prepare a set of sterile microcentrifuge tubes containing 0.5 mL of MEM-2.5 (preferably screw-cap tubes). Using sterile, cotton-plugged Pasteur pipets and a rubber bulb (*see* Note 11), pick well-separated plaques by squeezing the bulb, piercing through the agarose to the bottom of the well, scraping the monolayer, and aspirating the agarose plug containing infected cells into the pipet. Transfer the plug to a tube containing 0.5 mL of MEM-2.5. The number of plaque isolates picked depends on the selection protocol utilized. For recombinant viruses encoding the *lacZ* or *Ecogpt* gene (for example: pSC11 or pTKgptF1s, respectively; *see* Table 1 and Fig. 2) at least 6–12 plaques should be picked and screened. For recombinant viruses having *tk* selection only (e.g., pSC59, *see* Table 1),15–30 plaques should be picked owing to the high rate of spontaneous tk^- mutations (*see* Note 12).
11. After picking the plaques, vortex to mix and perform three freeze-thaw cycles as described in step 9 of Section 3.1. Store the virus isolates at –70°C. If tk^- selection is used, then the virus isolates should be screened by one of the methods described in Section 3.6. or mentioned in Note 2. After identifying plaques containing the recombinant vaccinia virus, proceed to step 12 below. If β-gal screening or XGPRT selection is used, no further analysis of the plaques is required at this time.
12. Plaque purify the recombinant vaccinia virus isolates as follows. Prepare monolayers of an appropriate cell line as described in steps 1 and 2; one 6-well plate for each plaque isolate (note that as with β-gal or XGPRT selected isolates, only a few tk^- isolates need to be plaque purified at this point).
13. Thaw the virus isolates and sonicate in an ice-water-filled cup sonicator as described in step 9 of Section 3.1. Prepare three tenfold serial dilution (beginning at 10^{-1}) of each of the isolates. If XGPRT selection is used, preincubate cell monolayers with selective drugs and add selective drugs to serial dilutions of virus.
14. Aspirate the medium from the 6-well plates and infect the monolayers in duplicate with 1-mL aliquots of the 10^{-1}, 10^{-2}, and 10^{-3} dilutions from step 13. Incubate for 2 h at 37°C in a 5% CO_2 atmosphere, rocking by hand at 30-min intervals.
15. Repeat steps 6–10 for three rounds of plaque purification to ensure a clonally pure recombinant vaccinia virus. Store the final recombinant vaccinia virus at –70°C. Proceed to Section 3.7.

3.6. Amplification and Screening of Recombinant Vaccinia Virus Plaque Isolates

3.6.1. Amplification of Plaque Isolates

1. Amplify each plaque isolate on cell monolayers as follows: Prepare BS-C-1 cells (for XGPRT selection) or HuTK⁻1 43B cells (for tk^- selection) in 12- or 24-well tissue culture plates by seeding 1.25×10^5 or 2.5×10^5 cells/well, respectively.

Incubate at 37°C in a 5% CO_2 atmosphere until confluent (usually overnight). If XGPRT selection is used, preincubate the cell monolayers for 12–24 h in MEM-2.5 containing MPA, xanthine, and hypoxanthine (step 2, Section 3.5.). It is also recommended that a monolayer of cells be infected with the parental vaccinia virus, and a monolayer of cells be left uninfected. These samples will be useful negative controls during later screening processes.

2. Infect individual wells containing confluent cell monolayers with 0.25 mL of each sonicated plaque isolate. For XGPRT selection, carry out infection in the presence of MPA, xanthine, and hypoxanthine; for tk^- selection carry out infection in the presence of BrdU. Incubate the plates for 2 h at 37°C in a 5% CO_2 atmosphere, rocking by hand at 15 min intervals.
3. Overlay each well with 0.5 mL of MEM-2.5 containing the appropriate drugs and incubate the plates for 2–3 d at 37°C in a 5% CO_2 atmosphere or until cytopathic effect (cell rounding) is evident throughout the monolayer (*see* Note 13). At this point the treatment of the amplified plaque isolates will vary depending on what screening method will be employed. Four examples of methods for analysis of plaques are given in the next sections: additional examples are mentioned in Note 2.

3.6.2. Detection of Recombinant Vaccinia Virus by DNA Hybridization

1. After complete cytopathic effect is observed during the amplification of plaque isolates (*see* Note 13), as described in step 3, Section 3.6.1., harvest the cells in each well by scraping and transfer to microcentrifuge tubes. Centrifuge the cells at full speed in a microcentrifuge for 5 min and aspirate the medium. Resuspend the cell pellets in 0.5 mL PBS, perform three freeze-thaw cycles as described in step 9, Section 3.1., and place on ice.
2. Cut a section of the GeneScreen Plus membrane and two sections of Whatman 3MM filter paper to fit the dot- or slot-blotting apparatus, and soak in a tray containing 0.4*M* Tris-HCl, pH 7.5, for 30 min.
3. Transfer 100 µL of each lysate to a new microcentrifuge tube, and denature the DNA by addition of 5 µL of 5*N* NaOH (final concentration of 0.25*N* NaOH). Vortex to mix and incubate for 10 min at room temperature.
4. Chill the denatured DNA on ice.
5. Dilute the denatured DNA with 200 µL of 0.125*N* NaOH, 0.125X SSC.
6. Sonicate the diluted denatured DNA in an ice-water-filled cup sonicator, and store on ice.
7. Assemble the dot- or slot-blotting apparatus with the presoaked membrane and filter paper.
8. Add 100 µL of each DNA sample in duplicate to the wells of the apparatus.
9. Allow solutions to remain on the membrane without any suction for 30 min.
10. After 30 min, apply a slight suction to the apparatus until all liquid has passed through the membrane.
11. Remove the membrane and air dry at room temperature.
12. Denature an aliquot of the 5 mg/mL sheared salmon sperm DNA stock by heating for 3 min at 100°C and chilling on ice. Prehybridize the membrane for 30 min at

65°C by incubation in 10 mL of 1% SDS, 1*M* NaCl containing 200 µg/mL denatured salmon sperm DNA in a sealable plastic bag.

13. Add 0.5–1.0 mL of 1 % SDS, 1*M* NaCl containing 200 µg/mL denatured salmon sperm DNA and 100 ng of [^{32}P]-labeled probe DNA (approx 1–4 × 10^7 dpm/µg) (*see* Note 14). Reseal the bag and incubate with constant agitation for 6–24 h at 65°C.
14. Remove the membrane from the bag and wash as follows: (a) two times with 100 mL of 2X SSC at room temperature for 5 min; (b) two times with 200 mL of 2X SSC containing 1 % SDS at 65°C for 30 min; and (c) two times with 100 mL of 0.1 X SSC at room temperature for 30 min. All washes should be performed with constant agitation.
15. Place the membrane with the DNA face up on a sheet of filter paper to adsorb excess liquid, wrap in plastic wrap, and expose to X-ray film. Plaque isolates containing recombinant virus are identified by hybridization with the probe DNA.
16. When one or several plaque isolates are identified, proceed with the plaque purification steps starting at step 12, Section 3.5.

3.6.3. Detection of Recombinant Vaccinia Virus by Immunoblotting of the Recombinant Gene Product

1. After complete cytopathic effect is observed during the amplification of plaque isolates (*see* Notes 13 and 15), as described in step 3, Section 3.6.1., harvest the cells in each well by scraping and transfer to a set of microcentrifuge tubes. Centrifuge the cells at full speed in a microcentrifuge for 5 min and aspirate the medium (recover the supernatants if the protein of interest is secreted). Resuspend the cell pellets in 0.5 mL PBS and perform three freeze-thaw cycles, as described in step 9, Section 3.1., sonicate in a cup sonicator and place on ice.
2. Cut a section of nitrocellulose and 2 sections of Whatman 3MM filter paper and soak them in distilled water.
3. Assemble the dot or slot-blot apparatus and apply 50 µL of each lysate into individual wells (in duplicate). If the protein of interest is secreted, the medium from the infected cell monolayers can be substituted for the cell lysates.
4. Allow lysates to remain on the membrane without any suction for 30 min.
5. After 30 min. apply a slight suction to the apparatus until all liquid has passed through the membrane.
6. Soak the membrane in 50 mL of PBS/Tween containing 4% BSA or 1% hydrolyzed gelatin for 30 min to 1 h. Plastic lids from micropipet tip racks work well as washing trays.
7. Wash the membrane in 50–100 mL of PBS/Tween; dilute the antibody to the foreign protein in PBS/Tween (as appropriate for the antibody) using a minimal volume (just enough to cover the membrane). Pour off the wash solution, replace with the antibody solution, and incubate for at least 1 h at room temperature or overnight at 4°C with gentle rocking.
8. Wash the membrane with four changes of PBS/Tween (50–100 mL/wash; 15–20 min/wash); dilute [^{125}I]-labeled protein A, protein G, or appropriate second anti-

body in a minimal volume of PBS/Tween. Pour off the wash solution, replace with the radiolabeled solution, and incubate for at least 1 h at room temperature.
9. Pour off the radiolabeled solution and wash the membrane with four changes of PBS/Tween as in step 8. Blot the membrane on filter paper to remove excess liquid, wrap in plastic wrap, and expose to X-ray film. Develop the autoradiograph and determine which amplified plaque isolates contain recombinant virus producing the protein of interest.
10. When one or several recombinant virus plaque isolates are identified, proceed with the plaque purification steps starting at step 12, Section 3.5.

3.6.4. Detection of Recombinant Vaccinia Virus by Western Blotting of the Recombinant Gene Product

1. After complete cytopathic effect is observed during the amplification of plaque isolates (*see* Notes 13 and 15), as described in step 3, Section 3.6.1., harvest the cells in each well by scraping and transfer to a set of microcentrifuge tubes. Centrifuge the cells at full speed in a microcentrifuge for 5 min, and aspirate the medium. Alternatively, if the protein of interest is secreted the supernatants can be recovered and either analyzed directly or concentrated by immunoprecipitation or use of a microconcentrator. Lyse the cell pellets by resuspending in 200 µL of cell lysis buffer (*see* Section 2.4.3. and Note 16), vortex, and incubate on ice for 15 min. Centrifuge the cell lysates for 5 min at full speed to remove nuclei and debris, and transfer the supernatants to clean tubes.
2. Prepare an SDS-PAGE gel(s).
3. Aliquot 20 µL of each lysate into a new microcentrifuge tube, add 20 µL of 2X SDS protein gel sample buffer, and heat at 100°C for 3 min. Centrifuge for 1 min at top speed in a microcentrifuge. Load 25–30 µL of each sample into the wells of the gels and separate the proteins by electrophoresis.
4. Transfer the separated proteins electrophoretically onto a sheet of nitrocellulose membrane using a transfer apparatus.
5. Carry out steps 6–10, Section 3.6.3.

3.6.5. Detection of Recombinant Vaccinia Virus by Radioimmunoprecipitation of the Recombinant Gene Product

1. Amplify the plaque isolates as described in Section 3.6.1., steps 1 through 3 (*see* Notes 13 and 15). However, after 1–2 d postinfection, proceed as follows.
2. Aspirate the medium from each well and wash two times with 2 mL of methionine- and/or cysteine-free MEM containing 2.5% dialyzed FBS.
3. Remove the final wash and overlay each well with 0.5 mL of methionine- and/or cysteine-free MEM containing 2.5% dialyzed FBS and 50–100 µCi/mL of [^{35}S]methionine (>1000 Ci/mmol) and/or [^{35}S]cysteine (>600 Ci/mmol). Incubate for an additional 24 h at 37°C in a 5% CO_2 atmosphere.
4. Add 100 µL of MEM-2.5 to each well for 1 h at 37°C in a 5% CO_2 atmosphere. Aspirate the radioactive supernatants (recover and save the supernatants if the protein of interest is secreted), overlay each well with 0.5 mL PBS, scrape, and

transfer to a set of microcentrifuge tubes. Centrifuge the cells for 5 min at full speed, aspirate the PBS, and lyse the cell pellets in 200 µL of cell lysis buffer (*see* Section 2.4.3. and Note 16). Vortex and incubate on ice for 15 min. Centrifuge the cell lysates for 5 min at full speed to remove nuclei and debris and transfer the supernatants to new tubes.

5. Aliquot 10–50 µL of each metabolically labeled cell lysate (use 100–500 µL of each supernatant if the protein of interest is secreted) to a set of microcentrifuge tubes.
6. Add 100 µL of PBS-Triton X-100 containing the appropriate dilution of antibody to the foreign protein (as appropriate for the antibody). Incubate the tubes for 2 h at room temperature or 4°C overnight.
7. Add 50-µL of a 20% (v/v) suspension of immobilized protein A or protein G Sepharose CL-4B (or agarose) beads (as appropriate for the antibody *[48]*). If necessary, a second antibody with specificity to the species of the first antibody can be included as well. Rotate the tubes for 1 h at 4°C.
8. Centrifuge the tubes for 5 min at 500*g* (a swinging bucket rotor works best to create a small pellet; alternatively use a microcentrifuge). Aspirate the supernatants with a round gel-loading micropipet tip taking care not to touch the beads.
9. Wash the beads two times as follows: Add 1.0 mL of PBS-Triton X-100 to each tube, shake to mix, and repeat step 8 (*see* Note 17).
10. Resuspend the pellet in 20 µL of 2X SDS sample buffer and heat 100°C for 3 min. Centrifuge for 5 min at top speed in a microcentrifuge.
11. Prepare SDS-PAGE gel(s).
12. Load 20 µL of each sample into the wells of the gels and separate the proteins by electrophoresis. Process the SDS-PAGE gels for detection of labeled proteins by fixation, amplification, and fluorography. When one or several recombinant virus plaque isolates are identified proceed with the plaque purification steps starting at step 12, Section 3.5.

3.7. Final Amplification of a Recombinant Vaccinia Virus Plaque Isolate

1. Prepare a 25-cm^2 flask with the appropriate cell line: BS-C-1 cells (for XGPRT selection) or HuTK$^-$ 143B cells (for *tk*$^-$ selection) by seeding 1×10^6 cells and incubating at 37°C in a 5% CO_2 atmosphere until confluent (usually overnight). If XGPRT selection is used, preincubate the cell monolayer for 12 to 24 h in MEM-2.5 containing MPA, xanthine, and hypoxanthine (step 2, Section 3.5.).
2. Choose one or several of the plaque-purified recombinant vaccinia virus isolates, thaw and sonicate in an ice-water-filled cup sonicator.
3. Infect the cell monolayer as follows: Add 0.25 mL of one sonicated plaque isolate to a plastic centrifuge tube and add an additional 0.75 mL of MEM-2.5 and the appropriate selective drugs; remove the medium from the monolayer by aspiration; and overlay the monolayer with the diluted virus preparation. Incubate the flask for 2 h at 37°C in a 5% CO_2 atmosphere, rocking by hand at 15-min intervals.

4. Overlay the monolayer with 5 mL of MEM-2.5 containing the appropriate selective drugs and incubate the culture at 37°C in a 5% CO_2 atmosphere for 2–3 d or until cytopathic effect (cell rounding) is evident throughout the monolayer.
5. Harvest the cells by scraping, transfer to a sterile 15-mL screw-cap conical centrifuge tube, and centrifuge for 10 min at 1800*g* at 4°C. Aspirate the supernatant, resuspend the cell pellet in 0.5 mL of MEM-2 5, and perform three freeze-thaw cycles as described in step 9, Section 3.1. The amplified recombinant virus stock can be stored at –70°C if not used immediately.
6. Scale up the virus stock by infecting a 150-cm² tissue culture flask containing a confluent monolayer of the appropriate cell line. To prepare the cells, scale up the procedure described in step 1 above.
7. Infect the cell monolayer as follows: Add 0.25 mL of sonicated virus stock from step 5 to a plastic centrifuge tube and add an additional 2.75 mL of MEM-2.5 and the appropriate selective drugs; remove the medium from the monolayer by aspiration; overlay the monolayer with the diluted virus preparation. Incubate the flask for 2 h at 37°C in a 5% CO_2 atmosphere, rocking by hand at 15 min intervals.
8. Overlay the monolayer with 10 mL of MEM-2.5 containing the appropriate selective drugs and incubate the culture at 37°C in a 5% CO_2 atmosphere for 2–3 d or until cytopathic effect (cell rounding) is evident throughout the monolayer.
9. Harvest the cells by scraping; transfer to a sterile 15 mL screw-cap conical centrifuge tube, and centrifuge for 10 min at 1800*g* at 4°C. Aspirate the supernatant, resuspend the cell pellet in 2 mL of MEM-2.5, and perform three freeze-thaw cycles as described in step 9, Section 3.1. The amplified recombinant virus stock can be stored at –70°C if not used immediately.
10. Count the HeLa S3 spinner cell culture; for each 150-cm² flask to be infected remove 5×10^7 cells and centrifuge for 5 min at 1800*g* at room temperature. (Usually 5 flasks of cells are prepared at this stage).
11. Resuspend cells to a density of 2×10^6 cells/mL in MEM-10 equilibrated to 37°C and dispense 25 mL to each 150-cm² tissue culture flask. Incubate overnight at 37°C in a 5% CO_2 incubator.
12. Sonicate the virus stock from step 9 above. For each 150-cm² tissue culture flask to be infected, dispense 0.25 mL of the virus stock (step 9) and 2.75 mL MEM-2.5 into a 15-mL screw-cap conical centrifuge tube (selective drugs are not required at this stage).
13. Aspirate the medium from the flasks containing the HeLa S3 cells and overlay with 3 mL of the diluted virus suspension. Incubate at 37°C in a CO_2 incubator for 2 h, rocking the flasks by hand at 15–30-min intervals to prevent drying of the monolayer.
14. Overlay the cells with 25 mL of MEM-2.5/flask and incubate for 3 d at 37°C in a CO_2 incubator.
15. Harvest the cells by shaking, thumping, or scraping the flasks, and pipet into sterile plastic screw-cap centrifuge tubes. Centrifuge for 10 min at 1800*g* at 4°C. Aspirate the supernatant and resuspend the cell pellet in 2 mL of MEM-2.5/150-cm² flask. Disperse the cells by vortexing and lyse with at least three cycles of freeze-

thawing in a dry ice/ethanol bath and 37°C water bath. Vortex cells during each thaw. Sonicate the thawed lysate in an ice-water filled cup sonicator for 30 s.

16. Store the recombinant virus preparation at –70°C. This virus stock can now be titered as detailed in Section 3.3.

4. Notes

1. Contaminants in LMP agarose from some sources may be toxic to cells; we have found that the LMP agarose from GIBCO/BRL has been consistently suitable.
2. Variations in the DNA and immunobased assays described in this chapter can be performed. For example, utilization of other membranes for immobilization of DNA and protein samples and utilization of nonradioactive-based detection mechanisms can be employed. Other methods for screening and analyzing recombinant vaccinia viruses include DNA analysis by Southern blot or PCR techniques and mRNA analysis by Northern blot techniques. For detailed protocols of these techniques, *see* ref. *8*. A method of *in situ* immunostaining of virus plaques can be performed for cell surface recombinant proteins *(49)*. Also, a recent streamlined procedure in which single *tk* recombinant plaque isolates can be obtained in 96-well cell culture plates directly, without the agarose overlays, has been described *(50)*.
3. Maintain the density of the HeLa S3 spinner culture at 1.5–5 × 10^5 cells/mL. Culture viability drops off dramatically at higher densities and the cells do not grow well at densities below 1 × 10^5 cells/mL.
4. As a general guideline, the yield of vaccinia virus from a cell lysate of either HeLa S3 suspension or HeLa monolayer cultures is approx 5 × 10^8–4 × 10^9 PFU/mL when each 150-cm^2 flask of infected cells is resuspended into 2 mL of MEM-2.5. After purification of vaccinia virus by banding in sucrose, each liter of 5 × 10^8 infected HeLa S3 cells yields 0.5–1 mL with a titer of approx 1–5 × 10^9 PFU/mL.
5. Stocks of vaccinia virus can be prepared using the HeLa monolayer cell line in place of the HeLa S3 suspension cell line. This is convenient when smaller stocks of virus are required (20–40 mL of stock with titers of about 10^9 PFU/mL) or if equipment for growing spinner cells is not available.
6. Always perform trypsinization of vaccinia virus stocks just prior to use. Never store trypsinized viruses as this results in major losses in virus titer even at –70°C.
7. When purifying vaccinia virus and a probe sonicator is unavailable, split the cell lysate into 3-mL aliquots and sonicate each separately in an ice-water-filled cup sonicator at full power for 1 min. Repeat sonication four times, with at least a 30-s interval of incubation on ice each time. Replenish the ice in the cup as required to maintain cold temperature.
8. It is important that the density of cells in the monolayer not be too high when plaqueing virus, as this may result in small plaque size and/or deterioration of the cell monolayer. It is best to use monolayers of cells in that have just reached confluence (10^6 cells/35-mm well tissue culture dish) for $HuTK^-$, CV-1, or BS-C-1 cells. This can usually be achieved by seeding 5 × 10^5 cells/well (35-mm diameter) in a total volume of 2 mL of medium the morning of the day prior to use.

9. The inclusion of wild-type vaccinia DNA in the transfection preparation yields a higher efficiency of recombination. This is accomplished by adding 1 μg of wild-type vaccinia DNA with the transfer vector DNA containing the cloned gene of interest. Also, alternative DNA transfection protocols such as those using Lipofectin (GIBCO/BRL), DOTAP (Boehringer Mannheim, Indianapolis, IN), or Transfectam (Promega, Madison, WI) can be utilized.
10. After thawing the BrdU stock, a 5–10-min incubation at 37°C followed by vortexing is required to ensure that the BrdU is in solution.
11. When picking plaques, move the Pasteur pipet tip in a circular motion covering an area just slightly larger than the size of the plaque while maintaining contact with the bottom of the well. This will ensure good recovery of the infected cells in the plaque area. Also, the use of screw-cap microcentrifuge tubes ensures tight seals and prevents sample loss and contamination during the freeze-thaw cycles and manipulations.
12. When the *tk* phenotype is used for selection without a concomitant screening protocol, it is important to pick 15–30 plaques because up to 80–90% of the plaques can be the result of spontaneous tk^- vaccinia virus mutants. It is critical to screen plaque isolates at this stage by one of the methods described in Section 3.6., or mentioned in Note 2, to identify positive recombinant viruses. After this initial recombinant virus identification is performed, only 6–8 plaques need be picked during the second and third rounds of plaque purification. It is usually necessary to purify only 1 or 2 of these isolates. Save the others until the final virus preparation has been made. If screening for the production of β-gal is used in conjunction with *tk* selection, or if XGPRT selection is employed, then only 6-12 plaques need be picked in the initial plaque purification step. A few of these can be immediately plaque purified. The presence of β-gal activity or MPA resistance is a very good indication that the virus isolate contains the inserted gene of interest.
13. When amplifying a series of plaque isolates (especially in the first round of tk^- selection) allow sufficient time (up to 3–4 d) for all or most of the individual monolayers to achieve a high degree of infection. Make note of which wells, if any, have little cytopathic effect; this will aid in assessing the positive signals obtained with DNA or immunobased analyses. If minimal cytopathic effect is observed after 2 d, the cultures can be supplemented with fresh medium containing appropriate selective drugs and incubated further.
14. When screening amplified plaque isolates by DNA hybridization, prepare the [^{32}P]-labeled probe from DNA containing the gene of interest and not the flanking vaccinia virus sequences, as the latter will hybridize with all vaccinia virus samples. The DNA probe can be prepared via any commercially available nick-translation or random-priming kit.
15. The Western blot, immunoblot, and radioimmunoprecipitation assays outlined in Sections 3.6.3. through 3.6.5 are intended for screening many plaque isolates in order to identify recombinant vaccinia viruses producing the protein of interest. When utilizing the hybrid vaccinia virus/T7 system, the cell monolayers must be

coinfected with a recombinant vaccinia virus expressing T7 RNA polymerase such as vTF7-3 *(16)* (moi of 1). Foreign genes controlled by early vaccinia promoters may yield a weaker signal in the immunobased detection assays. However, all these assays can be used to characterize the protein produced by a recombinant vaccinia virus after it has been plaque purified and grown up as a working stock of virus.

16. When preparing cell lysates for analysis of the protein of interest, it may be necessary to add one or several protease inhibitors to the cell lysis buffer. Phenylmethylsulfonyl fluoride is commonly used at a final concentration of 0.2 m*M* (20 m*M* stock). Also, virus isolates producing a secreted recombinant protein can be screened by harvesting the medium from cells infected at any stage in the plaque-amplification process. However, if the medium will be concentrated with microconcentrators, it is important to use serum-free medium or only 1% serum, as a high concentration of serum proteins interfere with SDS-PAGE analysis.
17. During the immunoprecipitation assays, an additional wash with PBS containing 0.1% deoxycholic acid and 0.1% SDS can be performed to reduce nonspecific background. If background remains problematic, then the cell lysates can be precleared by performing a mock immunoprecipitation using preimmune sera or irrelevant antibody with the protein A or protein G beads. After pelleting these beads by centrifugation, the precleared lysate is recovered and then used in the immunoprecipitation assay.

Acknowledgments

We thank Stuart N. Isaacs and Paul E. Kennedy for critical reading of this manuscript.

References

1. Moss, B. (1991) Vaccinia virus: a tool for research and vaccine development. *Science* **252,** 1662–1667.
2. Smith, G. L. (1991) Vaccinia virus vectors for gene expression. *Curr. Opin. Biotechnol.* **2,** 713–717.
3. Cox, W. L., Tartaglia, J., and Paoletti, E. (1992) Poxvirus recombinants as live vaccines, in *Recombinant Poxviruses* (Binns, M. M. and Smith, G. L., eds.), CRC, Boca Raton, pp. 123–162.
4. Fenner, F. (1992) Vaccinia virus as a vaccine, and poxvirus pathogenesis, in *Recombinant Poxviruses* (M. M. Binns and G. L. Smith, eds.), pp. 1–43. CRC, Boca Raton.
5. Smith, G. L. and Mackett, M. (1992) The design, construction, and use of vaccinia virus recombinants, in Recombinant Poxviruses" (Binns, M. M. and Smith, G. L., eds.), CRC, Boca Raton, pp. 81–122.
6. Moss, B. (1993) Poxvirus vectors: cytoplasmic expression of transferred genes. *Curr. Opin. Gen. Dev.* **3,** 86–90.
7. Moss, B. (1994) Replicating and host-Restricted non-replicating vaccinia virus vectors for vaccine development. *Dev. Biol. Stand.* **82,** 55–63.

8. Earl, P. L., Cooper, N., and Moss, B. (1991) Expression of proteins in mammalian cells using vaccinia viral vectors, in *Current Protocols in Molecular Biology* (Ausubel, F. M., Brent, R., Kingston, R. E., Moore, D. D., Seidman, J. G., Smith, J. A., and Struhl, K., eds.), Greene; Wiley Interscience, New York, pp. 16.15.1–16.18.10.
9. Mackett, M. (1991) Manipulation of vaccinia virus vectors. *Gene Transfer Expression Protocols* **7,** 129–146.
10. Talavera, A. and Rodriguez, J. M. (1991) Isolation and handling of recombinant vaccinia viruses. *Practical Mol. Virol.* **8,** 235–248.
11. Moss, B. (1990) Poxviridae and their replication, in *Virology* (Fields, B. N., Knipe, D. M., Chanock, R. M., Hirsch, M. S., Melnick, J., Monath, T. P., and Roizman, B., eds.), Raven, New York, pp. 2079–2112.
12. Moss, B. (1992) Molecular biology of poxviruses, in *Recombinant Poxviruses* (Binns, M. M. and Smith, G. L., eds.), CRC, Boca Raton, pp. 45–80.
13. Moss, B. (1994) Vaccinia virus transcription, in *Transcription: Mechanisms and Regulation* (Conaway, R. C. and Conaway, J. W., eds.), Raven, New York, pp. 185–205.
14. Mackett, M., Smith, G. L., and Moss, B. (1982) Vaccinia virus: a selectable eukaryotic cloning and expression vector. *Proc. Natl. Acad. Sci. USA* **79,** 7415–7419.
15. Panicali, D. and Paoletti, E. (1982) Construction of poxviruses as cloning vectors: insertion of the thymidine kinase gene from herpes simplex virus into the DNA of infectious vaccinia virus. *Proc. Natl. Acad. Sci. USA* **79,** 4927–4931.
16. Fuerst, T. R., Niles, E. G., Studier, F. W., and Moss, B. (1986) Eukaryotic transient-expression system based on recombinant vaccinia virus that synthesizes bacteriophage T7 RNA polymerase. *Proc. Natl. Acad. Sci. USA* **83,** 8122–8126.
17. Fuerst, T. R., Earl, P. L., and Moss, B. (1987) Use of a hybrid vaccinia virus-T7 RNA polymerase system for expression of target genes. *Mol. Cell. Biol.* **7,** 2538–2544.
18. Elroy-Stein, O. and Moss, B. (1992) Gene expression using the vaccinia virus/T7 RNA polymerase hybrid system, in *Current Protocols in Molecular Biology* (Ausubel, F. M., Brent, R., Kingston, R. E., Moore, D. D., Seidman, J. G., Smith, J. A., and Struhl, K., eds.), Greene;Wiley Interscience, New York, pp. 16.
19. Rodriguez, D., Zhou, Y., Rodriguez, J.-R., Durbin, R. K., Jiminez, V., McAllister, W. T., and Esteban, M. (1990) Regulated expression of nuclear genes by T3 RNA polymerase and lac repressor, using recombinant vaccinia virus vectors. *J. Virol.* **64,** 4851–4857.
20. Usdin, T. B., Brownstein, M. J., Moss, B., and Isaacs, S. N. (1993) SP6 RNA polymerase containing vaccinia virus for rapid expression of cloned genes in tissue culture. *Biotechniques* **14,** 222–224.
21. Fuerst, T. R., Fernandez, M. P., and Moss, B. (1989) Transfer of the inducible lac repressor/operator system from *Escherichia coli* a vaccinia virus expression vector. *Proc. Natl. Acad. Sci. USA* **86,** 2549–2553.
22. Rodriguez, J. F. and Smith, G. L. (1990) Inducible gene expression from vaccinia virus vectors. *Virology* **177,** 239–250.

23. Alexander, W. A., Moss, B., and Fuerst, T. R. (1992) Regulated expression of foreign genes in vaccinia virus under the control of bacteriophage T7 RNA polymerase and the Escherichia colilac repressor. *J. Virol.* **66,** 2934–2942.
24. Cullen, B. R. and Garrett, E. D. (1992) A comparison of regulatory features in primate lentiviruses. *AIDS Res. Hum. Retroviruses* **8,** 387–393.
25. Stomatos, N., Chakrabarti, S., Moss, B., and Hare, D. J. (1987) Expression of polyomavirus virion proteins by a vaccinia virus vector: association of VP1 and VP2 with the nuclear framework. *J. Virol.* **61,** 516–525.
26. Merchlinsky, M. and Moss, B. (1992) Introduction of foreign DNA into the vaccinia virus genome by in vitro ligation: recombination-independent selectable cloning vectors. *Virology* **190,** 522–526.
27. Scheiflinger, F., Dorner, F., and Falkner, F. G. (1992) Construction of chimeric vaccinia viruses by molecular cloning and packaging. *Proc. Natl. Acad. Sci. USA* **89,** 9977–9981.
28. Davison, A. J. and Moss, B. (1990) New vaccinia virus recombination plasmids incorporating a synthetic late promoter for high level expression of foreign proteins. *Nucleic Acids Res.* **18,** 4285,4286.
29. Broder, C. C. and Berger, E. A. (1993) CD4 molecules with a diversity of mutations encompassing the CDR3 region efficiently support human immunodeficiency virus type 1 envelope glycoprotein-mediated cell fusion. *J. Virol.* **67,** 913–926.
30. Earl, P. L., Broder, C. C., Long, D., Lee, S. A., Peterson, J., Chakrabarti, S., Doms, R. W., and Moss, B. (1994) Native oligomeric human immunodeficiency virus type 1 envelope glycoprotein elicits diverse monoclonal antibody reactivities. *J. Virol.* **68,** 3015–3026.
31. Zhang, Y. and Moss, B. (1991) Inducer-dependent conditional-lethal mutant animal viruses. *Proc. Natl. Acad. Sci. USA* **88,** 1511–1515.
32. Yuen, L. and Moss, B. (1987) Oligonucleotide sequence signaling transcriptional termination of vaccinia virus early genes. *Proc. Natl. Acad. Sci. USA* **84,** 6417–6421.
33. Earl, P. L., Hugen, A. W., and Moss, B. (1990) Removal of cryptic poxvirus transcription termination signals from the human immunodeficiency virus type 1 envelope gene enhances expression and immunogenicity of a recombinant vacciniavirus. *J. Virol.* **64,** 2448–2451.
34. Broder, C. C., Kennedy, P. E., Michaels, F., and Berger, E. A. (1994) Expression of foreign genes in cultured human primary macrophages using recombinant vaccinia virus vectors. *Gene* **142,** 167–174.
35. Cook, D. G., Lee, V. M.-Y., and Doms, R. W. (1994) Expression of foreign proteins in a human neuronal system. *Methods Cell Biol.* **43,** 289–303.
36. Chakrabarti, S., Brechling, K., and Moss, B. (1985) Vaccinia virus expression vector: coexpression of β-galactosidase provides visual screening of recombinant virus plaques. *Mol. Cell. Biol.* **5,** 3403–3409.
37. Ramshaw, I. A., Andrew, M. E., Philips, S. M., Boyle, D. B., and Coupar, B. E. H. (1987) Recovery of immunodeficient mice from a vaccinia virus/IL2 recombinant infection. *Nature* **329,** 545–546.

38. Boyle, D. B. and Coupar, B. E. H. (1988) A dominant selectable marker for the construction of recombinant poxviruses. *Gene* **65,**123–128.
39. Falkner, F. G. and Moss, B. (1988) *Escherichia coli gpt* gene provides dominant selection for vaccinia virus open reading frame expression vectors. *J. Virol.* **62,** 1849–1854.
40. Isaacs, S. N., Kotwal, G. J., and Moss, B. (1990) Reverse guanine phosphoribosyltransferase selection of recombinant vaccinia viruses. *Virology* **178,** 626–630.
41. Falkner, F. G. and Moss, B. (1990) Transient dominant selection of recombinant vaccinia viruses. *J. Virol.* **64,** 3108–3111.
42. Franke, C. A., Rice, C. M., Strauss, J. H., and Hruby, D. E. (1985) Neomycin resistance as a dominant selectable marker for selection and isolation of vaccinia virus recombinants. *Mol. Cell. Biol.* **5,** 1918–1924.
43. Zhou, J., Crawford, L., Sun, X.-Y., and Frazer, I. H. (1991) The hygromycin-resistance-encoding gene as a selection marker for vaccinia virus recombinants. *Gene* **107,** 307–312.
44. Rodriguez, J. F. and Esteban, M. (1989) Plaque size phenotype as a selectable marker to generate vaccinia virus recombinants. *J. Virol.* **63,** 997–1001.
45. Blasco, R. and Moss, B. (1995) Selection of recombinant vaccinia viruses on the basis of plaque formation. *Gene* **158,** 157–162.
46. Shida, H., Tochikura, T., Sato, T. T. K., Hirayoshi, K., Seki, M., Ito, Y., Hatanaka, M., Hinuma, Y., Sugimoto, M., Takahashi-Nishimaki, F., Marayama, T., Miki, K., Suzuki, K., Morita, M., Sashiyama, H., and Hayami, M. (1987) Effect of the recombinant vaccinia viruses that express HTLV-1 envelope gene on HTLV-1 infection. *EMBO J.* **6,** 3379–3384.
47. Perkus, M. E., Limbach, K., and Paoletti, E. (1989) Cloning and expression of foreign genes in vaccinia virus, using a host range selection system. *J. Virol.* **63,** 3829–3836.
48. Boyle, M. D. P. and Reis, K. J. (1987) Bacterial Fc receptors. *Biotechnology* **5,** 697–703.
49. Sutter, G., Wyatt, L. S., Foley, P. L., Bennink, J. R., and Moss, B. (1994) A recombinant vector derived from the host range-restricted and highly attenuated MVA strain of vaccinia virus stimulates protective immunity in mice to influenza virus. *Vaccine* **12,** 1032–1040.
50. Chen, H. and Padmanabhan, R. (1994) A modified method for isolation of recombinant vaccinia virus. *BioTechniques* **17,** 41–42.
51. Mackett, M., Smith, G. L., and Moss, B. (1984) General method for production and selection of infectious vaccinia virus recombinants expressing foreign genes. *J. Virol.* **49,** 857–864.
52. Flexner, C., Hugin, A., and Moss, B. (1987) Prevention of vaccinia virus infection in immunodeficient nude mice by vector-directed IL-2 expression. *Nature* **330,** 259–262.

16

A Simplified Method of Screening for Isolation of Recombinant Vaccinia Virus

Haifeng Chen and R. Padmanabhan

1. Introduction

The utility of a recombinant vaccinia virus expression system for transient expression of genes was demonstrated in 1982 *(1,2)*. Among a number of useful characteristics of this expression system are the capacity of the vaccinia virus genome to accommodate large genes (>20 kb pairs), availability of insertional sites on the viral genome yielding viable recombinant viruses, ability of the vaccinia viruses to infect a wide range of mammalian hosts, cytoplasmic transcription and replication of the viral genome unlike many other expression systems, a relatively high level of recombinant proteins produced in infected cells, and the primary sequence-directed posttranslational modifications of recombinant proteins akin to native proteins (for a review, *see* ref. *3*). As a first step in the construction of a recombinant vaccinia virus, the gene encoding a protein in the form of a cDNA is cloned into a plasmid vector under the control of an early or late vaccinia virus promoter *(4–6)*, or a bacteriophage T7 promoter/the encephalomyocarditis virus 5'-untranslated leader sequence which would allow a cap independent mechanism of translation of the transcripts *(7)*. Flanking the gene chosen for expression are DNA sequences representing portions of a nonessential viral gene such as that encoding the thymidine kinase in order to allow homologous recombination at the locus of the viral gene in vivo *(6)*. The frequency with which the gene is integrated into the viral genome has been estimated to be about 0.1% (for a review, *see* ref. *3*). In the original protocol for the isolation of a recombinant vaccinia virus *(6)*, cell monolayers are infected with the wild-type virus (WR strain) and transfected with the plasmid DNA encoding the gene of interest. Since insertion of the gene of interest at the viral thymidine kinase (TK) gene locus results in recombinant viruses with the

From: *Methods in Molecular Biology, vol. 62: Recombinant Gene Expression Protocols*
Edited by: R. Tuan Humana Press Inc., Totowa, NJ

TK^- phenotype, plaque isolates of recombinant viruses are selected in the presence of 5-bromodeoxyuridine (5-BrdU). The conditions chosen for selection in the presence of this analog of thymidine are lethal to replication of the wild-type virus owing to incorporation of the phosphorylated form into the viral genome, but not to the recombinant virus deficient in TK. The cell lysates containing the wild-type and recombinant viruses at appropriate dilutions are used to infect cell monolayers of human TK^- cells, and then overlaid with growth medium containing agarose and 5-BrdU to prevent spreading of recombinant virus plaques and inhibit replication of the wild-type virus, respectively. A second agarose overlay containing the 5-BrdU and the neutral red stain is used to enable one to visualize recombinant virus plaques *(6)*. A single plaque results from a productively infected cell and the cell-to-cell spread of the virus, and an agarose overlay localizes each of these plaques on a monolayer. Since a plaque could consist of either a recombinant virus or a spontaneous TK^- mutant virus often present as a contaminant in the cell lysates, a suitable hybridization technique is employed in the screening protocol for isolation of recombinants *(6)*.

In the use of agarose overlay, pouring of the molten agarose to cell monolayers is a critical step since the temperature of the agarose must be not too high since that would kill the cells, and not too low since that would cause premature solidification and uneven spreading of agarose on cell monolayers. Moreover, the presence of two agarose overlays on cell monolayers poses a technical problem of isolating a single plaque by mild aspiration with a Pasteur pipet. The yield of virus particles recovered from an area of a plaque is variable and often poor, which would affect subsequent steps of amplification in TK^- cells and screening by hybridization. In addition, some lots of agarose are toxic to the cells, and hence are unsuitable for overlays. We describe a simple and rapid method for the isolation of a recombinant vaccinia virus applied to the gene encoding a temperature-sensitive adenovirus DNA polymerase. This method avoids the use of agarose overlays and is essentially a limiting dilution method which relies on infection of monolayers of cells in a 96-well cell culture plates with an input virus from a cell lysate diluted 10^2–10^4-fold so that only single plaques are formed in some wells.

2. Materials

1. Medium for cell culture: DMEM-10, DMEM-2.5, and DMEM-0 (Dulbecco's modified Eagle medium containing 10%, 2.5%, and 0% BCS).
2. BCS (defined/supplemented bovine calf serum, Hyclone, Logan, UT).
3. Cell lines: CV-1 cells (ATCC #CCL70), $HuTK^-$143B cells (ATCC #CRL8303).
4. Neutral red stock solution (5 mg/mL in sterile distilled water).
5. 5-Bromo-2'-deoxyuridine (BrdU) stock solution (5 mg/mL in sterile distilled water).

6. Wild-type vaccinia virus stock (Western Reserve, WR) strain of vaccinia virus (ATCC #VR1354).
7. Lipofectin (Gibco-BRL, Gaithersburg, MD).
8. 0.5X trypsin/EDTA solution: 0.125% trypsin/0.01% EDTA.
9. 0.25 mg/mL trypsin solution.
10. [α-^{32}P]dCTP.
11. 0.4*M* Tris-HCl (pH 7.5).
12. 5*N* NaOH.
13. Dilution solution: 0.125X SSC and 0.125*N* NaOH.
14. Hybridization solution: 1% SDS, 1*M* NaCl and 10% dextran sulfate.
15. Salmon sperm DNA solution: 10 µg/µL in TE buffer.
16. 20X SSC (1X = 0.15*M* sodium chloride and 0.015*M* sodium citrate).
17. 10% SDS in water.
18. 0.1% crystal violet in 20% ethanol.

3. Methods

3.1. Construction of Recombinant Vaccinia Virus

1. Subclone the gene of interest in pTM1 vector *(7)* or other suitable vectors (ref. *6*; *see* Note 1).
2. Seed monkey kidney (CV-1; 1×10^6) cells in a 25-cm^2 flask in DMEM-10 and grow in a CO_2 incubator at 37°C until cells are about 90% confluent. One 25-cm^2 flask contains approx 3×10^6 cells.
3. Treat wild-type vaccinia virus just prior to use with trypsin by mixing an equal volume of a virus stock and 0.25 mg/mL trypsin. Vortex vigorously and incubate for 30 min in a 37°C water bath.
4. Dilute 10 µg of an expression plasmid containing the gene of interest with 100 µL of DMEM-0 in a microfuge (1.5 mL capacity) tube, and 30 µL of Lipofectin with 100 µL DMEM-0 in a 15-mL polystyrene tube.
5. Transfer the diluted plasmid DNA to the polystyrene tube and gently mix the contents. Incubate at room temperature for 15 min.
6. Dilute trypsinized virus in DMEM-0 to 1.5×10^5 PFU/mL and transfer 1 mL of diluted virus (0.05 PFU/cell) to the polystyrene tube containing the plasmid DNA and lipofectin. Mix by gentle vortexing.
7. Aspirate medium from confluent monolayer of CV-1 cells and wash the cells once with 4 mL of DMEM-0. Aspirate the medium again and add the DNA-lipofectin-virus mixture to the cell monolayer. Spread the solution evenly to cover the cells and incubate at 37°C in a CO_2 incubator for 3–5 h.
8. Add 4 mL of DMEM-2.5 and incubate at 37°C in a CO_2 incubator for 2 d (*see* Note 2).
9. Collect infected cells by using a disposable cell scraper or a sterile rubber policeman and transferring the cell suspension to a 15-ml centrifuge tube. Centrifuge for 10 min in a HB-4 rotor (DuPont/Sorvall) at 659*g*) at 4°C and discard the supernatant.

10. Resuspend cells in 0.5 mL DMEM-2.5 by pipeting and mixing several times. Lyse cells by three freeze-thaw cycles consisting of freezing in dry ice, thawing in a 37°C water bath, and vortexing.
11. Store the lysate at –70°C until use in selection and screening protocols.

3.2. Isolation of the Recombinant Vaccinia Virus

1. Grow HuTK$^-$ 143B cells in a 75-cm^2 flask in DMEM-10 containing 50 μg/mL 5-BrdU to near confluency. One confluent 75-cm^2 flask contains approx 1×10^7 cells.
2. Trypsinize the cells as follows: Aspirate medium from the cell monolayer and add 3 mL of 0.5X trypsin/EDTA to cover the cell monolayer. Incubate for 1 min at room temperature and aspirate the trypsin/EDTA solution. Gently tap and shake the flask to completely dislodge cells. Add 40 mL DMEM-10 and pipet the cell suspension several times to disrupt clumps of cells.
3. Transfer 100 μL of cell suspension (2.5 X 10^5 cells/mL) to each well of three 96-well plates and incubate at 37°C in a CO_2 incubator until confluent (usually overnight).
4. Sonicate the cell lysate from step 11 (*see* Section 3.1.) in a capped sterile tube kept in ice water for 30 s. Dilute lysate serially to 10^{-2}, 10^{-3}, and 10^{-4} in a final volume of 10 mL DMEM-2.5 containing 50 μg/mL of 5-BrdU (*see* Note 3).
5. Aspirate medium from the 96-well plates two rows at a time and transfer 100 μL of the diluted lysate from Step 4 to each well (*see* Note 4). When diluted virus suspension is added to all wells, incubate the plates at 37°C in a CO_2 incubator for 2 d.
6. Aspirate the virus inoculum from the 96-well plates two rows at a time and add 100 μL DMEM-10 containing 50 μg/mL of 5-BrdU and 50 μg/mL of neutral red. Incubate at 37°C in a CO2 incubator overnight. At the same time, seed each well of 12-well plates with 1×10^5 HuTK$^-$ 143B cells suspended in a total volume of 2 mL DMEM-10 containing 50 μg/mL 5-BrdU and grow the cells until confluent.
7. Place the 96-well plates on a light-box and mark the wells that contain single plaques *(8)*. Aspirate medium from 96-well plates and add 50 μL DMEM-2.5 containing 50 μg/mL of 5-BrdU to the wells containing single plaques.
8. Freeze the 96-well plates in dry ice and thaw in a 37°C water bath for a total of three cycles to dislodge and lyse cells. Pipet forcefully (without splashing to adjacent wells) several times to dislodge all cells. Store at –70°C until needed or use it immediately in the following step (*see* Note 5).
9. Sonicate lysates for 30 s and add 25 μL to a microfuge tube containing 0.5 mL DMEM-2.5 and 50 μg/mL of 5-BrdU. Store the remaining lysates at –70°C. Aspirate medium from cell monolayers grown in 12-well plates and add diluted cell lysates to cells. Incubate cells for 1 h in a 37°C CO_2 incubator. Add 1.5 mL of DMEM-10 containing 50 μg/mL 5-BrdU and incubate the plates in a 37°C CO_2 incubator for 2 d.
10. Aspirate medium from cell monolayers and add 500 μL DMEM-2.5 containing 50 μg/mL of 5-BrdU to each well. Subject cells to three freeze-thaw cycles as in step 8. Collect the lysate and sonicate for 30 s on ice water. Use 40 μL for DNA hybridization and store the rest at –70°C.

3.3. Identification of Recombinant Vaccinia Virus Containing the Target Gene by DNA Hybridization

1. Isolate 50–100 ng of a DNA fragment from the gene of interest and label the DNA with [α-^{32}P]dCTP using an in vitro labeling kit following the manufacturer's instructions (*see* Note 6).
2. Remove unincorporated nucleotides by using a NENSORB affinity matrix following the instructions of the manufacturer (*see* Note 7).
3. Cut one piece of 3-mm filter paper and one piece of hybridization transfer membrane (GeneScreen Plus) according to the size of a dot-blot apparatus and soak them in 0.4*M* Tris-HCl (pH 7.5) buffer. Assemble the dot-blot apparatus according to the manufacturer's instructions.
4. Mix 40 µL of each sample with 2 µL of 5*N* NaOH and leave at room temperature for 10 min. Chill samples on ice and mix with 40 µL of a diluent containing 0.125X SSC and 0.125*N* NaOH.
5. Load the samples on the dot-blot apparatus and incubate for 30 min. Apply gentle suction until samples pass through the membrane (*see* Note 8). Disassemble the apparatus and air-dry the membrane.
6. Put the membrane into a sealable bag and add 10 mL of hybridization solution containing 1% SDS, 1*M* NaCl, and 10% dextran sulfate. Incubate the membrane at 65°C for 15 min.
7. Mix 100 µg salmon sperm DNA with the labeled DNA probe (about 1×10^6 cpm) in 0.5 mL deionized distilled H_2O. Boil the probe for 5 min and chill it on ice water quickly. Cut a small hole in the sealable bag containing the membrane and transfer the DNA probe into the bag. Seal the bag and mix the probe by inverting the bag several times. Incubate the bag at 65°C for 4–24 h.
8. Take out the membrane and wash the membrane with 2 X 100 mL of 2 X SSC at room temperature for 5 min per wash.
9. Wash the membrane with 2X 200 mL of 2X SSC containing 1.0% SDS at 65°C for 30 min each time.
10. Wash the membrane with 2X 100 mL of 0.1X SSC at room temperature for 30 min each time.
11. Air-dry the membrane and wrap it with a plastic wrap. Expose the membrane to an X-ray film at –70°C for a few hours or overnight depending on the amount of radioactivity on the membrane.

3.4. Amplification of Recombinant Vaccinia Virus

1. Culture Hu143TK$^-$ cells in a 25-cm^2 flask with DMEM-10 containing 50 µg/mL of 5-BrdU until confluent.
2. Sonicate the remainder of cell lysate of a positive plaque as verified by DNA hybridization and dilute 250 µL of the lysate with 750 µL DMEM-2.5 containing 50 µg/mL of 5-BrdU.
3. Remove medium from the cell monolayer and infect the cells with the diluted lysate for 1 h at 37°C in a CO_2 incubator. Add 4 mL DMEM-2.5 containing 50 µg/mL of 5-BrdU and incubate for 2 d.

4. Dislodge the cells using a disposable cell scraper or a rubber policeman and transfer the cells to a 15-mL centrifuge tube. Centrifuge at 4°C in a HB-4 rotor at 659*g* for 10 min to collect the cells. Discard the supernatant and resuspend the cells in 0.5 mL DMEM-2.5. Lyse the cells by three freeze-thaw cycles followed by a 30 s sonication. Store the lysate at –70°C until needed.
5. Culture HeLa cells with DMEM-10 in two 75-cm^2 flasks until confluent. Dilute 0.25 mL lysate with 5.75 mL DMEM-2.5. Remove the medium from cells and add 3 mL of the diluted lysate to each flask. Incubate at 37°C in CO_2 incubator for 1 h, rocking the flasks at 15- to 30-min intervals.
6. Add 12 mL DMEM-2.5 to each flask and incubate at 37°C in CO_2 incubator for 2 d.
7. Collect the infected cells in two 15-mL centrifuge tubes using a disposable cell scraper or a rubber policeman and centrifuge at 4°C in HB-4 rotor at 659*g* for 10 min. Aspirate and discard the supernatant.
8. Add 1 mL DMEM-2.5 to each tube, resuspend, and combine cells in one tube. Lyse the cells by three freeze-thaw cycles followed by a 30-s sonication on ice water. Keep the viral stocks frozen at –70°C until needed.

3.5. Determination of the Recombinant Virus Titer by Plaque Assay

1. Seed CV-1 cells (5×10^5) in each well of a six-well plate and culture in a CO_2 incubator at 37°C until confluent (usually overnight).
2. Trypsinize the virus as described in step 3 of Section 3.1. and sonicate for 30 s. Make nine 10-fold serial dilutions of the trypsinized virus in DMEM-2.5, using a fresh pipet tip for each dilution.
3. Remove medium from CV-1 cells and infect cells in duplicate wells with 0.5 mL of the 10^{-7}, 10^{-8}, and 10^{-9} dilutions. Incubate at 37°C in a CO_2 incubator for 1 h. Add 2.5 mL DMEM-2.5 to each well and continue the incubation for 2 d.
4. Remove the medium and add 0.5 mL of 0.1% crystal violet in 20% ethanol to each well and incubate at room temperature for 5 min.
5. Aspirate the crystal violet solution and allow the wells to dry. Determine the titer by counting the plaques in duplicate wells and multiplying by the dilution factor.

4. Notes

1. There are several vaccinia virus transfer vectors described by Earl and Moss *(6)*. Suitable vector can be chosen depending on the cloning strategy. We use the pTM1 vector *(7)* and clone the gene of interest into the NcoI site of the vector.
2. In the standard procedure described by Earl and Moss *(6)*, transfection is performed using the calcium-phosphate method one half hour postinfection. We use liposome-mediated transfection protocol (Lipofectin from Gibco-BRL), and combine the infection and transfection steps into one, which not only saves time but also in our experience increases the efficiency of transfection.
3. To begin the isolation of a recombinant vaccinia virus, generally three dilutions (10^{-2}, 10^{-3}, and 10^{-4}) of a cell lysate obtained from an infection-transfection experiment are sufficient to give rise to an appropriate dilution containing less

than 100 PFU/mL of the recombinant vaccinia virus so that some wells of the 96-well plates would contain single plaques. If no well contains a single plaque, then further dilution of infection-transfection lysate may be necessary.

4. While aspirating the medium from the wells of 96-well plates, do not remove medium from all wells at the same time. This will result in drying of some cell monolayers. In addition, do not let the Pasteur pipet tip touch a cell monolayer which may result in a localized loss of cells that can give rise to a false plaque morphology subsequent to the neutral red staining.
5. It is best to harvest as many single plaques as possible. Some single plaques may be derived from inactivation of the TK gene of the virus without insertion of the target gene. In addition, great care must be taken not to let water get into the wells during freeze-thaw cycles.
6. There are several methods for preparation of DNA probes as described *(9)*. We use the Amersham's Multiprime DNA labeling system for the preparation of labeled probes which gives satisfactory results. To eliminate false-positive virus plaques having TK^- phenotypes, a DNA hybridization step is necessary to identify the insertion of the target gene into the viral genome.
7. Different methods are available to remove unincorporated nucleotide. We have used QIAEX II Gel Extraction Kit, Sephadex G-50 gel filtration, and NENSORB column to purify the labeled probe. In our hands, the purification of labeled probes using NENSORB column has given consistently reliable results.
8. Sometimes the samples may pass through the membrane more slowly requiring longer time of aspiration.

Acknowledgments

This work was supported from a grant from NIH (CA33099 and AI32078) to R. P., and H. C. was supported by a postdoctoral training grant from Marion Merrell Dow Foundation.

References

1. Panicali, D. and Paoletti, E. (1982) Construction of poxviruses as cloning vector: insertion of the thymidine kinase gene from herpes simplex virus into the DNA of infectious vaccinia virus. *Proc. Natl. Acad. Sci. USA* **79,** 4927–4931.
2. Mackett, M., Smith, G. L., and Moss, B. (1982) Vaccinia virus: a selectable eukaryotic cloning and expression vector. *Proc. Natl. Acad. Sci. USA* **79,** 7415–7419.
3. Moss, B. (1991) Vaccinia virus: A tool for research and vaccinie development. *Science* **252,** 1662–1667.
4. Coupar, B. E. H., Andrew, M. E., and Boyle, D. B. (1988) A general method for the construction of recombinant vaccinia viruses expressing multiple foreign genes. *Gene* **68,** 1–10.
5. Perkus, M. E., Limbach, K., and Paoletti, E. (1989) Cloning and expression of foreign genes in vaccinia virus, using a host range selection system. *J. Virol.* **63,** 3829–3836.

6. Earl, P. L. and Moss, B. (1991) Expression of proteins in mammalian cells using vaccinia viral vectors, in *Current Protocols in Molecular Biology* (Ausubel, F. M., Brent, R., Kinston, R. E., Moore, D. D., Smith, J. A., Seidman, J. C., and Struhl, K., ed.), Wiley, New York, pp. 16.15.1–16.18.10.
7. Moss B., Elroy-Stein, T., Mizukami, T., Alexander, W. A, and Fuerst, T. R. (1990) New mammalian expression vectors (Product review). *Nature* **348,** 91–92.
8. Chen, H. and Padmanabhan, R. (1994) A modified method for isolation of recombinant vaccinia virus. *Biotechniques* **17,** 41–42.
9. Sambrook, J., Fritsch, E. F., and Maniatis, T. (1989). *Molecular Cloning: A Laboratory Manual.* Cold Spring Harbor Laboratory, Cold Spring Harbor, NY.

17

Design of Retroviral Expression Vectors

Kenine E. Comstock, Nathaniel F. Watson, and John C. Olsen

1. Introduction

Several characteristics of amphotropic murine retroviruses have made them useful as vectors for gene transfer with some distinct advantages over other methods of transduction. First, the normal replication cycle of retroviruses includes integration of the viral genome into the host's chromosomal DNA, and these viruses have evolved a very efficient mechanism of stable gene transfer for this purpose. Second, because the integration machinery uses the termini of the viral DNA as substrate, insertion of viral DNA takes place in a predictable manner with the long terminal repeat (LTR) sequences flanking the genes they carry. Third, retroviruses have a broad host range, which allows gene transfer and expression of foreign genes in many cell types, even including some cell types that are refractory to gene transfer by other means. Fourth, a cytopathic effect on infected cells is lacking, which is especially important for gene expression studies and the generation of stable cell lines. Fifth, all retroviral proteins required for the assembly of infectious virions can be supplied *in trans*, thereby enabling the expression of exogenous genes up to approx 8 kbp. Finally, retroviral vectors can be conveniently manipulated in plasmid form and, as discussed below, are somewhat flexible in terms of vector design.

There are, however, some limitations to the use of retroviral vectors. One limitation is that gene transfer only occurs when cells are actively dividing *(1)*. A second limitation is that expression levels are only moderate compared to those achieved by other methods of gene transfer, owing in part to the low copy numbers of the integrated provirus generally achievable by these vectors.

The first component of the retrovirus gene transfer system is the retroviral packaging cell line. Packaging cells provide all the retroviral proteins required for the encapsidation of the retroviral vector (in RNA form) into infectious

From: *Methods in Molecular Biology, vol. 62: Recombinant Gene Expression Protocols*
Edited by: R. Tuan Humana Press Inc., Totowa, NJ

virus particles including the reverse transcriptase required for the synthesis of double-stranded DNA from the viral RNA and the integrase protein, which is required for the integration of DNA product into the host genome. The packaging cell line also produces the viral envelope glycoproteins required for receptor recognition and fusion of the virus with the target cell. Since the nature of the viral envelope glycoproteins determines the host range of the retroviral vector, the type of packaging cell line is an important choice. Packaging cells expressing the envelope glycoproteins of amphotropic murine retroviruses are useful because these viruses infect many cell types from a number of species including mouse and human, although there are some important exceptions *(2)*. Commonly used amphotropic packaging cell lines include PA317 *(3)*, ΨCRIP *(4)*, and GP+*env*Am12 *(5)*, all of which are derived from mouse fibroblasts and produce viral proteins from murine leukemia virus (MuLV). Since these cell lines have been constructed to drastically minimize the frequency of helper virus production, gene transfer using virus obtained from these packaging cells does not spread beyond the cells initially infected.

Other packaging cells express viral envelope glycoproteins that use other receptors. For example, infection with retroviral vectors produced by ecotropic packaging cells such as PE501 *(6)*, BOSC *(7)*, ΨCRE *(4)*, or GP+E-86 *(8)* is limited to cells from mouse and rat. Xenotropic packaging cells *(9)* produce viruses that infect cells of many species, but these viruses do not infect mouse cells. The PG13 cells produce viruses having the broad host range of gibbon ape leukemia virus *(10)* and have been shown to deliver genes more efficiently than amphotropic viruses to some primate cells *(11)*. Virions pseudotyped with the G envelope glycoprotein from vesicular stomatitis virus (VSV) have perhaps the broadest host range of any retrovirus vector, owing to the utilization of a phospholipid component of the cell membrane as a receptor. The VSV-G glycoprotein also confers stability to retrovirus particles, so that virus(es) containing this envelope protein can be concentrated to very high titers using ultracentrifugation methods *(12)*. Whereas many packaging cells produce viruses having a broad host range, a current area of research is aimed at designing retroviruses that utilize specific receptors *(13)*. This approach may be useful for gene transfer to specific in vivo targets.

The second component of retrovirus-mediated gene transfer is the retroviral vector. Figure 1 summarizes the *cis*-acting sequence elements in retroviral vectors and how they contribute to replication as a retroviral genome. These elements include the long terminal repeat (LTR) sequences. The LTR is further divided into sequence domains referred to as U3, R, and U5. The U3 domain contains the viral enhancer and promoter sequences. Transcription by cellular RNA polymerase II begins at the U3-R boundary in the 5' LTR (as indicated by the arrow in Fig. 1). For murine retroviruses, a polyadenylation signal is present

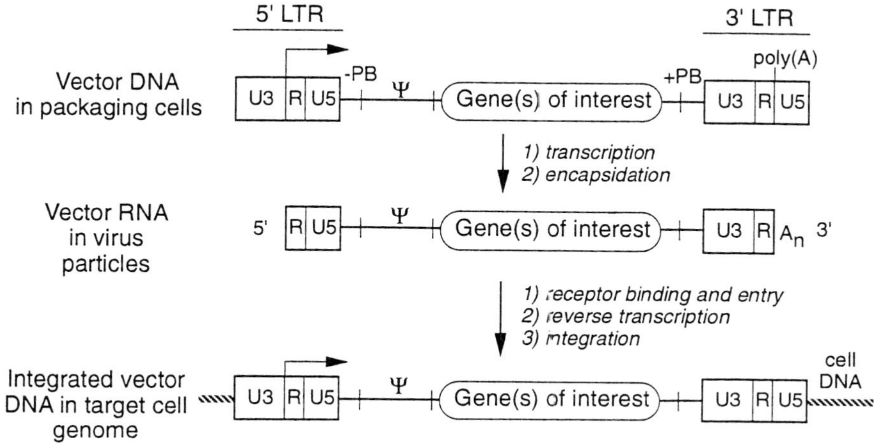

Fig. 1. The flow of genetic information during retrovirus-mediated gene transfer.

in the R domain and RNA transcripts are polyadenylated at the R-U3 boundary in the 3' LTR. The sequence referred to as Psi "Ψ" is required for encapsidation, or packaging, of the vector RNA into retrovirus particles. Newly formed retrovirus particles bud from the producer cell and are harvested by collecting the medium. Gene transfer is accomplished by using the virus-containing medium to infect the target cell. Shortly after infection, viral reverse transcription is initiated in the target cell. Sequences required for reverse transcription of the viral RNA into double-stranded DNA include the primer binding sequences required for the initiation of minus and plus strand DNA synthesis. These sequences, (–PB) and (+PB), are located adjacent to the 5' LTR and 3' LTR, respectively (Fig. 1). During reverse transcription, the U3 and U5 sequences in the RNA are copied twice to regenerate the 5' and 3' LTR sequences. This maneuver serves to place the U3 promoter sequences in an upstream position relative to the viral genes and also generates short inverted sequences at the termini of the viral DNA that are used by the viral integrase to covalently join the linear viral DNA with the host cell chromosomal DNA (Fig. 1).

In terms of retroviral vector design, all the cis-acting sequences required for propagation of the vector as a retroviral genome are essentially on the outside portions of the vector. All of the internal sequences can be manipulated as the investigator wishes taking into account the following considerations:

1. As the size of the vector increases, the efficiency of gene transfer and expression decreases. As the size of the RNA exceeds about 10 kb, gene transfer efficiency falls drastically. Since the *cis*-acting sequences use about 1.5 kb of a retroviral vector, this leaves a total of about 8.5 kb that can be devoted for other sequences.

This 8.5 kb includes the gene of interest as well as other elements that may be useful to have in the vector including selectable markers or additional promoters etc.

2. Since the *cis*-acting sequences required for replication are present near each end of the vector, the inclusion of poly(A) signals may result in truncation of the retroviral transcription unit and may be a cause of decreased retroviral gene transfer efficiency. Thus, when cloning cDNAs into retroviral vectors, the inclusion of the poly(A) signal sequences found in 3' untranslated regions should be avoided.
3. Since the flow of retroviral gene transfer includes an RNA intermediate, introns from chromosomal genes or present in some nonretroviral vectors to increase expression can be removed from viral RNA transcripts by the normal cellular splicing machinery. This later property can be exploited to use the retroviral vector/packaging cell system to remove introns from cloned chromosomal DNA to prepare intronless cDNA-like sequences *(14)*.

The choice of a retroviral vector design depends on the intended application of the technology. All modern retroviral vectors contain extended packaging sequences required for efficient encapsidation of viral RNA into virions and have been engineered to not synthesize retroviral-related polypeptides. In addition, these vectors have minimal homology to virus-related sequences in packaging cells to minimize the likelihood of recombination events leading to the production of replication competent helper viruses. The following vectors, described in Fig. 2, are all based on MuLV and are intended to give the reader a sampling of the various types available:

1. Vectors with internal promoters include those similar to the LXSN and LNCX vectors developed by Miller and Rosman *(6)*. These vectors are designed such that one gene is expressed from the retroviral LTR and a second gene is expressed from an internally located promoter. These vectors may include one of a variety of selectable markers *(15–17)* and further modifications have been made to optimize expression in embryonal cells *(17)*. In many of these vectors, the internal promoter sequence is a promoter known to be active in a wide variety of cell types. Internal promoters may include strong viral promoter such as the SV40 early promoter or the CMV immediate early promoter or, alternatively, housekeeping promoters such as the phosphoglycerate kinase promoter. The disadvantage of these vectors is that expression from one promoter within a retroviral vector can alter the expression from other promoters within the vector *(18,19)*. Also, this two promoter arrangement can sometimes lead to selection for proviruses having deletions *(20)*.
2. Vectors with internal ribosomal entry sites (IRES) circumvent the above problems by eliminating the downstream promoter and allowing cap-independent translation *(21–23)*. In this configuration, two (or more) genes are translated from a single bicistronic (or polycistronic) mRNA using the IRES sequence to direct the ribosome to begin translation at internal sites.

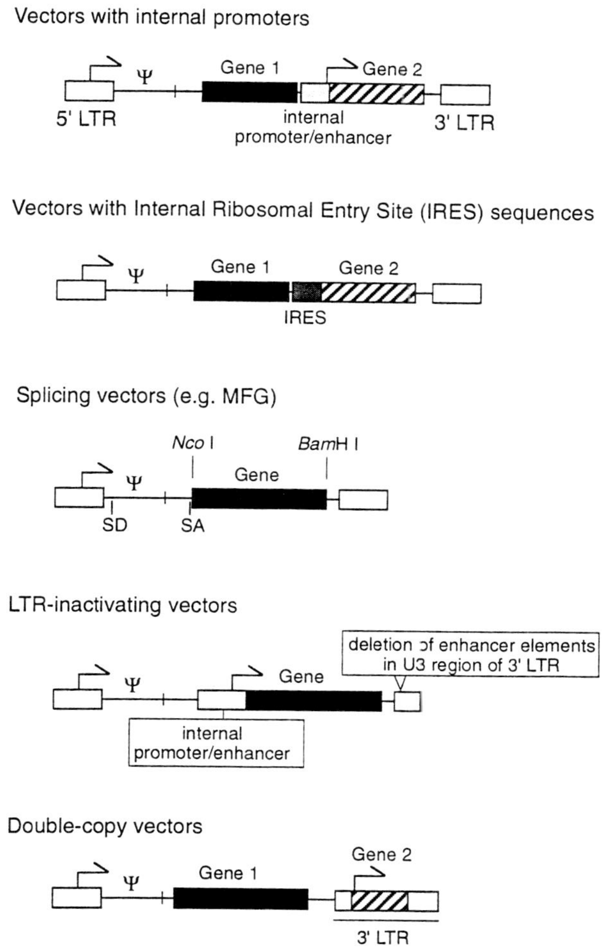

Fig. 2. Examples of various types of retroviral vectors.

3. Splicing vectors use splice donor (SD) and splice acceptor (SA) elements to express the gene of interest. The MFG vector was designed so that the normal viral splice sites used in expression of the viral envelope gene are retained and the gene of interest is expressed from this spliced transcript *(24)*. In an attempt to optimize expression, the use of an *Nco* I cloning site in the vector allows the insertion of the gene of interest such that the start codon for the gene of interest is in the same sequence context as the viral Env protein. To make full use of this strategy it is necessary to engineer an *Nco* I cloning site at the translation initiation site of the gene of interest. The advantage of this type of vector is that the gene of interest may be translated at high efficiency. MFG vectors with IRES sequences have also been constructed to allow the expression of more than one gene *(25)*.

4. The LTR-inactivating vectors are constructed so that deletion of the enhancer elements in the LTR sequence occurs during viral replication, which permits transcription to be controlled by enhancer/promoter sequences of the investigator's choosing *(26–28)*. This type of vector is useful if it is desirable to have tissue-specific or inducible expression of the gene of interest. An examination of Fig. 1, which summarizes the replication of retroviral sequences during a round of gene transfer, will explain how this vector works. A deletion of the enhancer sequence present in the U3 domain of the 3' LTR is not only maintained in the 3' LTR, but also transferred to the 5' LTR during reverse transcription. Thus, transcription in the target cell can be regulated by enhancer/promoter sequences placed upstream of the gene of interest (Fig. 2).
5. Double-copy vectors are constructed so that the integrated provirus will have identical transcription units (promoter and gene of interest) within both 5' and 3' LTRs, theoretically increasing the level of expression *(28–30)*. The principle behind the design of double-copy vectors is similar to the strategy described for LTR-inactivating vectors, except that for double-copy vectors, the cloning of sequences into the U3 domain of the 3' LTR will result in the duplication of these sequences following a round of retroviral replication. This type of strategy has been successfully used to construct vectors using RNA polymerase III promoters to express high levels of RNA for antisense and ribozyme applications *(30–33)*.

The retroviral vectors described above can be customized for the individual investigator's particular application. RNA polymerase II promoters with activity in a wide range of cell types, such as the SV40 immediate early promoter and MuLV LTR, can nonetheless vary in levels of expression in different cell types, so the ideal promoter for a particular cell type and amount of protein expression necessary for the desired effect must be determined empirically. In another approach, it is possible to construct vectors with hybrid enhancer/promoter sequences in the LTR without disrupting *cis*-acting sequences required for retroviral replication *(34,35)*.

In addition to their usefulness as expression vectors for gene transfer to cultured cells, retroviral vectors can be used in a number of other applications. For example, since retroviral DNA integrates at essentially random sites in host cell chromosomal DNA, integration of retroviral DNA can inactivate functional genes by insertional mutagenesis *(36,37)*. This property can be exploited in gene trap approaches that use the ability of an endogenous promoter to drive the expression of a reporter gene in a retroviral vector lacking promoter/enhancer sequences *(38–40)*. Retrovirus vectors containing reporter genes have also been used for cell lineage analysis during cellular differentiation *(41,42)*.

This chapter describes a method for rapidly generating amphotropic retroviral vectors for gene transfer experiments. The method involves the use of sodium butyrate to maximize transient production of virus following DNA-mediated transfection of the PA317 packaging cell line. The virus that is produced can be used in gene transfer experiments to generate cell lines stably expressing an exogenous gene or can be used in the development of retroviral vectors with new design features. This method is not as laborious as those procedures involving the generation of clonal cell lines, although the titers achieved using our method can be nearly as high. An important determinant of the successful application of the method is the purity of the plasmid DNA used for transfection. Preparation of plasmid DNA by isopycnic centrifugation using cesium chloride or by commercially available plasmid purification kits yields plasmid DNA of acceptable quality, but is time consuming or expensive. Therefore, an alternate procedure is described using polyethylene glycol precipitation that results in plasmid DNA of sufficient purity to be used in DNA transfection experiments.

2. Materials

2.1. Purification of Plasmid DNA Using Polyethylene Glycol Precipitation

1. Terrific Broth (Life Technologies, Gaithersburg, MD). Dissolve 47 g in 1 L of H_2O. Add 8 mL 50% (v/v) glycerol and sterilize by autoclaving. Store at room temperature.
2. Tris/glucose/EDTA solution: 25 m*M* Tris-HCl, pH 8.0, 50 m*M* glucose, 10 m*M* EDTA (pH 8.0). Store at 4°C.
3. 3*M* potassium acetate (pH 4.8): To 120 mL of 5*M* potassium acetate, add about 25 mL of glacial acetic acid to pH 4.8 and add H_2O to a final volume of 200 mL. Store at 4°C.
4. NaOH/SDS solution: To 167 mL H_2O, add 13 mL of 3*N* NaOH and 20 mL of 10% SDS. Store at room temperature in a polypropylene container. Do not store in glass.
5. $T_{10}E_{10}$: 10 m*M* Tris-HCL, pH 8.0, 10 m*M* EDTA.
6. $T_{10}E_{0.1}$: 10 m*M* Tris-HCL, pH 8.0, 0.1 m*M* EDTA.
7. RNase, DNase-free (Boehringer-Mannheim, Indianapolis, IN).
8. Phenol:chloroform:isoamyl alcohol (25 vol:24 vol:1vol; made with buffer equilibrated phenol; U. S. Biochemical, Cleveland OH). After opening the bottle for the first time, store for periods of up to several weeks at 4°C, or at –20°C for longer periods.
9. Chloroform:isoamyl alcohol (24 vol:1 vol). Store at room temperature.
10. 13% (w/v) polyethylene glycol: Dissolve 13 g PEG 8000 (Carbowax; Fisher Scientific, Pittsburgh, PA) in H_2O to a final volume of 100 mL.
11. 5*M* NaCl: Dissolve 29.2 g NaCl in H_2O to a final volume of 100 mL.

2.2. Culture of PA317 Retroviral Packaging Cells

1. PA317 cells (American Type Culture Collection, Rockville MD).
2. Dulbecco's Modified Eagle Medium, High glucose (DMEM-H), with 4500 g/L glucose and L-glutamine (Life Technologies). Supplement with 10% fetal bovine serum (Life Technologies). Penicillin and streptomycin can be added as antibiotics.

2.3. Transfection of PA317 Cells

1. 0.5*M* HEPES, pH 7.1: Dissolve 12 g HEPES (4-[2-hydroxyethyl]-1-piperazine ethanesulfonic acid, Boehringer Mannheim) in H_2O to a final volume of 100 mL. Adjust pH to 7.1 ± 0.05 with 3*N* NaOH. Sterilize by filtration using a 0.45-µm filter.
2. 2*M* NaCl: Dissolve 11.7 g NaCl in H_2O to a final volume of 100 mL. Sterilize by filtration using a 0.45-µm filter.
3. 2*M* $CaCl_2$: Dissolve 29.4 g $CaCl_2 \cdot 2H_2O$ in H_2O to a final volume of 100 mL. Sterilize by filtration using a 0.45-µm filter.
4. 150 m*M* sodium phosphate buffer, pH 7.0: Dissolve 2.01 g $Na_2HPO_4 \cdot 7H_2O$ in H_2O to a final volume of 50 mL. Dissolve 1.04 g $NaH_2PO_4 \cdot H_2O$ in H_2O to a final volume of 50 mL. Adjust the 50 mL NaH_2PO_4 solution to pH 7.0 with about 40 mL of the Na_2HPO_4 solution. Sterilize by filtration using a 0.45-µm filter.
5. 2X HBS solution: Mix 1.53 mL sterile H_2O, 200 µL 0.5*M* HEPES (pH 7.1), 250 µL 2*M* NaCl, and 20 µL 150 m*M* sodium phosphate buffer, pH 7.0. Make up fresh 2X HBS solution on the day of transfection.
6. 500 m*M* sodium butyrate: Dissolve 0.55 g (Sigma, St. Louis, MO) in H_2O to a final volume of 10 mL. Sterilize by filtration using a 0.45-µm filter. Store at –20°C.

2.4. Infection of Adherent Target Cells with Retroviral Vectors

1. 4 mg/mL Polybrene (hexadimethrine bromide, Sigma). Sterilize by filtration using a 0.45-µm filter. Store at 4°C.
2. 20 mg/mL G418 (Geneticin, Life Technologies). Dissolve in 0.1*M* HEPES (pH 7.1) to a final concentration of 20 mg of the active component of the drug per mL. Sterilize by filtration using a 0.45-µm filter.

2.5. Determination of Viral Titers

1. NIH 3T3 cells (American Type Culture Collection, Rockville MD).
2. Dulbecco's Modified Eagle Medium, High glucose (DMEM-H), with 4500 g/L glucose, L-glutamine (Life Technologies). Supplement with 10% calf serum (Life Technologies). Penicillin and streptomycin can be added as antibiotics.
3. Crystal violet staining solution: Dissolve 4 g crystal violet (Sigma) in 475 mL H_2O. Add 525 mL 95% ethanol. Filter solution if there is any insoluble material. Store at room temperature. The solution may be reused many times.

3. Methods

3.1. Purification of Plasmid DNA Using Polyethylene Glycol Precipitation

This method of plasmid DNA purification yields a minimum of 100–300 µg of DNA and is suitable for transfection into cultured cells using the calcium phosphate method.

1. Inoculate 100 mL of Terrific Broth medium containing the appropriate antibiotic with a single colony of bacteria (*see* Note 1) containing the retroviral vector on plasmid DNA. Alternatively, if the starting material is a liquid culture (e.g., a portion of an overnight culture used for a small-scale culture) inoculate medium with 0.5 mL. Grow bacteria overnight (16–20 h) at 37°C with shaking at 250 rpm using an orbital shaker.
2. Harvest the cells by centrifugation at 4000*g* (all values given are for r_{max}) for 30 min at 4°C.
3. Suspend the bacterial pellet in 6 mL of ice-cold Tris/glucose/EDTA solution. A homogeneous suspension is important, therefore vigorous pipeting (without making bubbles) may be necessary.
4. Add 12 mL of prechilled, ice-cold NaOH/SDS solution. Mix *gently* by inversion and incubate on ice for 5 min (*see* Note 2).
5. Add 9 mL of 3*M* potassium acetate (pH 4.8). Mix initially by gentle inversion and then more vigorously by shaking until a very particulate precipitate is observed. Centrifuge at 4000*g* for 20 min at 4°C.
6. Remove the supernatant to a 50 mL tube containing 20 mL of isopropanol, carefully avoiding the debris floating on the surface and pelleted at the bottom of the tube. Incubate at –20°C for several hours or overnight until a precipitate collects at the bottom of the tube.
7. Centrifuge at 4000*g* for 30 min at 4°C. Decant the supernatant and rinse the pellet with 10 mL of 70% ethanol followed by centrifugation for 5 min. Aspirate the ethanol and allow the pellet to air dry for 5–10 min at room temperature.
8. Dissolve the pellet in 1 mL of $T_{10}E_{10}$ and transfer to a 15 mL tube. Any insoluble material will be removed in subsequent steps.
9. Add 4 µL of 0.5 mg/mL RNase and incubate at 37°C for 20 min.
10. Extract the solution twice with 1 mL phenol:chloroform:isoamyl alcohol, carefully avoiding the insoluble material at the interface between the lower organic phase and upper aqueous phase.
11. Extract the aqueous phase once with an equal volume of chloroform:isoamyl alcohol to remove phenol.
12. Transfer the aqueous phase to a 15 mL tube and precipitate the DNA by adding 1/10th vol 5*M* NaCl and 2 vol 95% ethanol. Incubate on ice or at –20°C until a precipitate collects at the bottom of the tube.
13. Centrifuge at 4000*g* for 30 min at 4°C. Decant the supernatant and wash the pellet with 5 mL 70% ethanol. Centrifuge at 4000*g* for 5 min. Remove the supernatant using an aspirator.
14. Dissolve the pellet in 462 µL of $T_{10}E_{0.1}$. Transfer the solution to a microcentrifuge tube and add 88 µL of 5*M* NaCl and 550 µL of 13% polyethylene glycol. Incubate for 60 min on ice.
15. Pellet the fine white precipitate in a microcentrifuge at maximum speed (>12,000*g*) for 20 min at 4°C. Without disturbing the pellet, which will be difficult to see, aspirate the polyethylene glycol supernatant down to ~50–100 µL above the pellet. Add 1 mL 70% ethanol and mix by inversion. The translucent

pellet will turn white within a few minutes. Centrifuge for 5 min using the microcentrifuge and remove the supernatant. Rinse the pellet a second time with 1 mL 70% ethanol and then centrifuge for 5 min.

16. Following the second wash, carefully remove the supernatant with an aspirator. Remove as much ethanol as possible as excessive amounts can interfere with subsequent enzymatic manipulations of the DNA. Allow the pellet to air dry 5 min, then dissolve the purified DNA plasmid pellet in 300 µL of $T_{10}E_{0.1}$. Determine the yield by subjecting a portion of plasmid DNA to electrophoresis in an agarose gel, stain with ethidium bromide, and compare the intensity of staining to that of known amounts of a DNA standard run in parallel (*see* Note 3). Store plasmid DNA at –20°C.

3.2. Culture of PA317 Retroviral Packaging Cells

PA317 cells are an adherent, amphotropic retrovirus packaging cell line. This cell line was derived from NIH3T3 TK^- cells by cotransfection of a retrovirus packaging construct DNA with mutations designed to prevent transfer of packaging function and the herpes simplex virus thymidine kinase (TK) gene. Passage the cells 1:8 as the culture becomes 90% confluent. With time, the cells can lose the ability to package retrovirus vectors. For performing a series of gene expression studies, we recommend freezing a number of vials when the cells are initially obtained and thawing out a fresh vial every 2–3 mo.

3.3. Transfection of PA317 Cells

The introduction of retroviral vector plasmid DNA into PA317 cells results in the production of retrovirus virions with an amphotropic host range capable of gene transfer to many species, including those of mouse, rat, dog, and human origin (*see* Note 4).

1. Seed PA317 cells at 5×10^5 cells/60 mm polystyrene dish. DNA transfer will be most efficient if the cells are about 70–80% confluent at the time of transfection. Incubate overnight at 37°C in a humidified incubator with 5% CO_2.
2. Remove medium from cells and replace with 4 mL fresh medium. Make up the 2X HBS solution (*see* Note 5).
3. In polystyrene tubes, mix 20 µg DNA with H_2O to give a volume of 175 µL. To this add 25 µL 2*M* $CaCl_2$ (final concentration of 250 m*M* $CaCl_2$).
4. For each transfection, aliquot 200 µL 2X HBS solution into a polystyrene tube. To this add the 200 µL DNA/$CaCl_2$ mixture drop-wise, then bubble air through the solution 5–10 times with the pipetor. This should help prevent formation of coarse precipitates which will decrease the efficiency of the transfection.
5. Incubate the mixture 20–30 min at room temperature (*see* Note 6).
6. Pipet the solution up and down several times to mix, then add the 400 µL sample to the cells, drop-wise. Swirl the plate gently to mix. Incubate overnight at 37°C in a humidified incubator with 5% CO_2.

7. Remove the medium. Add 3 mL of fresh growth medium per plate. Add 30 µL 500 m*M* sodium butyrate per plate. Incubate 48 h at 32°C in a humidified incubator with 5% CO_2 (*see* Notes 7 and 8).
8. Swirl the plates to mix, then remove virus with a 5-cc syringe. Filter virus through an 0.45 µm syringe filter. The virus can be used in gene transfer experiments immediately or can be aliquoted and stored for years at –70°C.

3.4. Infection of Adherent Target Cells with Retroviral Vectors

1. Seed cells of interest at approx 2.5×10^5 cells per 35-mm dish. Incubate overnight at 37°C in a humidified incubator with 5% CO_2 (*see* Note 9).
2. Remove medium and add 0.5 mL retrovirus containing 8 µg/mL polybrene (1 µL of a 4 mg/mL stock).
3. Return cultures to the 37°C incubator for 2 h.
4. Replace the virus with fresh growth medium without polybrene. Incubate at 37°C in a humidified incubator with 5% CO_2.
5. After 48 h analyze cells for gene transfer and expression and/or passage cells at 1:10 to 1:20 into selective growth medium, if applicable. Replace selective medium every 2–3 d for 10–14 d until all cells have disappeared from a mock-infected control culture (*see* Note 10).

3.5. Determination of Viral Titers

Amphotropic retrovirus vectors can be titered very effectively using the murine cell line NIH 3T3. For routine maintenance of the cell line, passage the cells 1:8 as the culture becomes 90% confluent.

1. Seed NIH 3T3 cells at 2.5×10^5 per 35-mm well of a six-well plate. Incubate overnight at 37°C in a humidified incubator with 5% CO_2.
2. Remove medium and add 1 mL fresh medium containing 8 µg/mL polybrene per well.
3. Add the virus at several dilutions (typically 1 µL, 3 µL, and 10 µL of a 1:10 dilution). Incubate 2 h at 37°C in a humidified incubator with 5% CO_2.
4. Replace the virus with 1.5 mL of fresh medium (without polybrene). Return to the incubator overnight.
5. Passage cells from one 35-mm well into two 100-mm dishes (1:20) containing medium with the selective agent. For G418 selection of NIH 3T3 cells expressing the *neo* gene, use 0.5 mg active G418 per mL.
6. Replace the medium every 2–3 d until all the cells have disappeared from a mock-infected culture, usually about 10–14 d. Colonies of cells should be clearly visible in infected cultures.
7. To stain the colonies, remove the selective medium and add approx 5 mL of crystal violet staining solution. Let sit at room temperature 10–20 min, then remove the crystal violet solution. This solution can be reused several times. Very gently rinse the plates with H_2O until clear. Let dry at room temperature overnight.

8. Count the number of colonies on each plate. Use the following formula to determine the viral titer: Number of total colonies on the two plates x dilution factor = colony forming units (CFU) per mL.

4. Notes

1. Bacteria for plasmid propagation: We use the DH5α strain of *Escherichia coli* for routine plasmid preparation, although it is likely that the method described here will work for other *recA*⁻, *endA*⁻ *E. coli* strains. For plasmids requiring preparation in a *dam*⁻, *dcm*⁻ host, good yields of plasmid have been achieved using the DM1 strain (Life Technologies).
2. Prechilling of NaOH/SDS: Although the NaOH/SDS solution is stored at room temperature portions are removed and prechilled on ice prior to use. The NaOH/SDS should be cold enough such that a visible precipitate is apparent prior to adding to the bacteria. The NaOH/SDS will cause lysis of the bacteria. The prechilling step is done to prevent alkaline denaturation of nicked circular DNA. Gentle mixing by inversion is done to prevent shearing of high mol wt bacterial chromosomal DNA.
3. Factors affecting the yield of DNA: The problem of a low yield can be multifactorial: (a) Many of the readily available retrovirus vectors are cloned into pBR322, which replicates at a relatively low copy number per *E. coli* cell. For performing a series of experiments, it may be worthwhile to subclone the retroviral vector sequences into plasmids (such as those derived from pUC) capable of replicating to higher copy number. (b) Growth of plasmids using selection for the ampicillin resistance gene can result in the outgrowth of bacteria lacking plasmid, as β-lactamase secreted into the medium can inactivate ampicillin which is already relatively unstable. We have found that use of the antibiotic carbenicillin at 100 μg/mL can help increase yields. (c) Protein can trap DNA at the interface of the organic layer during phenol/chloroform extractions. We have found that yields can be increased substantially (30–50%) by back-extracting plasmid DNA from the pooled phenol/chloroform phases with 1 mL $T_{10}E_{0.1}$.
4. Safety while working with amphotropic retroviruses: Several precautions should be observed while working with retroviral vectors in the laboratory. Foremost, always wear gloves when handling viruses or virus-infected cultures. Dispose of the gloves in an autoclave bag before leaving the tissue culture work area. Plasticware used in contact with retroviruses should be autoclaved prior to disposal and medium containing virus should be inactivated by treating with ≥15% bleach for 15 min at room temperature prior to disposal. Additional containment procedures may be required for amphotropic retroviral vectors expressing certain genes, including known or suspected oncogenes. We advise that an institutional or facility safety officer be consulted prior to initiating potentially biohazardous experiments.
5. Preparation of reagents used for transfections: Calcium phosphate transfections are very sensitive to deviations from optimal conditions. In particular, the HEPES

buffer should be pH 7.1 ± 0.05. Also, the plasmid DNAs should be of high purity which can be attained by using the polyethylene glycol method described above, by banding plasmids two or more times in CsCl, or by using commercially-available plasmid purification kits, e.g., the Qiagen Plasmid Kit (Qiagen, Chatsworth, CA).

6. Appearance of calcium/phosphate-DNA co-precipitate: After the room temperature incubation, a fine white precipitate should only be faintly visible. Very course precipitates invariably result in poor transfection efficiencies. A good way to monitor transfection efficiencies is to use a retrovirus vector [e.g., BAG *(42)* or LNPOZ *(21)*] containing an *E. coli lacZ* reporter gene and to stain the cells with X-gal *(41)* when virus is harvested. Using PA317 cells, a minimum of 20–30% the cells in the culture should stain positive. If this efficiency is not obtained, and it is important to generate as high a viral titer as possible, it may be necessary to purify the DNA further, perhaps by performing an additional polyethylene glycol precipitation. Alternatively, higher titers may be obtained if PA317 cells of an earlier passage are used. Rarely, certain lots of fetal bovine serum have been known to adversely affect virus production.
7. Sodium butyrate: Sodium butyrate treatment appears to increase the steady-state levels of RNA transcribed from the viral LTR resulting in increased vector production, and therefore, higher titer supernatants *(43)*. The extent of this effect varies depending on the vector but generally results in a 5- to 25-fold increase in vector production. The titers that one can expect using this method should be in the range of 2×10^5 to 5×10^6 infectious units/mL.
8. Temperature during retroviral vector production: We have confirmed that vector production can be increased by incubating packaging cells at 32°C and collecting virus 48 h after transfection, as suggested by a previous report *(44)*. If a 32°C incubator is not available, cells are incubated at 37°C and virus is harvested 24 h later.
9. Cell density at time of infection: For optimal retrovirus-mediated gene transfer efficiency, cultures should be healthy and proliferating at the time of exposure to virus. Considerations of the optimal seeding density prior to infection include the rate of cell division and the size of the cells. For many cell types, a seeding density somewhere in the range of 5×10^4 to 3×10^5 cells per 35-mm well is usually optimal.
10. Concentration of selective agent: The optimal concentration of selective agent varies somewhat between cell lines, so the minimum level sufficient to kill uninfected cells in 7–10 d should be determined. For G418, the appropriate concentration usually falls between 0.1 mg/mL and 1 mg/mL active G418.

References

1. Miller, D. G., Adam, M. A., and Miller, A. D. (1990) Gene transfer by retrovirus vectors occurs only in cells that are actively replicating at the time of infection. *Mol. Cell. Biol.* **10,** 4239–4242.
2. Miller, A. D., Miller, D. G., Garcia, J. V., and Lynch, C. M. (1993) Use of retroviral vectors for gene transfer and expression, in *Methods in Enzymology*, Vol. 217 (Wu, R., ed.), Academic, New York, pp. 581–599.

3. Miller, A. D. and Buttimore, C. (1986) Redesign of retrovirus packaging cell lines to avoid recombination leading to helper virus production. *Mol. Cell. Biol.* **6,** 2895–2902.
4. Danos, O. and Mulligan, R. C. (1988) Safe and efficient generation of recombinant retroviruses with amphotropic and ecotropic host ranges. *Proc. Natl. Acad. Sci. USA* **85,** 6460–6464.
5. Markowitz, D., Goff, S., and Bank, A. (1988) Construction and use of a safe and efficient amphotropic packaging cell line. *Virology* **167,** 400–406.
6. Miller, A. D. and Rosman, G. J. (1989) Improved retroviral vectors for gene transfer and expression. *BioTechniques* **7,** 980–990.
7. Pear, W. S., Nolan, G. P., Scott, M. L., and Baltimore, D. (1993) Production of high-titer helper-free retroviruses by transient transfection. *Proc. Natl. Acad. Sci. USA* **90,** 8392–8396.
8. Markowitz, D., Goff, S., and Bank, A. (1988) A safe packaging cell line for gene transfer: Separating viral genes on two different plasmids. *J. Virol.* **62,** 1120–1124.
9. Adams, R. M., Soriano, H. E., Wang, M., Darlington, G., Steffen, D., and Ledley, F. D. (1992) Transduction of primary human hepatocytes with amphotropic and xenotropic retroviral vectors. *Proc. Natl. Acad. Sci. USA* **89,** 8981–8985.
10. Miller, A. D., Garcia, J. V., von Suhr, N., Lynch, C. M., Wilson, C., and Eiden, M. V. (1991) Construction and properties of retrovirus packaging cells based on gibbon ape leukemia virus. *J. Virol.* **65,** 2220–2224.
11. Bayle, J.-Y., Johnson, L. G., George, J. A. S., Boucher, R. C., and Olsen, J. C. (1993) High efficiency gene transfer to primary monkey airway epithelial cells with retrovirus vectors using the GALV receptor. *Human Gene Therapy* **4,** 161–170.
12. Yee, J. K., Miyanohara, A., LaPorte, P., Bouic, K., Burns, J. C., and Friedmann, T. (1994) A general method for the generation of high–titer, pantropic retroviral vectors: Highly efficient infection of primary hepatocytes. *Proc. Natl. Acad. Sci. USA* **91,** 9564–9568.
13. Kasahara, N., Dozy, A. M., and Kan, Y. W. (1994) Tissue-specific targeting of retroviral vectors through ligand-receptor interactions. *Science* **266,** 1373–1376.
14. Bates, P., Young, J. A., and Varmus, H. E. (1993) A receptor for subgroup A Rous sarcoma virus is related to the low density lipoprotein receptor. *Cell* **74,** 1043–1051.
15. Morgenstern, J. P. and Land, H. (1990) Advanced mammalian gene transfer: High titre retroviral vectors with multiple drug selection markers and a complementary helper-free packaging cell line. *Nucleic Acids Res.* **18,** 3587–3596.
16. Stockshlaeder, M. A., Storb, R., Osborne, W. R., and Miller, A. D. (1991) L-histidinol provides effective selection of retrovirus-vector-transduced keratinocytes without impairing their proliferative potential. *Human Gene Therapy* **2,** 33–39.
17. Hawley, R. G., Lieu, F. H. L., Fong, A. Z. C., and Hawley, T. S. (1994) Versatile retroviral vectors for potential use in gene therapy. *Gene Therapy* **1,** 136–138.
18. Emerman, M. and Temin, H. M. (1984) Genes with promoters in retrovirus vectors can be independently suppressed by an epigenetic mechanism. *Cell* **39,** 459–467.
19. Emerman, M. and Temin, H. M. (1986) Comparison of promoter suppression in avian and murine retrovirus vectors. *Nucleic Acids Res.* **14,** 9381–9396.

20. Olsen, J. C., Johnson, L. G., Wong-Sun, M. L., Moore, K. L., Swanstrom, R., and Boucher, R. C. (1993) Efficient retrovirus-mediated gene transfer with long term expression in cystic fibrosis airway epithelial cells. *Nucleic Acids Res.* **21,** 663–669.
21. Adam, M. A., Ramesh, N., Miller, A. D., and Osborne, W. R. (1991) Internal initiation of translation in retroviral vectors carrying picornavirus 5' nontranslated regions. *J. Virol.* **65,** 4985–4990.
22. Morgan, R. A., Coulture, L., Elroy-Stein, O., Ragheb, J., Moss, B., and Anderson, W. F. (1992) Retroviral vectors containing putative internal ribosome entry sites: Development of a polycistronic gene transfer system and applications to human gene therapy. *Nucleic Acids Res.* **20,** 1293–1299.
23. Aran, J. M., Gottesman, M. M., and Pastan, I. (1994) Drug-related coexpression of human glucocerebrosidase and P-glycoprotein using a bicistronic vector. *Proc. Natl. Acad. Sci. USA* **91,** 3176–3180.
24. Dranoff, G., Jaffee, E. M., Lazenby, A., Golumbeck, P., Levitsky, H., Brose, K., Jackson, V., Hamada, H., Pardoll, D. M., and Mulligan, R. C. (1993) Vaccination with irradiated tumor cells engineered to secrete murine GMCSF stimulates potent, specific and long lasting antitumor immunity. *Proc. Natl. Acad. Sci. USA* **90,** 3539–3543.
25. Zitvogel, L., Tahara, H., Cai, Q., Storkus, W. J., Muller, G., Wolf, S. F., Gately, M., Robbins, P. D., and Lotze, M. T. (1994) Construction and characterization of retroviral vectors expressing biologically active human interleukin-12. *Hum. Gene Ther.* **5,** 1493–1506.
26. Yu, S.-F., von Rüden, T., Kantoff, P. W., Garber, C., Seiberg, M., Rüther, U., Anderson, W. F., Wagner, E. F., and Gilboa, E. (1986) Self-inactivating retroviral vectors designed for transfer of whole genes intc mammalian cells. *Proc. Nat. Acad. Sci. USA* **83,** 3194–3198.
27. Guild, B. C., Finer, M. H., Housman, D. E., and Mulligan, R. C. (1988) Development of retrovirus vectors useful for expressing genes in cultured murine embryonal cells and hematopoietic cells in vivo. *J. Virol.* **62,** 3795–3801.
28. Faustinella, F., Kwon, H., Serrano, F., Belmont, J. W., Caskey, C. T., and Aguilar-Cordova, E. (1994) A new family of murine retroviral vectors with extended multiple cloning sites for gene insertion. *Hum. Gene Ther.* **5,** 307–312.
29. Hantzopoulos, P. A., Sullenger, B. A., Ungers, G., and Gilboa, E. (1989) Improved gene expression upon transfer of the adenosine deaminase minigene outside the transcriptional unit of a retroviral vector. *Proc. Natl. Acad. Sci. USA* **86,** 3519–3523.
30. Sullenger, B. A., Lee, T. C., Smith, C. A., Ungers, G. E., and Gilboa, E. (1990) Expression of chimeric tRNA-driven antisense transcripts renders NIH 3T3 cells highly resistant to Moloney murine leukemia virus replication. *Mol. Cell. Biol.* **10,** 6512–6523.
31. Sullenger, B. A., and Cech, T. R. (1993) Tethering ribozymes to a retroviral packaging signal for destruction of viral RNA. *Science* **262,** 1566–1569.
32. Chuah, M. K. L., Vandendriessche, T., Chang, H. K., Ensoli, B., and Morgan, R. A. (1994) Inhibition of human immunodeficiency virus type-1 by retroviral vectors expressing antisense-TAR. *Hum. Gene Ther.* **5,** 1467–1475.

33. Lee, S. W., Gallardo, H. F., Gilboa, E., and Smith, C. (1994) Inhibition of human immunodeficiency virus type 1 in human T cells by a potent rev response element decoy consisting of the 13-nucleotide minimal rev-binding domain. *J. Virol.* **68,** 8254–8264.
34. Wilke, M., Bout, B., Verbeek, B., Kappers, W., Verkerk, T., Valerio, D., and Scholte, B. (1992) Amphotropic viruses with a hybrid long terminal repeat as a tool for gene therapy of cystic fibrosis. *Biochem. Biophys. Res. Comm.* **187,** 187–194.
35. van den Wollenberg, D. J., Hoeben, R. C., van Ormondt, H., and van der Eb, A. J. (1994) Insertion of the human cytomegalovirus enhancer into a myeloproliferative sarcoma virus long terminal repeat creates a high-expression retroviral vector. *Gene* **144,** 237–241.
36. King, W., Patel, M. D., Lobel, L. I., Goff, S. P., and Nguyen-Huu, M. C. (1985) Insertion mutagenesis of embryonal carcinoma cells by retroviruses. *Science* **228,** 554–558.
37. Hubbard, S. C., Walls, L., Ruley, H. E., and Muchmore, E. A. (1994) Generation of chinese hamster ovary cell glycosylation mutants by retroviral insertional mutagenesis. Integration into a discrete locus generates mutants expressing high levels of N-glycolylneuraminic acid. *J. Biol. Chem.* **269,** 3717–3724.
38. von Melchner, H. and Ruley, H. E. (1989) Identification of cellular promoters by using a retrovirus promoter trap. *J. Virol.* **63,** 3227–3233.
39. Chen, Z., Friedrich, G. A., and Soriano, P. (1994) Transcriptional enhancer factor 1 disruption by a retroviral gene trap leads to heart defects and embryonic lethality in mice. *Genes Dev.* **8,** 2293–2301.
40. Chang, W., Hubbard, S. C., Friedel, C., and Ruley, H. E. (1993) Enrichment of insertional mutants following retrovirus gene trap selection. *Virology* **193,** 737–747.
41. Fields-Berry, S. C., Halliday, A. L., and Cepko, C. L. (1992) A recombinant retrovirus encoding alkaline phosphatase confirms clonal boundary assignment in lineage analysis of murine retina. *Proc. Natl. Acad. Sci. USA* **89,** 693–697.
42. Price, J., Turner, D., and Cepko, C. (1987) Lineage analysis in the vertebrate nervous system by retrovirus-mediated gene transfer. *Proc. Natl. Acad. Sci. USA* **84,** 156–160.
43. Olsen, J. C., and Sechelski, J. (1995) Use of sodium butyrate to enhance production of retroviral vectors expressing CFTR cDNA. *Hum. Gene Ther.* **6,** 1195–1202.
44. Kotani, H., Newton III, P. B., Zhang, S., Chiang, Y. L., Otto, E., Weaver, L., Blaese, R. M., Anderson, W. F., and McGarrity, G. J. (1994) Improved methods of retroviral vector transduction and production for gene therapy. *Hum. Gene Ther.* **5,** 19–28.

18

Use of Defective Herpes-Derived Plasmid Vectors

Filip Lim, Dean Hartley, Philip Starr, Song Song, Phung Lang, Linda Yu, Yaming Wang, and Alfred I. Geller

1. Introduction

Genetic intervention is becoming increasingly useful in elucidating the molecular basis of various biological processes, including those of the brain. Many genes have now been isolated which encode key regulatory molecules such as signal transduction components or transcription factors, but relatively little is known about their effects on neuronal physiology or higher order brain function. Gene transfer experiments offer the potential to study the effects of highly specific alterations in gene products in their normal neuronal environment. In addition, treatments of nervous disorders may also be devised using the same technology. Gene therapy protocols may be targeted towards specific diseases (e.g., the treatment of Parkinson's disease by expression of tyrosine hydroxylase in the striatum), or directed towards the treatment of general neuronal damage which may occur in many diseased conditions (e.g., prevention of cell death by expression of neurotrophins or neurotrophin receptors). Further knowledge of the molecular mechanisms underlying neuronal physiology may lead to insight into the nature of the defects in diseased conditions. The ability to manipulate normal physiological pathways specifically may also result in new treatments. For example, the regulation of neurotransmission is altered in many disorders such as Alzheimer's disease, Parkinson's disease, and epilepsy, and may be responsive to genetic intervention such as expression of signal transduction components controlling long-term changes in neurotransmitter release.

Two properties of neurons make them refractory to most gene transfer techniques such as transfection or retroviral transduction: first, they are highly sensitive to membrane or environmental perturbations, and second, neurons are

From: *Methods in Molecular Biology, vol. 62: Recombinant Gene Expression Protocols*
Edited by: R. Tuan Humana Press Inc., Totowa, NJ

15. Crystal violet stain. Prepare a solution containing 0.5% crystal violet and 0.2% sodium acetate, adjusting the pH to 3.6 with acetic acid before making it up to final volume. Filter through 1-mm Whatman paper. The stain can be reused several times.
16. Phosphate-buffered saline (PBS). Prepare a 10X PBS stock by mixing 1 g KH_2PO_4, 10.8 g $Na_2HPO_4 \cdot 7\ H_2O$, 1 g KCl, 40 g NaCl, and water to 500 mL final volume. The pH should be approx 7.0. Filter-sterilize. Dilute 10-fold before use.
17. Solutions of 60% sucrose, 30% sucrose, and 10% sucrose in PBS. These solutions can be sterilized by filtration or by autoclaving for no more than 20 min at 120°C. (Heating for longer periods leads to caramelization.)
18. Poly-D-lysine (PDL). Prepare 1 mg/mL PDL in water, filter sterilize, and store at –20°C. This is a 50X stock. Dilute to 20 μg/mL in water immediately prior to use.
19. Fe solution. Mix 200 mL PBS, 0.332 g potassium ferricyanide, 0.424 g potassium ferrocyanide, 0.2 mL 1*M* $MgCl_2$, 0.2 mL 20% Nonidet P40, and 0.2 mL 10% sodium deoxycholate. Store at 4°C in a darkened container since this reagent is light sensitive.
20. 5-bromo-4-chloro-3-indolyl-β-D-galactoside (X-gal). A 50 mg/mL stock in dimethyl sulfoxide can be stored in aliquots at –20°C.
21. 4% paraformaldehyde pH 7.0. Add 20 g paraformaldehyde to 300 mL H_2O and heat to 55–60°C. Slowly add 1*M* NaOH dropwise over about 10 min until the solution becomes clear. Cool the solution to room temperature. Use pH paper to check that the pH is 7.0–7.5 (add more NaOH if necessary). Add 100 mL of 0.5*M* sodium phosphate buffer pH 7.0 and then water to a final volume of 500 ml. The final pH should be 7.0–7.5. Store at 4°C.

3. Methods

3.1. Packaging

3.1.1. Cell Preparation

The 2-2 VERO cell line is grown in DMEM + 10% FBS at 37°C in the presence of 5% CO_2. Several vials of the cell line should be frozen at an early passage and a fresh vial thawed once every 4–6 mo (up to 50 passages). The cells are maintained under G418 selection (0.5 mg/mL), and passaged once in normal (nonselective) medium before use in experiments. Figure 2 shows a convenient schedule for plating cells. Two days before transfection, trypsinize the stock cells and plate at 3×10^5 per 60-mm dish in 5 mL DMEM + 10% FBS. The growth state of the stock cells is important for ensuring both good transfection and infectivity resulting in efficient packaging so trypsinization of the cells two days in succession should be avoided.

3.1.2. Transfection

1. Dilute 2 μg plasmid vector DNA with 100 μL OPTIMEM in a microcentrifuge tube. Mix 12 μL lipofectamine with 100 μL OPTIMEM in another tube and then

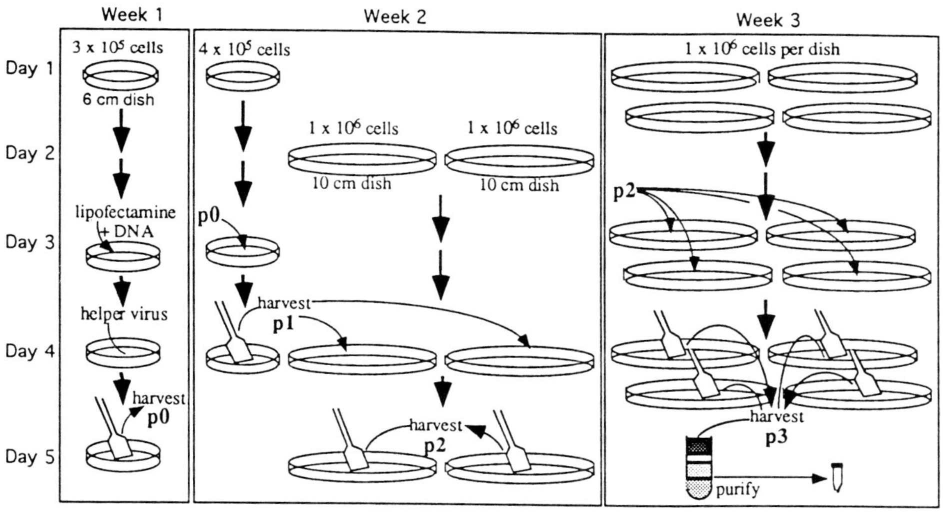

Fig. 2. Overview of the packaging procedure. During the first week, the packaging cells are transfected with vector DNA using lipofectamine and subsequently infected with helper virus to generate an initial virus stock (p0). During the second and third weeks three consecutive passages of the p0 supernatant result in an increase in the total amount of virus as well as the vector:helper ratio. In the final step, the p3 supernatant is purified by sucrose gradient centrifugation.

add to the 100 μL DNA in the first tube. Leave the solution for 20 min at room temperature to allow liposomes to form.

2. Remove the medium from the plates and wash each plate once with 2 mL OPTIMEM. Remove the wash and replace with 2 mL fresh OPTIMEM.
3. Add 800 μL OPTIMEM to the DNA/lipofectamine mix from step 1. Add this 1 mL to the cells dropwise to distribute the solution evenly over the entire plate. Incubate at 37°C for 5 h.
4. Wash the cells thoroughly to remove excess lipofectamine which will otherwise inactivate virus particles. Prewarm PBS and DMEM + 10% FBS for at least 20 min at 37°C. Remove the medium from the plates and wash three times with 2 mL of the prewarmed PBS. Remove the final wash and add 5 mL of prewarmed DMEM + 10% FBS. Incubate the cells overnight at 37°C.

3.1.3. Infection

Allow the cells to recover for at least 20 h after the transfection (from the time of the last wash). Prewarm DMEM + 5% FBS for at least 20 min at 37°C. Remove the medium from the plates and add 5 mL of prewarmed DMEM + 5% FBS. Add 6×10^5 PFU helper virus and incubate overnight at 37°C.

3.1.4. Harvesting HSV-1 Virus (see Notes 1–5)

1. Check the cells one day after infection for cytopathic effects: They should round up but remain stuck to the plate. Harvest the cells by scraping them off the plate and into the medium with a plastic cell scraper. Place 5 mL of the cell suspension into a 15 mL screw-top centrifuge tube.
2. Freeze thaw three times using a dry ice-ethanol bath and a 37°C water bath.
3. Sonicate in a cup sonicator cycling between on and off such that the sum of the sonication bursts is 2 min. The length of the on/off cycle can vary from 5 s to at least 30 s. The cup in the sonicator should be cooled with circulating (e.g., using a submersible pump) water from an ice water bath.
4. Remove the cellular debris by centrifugation at 1500*g* for 5 min.
5. Transfer the supernatant to a new tube - this is designated p0 (passage 0). This virus stock may be frozen in a dry ice-ethanol bath, and stored at –70°C before proceeding further. Titer the virus stock for the vector (Method 3.3.) and continue with the packaging protocol only if the titer is $\geq 1 \times 10^4$ infectious vector units (ivu)/mL.

3.1.5. Amplification

1. First amplification (p1): Plate the cells at 4×10^5 per 60-mm dish in 5 mL DMEM + 10% FBS. Two days later, replace the medium with 4 mL DMEM + 5% FBS and add 4 mL of the p0 preparation. The next day, check the cells and harvest the virus (*see* Section 3.1.4.) when the cells have rounded up. To facilitate assessment of cytopathic effects, include an uninfected (mock) plate as a negative control for each passage.
2. Second amplification (p2): Plate the cells at 1×10^6 per 100-mm dish (2 dishes per sample) in 10 mL DMEM + 10% FBS. Two days later, replace the medium in each plate with 6 mL DMEM + 5% FBS and add 4 mL of the p1 preparation. On the next day, check the cells and harvest the virus (*see* Section 3.1.4.) when the cells have rounded up.
3. Third amplification (p3): Plate the cells at 1×10^6 per 100-mm dish (4 dishes per sample) in 10 mL DMEM + 10% FBS. Two days later, replace the medium in each plate with 6 mL DMEM + 5% FBS and add 4 mL of the p2 preparation (from step 10). On the next day, check the cells and harvest the virus (*see* Section 3.1.4.) when the cells have rounded up.

3.1.6. Purification of Virus

1. Prepare sucrose step gradients in 40 mL Beckman SW28 ultracentrifuge tubes or their equivalent. Layer the following sucrose solutions into the tube in the following order: 7 mL 60% sucrose; 6 mL 30% sucrose; and 3 mL 10% sucrose.
2. Gently load 20 mL of p3 supernatant onto the gradient.
3. Centrifuge for 1 h at 125,000*g* (e.g., Beckman SW28 rotor). The virus bands at the interface between the 30% and 60% sucrose layers, and can be best visualized against a dark background using side illumination of the tube with a lamp.

4. Carefully remove the top layers of sucrose and then pipet off the virus which should be contained in about 2 mL.
5. Dilute the purified virus with PBS (and mix) to fill a 20 mL Beckman SW28 tube (or equivalent) and centrifuge at 125,000*g* for 1 h.
6. Carefully pipet off the supernatant, resuspend the viral pellet in 200 μL 10% sucrose in PBS, and dispense into 30 μL aliquots. Store at –70°C.

3.2. Plaque Assays

1. Plate 2-2 cells in 60-mm tissue culture plates (1×10^6 per plate) in 5 mL DMEM + 10% FBS and grow for 1 d. Prepare one plate for each dilution of virus stock to be titered (*see* step 2).
2. Infect the cells by removing 3 mL of medium from the plate and adding 100 μL virus dilution to the plate. Allow the virus to adsorb for at least 90 min, but less than 6 h.
3. Remove the medium and add 3 mL plaque agar (*see* Section 3.). Leave to solidify before returning to the 37°C incubator. The following day, add 2 mL DMEM + 5% FBS on top of the agarose. Repeat on the next day.
4. Three days after infection, check the plates with a microscope for plaques. If they are too small to reveal clear holes in the cell monolayer, the plates may be left for another day before staining.
5. Remove the medium and fix the cells with 3 mL 5% methanol, 10% acetic acid for at least 15 min. Remove the fixing solution and add 1 mL crystal violet stain. Remove the stain (which can be reused), wash with 1 mL water, and air dry. Plaques should be visible to the naked eye as holes.
6. Count the number of plaques and using the dilution factor of the virus stock tested, calculate the virus titer in PFU/mL.

3.3. Vector Assays

1. The surfaces of 24-well plates should be first coated with PDL to enable PC12 cells to adhere properly. Use 0.5 mL 20 μg/mL PDL for at least 5 min at room temperature. Aspirate the PDL completely before plating the cells.
2. Plate the PC12 cells at 3 x 10^5/well in DMEM (or RPMI 1640) + 5% FBS and 10% horse serum. Incubate overnight at 37°C.
3. The following day add the test virus (up to 100 μL of unpurified virus preparation). Always include an uninfected culture in a well for a negative control to detect the background level of staining.
4. One day after infection, remove the medium and fix the cells with either 4% paraformaldehyde, pH 7.0 for 20–60 min at room temperature.
5. Wash the fixed cells 3X with 0.5 mL PBS.
6. Dispense 0.5 mL Fe solution for each well of cells to be stained and warm in a 37°C water bath. Dilute the X-gal stock 50-fold into the warm Fe solution to a final concentration of 1 mg/mL. If the solutions are not warm, a precipitate will form.
7. Remove the final wash from the cells and add the X-gal/Fe solution (0.5 mL/well).

8. Incubate the cells at 37°C. Depending on the cell type and the promoter in the vector, the staining will develop in a few h to overnight.
9. To stop the staining reaction, remove the X-gal solution and wash the wells 2–3X with PBS. The plates can be stored in PBS at room temperature for several weeks.

3.4. Preparing Helper Stocks

1. Seed 60 mm tissue culture plates with 1×10^6 2-2 cells per plate in 5 mL DMEM + 10% FBS.
2. On the next day, remove 3 mL of the medium and add an amount of virus that will give well-separated plaques (about 50 PFU/plate). Incubate at 37°C for 1.5–6 h.
3. Remove the medium and add 3 mL plaque agar (*see* Section 3.). Leave to solidify before returning to the 37°C incubator. The following day, add 2 mL DMEM + 5% FBS on top of the agarose. Repeat on the next day.
4. Two to four days after infection, when plaques are visible, pick a single plaque by using a Pasteur pipet to stab through the agar and into the cells, and then transfer the plug plus the cells to a 60-mm plate of cells plated as in step 1. It is advisable to pick several different plaques to ensure that at least one isolate has the appropriate phenotype for the virus.
5. After one or two days (depending on the amount of virus obtained from the plaque), the cells should show cytopathic effects (CPE). The cells should round up but remain stuck to the plate. Harvest the cells and prepare virus (*see* Section 3.1.4.). These plaque-purified stocks should be stored at –70°C and used only as seed stocks for innoculation for large-scale virus preparations (steps 6–8).
6. Seed four 100 mm plates with 1×10^6 2-2 cells/plate in 10 mL DMEM + 10% FBS and grow for 2 d.
7. Change the medium to 10 mL DMEM + 5% FBS and add 50 μL of seed stock virus. The cells should be infected at a m.o.i. of 0.1 or lower. At high m.o.i., spontaneously arising defective viruses are able to grow and interfere with virus replication, resulting in lower pfu/mL. The 2-2 cells should reach a density of $0.5–2 \times 10^7$ cells per plate by the second day. Seed stocks prepared as outlined in steps 1–5 usually have titers of $0.5–1 \times 10^7$/mL.
8. Harvest the cells when they have rounded up but still remain stuck to the plates. Prepare virus (*see* Section 3.1.4) and store in aliquots at –70°C (*see* Section 4.). on how to determine the optimum amount of helper virus stock to use for packaging.

4. Notes

1. As a control for the transfection procedure, include a tube with pHSVlac and one day later assay expression of β-galactosidase by staining with X-gal (*see* Section 3.3.). To monitor the packaging efficiency during the subsequent steps, include a second tube with pHSVlac and use the X-gal assay (*see* Section 3.3.) to monitor vector titer at each step.
2. We have found the packaging procedure to have an optimum efficiency when the m.o.i. of the helper virus is between 0.2–0.6. It is advised that a large amount of helper virus be kept aside and stored in aliquots specifically for packaging. An

initial experiment to titrate for the optimum amount of this helper stock can then be set up by transfecting several identical plates with pHSVlac and varying the volume of helper virus (to cover a range from 10^5–10^6 PFU) added in the infection step. The cells are harvested on the next day, virus stocks are prepared as described in Section 3.1.4. and assayed for vector by X-gal staining (*see* Section 3.3.). The optimum volume of helper stock which results in the highest number of blue-staining cells can then be recorded and used for subsequent packaging.

3. High transfection efficiency is important to obtain good vector titers. In our hands, transfection with lipofectamine results in higher packaging efficiencies than other reagents such as lipofectin or calcium phosphate.
4. Different helper viruses and the appropriate complementary cell lines may be used. We have also used the IE3 deletion mutant d120 *(11)* grown on E5 cells *(12)* and the IE3 deletion mutant D30EBA *(13)* grown on RR1 cells *(14)*, but packaging with either of these two systems is less efficient than using the 5dl1.2/2-2 system described herein. The RR1 cells have a faster growth rate than 2-2 or E5 cells, and should be plated at approx 0.3X the density.
5. Virus particles are thermolabile and should be kept on ice during manipulations. Approximately 50% loss in titer can be expected after storage of unpurified virus at –70°C for one year. Concentrated virus in sucrose solutions has higher stability.

References

1. Federoff, H. J., Geschwind, M. D., Geller, A. I., and Kessler, J. A. (1992) Expression of nerve growth factor in vivo from a defective herpes simplex virus 1 vector prevents effects of axotomy on sympathetic ganglia. *Proc. Natl. Acad. Sci. USA* **89,** 1636–1640.
2. Geschwind, M. D., Kessler, J. A., Geller, A. I., and Federoff, H. J. (1994) Transfer of the nerve growth factor gene into cell lines and cultured neurons using a defective herpes simplex virus vector. Transfer of the NGF gene into cells by a HSV-1 vector. *Mol. Brain Res.* **24,** 327–335.
3. Ho, D. Y., Mocarski, E. S., and Sapolsky, R. M. (1993) Altering central nervous system physiology with a defective herpes simplex virus vector expressing the glucose transporter gene. *Proc. Natl. Acad. Sci. USA* **90,** 3655–3659.
4. Battleman, D. S., Geller, A. I., and Chao, M. V. (1993) HSV-1 vector-mediated gene transfer of the human nerve growth factor receptor p75hNGFR defines high-affinity NGF binding. *J. Neurosci.* **13,** 941–951.
5. Bergold, P. J., Casaccia-Bonnefil, P., Zeng, X. L., and Federoff, H. J. (1993) Transsynaptic neuronal loss induced in hippocampal slice cultures by a herpes simplex virus vector expressing the GluR6 subunit of the kainate receptor. *Proc. Natl. Acad. Sci. USA* **90,** 6165–6169.
6. Geller, A. I., During, M. J., Haycock, J. W., Freese, A., and Neve, R. (1993) Long-term increases in neurotransmitter release from neuronal cells expressing a constitutively active adenylate cyclase from a herpes simplex virus type 1 vector. *Proc. Natl. Acad. Sci. USA* **90,** 7603–7607.

7. Xu, H., Federoff, H., Maragos, J., Parada, L. F., and Kessler, J. A. (1994) Viral transduction of trkA into cultured nodose and spinal motor neurons conveys NGF responsiveness. *Dev. Biol.* **163,** 152–161.
8. During, M. J., Naegele, J. R., O'Malley, K. L., and Geller, A. I. (1994) Long-term behavioral recovery in Parkinsonian rats by an HSV vector expressing tyrosine hydroxylase. *Science* **266,** 1399–1403.
9. McCarthy, A. M., McMahan, L., and Schaffer, P. A. (1989) Herpes simplex virus type 1 ICP27 deletion mutants exhibit altered patterns of transcription and are DNA deficient. *J. Virol.* **63,** 18–27.
10. Smith, I. L., Hardwicke, M. A., and Sandri-Goldin, R. M. (1992) Evidence that the herpes simplex virus immediate early protein ICP27 acts post-transcriptionally during infection to regulate gene expression. *Virology* **186,** 74–86.
11. DeLuca, N. A., McCarthy, A. M., and Schaffer, P. A. (1985) Isolation and characterization of deletion mutants of herpes simplex virus type 1 in the gene encoding immediate-early regulatory protein ICP4. *J. Virol.* **56,** 558–570.
12. DeLuca, N. A. and Schaffer, P. A. (1987) Activities of herpes simplex virus type 1 (HSV-1) ICP4 genes specifying nonsense peptides. *Nucleic Acids Res.* **15,** 4491–511.
13. Paterson, T. and Everett, R. D. (1990) A prominent serine-rich region in Vmw175, the major transcriptional regulator protein of herpes simplex virus type 1, is not essential for virus growth in tissue culture. *J. Gen. Virol.* **71,** 1775–1783.
14. Johnson, P. A., Miyanohara, A., Levine, F., Cahill, T., and Friedmann, T. (1992) Cytotoxicity of a replication-defective mutant of herpes simplex virus type 1. *J. Virol.* **66,** 2952–2965.

V

Expression in Nonmammalian Higher Eukaryotes

19

Use of Baculovirus Expression Vectors

David R. O'Reilly

1. Introduction

The baculovirus expression vector system is a helper-independent system that has found extensive use in the past decade for the expression of heterologous genes. Its popularity stems from a combination of high levels of expression with the ability to carry out most eukaryotic posttranslational modifications in an authentic manner. As a consequence, the system has an excellent track record for the production of functional eukaryotic gene products. (*See* ref. *1* for a detailed recent review of the expression vector system and the basic biology that underlies it.) Baculoviruses are a large family of DNA viruses that have only been isolated from arthropods, with the majority found in insects. *Autographa californica* nuclear polyhedrosis virus (AcMNPV), the prototype baculovirus strain and the virus most commonly used in expression vector systems, infects a number of lepidopteran species. Like most baculoviruses, it has two distinct morphological types. The budded virus form comprises single virus particles surrounded by a lipid envelope, whereas the occluded virus form comprises multiple virus particles embedded in a matrix composed almost exclusively of a protein called polyhedrin. The bodies containing the occluded virus particles are known as occlusion bodies or polyhedral inclusion bodies. Up to 100 such occlusion bodies may form in the nucleus of an infected cell.

These two morphotypes are formed by a unique biphasic mode of replication that, historically, was the basis of the development of the expression system. Following infection of a cell, a subset of viral genes, known as early genes, is expressed. These encode (among others) proteins required for viral DNA replication. Concomitant with the onset of DNA replication, a second subset of genes, the late genes, is expressed. Late genes encode the viral structural proteins. The structural proteins combine with the replicated viral DNA molecules

From: *Methods in Molecular Biology, vol. 62: Recombinant Gene Expression Protocols*
Edited by: R. Tuan Humana Press Inc., Totowa, NJ

to form the progeny virus particles. Progeny virus first buds from the infected cell (hence the term budded virus) into the extracellular fluid. This budded virus is infectious for individual cells and is responsible for cell-to-cell spread of the virus. However, budded virus production is transient. A third subset of genes, the very late genes, is then expressed. Very late genes include the polyhedrin gene, which is expressed at very high levels as the polyhedrin protein worms the matrix of the occlusion bodies. Progeny virus particles are no longer released from the cell, but rather become embedded within occlusion bodies forming in the nucleus. These occlusion bodies function in transmission of the virus from one insect to another. They are not required for infection of individual cells, either in culture or in the insect.

The earliest baculovirus expression vectors were based on the replacement of polyhedrin coding sequences with the heterologous gene of interest *(2,3)*. Such recombinant viruses are not able to form occlusion bodies because they lack the polyhedrin gene. However, they do form budded virus and can be propagated in cell culture. High levels of foreign gene expression are achieved because the gene is under the control of the powerful polyhedrin promoter. Many variations of this basic approach have been developed in the past decade, including the use of other promoters, the production of occlusion-positive recombinant viruses, and the expression of two or more foreign genes simultaneously (reviewed in ref. *1*). However, the majority of vectors currently in use are based on the replacement of the polyhedrin gene with the foreign gene of interest.

Because of the large size of the AcMNPV genome (133 kbp), it is difficult to simply digest the viral DNA with restriction endonucleases and ligate in the heterologous DNA. Instead, most recombinant baculoviruses are generated by recombination in vivo. This is illustrated in Fig. 1A. The recombinant virus is identified based on the formation of occlusion-negative plaques, i.e., plaques composed of cells lacking occlusion bodies. The principle drawback to this approach is that a certain amount of practice is required to be able to spot occlusion-negative plaques reliably. Many workers find it difficult to identify such plaques amongst the very large background of nonrecombinant occlusion-positive plaques. Several approaches have been devised to circumvent this problem. The most effective of these is based on the use of parental viral DNA that incorporates a lethal deletion. The gene downstream of polyhedrin (ORF 1629) encodes a capsid protein that is essential for virus viability *(4,5)*. Kitts and Possee *(6)* have produced a recombinant AcMNPV derivative (BacPAK6) that includes target sites for the restriction endonuclease Bsu36I upstream of the polyhedrin gene and within ORF 1629. Bsu36I does not cleave elsewhere in the genome. Digestion of BacPAK6 with Bsu36I thus deletes part of ORF1629 and renders the virus nonviable. However, the

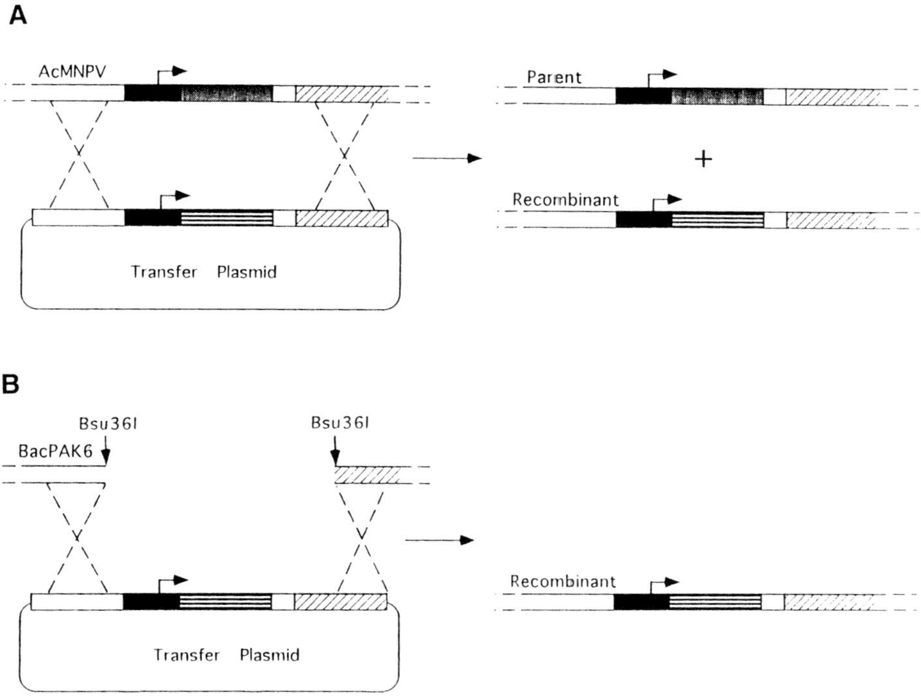

Fig. 1. Illustration of the generation of recombinant baculoviruses by in vivo recombination. **(A)** Following cotransfection of the transfer plasmid with wild-type virus, the vast majority of the progeny virus is wild-type, so that extensive screening is required to isolate the recombinant in a pure form. **(B)** The Bsu36I-cut BacPAK6 DNA is nonviable, so that the only progeny obtained after cotransfection of the transfer plasmid and digested BacPAK6 is the recombinant. The solid black box represents the polyhedrin promoter, with the arrow indicating the transcriptional start site. Polyhedrin coding sequences are represented by the shaded box. The hatched box represents ORF1629, whereas the lined box denotes the heterologous gene. Open boxes are other AcMNPV sequences and the single line represents plasmid vector sequences.

lethal deletion can be rescued by recombination with a plasmid that includes the deleted DNA (Fig. 1B). When Bsu36I-digested BacPAK6 is used as the parental viral DNA for the generation of a recombinant expression vector, virtually all the progeny virus obtained are recombinant, so that the difficulties in screening mentioned previously are avoided.

Over the past decade, a huge range of approaches for the generation of baculovirus expression vectors have been developed. A large number of transfer plasmids are available that enable the use of alternate screening methods, the use of different promoters, the generation of occlusion-positive recombinants, etc. A variety of parent viruses are available to complement these plasmids. Similarly, procedures involving different cell lines, or gene expression in the insect, have been described. Many of these alternative approaches are of value in specific circumstances. However, this chapter is confined to a description of the use of BacPAK6 for the expression of foreign genes under the control of the polyhedrin promoter in SF9 or SF21 cells. This basic approach will be suitable for the majority of expression projects. Researchers interested in alternative approaches or in very large scale expression ("industrial scale") are directed to refs. *1* and *7* for further information.

2. Materials

1. SF cells. The cells most commonly used with AcMNPV expression vectors are SF 9 and SF-21 AE cells. Both originate from IPLB-SF-21 cells which were derived from *Spodoptera frugiperda* pupal ovarian tissue *(8)*. The SF-9 cell line was cloned from SF-21 AE cells. In this chapter, both lines are referred to as SF cells. They are both widely available from workers using the expression system, and are often provided in commercially supplied expression vector kits. In addition, SF-9 cells are available from the ATCC (ATCC CRL 1711).
2. Virus and viral DNA. BacPAK6 is available from Clontech (Palo Alto, CA) either alone or as part of a kit. Clontech also provide Bsu36I-digested BacPAK6 DNA. Alternatively, the virus can be propagated and the DNA prepared as described in Sections 3.3. and 3.4.
3. Complete TC100. TC100 insect cell culture medium (Gibco-BRL, Gaithersburg, MD) supplemented with 10% fetal calf serum and 2.6 g/L tryptose broth (Sigma, St. Louis, MO) (*see* Note 1). Antibiotics (100X antibiotic antimycotic solution, Sigma) may be added as required.
4. Tissue culture ware. Any standard brand of tissue culture plastic ware may be used. If possible, it is best to use the brand the cells are accustomed to. Do not use bacterial Petri dishes.
5. 0.4% Trypan blue in TC100.
6. SeaKem ME agarose (FMC).
7. 20 mg/mL X-Gal in dimethylformamide.
8. 1 mg/mL Neutral red in TC100.
9. Sucrose cushion: 25% sucrose (w/w) in 5 m*M* NaCl, 10 m*M* EDTA.
10. Virus disruption buffer: 10 m*M* Tris-HCl pH 7.6, 10 m*M* EDTA, 0.25% SDS.
11. 10 mg/mL Proteinase K in H_2O.
12. TE: 10 m*M* Tris-HCl pH 8.0, 1 m*M* EDTA.
13. Phenol saturated in TE.
14. Phenol-chloroform-isoamylalcohol (IAA) (25:24:1).
15. Chloroform-IAA (24:1).
16. Bsu36I (New England Biolabs, Beverly, MA).
17. Complete Grace's medium. Grace's insect cell culture medium (Gibco-BRL) supplemented with 10% fetal calf serum.

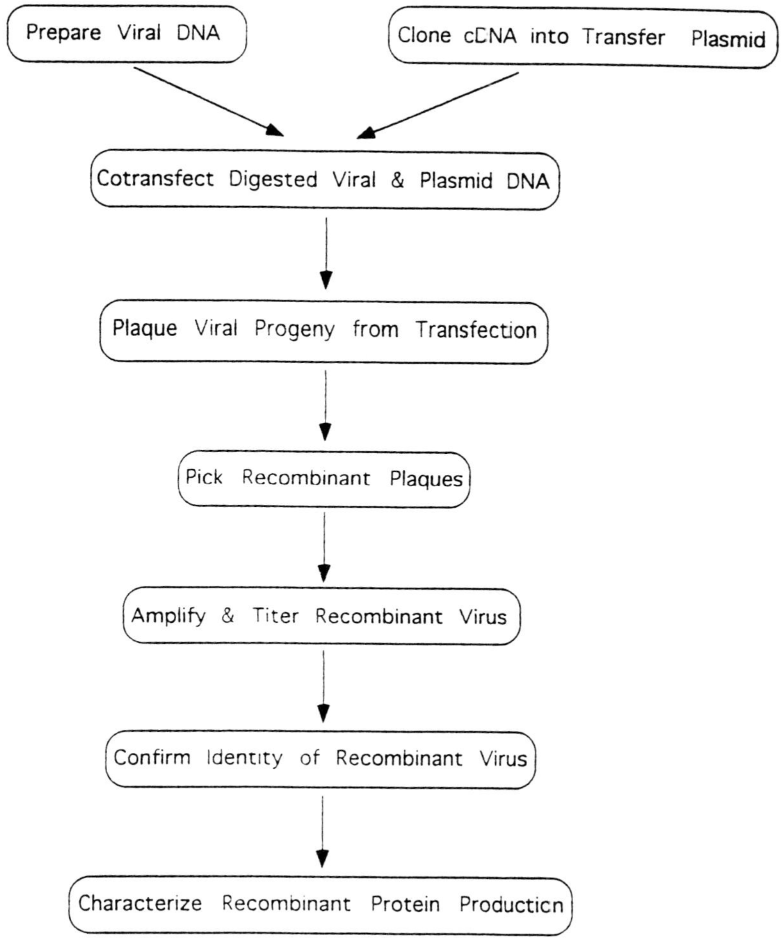

Fig. 2. Flow chart outlining the generation of recombinant baculoviruses.

18. Transfection buffer: 25 m*M* HEPES pH 7.1, 140 m*M* NaCl, 125 m*M* $CaCl_2$ · Filter sterilize and store at 4°C (*see* Note 2 in Section 4.).
19. Phosphate-buffered saline (PBS): 1 m*M* $Na_2HPO_4.7H_2O$, 10.5 m*M* KH_2PO_4, 140 m*M* NaCl, 40 m*M* KC1, pH 6.2.
20. SDS-gel loading buffer: 50 m*M* Tris-HCl pH 6.8, 2% SDS, 0.1% bromophenol blue, 10% glycerol. Before use add β-mercaptoethanol to a concentration of 2% (v/v).

3. Methods

A flow chart showing the stages in the generation of a baculovirus expression vector is presented in Fig 2. Briefly, a cDNA encoding the gene of interest is first subcloned into a suitable transfer plasmid. (Most AcMNPV genes are not spliced and splicing of heterologous genes is not expected to be efficient. Thus, it is best to use a cDNA clone or other DNA that lacks introns.) The transfer plasmid DNA is then cotransfected into insect cells with digested viral DNA. Progeny virus is harvested 4–5 d later and individual viral clones are identified by plaque assay. Several plaques are picked and the recombinant virus is amplified. Following confirmation of the identity of the selected recombinants, the protein of interest can be overproduced by infection of insect cell cultures. The methods that follows describe insect cell culture, plaque assay and amplification of viral stocks, preparation of viral DNA, cotransfection of viral and plasmid DNA, selection and amplification of recombinants, and overexpression of the heterologous gene. The protocols will not necessarily be followed in the order given. For example, researchers who purchase Bsu36I-digested viral DNA will not need to refer to the protocols in Sections 3.2., 3.3., and 3.4. until after cotransfection (Section 3.5.). Subcloning the gene into the transfer plasmid or preparation of plasmid DNA are not described; these are common molecular biology protocols and are well described elsewhere (e.g., ref. *9*).

3.1. Insect Cell Culture

SF cells will grow either as monolayer cultures or in suspension. It is best to maintain stock cultures in tissue culture flasks and use these to seed suspension cultures when necessary. Stock cultures are generally subcultured twice a week. They are maintained at 27°C in a humidified incubator with the caps of the flasks open one quarter turn. A typical cell passage schedule is as follows.

1. Subculture the cells when they are 80–90% confluent. Remove the cells from the surface of a 25-cm^2 tissue culture flask by gentle pipeting or scraping with a rubber policeman.
2. Take a small aliquot of cells and add 0.1 vol of Trypan blue solution. Examine immediately on a hemocytometer using a tissue culture microscope. Determine the concentration of viable cells (i.e., cells that exclude Trypan blue)/mL and the proportion of viable cells. Ideally, viability should be ≥98%.
3. Put 5 mL of prewarmed complete TC100 into each of two 25-cm^2 flasks (*see* Note 3). Inoculate one 25-cm^2 flask with 1×10^6 cells (the stock flask) and the other with 5×10^5 cells (the backup flask). If required, the remaining cells can be used to inoculate a large flask in preparation for setting up a suspension culture (step 6). Incubate at 27°C.
4. The stock flask should be ready to subculture again in 3–4 d. Do not subculture the backup flask.
5. The day after subculturing, examine the newly passaged cells for signs of contamination. If the cells are contaminated, the backup flask can be passaged to maintain the culture.

6. To set up a suspension culture, inoculate a 75-cm^2 flask with the cells remaining after subculture of the stock flask (≈5 × 10^6 cells; step 3). Incubate at 27°C for 3–4 d.
7. Scrape all the cells from the confluent 75-cm^2 flask (≈4 × 10^7 cells) and transfer to a spinner bottle or flask (*see* Note 4). Feed with 100–200 mL of prewarmed complete TC100 + antibiotics and incubate at 27°C with stirring at 60–80 rpm.
8. Use the cells from the suspension culture when they reach a density of ≈1 × 10^6 cells/mL (3–4 d after inoculation).

3.2. Plaque Assay of Virus Stocks

Prior to use of a virus stock, it will be necessary to determine the titer of the stock by plaque assay. Provided the virus is stored at 4°C in the dark, the titer should remain stable for several months.

1. Put 2 mL of complete TC100 into each of four 60-mm tissue culture dishes. Inoculate each dish with 2 × 10^6 cells (*see* Note 5) and incubate at 27°C for 30 min to 1 h to allow the cells to attach (*see* Note 6).
2. Dilute the virus stock in TC100. Dilutions of 10^{-3}, 10^{-4}, 10^{-5}, and 10^{-6} should be appropriate for most stocks.
3. Aspirate the tissue culture fluid from the plates and infect with 0.5 mL of each dilution. Incubate at room temperature for 1 h with gentle rocking.
4. While the cells are rocking, prepare the agarose overlay.
   a. Prepare 3 mL of a 5% agarose solution in distilled H_2O in a flask or bottle capable of accommodating at least 30 mL. Melt by autoclaving.
   b. Heat 27 mL of complete TC100 to 60°C and add to the melted agarose. This gives a solution of 0.5% agarose in TC100. Cool to 40–42°C before use. Add 180 µL X-gal stock (final concentration 120 µg/mL) and 300 µL 100X antibiotic solution just before overlaying.
5. Aspirate the virus inoculum from the cells and add 4 mL of overlay to each plate. Let the agarose set at room temperature for 15–20 min; then incubate at 27°C for 4–5 d.
6. β-galactosidase-producing viruses, such as BacPAK6, form easily visible blue plaques in the presence of X-gal. Plaques of viruses that do not produce β-galactosidase (e.g., the recombinants) can be visualized by staining with neutral red as follows. Prepare 20 mL of a 0.5% solution of agarose in tissue culture fluid as described in step 4. Cool to 42°C and add 1 mL neutral red stock (final concentration 50 µg/mL). Add 3 mL to each plate (on top of the existing overlay), let set and incubate at 27°C overnight.
7. After neutral red staining, plaques will appear as clear, circular areas 0.5–3 mm in diameter against a red background. Determine the titer of the stock by counting the number of plaques present on each plate. The most reliable counts will be obtained from plates that have between 10–100 plaques. To calculate the titer of the stock in plaque-forming units (PFU) per mL, multiply the number of plaques by the dilution factor and by 2 (only 0.5 mL were inoculated onto each plate). For example, if 30 plaques are counted on the plate infected with the 10^{-6} dilution, the titer is 30 × 2 × 10^6 = 6 × 10^7 PFU/mL.

3.3. Amplification of Virus

Large stocks of budded virus can be prepared simply by infecting cell cultures and harvesting the extracellular fluid 4–5 d later. The protocol below describes two rounds of amplification to yield 40 mL of virus. If your initial inoculum is reasonably large, you may be able to skip the first round of amplification, i.e., start at step 6 (*see* Note 7 in Section 4.).

1. Put 1 mL of complete TC100 into a 35-mm tissue culture dish. Inoculate with 1×10^6 cells and incubate at 27°C for 30 min to 1 h to allow the cells to attach (*see* Note 6 in Section 4.).
2. Aspirate the tissue culture fluid from the plates and add in 1×10^5 PFU of virus. (the determination of the titer of your virus stock is described in Section 3.2.) This gives a multiplicity of infection (m.o.i.) of 0.1 PFU/cell. Adjust the volume in the plate to 250 µL with TC100, and incubate at room temperature for 1 h with gentle rocking.
3. Feed the cells with 1.75 mL complete TC100 + antibiotics and incubate at 27°C for 4–5 d.
4. Collect the extracellular fluid and centrifuge at 1000*g* for 5 min to remove cell debris. Store the supernatant at 4°C in the dark.
5. Determine the titer of this virus stock by plaque assay as described in Section 3.2.
6. Put 5 mL of complete TC100 into each of four 100-mm tissue culture dishes. Inoculate each dish with 5×10^6 cells and incubate at 27°C for 30 min to 1 h to allow the cells to attach (*see* Note 6).
7. Aspirate the tissue culture fluid from the plates and add in viral inoculum to give an m.o.i. of 0.1 PFU/cell. Adjust the volume in the plate to 1 mL with TC100, and incubate at room temperature for 1 h with gentle rocking.
8. Feed the cells with 9 mL complete TC100 + antibiotics and incubate at 27°C for 4–5 d.
9. Collect the extracellular fluid and centrifuge at 1000*g* for 5 min to remove cell debris. The supernatant is the virus stock. Store at 4°C in the dark.
10. Titer the stock by plaque assay (*see* Section 3.2.) before use.

3.4. Preparation of Viral DNA

1. Centrifuge 10 mL of high titer viral stock (e.g., from Section 3.3.) through a 2 mL sucrose cushion at 80,000*g* for 75 min at 4°C.
2. Decant the supernatant and remove traces of sucrose. The viral pellet should be translucent white with a faint, blue tinge around the edges when viewed against a dark background.
3. Add 1 mL disruption buffer and resuspend the pellet carefully by pipeting gently using a pipet with a cutoff tip (*see* Note 8).
4. Add 50 µL proteinase K and incubate at 37°C overnight.
5. Add 1 mL of phenol and extract by inverting repeatedly until an emulsion is formed. Do not vortex! Centrifuge for 5 min at 4°C to separate the phases.

6. Transfer the aqueous phase to a fresh tube. Add 100 µL TE to the organic phase and extract again as in step 5. Combine the two aqueous phases.
7. Extract the aqueous phases with an equal volume of phenol-chloroform-IAA as in step 5. Centrifuge for 3 min at 4°C to separate the phases.
8. Extract the aqueous phase with an equal volume of chloroform-IAA, centrifuge for 1 min, and collect the aqueous phase.
9. Dialyze the DNA at 4°C against 500 vol of TE with 2 changes for at least 4 h each time (*see* Note 9). Store the DNA at 4°C.
10. Determine the DNA concentration by measuring the OD_{260} (1 OD unit = 50 µg/mL in a 1 cm path cell). Typically, the protocol above will yield approx 20–30 µg of DNA at a concentration of 10–20 µg/mL.
11. Digest 1 µg of viral DNA with 12 U of Bsu36I for 5–6 h at 37°C. (Keep the volume as small as possible.) It is best to add the restriction enzyme in 3 aliquots (4 U every 2 h).
12. Inactivate the enzyme by heating the DNA at 70°C for 15 min. Store at 4°C.

3.5. Cotransfection

1. Put 1 mL of complete TC100 into each of two 35-mm tissue culture dishes. Inoculate each dish with 1×10^6 cells and incubate at 27°C for 30 min to 1 h to allow the cells to attach (*see* Note 6).
2. Aspirate the tissue culture fluid from the plates and replace with 375 µL complete Grace's medium (*see* Note 10). Incubate at room temperature.
3. Mix 2 µg plasmid DNA and 1 µg digested viral DNA in a tube and add transfection buffer to a final volume of 750 µL. Mix gently.
4. Add 375 µL of the DNA plus transfection buffer mix dropwise to each dish. Incubate at 27°C for 4 h.
5. Aspirate the Grace's medium plus transfection buffer; rinse the cells with TC100, and refeed with 2 mL complete TC100 plus antibiotics. Incubate at 27°C for 4 d.
6. Collect the extracellular fluid (which will contain any recombinant virus formed) from both plates of transfected cells. Keep these stocks separate. Centrifuge at 1000*g* for 5 min to pellet the cell debris. Store the supernatants at 4°C in the dark.

3.6. Plaque Purification and Amplification of Recombinant Virus

1. Plaque out 10^{-1}, 10^{-2}, and 10^{-3} dilutions of both recombinant virus stocks obtained from the cotransfection as described in Section 3.2. Stain with neutral red after 4–5 d.
2. Residual parent BacPAK6 virus will form blue plaques whereas recombinants will form colorless plaques. Select two, well separated, colorless plaques from each recombinant virus stock (*see* Note 11).
3. Pick each plaque by placing the tip of a sterilized Pasteur pipet or glass micropipet directly onto the plaque. Carefully apply very gentle suction to draw a small plug of agarose into the pipet.
4. Place each agarose plug in 1 mL TC100. Vortex well to release the virus particles and store at 4°C in the dark.

5. Prepare large stocks of each recombinant as described in Section 3.3. In this case, the precise titer of the infecting virus is not known. However, we have found that for the first round of amplification, each 35-mm dish can be infected with 0.5 mL of the plaque-purified inoculum from step 4. 250 μL of the (passage 1) stock obtained from this infection can then be used to infect each 100-mm dish.
6. Titer the passage 2 stock of virus obtained (*see* Section 3.2.) and store at 4°C in the dark.

3.7. Characterizing Recombinant Protein Production

Once you have prepared your amplified stocks of recombinant virus, you will want to confirm that the expected virus has been obtained and investigate whether it expresses the gene of interest. Obviously, demonstration of expression of the required foreign gene is a strong indication that the recombinant virus has the appropriate structure. However, we strongly recommend that, in addition, you characterize the genome structure of the virus in each stock by PCR, restriction enzyme analysis, and/or Southern blotting to confirm that the gene has been inserted into the polyhedrin locus and that no other alterations have taken place. Viral DNA for analysis can be prepared as described in Section 3.4. omitting the Bsu36I digestion (step 11). For information on strategies for confirming the genome structure and details of the expected restriction profiles, *see* ref. *1*.

The method chosen to investigate protein production will obviously depend on the reagents available for each individual protein of interest (antibodies, enzymatic assays, etc.). In addition, the procedure for processing infected cells to extract the overproduced protein will also depend to a large extent on the nature of the protein concerned. Thus, it is up to each individual researcher to choose the best approach for extraction and analysis of the infected cells. In general, protocols that work with mammalian cells should also work with insect cells. The protocol presented below is a typical one used to investigate the time course of protein production by SDS-PAGE. Each recombinant virus clone obtained should be characterized using a protocol such as this to determine which clone yields the highest levels of the required protein.

1. For each viral clone to be analyzed (each recombinant clone and the parent virus), seed three 35-mm tissue-culture dishes with 1×10^6 cells and incubate at 27°C for 30 min to 1 h to allow the cells to attach (*see* Note 6). Seed one additional dish to be used as an uninfected control.
2. Aspirate the tissue culture fluid from the plates and infect with 2×10^7 PFU of virus per plate (m.o.i. = 20; 3 plates per virus clone) in a final volume of 500 μL. Incubate at room temperature for 1 h with gentle rocking.
3. Aspirate the virus inoculum and feed the cells with 2 mL complete TC100 plus antibiotics. Incubate at 27°C.

4. At 24, 48, and 72 h postinfection, take one dish infected by each virus and remove the extracellular fluid (the uninfected cells can be processed with the 24 h samples). If the protein of interest is secreted, the extracellular fluid may be saved for analysis by SDS-PAGE (*see* Note 12). Rinse the cell monolayer twice with 2 mL ice-cold PBS (*see* Note 13). After the second rinse, stand the dishes on edge for a few minutes, then remove any residual PBS.
5. Lyse the cells by addition of 50 µL 1X SDS gel-loading buffer to each dish, scrape the lysate off the dish with a rubber policeman and store at –20°C or –80°C. 5–10 µL of each lysate can now be analyzed by SDS-PAGE followed by Coomassie blue staining and/or immunoblotting.

4. Notes

1. TC100 is only one of a range of cell culture media, some not requiring supplementation with serum. If possible, it is best to culture the cells in the medium used by the person supplying the cells.
2. The pH of the transfection buffer is critical for efficient transfection.
3. SF cells attach very rapidly to the tissue culture plastic. To ensure the cells are evenly spread on the surface, it is always better to add the medium before the cells. Rock the flask or dish gently as soon as the cells have been added.
4. A wide variety of systems is commercially available for the suspension culture of eukaryotic cells. These are generally suitable but quite expensive. A more economical alternative is to simply use a 500-mL sterile glass bottle containing a stir bar. This will be perfectly adequate as long as a stir plate capable of operating reliably at slow speeds without generating excess heat is used.
5. Proper cell density is critical for the success of a plaque assay. Generally, this will be around 2×10^6 cell per 60-mm dish, but it is best to try a range of cell densities to find the optimum for your conditions.
6. Alternatively, the cells may be inoculated at a density of 5×10^5 cells per dish the day before they are required.
7. It is important not to passage the virus too many times in cell culture as this can lead to the accumulation of variants within the stock. Make a large, low passage number stock (<5 passages in cell culture), and use this as inoculum for the generation of working stocks of virus.
8. Because of the large size of the AcMNPV genome (133 kbp), great care must be taken in this protocol to avoid shearing the DNA.
9. The DNA may be ethanol-precipitated instead of dialyzed if preferred. This has the advantage that it both sterilizes and concentrates the DNA. However, the DNA is often extremely difficult to redissolve.
10. Do *not* use TC100 at this step.
11. Theoretically, almost all the plaques obtained should represent recombinant virus. However, it is prudent to select several plaques at this stage (some from each cotransfection to be certain that independent clones are being examined), to ensure that at least one represents the required recombinant.

12. In the case of extracellular proteins, dilution of the protein into the tissue-culture fluid can make it extremely difficult to detect by Coomassie blue staining. Furthermore, the large amount of bovine serum albumin in the extracellular fluid can hinder detection of the protein of interest. Such extracellular proteins can generally be detected by immunoblotting with a suitable antibody. If no antibody is available to the protein, it should be possible to detect it by metabolic labelling of infected cells *(1)*.
13. Late in infection, the cells will become detached from the substratum very easily. Thus, extreme care must be taken during the rinse procedure to avoid excessive loss of cells. If it appears that significant numbers of cells are being lost at this stage, collect the cells by gentle scraping followed by centrifugation at 1000*g* for 5 min at 4°C. Carry out the rinses by resuspending the pellet in PBS and recentrifuging.

Acknowledgments

I would like to thank O. P. Evans for critical reading of the manuscript. I apologize to colleagues whose work I have not cited due to lack of space.

References

1. O'Reilly, D. R., Miller, L. K., and Luckow, V. A. (1992) *Baculovirus Expression Vectors—A Laboratory Manual.* Oxford University Press, New York.
2. Pennock, G. D., Shoemaker, C., and Miller, L. K. (1984) Strong and regulated expression of *Escherichia coli* β-galactosidase in insect cells with a baculovirus vector. *Mol. Cell. Biol.* **4,** 399–406.
3. Smith, G. E., Summers, M. D., and Fraser, M. J. (1983) Production of human β-interferon in insect cells infected with a baculovirus expression vector. *Mol. Cell. Biol.* **3,** 2156–2165.
4. Possee, R., Sun, T.-P., Howard, S., Ayres, M., Hill-Perkins, M., and Gearing, K. (1991) Nucleotide sequence of the *Autographa californica* nuclear polyhedrosis 9.4 kbp EcoRI-I and -R (polyhedrin gene) region. *Virology* **185,** 229–241.
5. Vialard, J. E. and Richardson, C. D. (1993) The 1,629-nucleotide open reading frame located downstream of the *Autographa californica* nuclear polyhedrosis virus polyhedrin gene encodes a nucleocapsid-associated phosphoprotein. *J. Virol.* **67,** 5859–5866.
6. Kitts, P. A. and Possee, R. D. (1993) A method for producing recombinant baculovirus expression vectors at high frequency. *BioTechniques* **14,** 810–817.
7. King, L. A. and Possee, R. D. (1992) *The Baculovirus Expression System: A Laboratory Guide.* Chapman and Hall, New York.
8. Vaughn, J. L., Goodwin, R. H., Tompkins, G. J., and McCawley, P. (1977) The establishment of two cell lines from the insect *Spodoptera frugiperda* (Lepidoptera: Noctuidae). *In Vitro* **13,** 213–217.
9. Sambrook, J., Fritsch, E. F., and Maniatis, T. (1989) *Molecular Cloning: A Laboratory Manual.* Cold Spring Harbor Laboratory, Cold Spring Harbor, NY.

20

Expression of Exogenous Genes in *Xenopus* Oocytes, Eggs, and Embryos

Koichiro Shiokawa, Chie Koga, Yuzuru Ito, and Mikihito Shibata

1. Introduction

Molecular biology of gene expression in early amphibian embryogenesis began in 1964, when undegraded, as opposed to alkaline-hydrolyzed, PCA (perchloric acid), or TCA (trichloroacetic acid)-degraded, RNAs were extracted by Brown and Littna *(1–3)*, Shiokawa and Yamana *(4,5)*, and Woodland and Gurdon *(6)* from amphibian embryos using phenol methods. From these early studies, especially from those of Don Brown, it was realized that amphibian embryos exhibit quite unusual RNA synthetic patterns: the pattern of active 4S RNA (mainly tRNA) synthesis with no rRNA synthesis in pregastrula stages and the pattern of gradually increasing rRNA synthesis in post-gastrular stages. It was then reported in 1982 by Newport and Kirschner *(7,8)* that in *Xenopus* embryogenesis, large changes called midblastula transition (MBT) take place at the 12th cleavage, which include the appearance of G1 phase in cell cycle, onset of gene expression from zygotic nuclei, and acquisition of cellular motility *(7,8)*. We then found in 1987 that *Xenopus* embryogenesis consists of three different phases with respect to the RNA synthetic pattern: The phase of a low level (on a per-embryo but not necessarily on a per-cell basis) of mRNA synthesis (pre-MBT stage), the phase of extremely active tRNA synthesis (both on per-cell and per-embryo bases) (MBT stage), and the phase of a nearly constant level (per cell) of rRNA synthesis (post-MBT stage) *(9–12)*.

While these studies on nuclear gene expression were in progress, John Gurdon injected exogenous genes into the *Xenopus* oocyte nucleus to test their functions *(13)*. Furthermore, Gurdon, in collaboration with Brown, injected for the first time, purified *Xenopus* genes (5S DNA, rDNA, and repetitive DNA)

From: *Methods in Molecular Biology, vol. 62: Recombinant Gene Expression Protocols*
Edited by: R. Tuan Humana Press Inc., Totowa, NJ

into both unfertilized and fertilized *Xenopus* eggs, with a finding that injected DNAs are expressed *(14)*. Since then, many workers started to analyze the behavior of normal or variously "reformed" genes, or their product RNAs, after introducing them into *Xenopus* fertilized eggs *(13,15)*.

In this chapter, we describe such DNA (and RNA, though only partly) microinjection studies in *Xenopus* oocytes, unfertilized eggs, and fertilized eggs, with an emphasis on the technical aspects.

2. Materials

2.1. Apparatus

For injection, a binocular microscope, a micromanipulator with a microinjector on it (Narishige Co., Japan), a pressure-generating apparatus, and glass needles are needed. The microinjecter is equipped with a syringe, to which a glass needle is directly attached, and this system is connected by flexible polytene tubing to the pressure-generating apparatus. Figure 1 shows such an assemblage (A), which is operated routinely in our laboratory (B).

To obtain glass needles, prepare glass capillary tubes (approx 1 mm in outer diameter) by heating a usual glass tube, preferably with a capillary tube-former operated by electricity. Cut the long capillary tubes in the length of 9–10 cm, and pull the capillary to obtain a pair of needles in a micropuller (Narishige Co., Japan) as in Fig. 2.

Since the tips of the needles obtained in this way are closed by the heat generated within the micropuller, cut off the tips with scissors. (Here, collect pieces of wasted needles in a covered glass bottle.) Take a look at the needles under a microscope, and select those with a sharp tip of approx 10 μm in diameter. If necessary, the tip of the needle can be further sharpened by touching it briefly to the heated element of a microforge. Use the needles after sterilizing them in a heated incubator.

2.2. Bacterial Cells

For preparation of the plasmid DNAs, it is recommended to use *E. coli* cells of Rec A- strain, such as HB101, DH5, and XL1-blue, in order not to obtain dimer DNAs.

2.3. Oocytes and Eggs

Oocytes, eggs, coenocytic egg cells, and embryos of *Xenopus laevis* are used for injection experiments. Animals can be raised in the laboratory or purchased from appropriate local places.

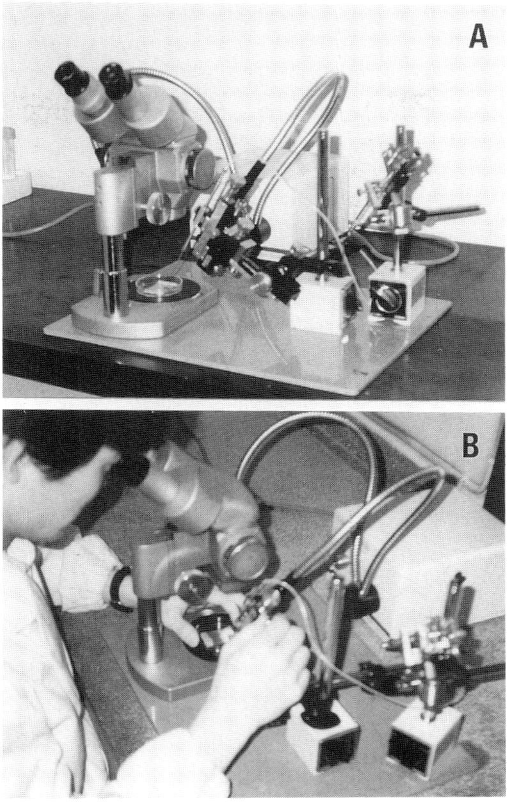

Fig. 1. An assemblage of the instruments for microinjection **(A)**, which is being operated **(B)**.

Fig. 2. A micropuller that is being operated for obtaining glass needles.

2.4. Culture Media and Other Reagents

Culture media to be used are as follows.

1. Modified Barth's solution (MBS) for culturing oocytes: 88 m*M* NaCl, 1.0 m*M* KCl, 0.41 m*M* $CaCl_2$, 0.33 m*M* $Ca(NO_3)_2$, 0.82 m*M* $MgSO_4$, 2.4 m*M* $NaHCO_3$, 10 m*M* HEPES-NaOH, pH 7.4–7.6.
2. De Beer solution for artificial fertilization: 0.1*M* NaCl, 1.3 m*M* KCl, 0.44 m*M* $CaCl_2$, and pH adjusted to 7.4 with a small amount of $NaHCO_3$.
3. Steinberg solution for culture of oocytes or embryos: 60 m*M* NaCl, 0.67 m*M* KCl, 0.34 m*M* $Ca(NO_3)_2$, 0.83 m*M* $MgSO_4$, 10 m*M* Hepes-NaOH, pH 7.4.
4. Modified Mark's solution (MMS) for culturing injected eggs: 0.1*M* NaCl, 2.0 m*M* KCl, 2.0 m*M* $CaCl_2$, 1.0 m*M* $MgCl_2$, 5.0 m*M* HEPES, pH 7.4.
5. TE for DNA stock solution: 10 m*M* Tris-HCl, 1 m*M* EDTA, pH 7.5. Tris-NaCl for RNA stock solution: 88 m*M* NaCl, 15 m*M* Tris-HCl, pH 7. 6.

3. Methods

3.1. Obtaining Oocytes, Eggs, Coenocytic Egg Cells

To obtain oocytes, chill a gravid female with ice, and excise the ovary using forceps and scissors. A fully mature *Xenopus* female contains more than 10^4 large oocytes. To ensure long survival of oocytes, wash the ovary thoroughly with the modified Barth's solution (MBS) shortly after the excision. For manual isolation, tear apart the ovarian lobes into several smaller clumps using forceps and scissors, grip a relatively small clump with one pair of forceps, and strip the oocytes off the clump one by one. Select full-grown stage VI oocytes ("white banded," so-called because they have a white belt around the marginal zone), unless younger oocytes are desired. The staging of oocytes is usually carried out according to Dumont *(16)*.

For enzymatic stripping of oocytes, gently swirl the clumps of ovary for 5–6 h at 20–24°C in MBS, containing 1 mg/mL collagenase (Sigma, St. Louis, MO) (Also, collagenase can be used at 2 mg/mL, but in this case treatment should be shorter). Figure 3 shows the oocytes (stages I–VI) obtained by this method. After the treatment, gently wash the oocytes with MBS several times by exchanging the medium. Incubate them in MBS, which contains 50 U/mL (or 50 μg/mL) penicillin and 50 μg/mL streptomycin.

To obtain unfertilized eggs, inject 100–250 U of a gonadotropic hormone (Gonatropin; Teikokozoki) into the dorsal lymph sac of female frogs, and after about 12–16 h, push their abdomen and gently squeeze out eggs into the Petri dish (10 cm in diameter, at least) containing a small volume of 0.5X De Beer solution (Fig. 4). Use a smaller amount of the hormone in summer at warmer temperature frogs ovulate eggs earlier. It is routine to inject the hormone into females in the evening (in winter) or late at night (in summer), and to collect eggs relatively early in the next morning.

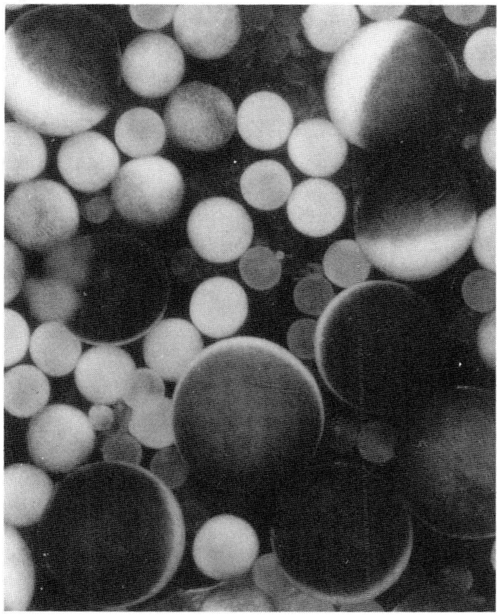

Fig. 3. The appearance of the oocytes obtained by collagenase treatment.

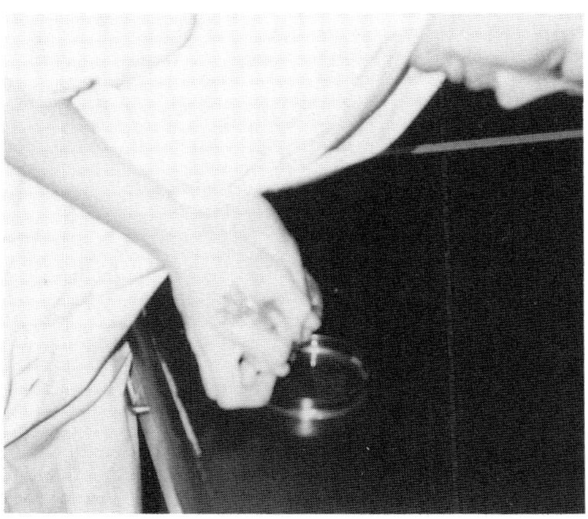

Fig. 4. Manually induced ovulation. From the female injected with a gonadotropic hormone, unfertilized eggs are being squeezed out.

To obtain fertilized eggs, isolate testes from a male frog chilled in ice. Add 1 mL of 0.5X De Beer solution to a testis, and mince it with scissors in a relatively large glass centrifugal tube, moving the scissors many times quickly from down to up in the direction from the top to the bottom of the tube. Meanwhile, the other testis could be kept in a Petri dish for 1–2 d in a refrigerator (do not forget to place water-soaked cotton in the Petri dish, lest the testis should dry up). The sperm suspension prepared as above contains fragments of testis, but they do not interfere with fertilization. For insemination, eliminate the 0.5X De Beer solution from the Petri dish using a pipet, and add 5–6 drops of the sperm suspension over the eggs. Then, mix the eggs and the sperm suspension well by gently shaking the dish for about 5 s. After waiting 10 min, add enough volume of 0.5X De Beer solution so that eggs are completely immersed in the solution. Then, wait for another 20 min. When fertilization takes place, perivitelline space is filled with fluid and the eggs rotate according to the gravity, and finally, black (or brown) animal pole comes to the top (Fig. 5). The efficiency of fertilization is usually 80–95%.

Before injection, both unfertilized and fertilized eggs are dejellied by treatment for about 5–10 min with either 3% cysteine-HCl solution (pH 7.9) or 2.5% sodium thioglycollate (pH 8.6–9.0). (The higher the pH, the shorter the time needed for dissolution of jelly. But, remember that higher pH is more harmful to eggs.) If the room temperature is relatively high, the treatment should accordingly be shorter, lest eggs be damaged. If the treatment is too long, eggs will be damaged (in that case, cleavage becomes irregular very soon). During the digestion, gently swirl the medium from time to time, watching the changes carefully, and when eggs start to contact to each other, stop the treatment. If jelly coat is not dissolved, eggs will not contact directly, because two layers of jelly coat separate the eggs (*see* Fig. 5). However, if some of the eggs stick together and form clumps, the digestion of jelly coat is not complete. In this case, repeat the digestion briefly.

It is worth mentioning here that even if the jelly coat appears to have been digested completely, a brief additional digestion just before using the eggs (or embryos, in most cases) helps eliminate bacteria that contaminate the surface of the eggs during culture. This is especially important when embryos are dissociated and the cells obtained are labeled with radioactively labeled precursors *(5,17)*.

Right after the digestion of jelly coat, rinse the eggs thoroughly by changing the medium with the fresh MBS or Steinberg solution at least several times. Oocytes, unfertilized eggs, and fertilized eggs obtained as above should be kept in MBS or Steinberg solution until used.

Coenocytic egg cells (or coenocytes) can be obtained by centrifuging fertilized eggs for a short period of time (~1 min) at a low speed (~500 rpm) immediately after fertilization. The centrifugation removes the spindle apparatus from its normal position, and totally inhibits cleavage. The coenocytic egg cells do not divide, but are viable for a relatively long period of time (~10 h), and

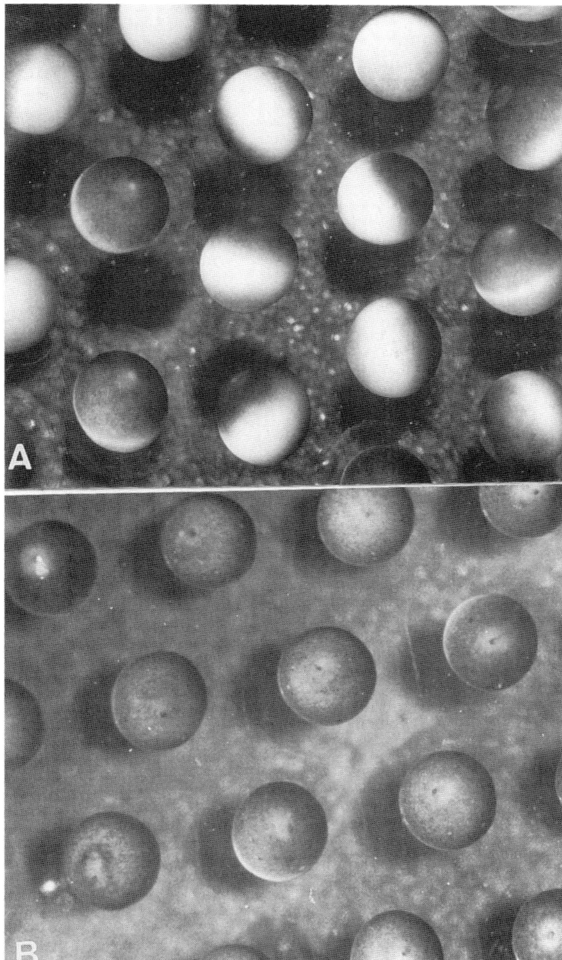

Fig. 5. The appearance of eggs before **(A)** and after **(B)** artificial fertilization. Top view. After fertilization, the animal pole comes to the top. Note that eggs cannot contact directly because of the presence of two jelly layers (~1.4 mm).

sometimes provide a useful system for studying the function of injected DNAs in the absence of cleavage *(8,18)*. However, interpretation of the results obtained has to be done carefully, because the synthesis of not only DNA but also tRNA and snRNA has been reported to be quite abnormal *(19)*.

Throughout these procedures, which involve transferring oocytes or eggs from one Petri dish to another, use a Pasteur pipet with a relatively large mouth, whose tip was heated briefly on a strong flame to avoid damaging the oocytes or eggs.

3.2. Before Starting the Injection

DNA may be dissolved in either sterile water or 0.1X TE (RNA may be dissolved in either sterile distilled water or Tris-NaCl). The concentration of the usual stock solution of DNA (or RNA) could be 1–2 mg/mL, but this has to be determined based on the amount of the DNA (or RNA) to be injected, taking the fact into account that the volume of the solution to be injected is in a range of 10–50 nL per oocyte or egg.

To fill the glass needle with the sample solution, place an appropriate volume of solution (5–10 μL) on a sheet of parafilm and suck it into the glass needle fixed directly to the syringe on a microinjector by operating the pressure-generating apparatus in a reverse direction. This method is used quite safely since the needle is stably fixed to the microinjector and its movement could be accurately controlled by operating the micromanipulator under the binocular microscope.

Figure 6 shows another simple device to do this. Here, instead of sucking the sample solution placed on a parafilm sheet, the sample solution is being sucked directly from the Eppendorf tube with the help of a big syringe (Fig. 6A), which could be fixed by adapting an appropriate stopper to generate negative pressure (Fig. 6B). If this method is used, the volume of the sample at the bottom of the tube should be 30 nL or more in order to guarantee the safe placing of the tip of the glass needle in the sample solution. Also, be careful not to damage the whole sample solution by using a needle that is not sterilized.

Before starting injection, confirm that all the necessary materials are ready for use. Do not prepare, however, several sample-loaded needles at a time; sometimes the tip of the needle dries in the air, rendering the needle clogged and useless. One way to overcome this is to keep such needles in a moisture chamber (a Petri dish with water-soaked cotton in it).

Assemble the injector and other equipment as shown in Fig. 1. Next, give the appropriate pressure to the injector, so that DNA (or RNA) solution comes out at a constant rate. The volume of the solution to be injected could be determined by measuring, under a binocular microscope with an internally set scaler, the approximate diameter of the droplet formed at the tip of the needle during a certain period of time.

3.3. During the Injection

After confirming the constant flow of the solution from the tip, start to prick the oocyte (or egg) under the binocular microscope. Unfertilized eggs usually have no strong surface tension and the membrane (surface coat) surrounding the egg is so elastic that the tip of the unsharpened needle sometimes only pushes the surface inward and does not penetrate the egg. By contrast, it is much easier to prick fertilized eggs with the needle, since they have relatively

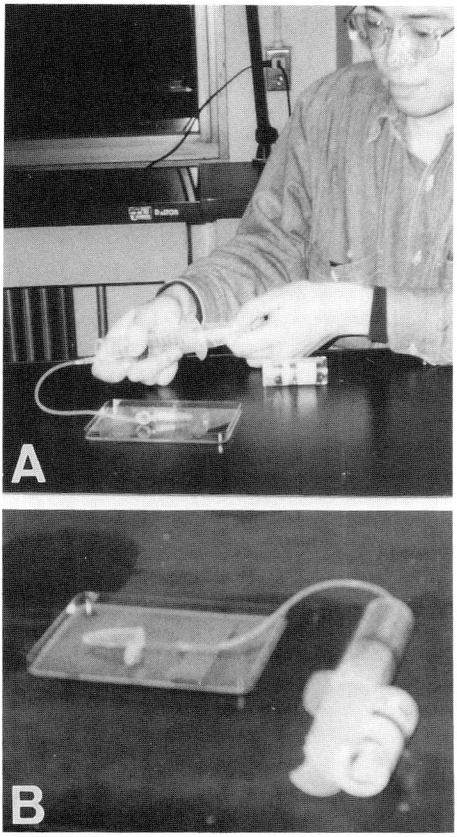

Fig. 6. A device to suck a DNA (or RNA) solution into the needle. The glass needle is directly connected to a large syringe via polytene tubing. The syringe is pooled **(A)**, and fixed with a small supporter to keep the negative pressure **(B)**. Note the tape that fixes the tubing.

high tension on their surfaces. Oocytes are sometimes difficult to prick, since they are surrounded by several layers of somatic cells even after the collagenase treatment. Therefore, it is a good idea to save needles with a sharp tip for injection experiments using oocytes and unfertilized eggs.

Since the oocyte nucleus is not visible, one has to inject the sample to the central portion of the animal hemisphere, where the nucleus (or germinal vesicle) is supposed to be. (When mRNAs are to be injected into the cytoplasm of oocytes, prick the egg, aiming at the vegetal hemisphere, lest the needle penetrate into the nucleus.) To learn how to inject DNA successfully into the oocyte nucleus, it is useful to practice in advance by injecting red ink to the stage VI oocyte, and to see the presence of the ink in the nucleus, which is easily pushed (squeezed) out of the oocyte after pricking (or cutting) it at the animal pole region.

It is also useful to use oocytes centrifuged for about 1 min at ~300–500 rpm to ensure successful injection, since the nucleus comes closer to the surface of the oocyte, and the site of the nucleus becomes visible as a less pigmented circular area under the binocular microscope. However, it is important to keep in mind that the centrifuged nucleus becomes a little flat (causing more leakage of the DNA into the cytoplasm) when the tip of the needle penetrates through the nucleus.

During the injection, oocytes are immersed in MBS solution, containing 5% Ficoll. Eggs are immersed in MBS solution, 1X MMR (Modified Mark's solution), or 1X Steinberg solution, all of which contain 5% Ficoll as above. It is essential that all the media used for injection contain penicillin (50 U/mL) and streptomycin (50 μg/mL) to suppress possible bacterial contamination. (We assume that at the moment the needle penetrates into the egg, bacteria could also penetrate into the egg along with a small amount of the medium.) When bacterial contamination kills the injected egg, the injected portion of the egg becomes white in 3–5 h and the white area expands as incubation lasts longer.

If the number of the oocytes (or eggs) to be injected is relatively small, gently hold the oocyte with a pair of forceps and inject the sample into it by operating the micromanipulator under a binocular microscope. An alternative method is to put the oocytes (or eggs) on a slide glass and to inject them after eliminating most of the medium, so that surface tension of the remaining medium holds each of the oocytes on the slide glass. (In this case, one needs a needle with a sharp tip to guarantee easy pricking as mentioned above.)

When a relatively large number of oocytes (or eggs) are to be injected, it is convenient to make another device to fix them underneath the medium. For instance, a slide glass with several short glass rods fixed by an appropriate adhesive could be immersed in the medium, with oocytes or eggs aligned along a ridge of the glass rods. By this method, the oocyte (or egg) is pushed by the tip of the needle to the ridge of the glass rod and fixed there, easily pricked by the needle. This injection can be done serially by operating the injector with one hand and moving the Petri dish bit by bit with the other hand (Fig. 1B). Also, it is a good way to carry out injection after placing oocytes (or eggs) in each of the wells in the microtiter plate, especially when one changes the dosage of the injected DNA serially or the position of the egg at which the DNA is injected in each group of eggs. Similarly, it is useful to minimize injury in the egg by using a Petri dish, whose bottom was covered by agar gel with small holes on its surface.

Throughout the injection, the sample solution must be coming out of the tip at a constant rate. Therefore, at each injection, draw the tip of the needle back into the air and confirm the flow. During the injection, one has to keep in mind that the length of the time that the tip of the needle remains in the egg must be constant. The shorter the time, the less solution wasted. In an ideal condition, the injection into several hundreds of eggs can be done rhythmically in 5 min or so.

It is possible to change the dosage of the injected DNA by simply changing the length of the period of injection. However, try not to use too small an amount of solution, since the volume control is usually difficult; furthermore, injection of too small a volume of the sample solution might confine the injected DNA (or RNA) to the place of the injection. In this connection, it has not yet been clarified to what extent a given DNA (or RNA) sample spread within the egg.

The amount of DNA to be injected should be 1 ng per egg or less (the amount of RNA could be as much as 30 ng, in the volume of 30–50 nL). In the authors' experience, 2 ng/egg is the upper limit to see no apparent toxic effect. When more than 2 mg/egg of DNA are injected, cleavage becomes abnormal dose-dependently, and large blastomeres, whose divisions are retarded or inhibited, are formed at the site of the injection relatively early during the cleavage. Nevertheless, it is sometimes intended to inject as much as 10 ng/egg or more of DNA *(20)*. Under these conditions, cleavage becomes abnormal, but embryos can sometimes develop beyond the blastula and/or gastrula stage, stopping the development at neurulation. If the dosage of the DNA were further increased to 24 ng/egg (the amount corresponding to the total nuclear DNA of a 4000-celled blastula as calculated from the value of 6 pg of nuclear DNA per cell *[21]*), development would stop at blastula or early gastrula stage. There is an experiment in which as much as 50 ng/egg of DNA were injected; in this case, information on its toxicity is not given *(19)*.

In addition to the injection into uncleaved fertilized eggs, it is also possible to inject DNA into embryos at the two-cell stage. In such a case, usually both blastomeres of the embryo are injected, unless it is desired to leave half of the embryo as an uninjected control. The injection into only half of the embryo has been done to test the function of the in vitro synthesized mRNA, as in the experiment by Melton and coworkers *(22)*, who invented the method of mRNA synthesis in an in vitro system.

3.4. After the Injection

Oocytes injected with DNA samples are cultured in 1X MBS, which contains penicillin and streptomycin (50 U each/mL). If necessary, it is possible to put the injected oocytes back to the body cavity of an uninjected female after making a small window at the abdomen with forceps and scissors. In such a case, implanted oocytes are matured and ovulated as fertilizable eggs together with normal eggs by injecting a gonadotropic hormone later. This method was used, for instance, in the experiment by Marc Kirschner and his coworkers *(23)*, using a combination of wild-type oocytes and an albino female (If albino animal was not available, the injected oocytes can be marked by vital staining using Nile blue or neutral red).

DNA- or RNA-injected fertilized eggs must be cultured for about 5 h in either MBS solution, 1X MMR, or Steinberg solution, all containing 5% Ficoll and penicillin and streptomycin as above. During this period, wound healing is expected to occur, and embryos reach stage 7 (early blastula stage). The staging of *Xenopus* embryos is carried out according to Nieuwkoop and Faber *(24)*. Then, the medium has to be changed with 0.1X MBS solution containing penicillin and streptomycin as above until the desired stage is reached. This medium change is absolutely necessary, since high salt conditions induce retardation of development or even malformation.

Before leaving the laboratory, carefully check the development of the injected embryos. If any embryos are injured, eliminate them or culture them separately under the same conditions as the apparently normally developing ones. (If damaged embryos are left with the normally developing ones, they will die overnight and leak substances that will seriously damage the normal embryos.) Also, it is important not to culture the embryos in a crowded condition.

4. Notes

4.1. General Points

It is true that *Xenopus* oocytes and eggs are suitable materials for injecting substances into the cytoplasm or the nucleus (in the case of oocytes), since these are so large (1.2 mm in diameter) and the oocyte has an extremely large nucleus (diameter ~0.4 mm), which could be isolated manually in a pure form by simple surgery. However, unlike mouse eggs, *Xenopus* eggs are not transparent, and the nucleus cannot be seen. Besides, in the cytoplasm of the unfertilized egg, there is no nucleus, only metaphase-arrested chromosomes. A fertilized egg, however, has a zygotic nucleus of about 25 μm in diameter (usually in the upper part of the animal hemisphere) *(25)*. Therefore, DNA has to be injected into the cytoplasm when unfertilized eggs or fertilized eggs are used.

It should also be pointed out that the fertilized egg cleaves so rapidly (approx every 30 min) that its nuclear envelope disappears quickly. This means that even when one succeeds in injecting DNA into the zygotic nucleus in the fertilized egg, DNA will probably be scattered promptly into the cytoplasm at the next division cycle. Therefore, it is necessary to know the fate of the DNA in the egg cytoplasm, which will be described briefly later, with an emphasis on the characteristic features of the *Xenopus* egg system.

4.2. Formation of Nucleus-Like Structures by Injected DNA

DNA injected into the oocyte cytoplasm is degraded *(13)*. However, when injected into the cytoplasm of unfertilized or fertilized eggs, DNA, both circular and linear, is bound to maternally-inherited histones *(26)*, and wrapped by the apparently normal nuclear envelope that contains lamin as demonstrated by Forbes et al. *(27)*. Thus, injected DNA forms mininuclei or nucleus-like structures whose size depends on the amount of the DNA injected *(18,25–29)*. In fertilized eggs, such nucleus-like structures always form in the animal hemisphere *(25)*, a phenomenon that may be relevant to the inability of the injected DNA to replicate in the vegetal hemisphere *(30)*. The nucleus-like structures are partitioned randomly into the nuclei of descendant cells *(18,26,28,29,31)*, probably by sticking to chromosomes *(31)*, albeit mostly as an episomal or nonintegrated form *(32)*. When the amount of injected DNA is less than 1 ng/egg, it is usually difficult to find nucleus-like structures cytologically on sectioned materials. However, the injected DNA appears to be stabilized and partitioned into the nuclei of the blastomeres. Thus, it is a fact that the DNA injected into the cytoplasm of unfertilized and fertilized eggs remain in a stable form.

4.3. Replication and Persistence of Injected DNA

DNA injected into *Xenopus* eggs undergoes remarkable changes, instead of just staying stably within the egg. When closed circular (cc) plasmid DNA is injected into unfertilized or fertilized eggs, it is immediately converted into open circular (oc) form. However, within an hour or so, cc DNA is reformed *(25)* owing to the action of topoisomerase. Then, the levels of cc and oc DNA are kept more or less constant, until most of them disappear at the tailbud stage. When circular DNA containing viral replication origin (pSV2CAT, for instance) is injected into embryos, it replicates considerably, especially in later stages. This is probably because, in later stages, the amount of maternally-inherited DNA polymerase is greatly reduced, and occurrence of the replication origin becomes critical for efficient replication *(33,34)*. In unfertilized eggs, injected circular DNA replicates only to a limited extent, regardless of the replication origin within the DNA *(35,36)*.

On the other hand, linear DNA injected into both unfertilized and fertilized eggs replicates by about 50- to 100-fold; at the same time, its size increases by 10- to 20-fold *(33,37)*. The increase in the size is owing neither to recircularization nor to catenation, since the large-sized DNA is insensitive to type II topoisomerase under the conditions in which kinetoplast DNA (catenated DNA) is completely decatenated (K. Shiokawa and T. Andow, unpublished data). The increase in size is owing to the formation of concatemers of all the

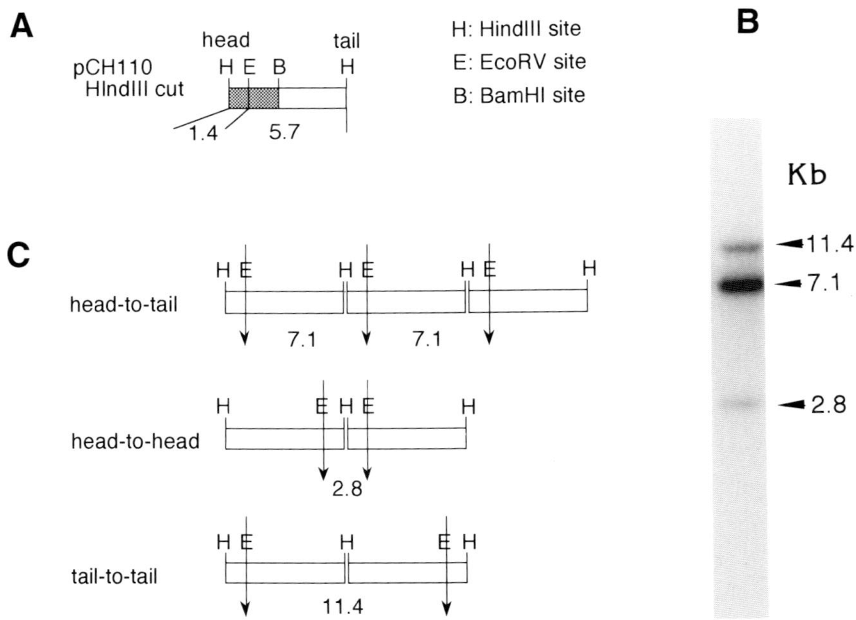

Fig. 7. A result showing the presence of all the possible combinations (head-to-tail, head-to-head, tail-to-tail) in the ligation of the linear DNA. A plasmid pCH110, linearized with *Hin*dIII was injected into the fertilized eggs, and DNA was extracted from the embryos at the gastrula stage. **(A)** Restriction maps of *Hin*dIII-digested pCH110 (7.1 kb) used for microinjection. **(B)** Southern blot analysis of the isolated DNA, whose molecular size was about 23 kb or more (not shown). The DNA was digested with *Eco*RV and analyzed using the *Bam*HI digest as a probe (shadowed box in A). **(C)** Restriction maps showing the fragments to be generated by digestion with *Eco*RV of the ligated DNA. In B, all the fragments of sizes expected to be generated in C are present (Ito, Tashiro, and Shiokawa, unpublished data).

possible combinations (head to head, head to tail, or tail to tail) *(38–40)*. Figure 7 shows a typical result to show this (Ito et al., unpublished data). During or prior to the ligation, both ends of linearized DNA appear to be degraded by exonuclease, since the large-sized DNA is not sensitive to the restriction enzyme used to linearize the DNA. They are digested by another restriction enzyme and give rise to the DNA of approximately monomer size *(33)*.

Most of the large-sized DNA disappears during the tailbud stage, with a relatively small amount of it being preserved even in 1-mo-old tadpoles *(33)*, probably owing to the integration into the host genome. It is said that linearized DNA tends to be integrated more efficiently than circular DNA *(31,33,41)*. In fact, Etkin et al. *(31)* showed that linearized pSV2CAT is transmitted to the genome of offsprings.

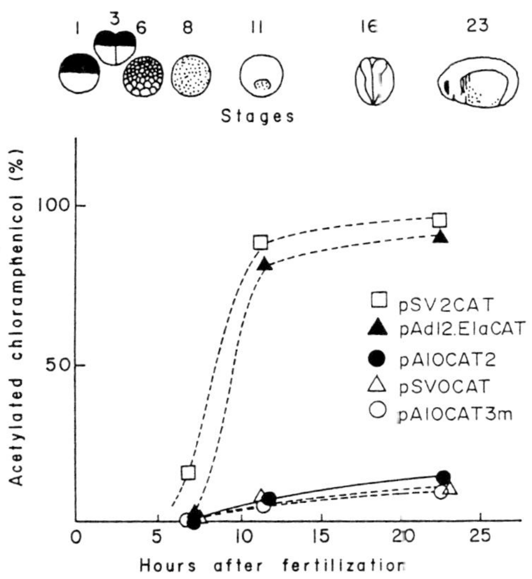

Fig. 8. Expression of CAT enzyme activity during development. Fertilized eggs were injected with circular plasmids (1 ng/egg), and embryos were collected at different stages. The relative strength of the CAT signal was determined by densitometry of the X-ray film. pSV2CAT, pAd12.E1aCAT, and pA10CAT2 contains the SV40 promoter, Adenovirus 12 promoter, and SV40 promoter without enhancer, respectively; pSV0CAT and pA10CAT3m contain no promoter. From Fu et al. *(33)*.

4.4. Expression of Injected DNA

In the oocyte nucleus, circular DNA is efficiently expressed but linearized DNA is not *(33,42–44)*, probably because they lack the supercoiled domain, hence, the torsional stress within the molecules *(44–46)*. In the oocyte nucleus, the expression is independent of the presence or absence of a promoter, probably owing to the occurrence of a large amount of RNA polymerase II *(47)*. As a result, the transcription in the oocyte nucleus is not necessarily from the correct initiation site *(42,48)*.

In both unfertilized and fertilized eggs, circular DNA is expressed mainly from the correct initiation site *(38,40,42)* but the level of the expression is much lower in unfertilized eggs than in embryos *(33,43)*. Also, in unfertilized eggs, the expression of linearized DNA takes place, but only to an extremely low level *(33)*. In embryos, the expression of both circular and linear DNA depends on the promoter. Figure 8 shows the expression of various circular CAT gene constructs in embryos during development. The number of injected embryos used here is the same throughout stages *(33)*. In embryos, the expres-

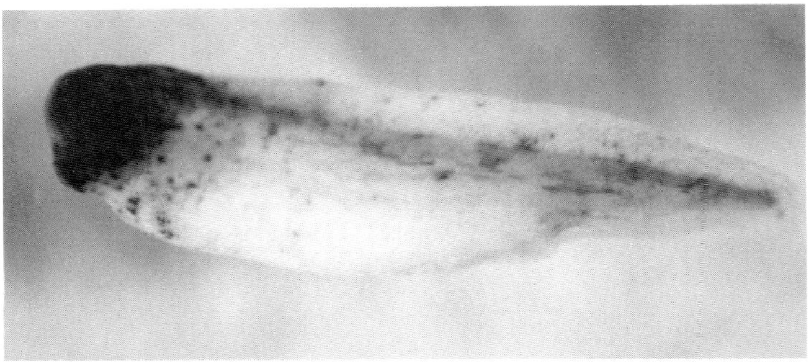

Fig. 9. A localized expression pattern of *Xenopus* Xdll gene promoter region connected to β-galactosidase gene. The injected gene was expressed in this case mainly in the head region. From Shiokawa et al. *(58)*.

sion of linear DNA is usually slightly lower than that of circular DNA. A peculiar observation in embryos is that, whereas circular pSV0CAT that does not contain a promoter is not expressed appreciably, its linear form is expressed at a high level *(33)*. Therefore, in embryos, linearized DNA is somehow endowed with the activity to serve as an active template independently of the presence or absence of a promoter. This may be related to the concatemerization and/or replication, which may generate a cryptic promoter on one hand, and also torsional stress by binding to nuclear structure such as the nuclear matrix, on the other *(33)*.

So far, many genes have been injected and expressed in *Xenopus* embryos in regionally and/or temporally regulated manners *(34,49–57)*. However, for the study of stage-dependent expression, use of circular DNA is recommended, since linearized DNA appears to be expressed temporally in an uncontrolled manner *(37)*. For the study of regionally regulated expression, β-galactosidase is often used as a reporter gene. For instance, Fig. 9 shows a localized expression of β-gal gene with *Xenopus* distal-less gene (*Xdll*) promoter *(58)*. Similarly, restricted expression of β-gal gene has been seen in our laboratory for *Xenopus* follistatin gene (Koga et al., unpublished observation) and type C (brain-type) aldolase (Yatsuki et al., unpublished observation).

However, these results are usually only qualitatively reliable. Therefore, to use β-galactosidase gene for quantitative studies, it is better to measure the activity with the embryo extract by colorimetric determination. Figure 10 gives such a result (Bannai et al., unpublished data).

Table 1 summarizes the behavior of exogenous DNA in the three kinds of *Xenopus* egg cells *(33)*.

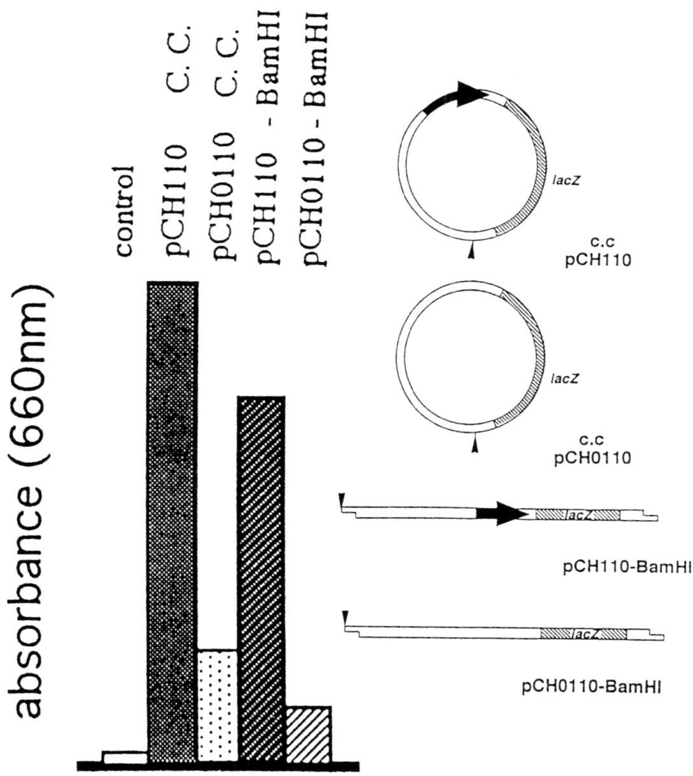

Fig. 10. A result of the colorimetric determination of the β-gal gene expression from circular and linear plasmid DNA. Fifteen embryos injected with different DNA (shown in the right; all at 0.5 pg/egg) were homogenized at the gastrula stage and β-galactosidase activity was determined in the crude embryo extract by measuring absorbancy at 660 nm (Bannai, Ito, Tashiro, and Shiokawa, unpublished data).

4.5. Expression of Injected DNA in Relation to MBT

When the extent of gene expression is compared at different stages, the difference in the number of cells per embryo should be taken into consideration in interpreting the data *(11,12,59,60)*. For instance, pSV2CAT (1 ng/egg) injected into fertilized eggs was once concluded to be expressed only at and after the MBT (*see* Fig. 8, and refs. *7,8,60*). Similar data were obtained also using CAT genes fused to the promoter of *Xenopus* elongation factor-1α (EF-1α) *(62)* and glucose-regulated protein (grp78) *(63)*.

However, when the amount of circular pSV2CAT was increased tenfold (10 ng/egg), CAT enzyme activity was detected at 3 h (early cleavage stage) and 5 h (late cleavage stage) after injection *(20)*. More importantly, even when the

Table 1
Expression and Molecular Structures of CAT genes in *Xenopus laevis* Embryos, Eggs, and Oocytes

Stages	CAT gene in circular form	CAT gene in linear form
Embryos	Actively expressed, depending on the strength of the promoter.	Actively expressed, regardless of the presence or absence of the promoter.
	DNA replicates moderately or only slightly; most DNA of monomer size.	DNA replicates actively and forms concatemers; occasionally integrated in the host DNA.
	Nucleus-like structures formed.	Nucleus-like structures formed.
Eggs	Very weakly expressed, depending on the promoter.	Very weakly expressed, irrespective of the presence or absence of the promoter.
	DNA replicates moderately; mostly circular DNA of monomer size.	DNA is ligated; forms concatemers and actively replicated.
	Nucleus-like structures formed.	Nucleus-like structures formed.
Oocytes	Actively expressed, even when the promoter is missing.	Expression undetectable even when a strong promoter is present.
	DNA stably preserved.	DNA stably preserved.
	DNA injected into the cytoplasm is degraded.	DNA injected into the cytoplasm is degraded.

number of embryos is increased tenfold (100 embryos/sample) without increasing the dosage of DNA (1 ng/egg), CAT enzyme signal is detectable at 4 h (middle of the cleavage stage) after fertilization (Fig. 11). In this relation, a CAT fusion gene #254, carrying the promoter of *Xenopus* α-actin gene, is expressed only at and after the neurula stage at both 1 ng/egg and 10 ng/egg. Therefore, it is not necessarily the MBT, but the nature of the promoter that determines the timing of the initiation of exogenous genes.

4.6. Advantages of Using Hybrid Embryos

When pXlr101A, a plasmid that contains *Xenopus laevis* rDNA single repeat, is injected into fertilized eggs of *Xenopus borealis*, it is expressed first at the late blastula stage *(64)* concurrently with the expression of endogenous rDNA *(10–12,60)*. Here, the "hybrid" combination is important, because the expression of *X. laevis* rDNA is dominant over that of the *X. borealis* rDNA *(65)*, and furthermore, *X laevis* rDNA new transcript is more easily detected in cells that contain *X. borealis* rRNA than in cells that contain *X. laevis* rRNA (4000 ng/egg) *(1,5)*. Thus, hybrid embryos are sometimes very useful.

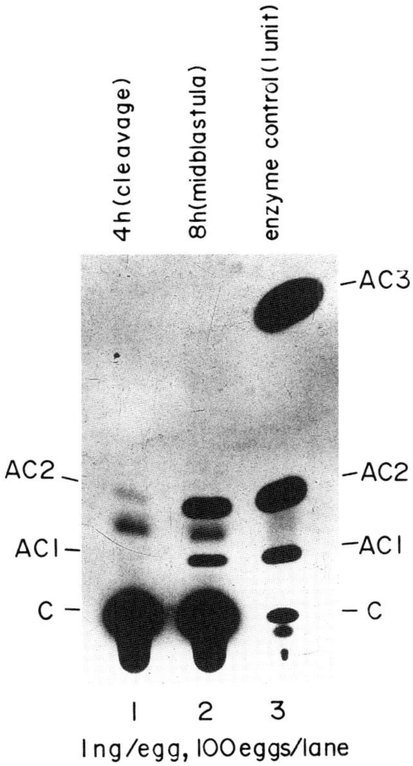

Fig. 11. CAT enzyme expression before MBT. Fertilized eggs (100 embryos each) were injected with circular pSV2CAT (1 ng/egg) and harvested at the indicated stages for CAT enzyme assay. Autoradiography was carried out for 2 mo. A relatively strong signal, which appeared between AC1 and AC2 is not owing to CAT enzyme expression because it was obtained even in uninjected embryos (not shown). From Shiokawa et al. *(37)*.

References

1. Brown, D. D. and Littna, E. (1964) RNA synthesis during the development of *Xenopus laevis*, the South African clawed toad *J. Mol. Biol.* **8,** 669–687.
2. Brown, D. D. and Littna, E. (1966) Synthesis and accumulation of DNA-like RNA during embryogenesis of *Xenopus laevis*. *J. Mol. Biol.* **20,** 81–94.
3. Brown, D. D. and Littna, E. (1966) Synthesis and accumulation of low molecular weight RNA during embryogenesis of *Xenopus laevis*. *J. Mol. Biol.* **20,** 95–112.
4. Shiokawa, K. and Yamana, K. (1965) Demonstration of "polyphosphate" and its possible role in RNA synthesis during early development of Rana japonica embryos. *Exptl. Cell Res.* **38,** 180–186.

5. Shiokawa, K. and Yamana, K. (1967) Pattern of RNA synthesis in isolated cells of *Xenopus laevis* embryos. *Dev. Biol.* **16,** 368–388.
6. Woodland, H. R. and Gurdon, J. B. (1968) The relative rates of synthesis of DNA. sRNA and rRNA in the endodermal region and other parts of *Xenopus laevis* embryos. *J. Embryol. Exp. Morph.* **19,** 363–385.
7. Newport, J. and Kirschner, M. (1982) A major developmental transition in early *Xenopus* embryos: I. Characterization and timing of cellular changes at the midblastula stage. *Cell* **30,** 675– 686.
8. Newport, J. and Kirschner, M. (1982) A major developmental transition in early *Xenopus* embryos: II. Control of the onset of transcription. *Cell* **30,** 687– 696.
9. Shiokawa, K., Misumi, Y., and Yamana, K. (1981) Demonstration of rRNA synthesis in pre-gastrular embryos of *Xenopus laevis*. *Dev. Growth Differ.* **23,** 579–587.
10. Shiokawa, K., Tashiro, K., Misumi, Y., and Yamana, K. (1981) Non-coordinated synthesis of RNAs in pre-gastrular embryos of *Xenopus laevis*. *Dev. Growth Differ.* **23,** 589–597.
11. Shiokawa, K. (1991) Gene expression from endogenous and exogenously-introduced DNAs in early embryogenesis of *Xenopus laevis*. *Dev. Growth Differ.* **33,** 1–8.
12. Shiokawa, K., Kurashima, R., and Shinga, J. (1994) Temporal control of gene expression from endogenous and exogenously-introduced DNAs in early embryogenesis of *Xenopus laevis*. *Int. J. Dev. Biol.* **38,** 249–255.
13. Gurdon, J. B. and Melton, D. A. (1981) Gene transfer in amphibian eggs and oocytes. *Ann. Rev. Genet.* **15,** 189–218.
14. Gurdon, J. B. and Brown, D. D. (1977) DNA microinjection, in *The Molecular Biology of the Genetic Apparatus*, vol. 2 (T'so, P., ed.), North-Holland, Amsterdam, pp. 111–123.
15. Asano, M. and Shiokawa, K. (1993) Behavior of exogenously-injected DNAs in early embryos of *Xenopus laevis*. *Zool. Sci.* **10,** 197–222.
16. Dumont, J. N. (1972) Oogenesis in *Xenopus laevis*. I. Stages of oocyte development in laboratory maintained animals. *J. Morphol.* **136,** 153–180.
17. Shiokawa, K., Nada, O., and Yamana, K. (1967) RNA synthesis in isolated cells from *Xenopus laevis* embryos. *Nature* **213,** 1027,1028.
18. Shiokawa, K., Fu, Y., Nakakura, N., Tashiro, K., Sameshima, M., and Hosokawa, K. (1989) Effects of the injection of exogenous DNAs on gene expression in early embryos and coenocytic egg cells of *Xenopus laevis*. *Roux's Arch. Dev. Biol.* **198,** 78–84.
19. Lund, E. and Dahlberg, J. E. (1992) Control of 4-8S RNA transcription at the midblastula transition in *Xenopus laevis* embryos. *Genes Dev.* **6,** 1097–1106.
20. Shiokawa, K., Yamana, K., Fu, Y., Atsuchi, Y., and Hosokawa, K. (1990) Expression of exogeneously introduced bacterial chloramphenicol acetyltransferase genes in *Xenopus laevis* embryos before the midblastula transition. *Roux's Arch. Dev. Biol.* **198,** 322–329.

21. Dawid, I. B. (1966) Deoxyribonucleic acid in amphibian eggs. *J. Mol. Biol.* **12,** 581–599.
22. Harvey, R. P. and Melton, D. A. (1988) Microinjection of synthetic Xhox-1A homeobox mRNA disrupts somite formation in developing *Xenopus* embryos. *Cell* **53,** 687–697.
23. Amaya, E., Musci, T. J., and Kirschner, M. W. (1991) Expression of a dominant negative mutant of the FGF receptor disrupts mesoderm formation in *Xenopus* embryos. *Cell* **66,** 257–270.
24. Nieuwkoop, P. D. and Faber, J. (1956) Normal Table of *Xenopus laevis* (Daudin). North Holland, Amsterdam.
25. Shiokawa, K., Tashiro, K., Yamana, K., and Sameshima, M. (1987) Electron microscopic studies of giant nucleus-like structure formed by lambda DNA introduced into the cytoplasm of *Xenopus laevis* fertilized eggs and embryos. *Cell Differ.* **20,** 253–261.
26. Shiokawa, K., Yoshida, M., Fukamachi, H., Fu, Y., Tashiro, K., and Sameshima, M. (1992) Cytological studies of large nucleus-like structures formed by exogenously-injected linear and circular DNAs in fertilized eggs of *Xenopus laevis*. *Dev. Growth Differ.* **34,** 79–90.
27. Forbes, D. J., Kirschner, M. W., and Newport, J. W. (1983) Spontaneous formation of nucleus-like structures around bacteriophage DNA microinjected into *Xenopus* eggs. *Cell* **34,** 13–23.
28. Shiokawa, K., Sameshima, M., Tashiro, K., Miura, T., Nakakura, N., and Yamana, K. (1986) Formation of nucleus-like structure in the cytoplasm of lambda-DNA-injected fertilized eggs and its partition into blastomeres during early embryogenesis in *Xenopus laevis*. *Dev. Biol.* **116,** 539–542.
29. Trendelenburg, M. F., Oudet, P., Spring, H., and Montag, M. (1986) DNA injections into *Xenopus* embryos: fate of injected DNA in relation to formation of embryonic nuclei. *J. Embryol. Exp. Morph.* **97(Suppl.),** 243–255.
30. Hofmann, A., Montag, M., Steinbeisser, H., and Trendelenburg, F. (1990) Plasmid and bacteriophage lambda-DNA show differential replication characteristics following injection into fertilized eggs of *Xenopus laevis*: dependence on period and site of injection. *Cell Differ. Dev.* **30,** 77–85.
31. Etkin, L. and Pearman, B. (1987) Distribution, expression and germ line transmission of exogenous DNA sequences following microinjection into *Xenopus laevis* eggs. *Development* **99,** 15–23.
32. Andres. A.-C., Muellner, D. B., and Ryffel, G. U. (1984) Persistence, methylation and expression of vitellogenin gene derivatives after injection into fertilized eggs of *Xenopus laevis*. *Nucleic Acids Res.* **12,** 2283–2302.
33. Fu, Y., Hosokawa, K., and Shiokawa, K. (1989) Expression of circular and linearized bacterial chloramphenicol acetyltransferase genes with or without viral promoters after injection into fertilized eggs, unfertilized eggs and oocytes of *Xenopus laevis*. *Roux's Arch. Dev. Biol.* **198,** 148–156.
34. Shiokawa, K., Yamazaki, T., Fu, Y., Tashiro, K., Tsurugi, K., Motizuki, M., Ikegami, Y., Araki, E., Andoh, T., and Hosokawa, K. (1989b) Persistence and

expression of circular DNAs encoding Drosophila amylase, bacterial chloramphenicol acetyltransferase, and others in *Xenopus laevis* embryos. *Cell Struct. Func.* **14,** 261–269.

35. Harland, R. M. and Laskey, R. A. (1980) Regulated replication of DNA microinjected into eggs of *Xenopus laevis*. *Cell* **21,** 761–771.
36. Mechali, M. and Kearsey, S. (1984) Lack of specific sequence requirement for DNA replication in *Xenopus* eggs compared with high sequence specificity in yeast. *Cell* **38,** 55–64.
37. Shiokawa, K., Fu, Y., Hosokawa, K., and Yamana, K. (1990) Temporally uncontrolled expression of linearized plasmid DNA which carries bacterial chloramphenicol acetyltransferase gene with *Xenopus* cardiac a-actin promoter after injection into *Xenopus* fertilized eggs. *Roux's Arch. Dev. Biol.* **199,** 171–180.
38. Bendig, M. M. (1981) Persistence and expression of histone genes injected into *Xenopus* eggs in early development. *Nature* **292,** 65–67.
39. Bendig, M. M. and Williams, J. G. (1984) Differential expression of the *Xenopus laevis* tadpole and adult b-globin genes when injected into fertilized *Xenopus laevis* eggs. *Mol. Cell Biol.* **4,** 567–570.
40. Rusconi, S. and Schaffner, W. (1981) Transformation of frog embryos with a rabbit beta-globin gene. *Proc. Natl. Acad. Sci. USA* **78,** 5051–5055.
41. Etkin, L. D., Pearman, B., Roberts, M., and Bektesh, S. (1984) Replication, integration and expression of exogenous DNA injected into fertilized eggs of *Xenopus laevis*. *Differentiation* **26,** 191–202.
42. Bendig, M. M. and Williams, J. G. (1984) Fidelity of transcription of *Xenopus laevis* globin genes injected into *Xenopus laevis* oocytes and unfertilized eggs. *Mol. Cell Biol.* **4,** 2109–2119.
43. Fu, Y., Sato, K., Hosokawa, K., and Shiokawa, K. (1990) Expression of circular plasmids which contain bacterial chloramphenicol acetyltransferase gene connected to the promoter of polypeptide IX gene of human adenovirus type 12 in oocytes, eggs and embryos of *Xenopus laevis*. *Zool. Sci.* **7,** 195–200.
44. Harland, R. M., Weintraub, H., and McKnight, S. L. (1983) Transcription of DNA injected into *Xenopus* oocytes is influenced by template topology. *Nature* **302,** 38–43.
45. Probst, E., Kressmann, A., and Birnstiel, M. L. (1979) Expression of sea urchin histone genes in the oocyte of *Xenopus laevis*. *J. Mol. Biol.* **135,** 709–732.
46. Pruitt, S. and Reeder, R. H. (1984) Effect of topological constraint on transcription of ribosomal DNA in *Xenopus* oocytes. Comparison of plasmid and endogenous genes. *J. Mol. Biol.* **174,** 121–139.
47. Roeder, R. G. (1974) Multiple forms of deoxyribonucleic acid-dependent ribonucleic acid polymerase in *Xenopus laevis*. Levels of activity during oocyte and embryonic development. *J. Biol. Chem.* **249,** 249–256.
48. Wickens, M. P., Woo, S., O'Malley, B. W., and Gurdon, J. B. (1980) Expression of a chicken chromosomal ovalbumin gene injected into frog oocyte nuclei. *Nature* **285,** 628–634.

49. Wilson, C., Cross, G. S., and Woodland, H. R. (1986) Tissue-specific expression of actin genes injected into *Xenopus laevis. Cell* **47,** 589–599.
50. Steinbeisser, H., Alonso, A., Epperlein, H.-H., and Trendelenburg, M. F. (1989) Expression of mouse histone H1(o) promoter sequences following microinjection into *Xenopus* oocytes and developing embryos. *Int. J. Dev. Biol.* **33,** 361–368.
51. Mohun, T. J., Garrett, N., and Gurdon, J. B. (1986) Upstream sequences required for tissue-specific activation of the cardiac actin gene in *Xenopus laevis* embryos. *EMBO J.* **5,** 3185–3193.
52. Steinbeisser, H., Hofmann, A., Stutz, F., and Trendelenburg, M. F. (1988) Different regulatory elements are required for cell-type and stage-specific expression of the *Xenopus laevis* skeletal muscle actin gene upon injection in *X. laevis* oocytes and embryos. *Nucleic Acids Res.* **16,** 3223–3238.
53. Brennan, S. M. (1990) Transcription of endogenous and injected cytoskeletal actin genes during early embryonic development in *Xenopus laevis*. *Differentiation* **44,** 111–121.
54. Jonas, E. A., Snape, A. M., and Sargent, T. D. (1989) Transcriptional regulation of a *Xenopus* embryonic epidermal keratin gene. *Development* **106,** 399–405.
55. Krieg, P. A. and Melton, D. A. (1987) An enhancer responsible for activating transcription at the mid-blastula transition in *Xenopus* development. *Proc. Natl. Acad. Sci. USA* **84,** 2331–2335.
56. Krone, P. H. and Heikkila, J. J. (1989) Expression of microinjected hsp70/CAT and hsp30/CAT chimeric genes in developing *Xenopus laevis* embryos. *Development* **106,** 271–281.
57. Winning, R. S., Bols, N. C., Wooden, S. K., Lee, A. S., and Heikkila, J. J. (1992) Analysis of the expression of a glucose-regulated protein (GRP78) promoter/CAT fusion gene during early *Xenopus laevis* development. *Differentiation* **49,** 1–6.
58. Shiokawa, K., Tashiro, K., Kurashima, R., and Amano, M. (1994) Frogs as the laboratory animal for exogenous gene introduction, in *Mouse as a Laboratory Animal* (Yamamura,K., Katsuki, M., and Aizawa, S., eds.), Nakayamashoten, Japan (In Japanese), p. 279.
59. Nakakura, N., Miura, T., Yamana, K., Ito, A., and Shiokawa, K. (1987) Synthesis of heterogeneous mRNA-like RNA and low-molecular-weight RNA before the midblastula transition in embryos of *Xenopus laevis*. *Dev. Biol.* **123,** 421–429.
60. Shiokawa, K., Misumi, Y., Tashiro, K., Nakakura, N., Yamana, K., and Oh-uchida, M. (1989) Changes in the patterns of RNA synthesis in early embryogenesis of *Xenopus laevis*. *Cell Differ. Dev.* **28,** 17–26.
61. Etkin, L. D. and Balcells, S. (1985) Transformed *Xenopus* embryos as a transient expression system to analyze gene expression at the midblastula transition. *Dev. Biol.* **108,** 173–178.
62. Johnson, A. D. and Krieg, P. A. (1994) pXeX, a vector for efficient expression of cloned sequences in *Xenopus* embryos. *Gene* **147,** 223–226.

63. Vezina, C., Wooden, S. K., Lee, A. S., and Heikkile, J. J. (1994) Constitutive expression of a microinjected glucose-regulated protein (grp78) fusion gene during early *Xenopus laevis* development. *Differentiation* **57,** 171–177.
64. Busby, S. J. and Reeder, R. H. (1983) Spacer sequences regulate transcription of ribosomal gene plasmids injected into *Xenopus* embryos. *Cell* **34,** 989–996.
65. Honjo, T. and Reeder, R. H. (1973) Preferential transcription of *Xenopus laevis* ribosomal RNA in interspecies hybrids between *Xenopus laevis* and *Xenopus mulleri*. *J. Mol. Biol.* **80,** 217–228.

21

Analysis of Heterologous Gene Expression in *Xenopus* Blastomeres

Sally A. Moody

1. Introduction

Like *Xenopus* oocytes, *Xenopus* cleavage blastomeres are ideal synthetic factories for the expression of heterologous gene products. For 6 h after fertilization, the embryos are transcriptionally quiescent *(1)*, but mRNA is recruited for translation at a rate greater than in the oocyte, and the protein products of exogenous mRNA and DNA, introduced by microinjection, are synthesized as efficiently as those of endogenous RNA *(2)*. This is an excellent system in which to examine the developmental function of heterologous genes, because one can express genes during developmentally inappropriate stages and in developmentally inappropriate regions. For example, to test whether a gene is involved in dorsal axis formation, researchers express the gene in a blastomere that produces ventral tissues and assay whether the blastomere's fate changes to a dorsal one *(3–6)*. Spatial misexpression is possible in *Xenopus* because:

1. The cardinal axes (anterior, posterior, dorsal, and ventral) can be distinguished by the two-cell stage;
2. Many embryos per clutch cleave in a reproducible pattern (Fig. 1); and
3. By using such embryos, very detailed fate maps have been constructed for the first five cell cycles *(7–11)*.

Thus, it is possible to target heterologous genes to blastomeres of the embryo that are the progenitors of the tissues in which one wants the gene to be expressed. Tissue origins are the most segregated at the 32-cell stage, and a fate map of that stage is provided in Fig. 2. A similar map for the 16-cell stage *(8)* and lineage diagrams for the major organ systems *(10)* also are available. *Xenopus* embryos also are ideal subjects in which to identify the elements in

From: *Methods in Molecular Biology, vol. 62: Recombinant Gene Expression Protocols*
Edited by: R. Tuan Humana Press Inc., Totowa, NJ

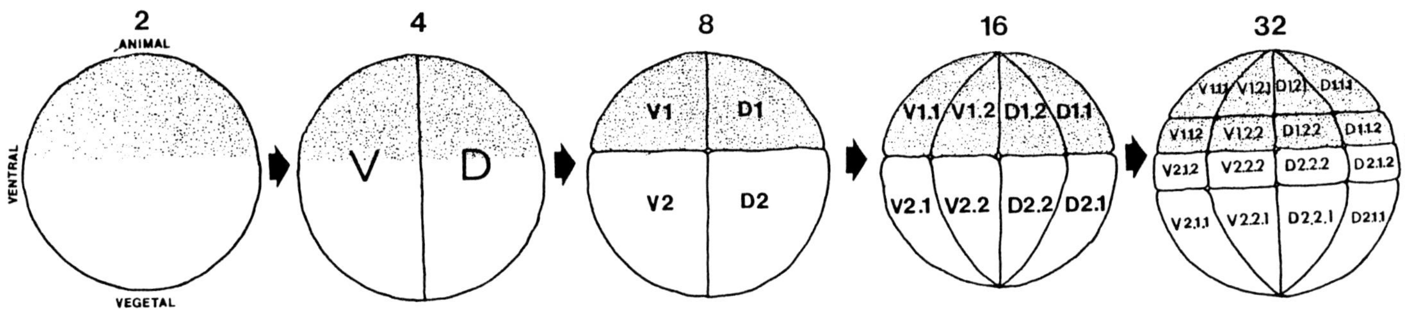

Fig. 1. The nomenclature of the cleavage stage blastomeres, right side view.

promoter/enhancer regions that control tissue-specific gene expression. By injecting promoter/reporter constructs into blastomere progenitors of the desired tissue, one can test for the presence of tissue-specific sequences *(12,13)*.

This chapter provides detailed protocols for preparing embryos, identifying unique blastomere progenitors, injecting nucleic acid constructs, and analyzing their expression.

2. Materials

2.1. Equipment

1. Refrigerated centrifuge for 1.5-mL tubes.
2. PLI-100 Pico-Injector (Medical Systems, Greenvale, NY) with pressurized N_2 tank attached.
3. Horizontal pipet puller, such as the Flaming/Brown Model P97 (Sutter Instruments, Novato, CA). Borosilicate capillary glass with outer diameter of about 0.8–1.0 mm. Glass should *not* contain a filament and should be siliconized and autoclaved before being pulled.
4. Dissection stereomicroscope placed on a steel plate in an area that is level and free of vibration. Illumination should be provided by a fiberoptic lamp, not by transillumination from the base of the microscope.
5. A micromanipulator with X, Y, and Z axes, mounted on a magnetic base secured to the steel plate.
6. Injection dish can be a 35-mm Petri dish filled with a base of blue, nontoxic, nondrying modeling clay (such as plasticene) in which 1.5-mm depressions have been made with a glass ball (obtained by melting the tip of a Pasteur pipet). Alternatively, affix a piece of Nitex mesh (Fisher Scientific, Pittsburgh, PA) to the bottom of a Petri dish with a few drops of chloroform. The grids of the mesh are about the same diameter as the embryos.
7. Cryostat with microtome blade.
8. Humidified chamber: A clear plastic box with tight fitting lid. Glue platforms to the floor to support microscope slides flat (such as empty coverslip boxes). Line remaining floor space with wet kimwipes during use.

2.2. Reagents

1. HCG: human chorionic gonadotropin made with sterile water at a concentration of 1000 IU/mL. Should be refrigerated and used within 1 mo.
2. Steinberg's solution: 60 m*M* NaCl, 0.67 m*M* KCl, 0.34 m*M* $Ca(NO_3)_2$, 0.83 m*M* $MgSO_4$, 4 m*M* Tris-HCl, 0.66 m*M* Tris Base, pH 7.4. Should be refrigerated and used within 1 mo.
3. MS222: 0.3% methanesulfonate salt (3-aminobenzcic acid ethyl ester, or tricaine).
4. Dejellying solution: 2% cysteine hydrochloride (aqueous) pH to 8.1 by adding 10*M* NaOH. Should be made fresh each day.
5. NA: 3*M* Na Acetate, pH 6.8. Can be stored at room temperature for months.
6. Ethanol: 100% EtOH: absolute (200 proof); 70% EtOH: mix 70 mL of absolute ethanol with 30 mL RNase DNase-free distilled water. Store both in freezer.

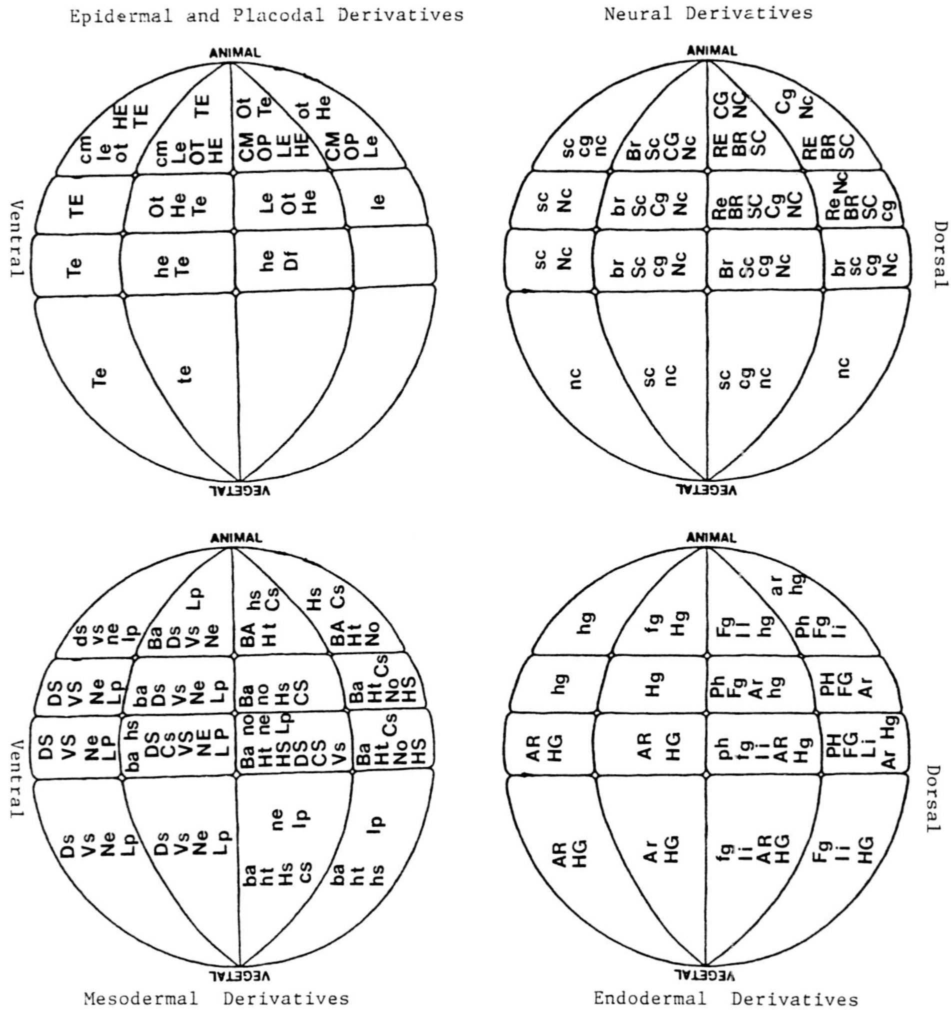

Fig. 2. Fate maps of the 32-cell blastomeres, summarized from data presented in ref. *9*. Only the 16 blastomeres of the right side of the embryo are shown. The same map applies to the left side. For clarity, the same embryo is shown four times with either epidermal, neural, mesodermal, or endodermal structures represented. If a structure appears entirely in capital letters, then it receives major contributions from the cell in which that structure is written. If a structure only has the first letter capitalized, then it is only moderately populated by descendants of the cell in which it is written. If a structure appears entirely in lower case, then it was populated by descendants of the cell in which it is written in only 50% of the embryos studied. Abbreviations: Epidermal and placodal derivatives: cm, cement gland; df, dorsal fin epidermis; he, head epidermis; le, lens; op, olfactory placode; ot, otocyst; and te, trunk epidermis. Neural derivatives: br, brain; cg, cranial ganglion; nc, neural crest; re, retina; sc, spinal cord.

(continued)

7. Ficoll solution: 3% Ficoll (type 400 DL) in Steinberg's solution. Solution should be filter-sterilized, stored cold, and used within a week.
8. PARA fixative: 4% paraformaldehyde powder in 50 mL H_2O. Heat to 60°C while stirring. Add a few drops 1*M* NaOH to clear. Cool on ice to RT, add 50 mL 2X PBS.
9. PBS: 0.1*M* phosphate buffered saline (81 mL 0.2*M* dibasic sodium phosphate; 19 mL 0.2*M* monobasic sodium phosphate; 100 mL H_2O; 1.8 g NaCl, pH 7.4).
10. Blocking solution: 5% bovine serum albumin, 5% normal serum of animal in which secondary antibody was made, in PBS.
11. Primary solution: Primary antibody diluted according to manufacturer; 10% blocking solution in PBS.
12. Secondary solution: 1% unlabeled secondary antibody (e.g., Gt α-Mo IgG); 10% blocking solution in PBS.
13. Tertiary solution: 1% peroxidase-antiperoxidase antibody (PAP, same species as primary antibody); 10% blocking solution in PBS.
14. DAB: 2.5 mL 1% diaminobenzidine; 6 µL 30% H_2O_2; 20 mL 0.2*M* monobasic sodium phosphate; 80 mL 0.2*M* dibasic sodium phosphate; 98 mL 20 m*M* sodium imidazole.

3. Methods

3.1. Micropipet Calibration

1. Pull capillary tubes into micropipets on a horizontal puller. For injecting embryos, the ideal shank length is 2 mm and taper length is about 1 mm (*see* Notes 1 and 2).
2. Break tip carefully under the microscope with sterile fine forceps, or bevel the edge at a 20–30° angle with a Narashige EG-3 grinder.
3. Attach micropipet to holder on Pico-injector.
4. Set input pressure at the gas regulator of the N_2 tank to 115 psi, balance pressure of the Pico-injector at 0.2–0.7 psi and output pressure of the Pico-injector between 15 and 50 psi.
5. Stretch clean Parafilm over a 35-mm Petri dish. Place a 2–4 µL droplet of sterile RNase-DNase-free water on Parafilm. With the micromanipulator, position micropipet tip in water and push "fill" button. Watch the process through the dissection microscope at all times to avoid getting air or debris into the micropipet.
6. Set injection duration pulse at 50 ms and eject 10 times onto the Parafilm to form a droplet. Pick up droplet in a 64-mm Drummond 1-µL microcap and measure length of column of water. Repeat at pulses of 100, 200, 400, and 600 ms. From the lengths of the water columns, one can calculate the average volume expressed by a single pulse at each duration (*see* Note 3):

Fig. 2. *(continued)* Mesodermal derivatives: ba, branchial arch; cs, central trunk somite; ds, dorsal trunk somite; hs, head somite; ht, heart; lp, lateral plate; ne, nephric tubules; no, notochord; vs, ventral trunk somite. Endodermal derivatives: ar, archenteron roof; fg, foregut; hg, hindgut; li, liver; and ph, pharynx. Orientation is the same as in Fig. 1.

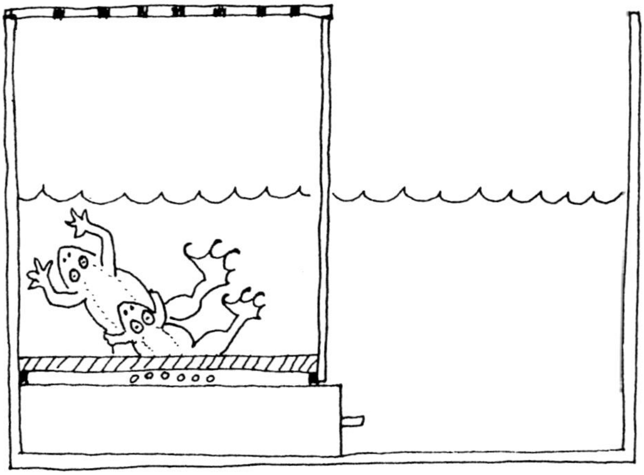

Fig. 3. Side view of plans for a tank designed to facilitate the collection of naturally fertilized eggs. The tank is constructed from opaque plexiglass. It is divided into two chambers. In one chamber (left side) the mating frogs sit on a stiff plastic grid floor (cross hatch) through which fertilized eggs (small dots) can fall. They fall into square Petri dishes that fit tightly into a drawer that occupies the space below the grid. A tightly fitting lid, with airholes drilled through, covers the top of this chamber to prevent the frogs from escaping. To collect eggs throughout the day without disturbing the frogs, one simply reaches into the chamber on the right side, pulls open the drawer, and removes the Petri dishes. By replacing the dishes each time, one can collect eggs multiple times.

$$X = (1000 \text{ nL}) (\text{length measured}) (0.1)/64 \text{ mm}^3$$

7. Store calibrated micropipets in a dust-free box until use (*see* Note 4).

3.2. Embryo Preparation

1. Two major methods are used to obtain fertilized eggs: natural matings and in vitro fertilization. For both methods, adult frogs are primed by hormone injections. Typically, males receive an injection of 300–500 IU of HCG 2 d before the experiment and again 12–14 h before the experiment. Females receive an injection of 600–800 IU 12–14 h before the experiment. Details for how to inject frogs can be found in ref. *14* (*see* Notes 5–7).
2. For natural matings, place the male and female frogs in a 15-gal tank filled with 8 gal of 0.1X Steinberg's solution 12 h prior to when fertilized eggs are desired. The bottom of the tank should contain square Petri dishes covered with a stiff plastic screen. The frogs should be left in the dark (we drape the chamber with black cloth) for the next 24 h. As eggs are laid, they drop through the plastic screen into square Petri dishes, and can be collected throughout the day. We use a specially-constructed plexiglass chamber for this purpose (Fig. 3).

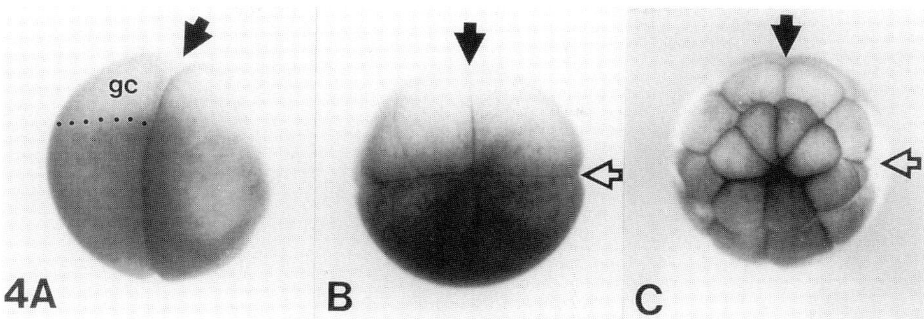

Fig. 4. Animal pole views of early cleavage stage embryos selected for consistency with the fate maps. **(A)** One should select embryos in which the grey crescent region is easily distinguished (gc, brackets) and is bisected by the first cleavage furrow (arrow). **(B)** At the 4-cell stage the second cleavage furrow (open arrow) should separate light dorsal cells from dark ventral cells. **(C)** Stereotypic cleavages at the 32-cell stage. Same orientation as in A. In all figures the closed arrow denotes the first cleavage furrow and the open arrow denotes the second furrow.

3. For in vitro fertilization, the mature eggs are gently squeezed from the hormone-treated female into Petri dishes. Males are anesthetized by submersion in an ice bath with MS222, sacrificed, and their testes are removed. The testes are minced and the released sperm are added to the eggs. Details for this method can be found in ref. *15*.
4. Remove the jelly coats from fertilized eggs that have just begun to cleave by gently swirling the eggs in 4X vol of dejellying solution for 5–15 min. After the jelly coats are free, immediately wash embryos in four changes of 1X Steinberg's solution (*see* Note 8).
5. Select two-cell embryos in which the first cleavage furrow bisects the gray crescent area (Fig. 4) (*see* Notes 9 and 10).
6. Transfer selected embryos to 1X Steinberg's solution in a clean Petri dish. Store at 18–20°C until they have reached the desired stage for injection.

3.3. Nucleic Acid Preparation

1. Isolate RNA or DNA preparations according to standard procedures and quantify.
2. Precipitate in 2.5 vol 100% EtOH and 0.1 vol NA in –20°C freezer.
3. Centrifuge 45 min, 4°C at 14,000*g*.
4. Remove supernatant, add 500 µL ice-cold 70% EtOH, and centrifuge for 10–15 min (4°C at 14,000*g*, *see* Note 11).
5. Remove supernatant, hold tube upside-down, and remove any residual ethanol with a sterile pipet tip or pledget of a Kimwipe.
6. Allow pellet to air dry on ice with cap open, covered with a Kimwipe, for about 30 min (*see* Note 12).

7. Using sterile RNase-free water, bring sample to a working concentration, which needs to be determined empirically. Generally, synthetic RNAs are injected in the range of 200 pg to 10 ng per blastomere, and plasmids are injected at 50–500 pg per blastomere. It is desirable to keep injection volumes at 1–4 nL per blastomere, although up to 10 nL per blastomere can be tolerated up to the 32-cell stage (*see* Notes 13–20).
8. Just before filling the injection micropipet, briefly (1–2 min) spin the tube to pellet any particulate material that might clog the tip.

3.4. Embryo Injection

1. Attach chosen micropipet to Pico-injector and set input, output, balance pressures, and duration pulse to match those used when that micropipet was calibrated.
2. Place a 0.5–1.0-µL drop of nucleic acid solution (spun briefly to pellet particulate debris) on clean parafilm surface. Watch under the dissecting scope to make sure there is no debris on the parafilm that could contaminate the material.
3. Place tip of micropipet in droplet and press "fill" button. Watch this process through the microscope so no air or debris are introduced into the micropipet. Clogged tips can be cleared by pressing the "inject" button or the "clear" button. Make sure the micropipet tip is not close to the parafilm surface when it is cleared, as a large pressure blast could cause it to move forward, stab the surface, and possibly break. With practice, 0.5–1.0 µL can be pulled into a single micropipet, which should be enough for hundreds of injections (*see* Note 21).
5. Once the micropipet is filled, test it once at the chosen duration pulse to make sure it is working, and immediately submerge it in the Ficoll solution filling the injection dish.
6. Place several embryos in the injection dish, one per well. With a hair loop or fine forceps, angle embryos so that the desired cell is facing the micropipet.
7. Using the micromanipulator, position the micropipet so its tip will puncture the chosen cell at a right angle. Drive the tip in with the Z axis control knob of the manipulator. There will be a little resistance when the tip hits the vitelline membrane, but the tip should slide in easily. Do not advance deeply into the cell. Using the foot pedal, inject the cell, count to 10, withdraw the micropipet, and move on to the next cell (*see* Note 22).
8. When all the cells in the dish have been injected, move the micropipet to the side of the dish, and transfer the embryos to a fresh Petri dish filled with 1X Steinberg's solution. If injection volumes are large, transfer embryos to Ficoll solution in a fresh Petri dish for another 1–2 h and then transfer to 1X Steinberg's solution (*see* Note 23).

3.5. Embryo Culture

1. Culture embryos in 1X Steinberg's solution for 5–6 h, through midblastula transition (stage 8; ref *16*). Embryos can be grown at room temperature or in an 18–20°C incubator; do not exceed 23°C (*see* Note 24).

2. Prior to the completion of neurulation (stage 20), they should be transferred to 0.5X Steinberg's solution.
3. At 3 d after fertilization (stage 38), the medium should be diluted to 0.1X Steinberg's solution.
4. Embryos should be fed tadpole powder (Nasco) starting at 4–5 d after fertilization.

3.6. Detection of Gene Products by Immunohistochemistry

1. Fix embryos by immersion in PARA fixative. If embryos are older than stage 40, tear a hole in the skin with forceps to enhance penetration to internal organs. Fix in a volume 20X that of embryos, in refrigerator for 6–12 h (*see* Notes 25 and 26).
2. Carefully note any gross morphological abnormalities in the embryos.
3. Wash in PBS; store in 5% sucrose in PBS for up to 7 d.
4. Embed in OCT medium (Tissuetek) and freeze on chuck in cryostat.
5. Section serially in ribbons at 14–20 µm with cryostat set at about –20°C. Collect ribbons on gelatin-coated slides.
6. Store slides at –20°C for up to 2 wk or at –70°C for many months.
7. Dry slides at 38°C for 20 min.
8. Wash off OCT with PBS (3 × 5 min).
9. Incubate slides in a Coplin jar filled with blocking solution for 1 h, agitating on a rotator.
10. Add about 200 µL of primary antibody solution to each slide. Store flat in a humidified chamber.
11. Incubate overnight in a refrigerator.
12. Transfer slides to Coplin jars. Wash 5 × 5 min in PBS on rotator.
13. Add about 200 µL of secondary antibody solution to each slide. Store flat in a humidified chamber.
14. Incubate for 2 h at room temperature.
15. Transfer slides to Coplin jars. Wash 5 × 5 min in PBS on rotator.
16. Add about 200 µL of tertiary antibody solution to each slide. Store flat in a humidified chamber (*see* Notes 27 and 28).
17. Incubate for 2 h at room temperature.
18. Transfer slides to Coplin jars. Wash 5 × 5 min in PBS on rotator.
19. Incubate in DAB solution for 5–20 min.
20. Stop reaction by washing twice in PBS.
21. Dehydrate in an ascending series of ethanols, clear in toluene, and coverslip.
22. After coverslips have dried for 6–12 h, it is safe to handle them.
23. There are several aspects of expression that should be checked:
    a. Confirm that the gene has been translated, as evidenced by dark brown HRP-DAB reaction product in cells.
    b. Confirm that spatial distribution has been preserved, as evidenced by reaction product in the tissues expected, based on the fate map of the injected blastomere (*see* Note 29).
    c. Determine the onset of expression and for how long expression can be detected by fixing embryos at several different developmental stages (*see* Note 30).

4. Notes

4.1. Micropipet Calibration

1. Micropipets must be calibrated prior to use, as each will have a different tip diameter. Always wear gloves while handling pipet glass to prevent RNase contamination.
2. The taper length needs to be short in order to pierce the vitelline membrane that surrounds the embryo.
3. For alternate methods of calibration and alternate injection equipment, *see* ref. *17*.
4. To hold micropipets in a storage box, press a strip of modeling clay onto the bottom of the box and make grooves in it with a capillary tube. Alternatively, glue a block of plastic foam (which will not shed small particles, as will rubber foam) to the bottom of the box and cut slits in it with a razor blade.

4.2. Embryo Preparation

5. In vitro fertilization is ideal for obtaining large numbers of embryos on demand, synchronized to the same stage of development. However, it requires sacrificing the male frog, and the embryos do not always cleave in regular patterns that match the fate maps. Natural fertilization frequently produces regular cleavage patterns and was used for all the fate maps from our lab *(7–10)*. Natural fertilization provides developmental stages of embryos spread out over a long time period, which is advantageous when complex manipulations are planned. However, frogs do not always mate successfully on a time frame convenient to the experimenter's schedule.
6. Allow female frogs to rest at least 6–8 wk between hormone treatment to prevent stress and to allow them to replenish egg supplies.
7. Many laboratories recommend injecting females with pregnant mare serum gonadotropin (30 IU) 3–4 d before HCG injection to obtain larger quantities of eggs *(18)*.
8. Dejellying must be performed carefully. Do not dejelly prior to the appearance of the first cleavage or the cysteine will disable sperm. Do not agitate the eggs, as this can cause polyspermy and very irregular cleavages; eggs should be gently swirled at intervals. Watch for signs that the jelly is falling off the eggs; the eggs will touch one another, rather than being separated by their coats. As soon as this happens, wash away the cysteine, as it can be quite toxic to the embryos.
9. If spatial localization of the heterologous gene is important, do not assume that the lightly pigmented area in the animal hemisphere is dorsal; this can lead to about 30% misidentification of blastomeres *(7,19)*. Rather, one should note the dorsal side of the embryos during the first cell cycle, either by marking the sperm entry point with a vital dye *(20)*, by tipping in vitro fertilized eggs and marking one side *(20)*, or by ensuring that the first cleavage furrow bisects the grey crescent (Fig. 3; refs. *7,8*). We routinely use the latter on naturally fertilized eggs at a reliability rate of greater than 95%. Each method has its virtues *(19,21)*.
10. After removing the jelly coats, observe embryos frequently until they reach the required cleavage stage to ensure that cleavage furrows are dividing the cyto-

plasm in a regular pattern. The stereotyped pattern used for the fate maps is illustrated in Fig. 1. If spatial localization is critical for the interpretation of the experiment, select embryos that adhere to this ideal pattern, at least on the side of the embryo to be injected. Each cleavage cycle takes 20–30 min, depending on the temperature at which they are stored.

4.3. Nucleic Acid Preparation

11. After the nucleic acid pellet is washed with 70% EtOH, if pellet is very white, large amounts of salt probably remain. In this case, repeat the 70% EtOH wash.
12. When drying nucleic acids, one can use a speed-vacuum dryer. However, while one wants all the excess liquid to evaporate, overdrying the pellet with a Speed-vac can render the pellet difficult to dissolve. This is especially true for RNA.
13. To confirm that injections were successful, it is helpful to add tracers to the nucleic acid solution. We have added a few crystals of horseradish peroxidase (Boerhinger-Mannheim, Mannheim, Germany) to native RNA solutions *(6)*. Fluorescent lineage tracers should not be mixed with the RNA, as they are highly charged and can bind to the RNA to reduce translation efficiency. We have used fluorescent tracers delivered by a second micropipet with success, but it doubles the number of injections one does. Frequently, synthetic mRNA for β-galactosidase is mixed with the test RNA as a tracer. Reporter plasmids that are transcribed efficiently are pSV_2CAT and pT109luciferase (American Type Culture Collection #37155 and 37583, ref. *17*). Plasmids using β-galactosidase as the reporter also are popular. Plasmids containing the gene for a fluorescent tracer, Fluorescent Green Protein (Clontech, Palo Alto, CA), should be considered if one wants to monitor expression in living cells.
14. Native RNA (total or poly A+) should be extracted and purified by methods recommended for the source tissue. After final precipitation, they can be dissolved in RNase-DNase free water or buffer (88 m*M* NaCl, 10 m*M* HEPES, pH 6.8) and directly injected into blastomeres. Avoid injection volumes of greater than 10 nL per cell at the 32-cell stage, as they can result in nonspecific fate changes *(6)* or mechanical damage. At earlier stages, larger volumes can be tolerated.
15. Functional RNA can be synthesized from cDNA cloned into expression vectors that contain SP6, T3, and T7 promoters. A most useful vector is pSP64T, which contains the 5' and 3' untranslated regions of the *Xenopus* β-globin mRNA *(23)*. These flanking sequences provide efficient translation of the mRNA. In addition, synthetic mRNA must contain a 5' terminal cap structure and 3' poly(A) tail to ensure efficient translation and mRNA stability. Detailed protocols for producing synthetic mRNA can be found in ref. *24*. It is important to consider that injected RNA does not diffuse throughout the injected blastomere; it tends to remain close to the injected site. Therefore, it tends to be expressed as a mosaic among the descendants of the injected blastomere. If widespread expression is required, multiple small injections are more effective.
16. Plasmids utilizing a variety of viral and tissue-specific promoters are very efficiently transcribed in *Xenopus* blastomeres *(13,17,25)*. However, plasmids can

be quite toxic to embryonic cells, frequently causing mitotic blockade. The concentration injected must be titrated for each construct to optimize for the highest level of expression, with the least amount of cell death.

17. While heterologous genes often are correctly regulated in developmental time and space, there are several examples where this is not the case *(13)*. One needs to carefully monitor the expression of one's construct.
18. DNA expression tends to be mosaic. Not all descendants of the injected blastomere will inherit sufficient copies of the gene so that protein product is detectable.
19. The level of transcription of injected DNA may depend on whether the DNA is in superhelical or linear form. The more efficiently transcribed form depends on the construct and needs to be individually determined.
20. Injected DNA that endogenously would be transcribed during development between midblastula and tadpole stages is more likely to be regulated in a tissue-specific manner.

4.4. Embryo Injection

21. Nucleic acids are viscous and can close fine micropipet tips that are exposed to the air. Keep the tip in an aqueous solution at all times after filling. Frequently check that tip is patent throughout the experiment by lifting tip from the solution and expelling a small amount of the nucleic acid solution.
22. If micropipet tip will not puncture the cell, it is too large or blunt. If tip bends when it touches the vitelline membrane, the taper is too long. In both cases, make adjustments on the pipet puller.
23. Injected embryos should remain in the Ficoll solution in the injection dish for at least 10 min to prevent leakage and the formation of blebs at the injection site. They can then be transferred to Ficoll medium in a Petri dish for up to 1–2 h, but will not develop normally if cultured longer in Ficoll.

4.5. Embryo Culture

24. An alternate solution for washing embryos and culturing them is Marc's Modified Ringers Solution (MMR): 100 m*M* NaCl, 2 m*M* KCl, 2 m*M* $CaCl_2$, 1 m*M* $MgCl_2$, 5 m*M* HEPES, pH 7.4. Embryos are cultured in 1X MMR up to stage 5 *(17,20)* and then in 0.1X MMR.

4.6. Detection of Gene Products by Immunocytochemistry

25. The choice of fixative depends on the ideal conditions for the primary antibody to be used. Consult antibody specification sheets.
26. Some antibodies may require unfixed tissue. In this case, anesthetize embryos in 0.1X MS222, embed in OCT, and quick-freeze in liquid N_2.
27. We prefer to use PAP methods for antibody detection in *Xenopus* because it is very sensitive, especially when enhanced with imidazole, and produces almost no nonspecific background staining. Always include a control slide, one not incubated in primary antibody, to detect nonspecific staining. Nonspecific staining most often occurs in the skin.

28. Using a fluorescent secondary antibody will eliminate steps 16–21. However, sometimes the protein is at too low a concentration to be detected above the high level of autofluorescence inherent to *Xenopus* tissue. Texas Red-tagged antibodies give the best signal-to-noise ratio, especially when using narrow band filter sets. Fluorescent antibodies are ideal if two proteins need to be simultaneously localized in the same tissue section. Combining Texas Red and fluorescein, and using dual-fluorochrome filter sets (Chromatec or Omega) are ideal for this. After washing slides in PBS (step 15), coverslip in FITC-Guard (Testog) or a comparable antifading aqueous medium.
29. It is important to demonstrate that injections have not simply damaged cells. The embryos should be visually inspected for large cells that have not divided and for dead cells that have sloughed off into the lumens of organs and into the perivitelline space. After antibody staining, abnormal cells can be identified as those that are much larger than normal, undifferentiated, or unincorporated into the surrounding tissue.
30. For processing large numbers of embryos, it is time-saving to perform whole-mount immunocytochemistry. An excellent protocol is provided in ref. *26*.

References

1. Newport, J. and Kirschner, M. (1982) A major developmental transition in early *Xenopus* embryos: I. Characterization and timing of cellular changes at the midblastula stage. *Cell* **30,** 675–686.
2. Hausen, P. and Riebesell, M. (1991) *The Early Development of* Xenopus laevis. Springer-Verlag, New York, p. 24.
3. Thompsen, G., Woolf, T., Whitman, M., Sokol, S., Vaughan, J., Vale, W., and Melton, D. A. (1990) Activins are expressed early in *Xenopus* embryogenesis and can induce axial mesoderm and anterior structures. *Cell* **63,** 485–493.
4. Cho, K. W. Y., Blumberg, B., Steinbesser, H., and De Robertis, E. M. (1991) Molecular nature of Spemann's Organizer: the role of the *Xenopus* homeobox gene goosecoid. *Cell* **67,** 1111–1120.
5. Smith, W. C. and Harland, R. M. (1992) Expression cloning of noggin, a new dorsalizing factor localized to the Spemann organizer in *Xenopus* embryos. *Cell* **70,** 829–840.
6. Hainski, A. M. and Moody, S. A. (1992) *Xenopus* maternal RNAs from a dorsal animal blastomere induce a secondary axis in host embryos. *Development* **116,** 347–355.
7. Klein, S. L. (1987) The first cleavage furrow demarcates the dorsal-ventral axis in *Xenopus* embryos. *Dev. Biol.* **120,** 299–304.
8. Moody, S. A. (1987) Fates of the blastomeres of the 16-cell stage *Xenopus* embryo. *Dev. Biol.* **119,** 560–578.
9. Moody, S. A. (1987) Fates of the blastomeres of the 32-cell stage *Xenopus* embryo. *Dev. Biol.* **122,** 300–319.
10. Moody, S. A. and Kline, M. J. (1990) Segregation of fate during cleavage of frog *(Xenopus laevis)* blastomeres. *Anat. Embryol.* **182,** 347–362.

11. Dale, L. and Slack, J. M. W. (1987) Fate map for the 32 cell stage of *Xenopus laevis*. *Development* **99,** 197–210.
12. Batni, S., Moody, S. A., and Knox, B. E. (1994) *Xenopus* rhodopsin promoter characterization by transient embryo transfection. *Mol. Biol. Cell* **5,** 110a.
13. Sargent, T. D. and Mathers, P. H. (1991) Analysis of Class II gene regulation. *Methods Cell Biol.* **36,** 347–365.
14. Etheridge, A. L. and Richter, S. M. A. (1978) Xenopus laevis: *Rearing and Breeding the African Clawed Frog*. Nasco, Ft. Atkinson, WI.
15. Heasman, J., Holwill, S., and Wylie, C. C. (1991) Fertilization of cultured *Xenopus* oocytes and use in studies of maternally inherited molecules. *Methods Cell Biol.* **36,** 213–230.
16. Nieuwkoop, P. D. and Faber, J. (1994) *Normal Table of* Xenopus Laevis (Daudin). Garland, New York.
17. Kay, B. K. (1991) Injections of oocytes and embryos. *Methods Cell Biol.* **36,** 663–669.
18. Wu, M. and Gerhart, J. (1991) Raising *Xenopus* in the laboratory. *Methods Cell Biol.* **36,** 3–18.
19. Masho, R. (1990) Close correlation between the first cleavage plane and the body axis in early *Xenopus* embryos. *Dev. Growth Diff.* **32,** 57–64.
20. Peng, H. B. (1991) Solutions and protocols. *Methods Cell Biol.* **36,** 657–662.
21. Moody, S. A., Bauer, D. V., Hainski, A. M., and Huang, S. (1995) Determination of *Xenopus* cell lineage by maternal factors and cell interactions, in *Current Topics in Development*, vol. 32 (Pedersen, R. and Schatten, G., eds.), Academic, New York, pp. 103–138.
22. Chalfie, M., Tu, Y., Euskirchen, G., Ward, W. W., and Prasher, D. C. (1994) Green fluorescent protein as a marker for gene expression. *Science* **263,** 802–805.
23. Krieg, P. A. and Melton, D. A. (1984) Functional messenger RNAs are produced by SP6 in vitro transcription of cloned cDNAs. *Nucleic Acids Res.* **12,** 7057–7070.
24. Wormington, M. (1991) Preparation of synthetic mRNAs and analyses of translational efficiency in microinjected *Xenopus* oocytes. *Methods Cell Biol.* **36,** 167–183.
25. Vize, P. D., Melton, D. A., Hemmati-Brivanlou, A., and Harland, R. M. (1991) Assays for gene function in developing *Xenopus* embryos. *Methods Cell Biol.* **36,** 367–387.
26. Klymkowsky, M. W. and Hanken, J. (1991) Whole-mount staining of *Xenopus* and other vertebrates. *Methods Cell Biol.* **36,** 420–441.

VI

Expression in Mammalian Cells

22

Overview of Vector Design for Mammalian Gene Expression

Randal J. Kaufman

1. Introduction

The technology of expressing foreign genes in mammalian cells has become increasingly important to study a number of biological questions and as a primary method for production of proteins for pharmaceutical use. Mammalian cells are frequently used as a host for expression of foreign genes because:

1. DNA cloned from higher eukaryotic cells (both cDNAs and genomic clones) is readily expressed since the signals for transcription, mRNA processing, and translation are conserved in higher eukaryotic systems;
2. Proteins are expressed in a stable functional form since the machinery to facilitate proper protein folding and assembly are conserved in higher eukaryotic cells;
3. Many post-translational modifications, especially for those proteins that transit the secretory pathway, are efficiently performed; and
4. Many proteins are readily secreted from mammalian cells providing the ability to isolate the protein from conditioned medium that contains low amounts of protein when cells are propagated under serum-free conditions.

Mammalian cells are used as a host for gene expression in order to:

1. Confirm that cloned genes can direct the synthesis of desired proteins;
2. Study protein structure-function relationships;
3. Isolate genes by direct screening or selecting transfected cells that express a desired protein;
4. Produce proteins that are available in limited quantity; and
5. Evaluate the physiologic consequences of expression of specific proteins in mammalian cells in order to study biological regulatory control mechanisms.

From: *Methods in Molecular Biology, vol. 62: Recombinant Gene Expression Protocols*
Edited by: R. Tuan Humana Press Inc., Totowa, NJ

The choice of a particular expression strategy is dependent upon the objectives of the study. The criteria in evaluating which expression system to employ include:

1. The method desired to introduce the foreign gene into the cell;
2. The particular requirement for a specific cell type in which to obtain expression;
3. The amount of protein expression required to achieve the goals of the study; and
4. The particular need for an inducible vector to obtain expression of proteins that are potentially toxic.

Two general methods for transfer of genetic material into mammalian cells are those mediated by virus infection and those mediated by direct DNA transfer. This chapter will discuss the advantages, utilities, and disadvantages of several vector systems that have proven most successful to obtain high level expression of heterologous genes in cultured mammalian cells.

Although there are significant advantages for the use of mammalian cells to express foreign genes, the size of the typical mammalian cell does limit the percentage of total cell protein that can be produced through mammalian cell expression systems. Table 1 shows that the amount of DNA, RNA, and protein per cell is roughly 100–1000 times greater for mammalian cells than for *Escherichia coli.* Thus, introduction of a single gene into mammalian cells would produce approximately 100–1000 times less of that gene product as the total cellular protein compared to a single gene in *E. coli.* A high level of expression in mammalian cells (approx 10 μg protein/10^6cells/d) represents only 2.5% of the total cellular protein.

2. Expression of Exogenous Genes Transfected into Mammalian Cells

2.1. Expression by Transient DNA Transfection

When cells take up DNA, they express it transiently over a period of several days to several weeks and eventually the DNA is lost from the population. The ability to express this DNA over a short period of time is called "transient expression." Transient expression is a convenient and rapid method to study expression of foreign genes in mammalian cells (Table 1). The efficiency of expression after transient transfection of plasmid DNA is dependent upon the number of cells that incorporate DNA, the gene copy number, and the expression level per gene. For several established cell lines it is possible to directly introduce plasmid DNA into 5–50% of the cells in the population. A variety of methods for introducing DNA have been reviewed *(1)*. The most convenient and reproducible methods are DNA transfection mediated by DEAE Dextran, electroporation *(2)*, and cationic phospholipids. Different labs find that one or another of these methods is generally more successful. Certain cell types may be more amenable to transfection by one procedure vs another. If one method

Table 1
The Main Macromolecular Components of *E. coli* and HeLa Cells

Component	Amount per HeLa cell	Amount per *E. coli* cell
Total DNA	15 pg[a]	0.017 pg[b]
Total RNA	30 pg	0.10 pg
Total protein	300 pg (5×10^9 mol, of average mw 40,000)	0.2 pg (3×10^6 mol, of average mw 40,000)
Cytoplasmic ribosomes	4×10^6	3×10^4
Cytoplasmic tRNA molecules	6×10^7	4×10^5
Cytoplasmic mRNA molecules[c]	7×10^5	4×10^3
Nuclear precursor rRNA molecules	6×10^4	
Heterogeneous nuclear RNA molecules[d]	1.6×10^5	
Total dry weight	400 pg	0.4 pg

[a]HeLa cells are hypotetraploid, i.e., they contain about four copies of each chromosome. The normal diploid human DNA complement is about 5 pg/cell.

[b]A rapidly growing *E. coli* cell contains, on the average, four DNA genomes. Each genomic DNA weighs 0.0044 pg.

[c]An average chain length of 1500 nucleotides is assumed.

[d]An average chian length of 6000 nucleotides is assumed; this group of molecules contains precursor mRNAs.

does not work effectively it is recommended to try an alternate procedure. Transient expression offers a convenient means to compare different vectors and ensure that an expression plasmid is functional before using the expression plasmid to establish a stable expressing cell line. Transient expression experiments obviate the effects of integration sites on expression and the possibility of selecting cells that harbor mutations in the transfected DNA. Expression vectors should be tested by transient transfection prior to the more laborious procedure of isolating and characterizing stably transfected cell lines.

A large variety of expression vectors for transient expression are described in the literature. Most useful vectors contain multiple elements that include:

1. An SV40 origin of replication for amplification to high copy number in COS-1 monkey cells;
2. An efficient promoter element for transcription initiation;
3. mRNA processing signals that include intervening sequences and mRNA cleavage and polyadenylation sequences;
4. Polylinkers that contain multiple endonuclease restriction sites for insertion of foreign DNA; and
5. Selectable markers that can be used to select cells that have stably integrated the plasmid DNA.

If a relatively efficient expression vector is used, then the expression level obtained from a particular insert is most dependent upon the particular gene insert and to a lesser extent upon the particular vector. Thus, it is not possible to generalize results obtained from one insert to another.

The most widely used and successful host-vector system for transient expression is based on the small DNA tumor virus, simian virus 40 (SV40). African green monkey kidney cells transformed with an origin-defective mutant of SV40 cells express high levels of the SV40 large tumor (T) antigen that is required to initiate viral DNA replication. The T-antigen-mediated replication can amplify the plasmid copy number to greater than 10,000 per cell. This large copy number results in high expression levels from the transfected DNA. Generally, with the high degree of replication mediated by SV40 T-antigen, the strength of the promoter is not a major factor in limiting expression of foreign genes. Most vectors for mammalian cells contain constitutive promoter elements such as the SV40 early promoter, the Rous sarcoma virus promoter, the adenovirus major late promoter, or the human cytomegalovirus immediate early promoter that are very active in a wide variety of cell types from many species.

pED is an efficient vector for transient expression in mammalian cells and can also be used to readily derive stably transfected cell lines (Fig. 1) *(3)*. pED can be obtained from Monique Davies at Genetics Institute (Cambridge, MA). It contains plasmid sequences from puc18 which allow for propagation and selection for ampicillin resistance in E. coli. It contains the SV40 origin of replication and enhancer element and utilizes the adenovirus major late promoter for transcription initiation. Contained within the 5' end of the mRNA is the tripartite leader from adenovirus late mRNA and a small intervening sequence. There are several cloning sites including for insertion of foreign DNA. In the 3' end of the transcript there is a cleavage and polyadenylation signal from the SV40 early region. This vector contains a DHFR coding region in the 3' end of the transcript. pED also contains the adenovirus virus-associated (VAI) gene, an RNA polymerase III transcription unit encoding a small RNA that inhibits the double-stranded RNA activated protein kinase (PKR). PKR is activated upon DNA transfection and it phosphorylates the alpha subunit of the eukaryotic initiation factor 2 to inhibit translation initiation *(4)*. The VAI gene improves translation of plasmid-derived mRNA by inhibiting PKR activation. pED encodes a dicistronic mRNA containing a dihydrofolate reductase (DHFR) coding region within the 3' end of the mRNA. Efficient translation of DHFR is mediated by the internal ribosomal entry site from encephalomyocarditis virus. Thus, expression of the dicistronic mRNA from this vector can be used to directly select for DHFR expression in Chinese hamster ovary cells that are deficient in DHFR *(5) (see* Section 2.3.*)*.

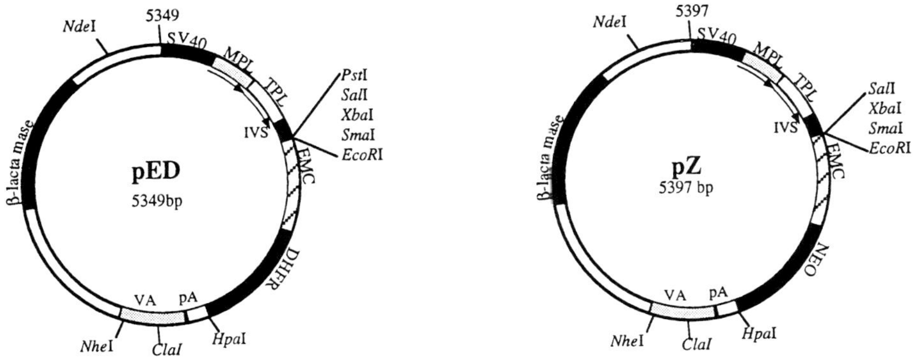

Fig. 1. Efficient cloning, selection, and expression dicistronic vectors for mammalian cells *(3)*. The vectors containing the selection markers for wild-type DHFR (pED) and neomycin phosphotransferase (pZ) are shown.

In many cases it is desireable to identify the transiently transfected cell within the total cell population. Several approaches have proven to be quite successful. It is possible to utilize a fluorescent derivative of methotrexate (MTX-F, obtained from Molecular Probes, Portland, OR) and stain DHFR in living cells that are transfected using the pED vector for analysis by fluorescence microscopy or by fluorescence-activated cell sorting. Fluorescence-activated cell sorting provides the ability to isolate the transfected subpopulation for direct study *(6)*. Alternatively, the expression vector pETF can be used. ETF produces a dicistronic mRNA encoding the membrane surface protein tissue factor that can be stained with specific monoclonal antibody *(7)*. In this manner, it is possible not only to identify the transfected cells, but it is also possible to isolate large numbers of transfected cells by "panning" with the specific antibody. It is possible to use cotransfection with an expression vector encoding the green fluorescence protein (GFP) derived from the bioluminescent jellyfish *Aequorea victoria* (psynGFPS65T can be obtained from Brian Seed, Mass. Gen Hospital [Boston, MA]) *(8)*. This protein fluoresceses green after excitation by 395-nm light. Frequently, it is desireable to cotransfect several different plasmid DNA molecules. By titration of the amount of different plasmid DNA molecules, we have shown that DEAE-mediated dextran transfection of COS-1 cells yields approximately 15-20 different individual plasmid DNA molecules expressed in a single cell. Thus, cotransfection of several different plasmid DNA molecules will yield a subpopulation of cells that coexpress each plasmid DNA.

2.2. Inducible Expression of Foreign Genes

Systems that permit stringent induction of gene expression offer unique advantages to study a diversity of biological questions as well as an approach

3. CHO cells adapt well to growth in the absence of serum and can grow either attached or in suspension; and
4. CHO cells can be scaled to greater than 5000 liters.

A variety of selection and coamplification vectors have been constructed in which the product gene and the selection gene are contained within the same transcription unit. These dicistronic mRNA expression vectors are based on the use of a picornaviral internal ribosomal entry site. Members of the picornavirus family, including poliovirus and EMC virus, have a long 5' untranslated region that mediates internal ribosome binding and cap-independent translation of mRNA *(20)*. The sequence required to promote internal initiation from encephalomyocarditis (EMC) virus extends from nucleotide 260 to 834 of the viral genome. Mammalian cell expression vectors were generated that utilize the 5' untranslated region from EMC virus to promote efficient internal translation initiation of selectable markers encoding DHFR (pED, Fig. 1), a methotrexate resistant DHFR (pEMC-MTX^r), neomycin phosphotransferase (pZ, Fig. 1), and adenosine deaminase (pEA) *(3)*. The use of pED is limited to DHFR-deficient cells. pZ, pEA, pEMC-MTX^r may be used as dominant markers in different cell types. These vectors permit the rapid derivation of stable cell lines that express high levels of the desired product. Since there is no selective advantage for deletion of the open reading frame contained in the 5' position, these vectors do not undergo deletion upon selection for further increases in expression. If deletion of the 5' open reading frame occurs, it is a good indication that the gene is detrimental to cell growth.

3. Vaccinia Virus Mediated Expression of Heterologous Genes in Mammalian Cells

Many viruses that infect mammalian cells have mechanisms to usurp the protein synthesis machinery of the host to produce their viral proteins. The ability to engineer the genetic material of these viruses enables insertion of foreign genes under the control of the viral expression elements and to produce infectious virus particles that direct high levels of foreign gene expression. Viral-mediated gene transfer provides a convenient and efficient means to introduce foreign DNA into a variety of different cell types. In addition, for many viruses, viral replication yields multiple copies of template DNA that can serve to amplify transcription of the foreign gene. Presently, one of the most convenient systems is based on vaccinia virus. For a more detailed review of the different eukaryotic viral vectors *see* ref. *21*.

Vaccinia virus is a member of the poxvirus family that replicates in the cytoplasm of mammalian or avian cells *(22)*. The genome is a linear double-stranded DNA molecule of 185 kb, packaged in the virus core. Vaccinia virus encodes its own transcription and RNA processing system within the viral particle. Upon infection, about 100 genes are expressed early and host protein synthesis is shut off. After DNA replication, at 6 h post infection, about 100 late genes are expressed. There is no evidence of splicing for any vaccinia virus transcript. As a consequence, it is necessary to express cDNAs, as opposed to genomic clones, in vaccinia virus vectors. Virus is formed about 6 h after infection and continues for about 2 d.

As a result of the large size of the vaccinia genome, DNA must be inserted by homologous recombination. Plasmid vectors were constructed so that foreign DNA is inserted under control of a strong promoter and flanked by DNA from a nonessential region of the vaccinia virus genome. After transfection of cells that have been infected with vaccinia virus, the foreign DNA recombines into the viral genome. Most commonly, insertion is directed into the thymidine kinase gene such that the TK^- phenotype can be used for selection of recombinants. Alternatively, inclusion of the β-galactosidase gene within the expression plasmid permits a selection for recombinant plaques by screening with an appropriate color indicator, or the *E. coli* xanthine guanine-phosphoribosyl transferase gene to select for recombinants by growth in mycophenolic acid. Recombinant viruses may also be screened by DNA hybridization or antibody binding.

A vaccinia virus/bacteriophage T7 promoter vector system utilizes the efficient and selective T7 bacteriophage RNA polymerase to express mRNA in a mammalian cell *(23–25)*. A recombinant vaccinia virus (vTF7-3) directs expression of the T7 RNA polymerase. Foreign DNA is cloned between two fragments of T7 DNA containing the 01O promoter and the T0 termination sequences into a plasmid vector containing cloning sites and flanking vaccinia virus thymidine kinase gene sequences for homologous recombination. Because the mRNA expressed from the T7 promoter was not efficiently capped, the mRNA was not efficiently translated since the 5' 7-methyl guanosine cap structure is important for efficient translation. The translational efficiency was improved by using an internal ribosomal binding site from EMC virus. Use of the 5' untranslated region from EMC virus improved translation sevenfold *(26)*. At 24 h post-infection, the marker gene product, chloramphenicol acetyltransferase, represented greater than 10% of the cell protein. The most convenient version of this vector system has a unique Nco1 restriction site at the position where translation initiation occurs (pTM-1, Fig. 2). cDNA clones should be engineered such that the initiator AUG codon is fused to the Nco1 site. pTM-1 contains a polylinker at the 3' end to facilitate insertion of foreign DNA. It also contains the vaccinia virus thymidine kinase sequences that are used for selection of homologous recombinants. Recently, a modified version of the vaccinia vector system, VOTE2 developed by Tom Fuerst, was derived that expresses the *lac* repressor. Thus, protein expression can be regulated by IPTG. This is especially beneficial for the expression of proteins that may be toxic for the cell.

4. 1000X Chloroquin (Sigma) 0.1*M*, stored –20°Cin dark.
5. 1X 10% DMSO Reagent, 1L.
6. 137 m*M* NaCl, 8 g.
7. 5 m*M* KCl, 0.37 g.
8. 0.7 m*M* NH_2HPO_4, 0.1g.
9. 6 m*M* D-glulcose, 1.08 g.
10. 21 m*M* HEPES, 5 g.

Then add 900 mL H_2O and pH to 7.1 and filter through 0.2 m. Then add 100ml 100% DMSO

5.2. Cells

COS-1 cells are grown in DME medium supplemented with 10% heat inactivated fetal calf serum. They are usually subcultured twice per week at a 1:4 to 1:8 split ratio depending on the rate of cell growth.

5.3. Growth Medium

Dulbecco's modified essential medium with 2 m*M* glutamine, 100 U/mL streptomycin, 100 µg/mL penicillin, and 10% heat inactivated fetal calf serum.

5.4. Transfection

1. Subculture cells 1:6 into 100 m*M* tissue culture plates at 12–24 h before transfection. The cells should be 60–80% confluent at the time of transfection.
2. Aspirate medium and wash 2× with 7 mL each of serum-free DME. Note: the cells will die if the DEAE dextran and the serum contact the cells at the same time.
3. Feed the cells the DNA-medium mix (4 mL/100-mm culture dish containing 8 µg of DNA); prepare as follows:
   a. Add DNA (generally prepared in sterile Tris-HCl [10 m*M* pH 8.0, EDTA 1 m*M*]) to 0.4 mL of Tris-HCl (final DNA concentration should be 2 µg/mL in the medium). Mix well.
   b. Add 0.4 ml of 10X DEAE dextran to DNA-Tris.HCl.
   c. Add 3.2 vol DME that contains 2 m*M* glutamine, 100 U/mL streptomycin, 100 µg/mL penicillin. Mix well.
4. Incubate 6–10 hrs at 37°C.
5. Rinse 1× with 7 mL serum-free DNA.
6. Add 2 mL of 10% DMSO reagent. Let sit on dish 2–3 min at T_r. Aspirate.
7. Add 5 mL/plate of DME + 10% fetal calf serum and 0.1 mM chloroquin for 2 h.
8. Remove chloroquin and rinse once with serum-free medium and add 10 ml DME growth medium per plate.
9. After 24–30 h aspirate medium and feed 10 mL fresh growth medium.
10. After 48–72 h, harvest medium and/or cells, as desired.

6. Notes

This technique results in significant cell death. In the optimal experiment, approximately 25% of the cells will die. Of the remaining cells, 20% usually acquire and express the DNA. The toxicity is most evident when the transfected cells are less than 50% confluent at the start of the transfection. It is recommended to use CsCl-banded DNA for this procedure. However, mini-plasmid DNA preparations may be used but yield significantly lower levels of expression.

References

1. Keown, W. A., Campbell, C. R., and Kucherlapati, R. S. (1990) Methods for introducing DNA into mammalian cells, in *Methods in Enzymology 185: Gene Expression Technology* (Goeddel, D., ed.), Academic, San Diego, CA, pp. 527–537.
2. Potter, H., Weir, L., and Leder, P. (1984) Enhancer-dependent expression of human g-immunoglobulin genes introduced into mouse pre-B lymphocytes by electroporation. *Proc. Natl. Acad. Sci. USA* **81,** 7161–7165.
3. Davies, M. V. and Kaufman, R. J. (1992) Internal translation initiation in the design of improved expression vectors. *Curr. Opin. Biotech.* **3,** 512–517.
4. Kaufman, R. J. (1994) Control of gene expression at the level of translation intiation. *Curr. Opin. Biotech.* **5,** 550–557.
5. Kaufman, R. J. (1990) Selection and coamplification of heterologous genes in mammalian cells, in *Methods in Enzymology 185: Gene Expression Technology* (Goeddel. D., ed.), Academic, San Diego, CA, pp. 537–566.
6. Kaufman, R. J., and Murtha, P. (1987) Translational control mediated by eucaryotic initiation factor 2 is restricted to specific mRNAs in transfected cells. *Mol. Cell. Biol.* **7,** 1568–1571.
7. Davies, M. V., Chang, H.-W., Jacobs, B. L., and Kaufman, R. J. (1993) The E3L and K3L vaccinia virus gene products stimulate translation through inhibition of the double-stranded RNA-dependent protein kinase by different mechanisms. *J. Virol.* **67,** 1688–1692.
8. Pines, J. (1995) GFP in mammalian cells. *Trends Genet.* **11,** 326–327.
9. Schweinfest, C. W., Jorcyk, C. L., Fujiwara, S., and Papas, T. S. (1988) A heat shock-inducible eukaryotic expression vector. *Gene* **71,** 207–210.
10. Israel, D. I. and Kaufman, R. J. (1989) Highly inducible expression from vectors containing multiple GRE's in CHO cells overexpressing the glucocorticoid receptor. *Nucleic Acids Res.* **17,** 4589–4604.
11. Ko, M. S. H., Sugiyama, N., and Takano, T. (1989) An auto-inducible vector conferring high glucocorticoid inducibility upon stable transformant cells. *Gene* **84,** 383–389.
12. Mulherkar, R. and Coulier, F. (1991) Overexpression of human Trk proto-oncogene in mouse cells using an inducible vector system. *Biochem. Biophys. Res. Commun.* **177,** 90–96.

13. Totzke, F., Marme, D., and Hug, H. (1992) Inducible expression of human phospholipase C-γ2 and its activation by platelet-derived growth factor A-chain homodimer in transfected NIH3T3 fibroblasts. *Eur. J. Biochem.* **203,** 633–639.
14. Daniels-McQueen, S., Goessling, L. S., and Thach, R. E. (1992) Inducible expression bovine papillomavirus shuttle vectors containing ferritin transitional regulatory elements. *Gene* **122,** 271–279.
15. Baim, S. B., Labow, M. A., Levine, A. J., and Shenk, T. (1991) A chimeric mammalian transactivator based on the *lac* repressor that is regulated by temperature and isopropyl-β-D-thiogalactopyranoside. *Proc. Natl. Acad. Sci.* **88,** 5072–5076.
16. Gossen, M., and Bujard, H. (1992) Tight control of gene expression in mammalian cells by tetracyline-responsive promoters. *Proc. Natl. Acad. Sci.* **89,** 5547–5551.
17. Gossen, M., Bonin, A., Freundlieb, S., and Bujard, H. (1994) Inducible gene expression systems for higher eukaryotic cells. *Curr. Opin. Biotech.* **5,** 516–520.
18. Gossen, M., Freundlieb, S., Bender, G., Muller, G., Hillen, W., and Bujard, H. (1995) Transcriptional activation by tetracyclines in mammalian cells. *Science* **268,** 1766–1769.
19. Urlaub, G., and Chasin, L. A. (1980) Isolation of Chinese hamster cell mutants deficient in dihydrofolate reductase activity. *Proc. Natl. Acad. Sci. USA* **77,** 4216–4220.
20. McBratney, S., Chen, C. Y., and Sarnow, P. (1993) Internal Initiation of Translation. *Curr. Opin. Cell Biol.* **5,** 961–965.
21. Muzyczka, N., ed. (1992) *Current Topics in Microbiology and Immunology: Eukaryotic Viral Expression Vectors*, Springer-Verlag Press, NY.
22. Moss, B. (1991) Vaccinia virus: a tool for research and vaccine development. *Science* **262,** 1662–1667.
23. Earl, P. L. and Moss, B. (1991) Characterization of recombinant vaccinia viruses and their products, in *Current Protocols in Molecular Biology* (Ausubel, F. M., Brent, R., Kingston, R. E., Moore, D. D., Seidman, J. G., Smith, J. A., and Struhl, K., eds.), Wiley, New York, pp. 16.18.1–16.18.10.
24. Earl, P. L. and Moss, B. (1991) Generation of recombinant vaccinia viruses, in *Current Protocols in Molecular Biology* (Ausubel, F. M., Brent, R., Kingston, R. E., Moore, D. D., Seidman, J. G., Smith, J. A., and Struhl, K., eds.), Wiley, New York, pp. 16.17.1–16.17.16.
25. Earl, P. L., Cooper, N., and Moss, B. (1991) Preparation of cell cultures and vaccinia virus stocks, in *Current Protocols in Molecular Biology* (Ausubel, F. M., Brent, R., Kingston, R. E., Moore, D. D., Seidman, J. G., Smith, J. A., and Struhl, K., eds.), Wiley, New York, pp. 16.16.1–16.16.7.
26. Elroy-Stein, O., Fuerst, T. R., and Moss, B. (1989) Cap-independent translation of mRNA conferred by encephalomycarditis virus 5' sequence improves the performance of the vaccinia virus/bacteriophage T7 hybrid expression system. *Proc. Natl. Acad. Sci. USA* **86,** 6126–6130.
27. Fuerst, T. R., Earl, P. L., and Moss, B. (1987) Use of hybrid vaccinia virus-T7 RNA polymerase system for expression of target genes. *Mol. Cell Biol.* **7,** 2538–2544.
28. Moss, B. (1992) Vaccinia and other poxvirus expression vectors. *Curr. Opin. Biotech.* **3,** 518–522.

23

Experimental Strategies in Efficient Transfection of Mammalian Cells

Calcium Phosphate and DEAE-Dextran

Gregory S. Pari and Wayne A. Keown

1. Introduction

Transfection of foreign DNA into mammalian cells is one of the most common procedures performed by investigators next to cell culture itself. Whether the goal is to express mammalian or viral genes transiently or by the generation of stable cells lines, researchers have a variety of transfection methods to choose from. Calcium phosphate coprecipitation and DEAE-dextran-mediated transfection are the oldest, and with many investigators, the most trusted of these methods. Paradoxically, the exact mechanism of DNA uptake for each of these methods remains unknown.

DEAE-dextran mediated DNA transfection is a favored approach for transient expression experiments The method has evolved somewhat since its original description *(1–4)*. In general this method is highly reproducible, more so than calcium phosphate transfection, and can be performed, with some optimization, on a wide variety of cell types. The efficiency of transfection can be as high as 80% with some cell types *(5)*. The principle drawback to DEAE-mediated DNA transfection is the poor efficiency of forming stably integrated expressing cell lines. Thus, DEAE-dextran transfection can only be used for transient expression assays. Other limitations include cellular toxicity, necessity for a DMSO (dimethyl sulfoxide) shock to improve efficiency, and the additional constraint that only plasmid DNA gives good transfection efficiencies. In addition, cotransfection efficiencies are not as good as those achieved with the calcium phosphate method. Notwithstanding these limitations, DEAE-dextran-mediated DNA transfection is a widely used procedure that can be made

From: *Methods in Molecular Biology, vol. 62: Recombinant Gene Expression Protocols*
Edited by: R. Tuan Humana Press Inc., Totowa, NJ

easy and efficient by preliminary optimizations. Variations of this procedure are batch transfection and transfection of cells in suspension, both of which are usually done to improve efficiency and to reduce variation within experiments.

Calcium phosphate coprecipitation is generally favored for generating stably expressing transfected cells. Although this approach is less reproducible than the DEAE-dextran method, it has several advantages over the DEAE-dextran procedure. First, the creation of stably intergrated constitutively expressing cell lines is efficiently achieved in a variety of cells. Second, cotransfection efficiencies are much greater; as many as 12 plasmids can be transfected at the same time *(6,7)*. Third, this method is not limited to the use of plasmid DNA, but also can be used to transfect genomic and/or linear DNA. Finally, cotransfection of more than one plasmid can occur in cells at the same ratio of plasmids that are present in the transfection mixture. One limitation of this method is the need for very high quality plasmid DNA; usually only double-banded CsCl DNA gives the highest efficiencies.

Two calcium phosphate methods are now routinely used, the first is the original HEPES based calcium phosphate transfection *(8,9)* and the second is a modification that uses BES buffer in which optimal pH is critical for achieving a high efficiency of transfection *(10)*. The HEPES buffer-based method is sufficient for achieving good results for transient and stable expression in many cell types. In this procedure, DNA is mixed with $CaCl_2$, buffer is added, then the mixture is added to cells, and DNA-calcium phosphate precipitate forms after a few hours. The BES method is essentially the same except that a precipitate forms slowly overnight in a 3% CO_2 atmosphere. This procedure leads to a much higher efficiency of the formation of stable expressing cell lines. In addition, the efficiency of cotransfection using the BES method is greater than that from the HEPES procedure. This is of particular importance when performing cotransfection-replication assays in which multiple plasmids are introduced into cells. In these assays, each plasmid may encode a single gene that is essential for activating an origin or effector plasmid, which is then assayed for amplification *(6)*. These types of assays are now widely used to ascertain essential factors required for viral replication. The BES calcium phosphate method is the preferred method for these transient replication assays.

Following are representative protocols for DEAE-dextran transfection of adherent of suspension cells, as well as a protocols for the BES method for calcium phosphate-mediated transfection.

2. Materials

2.1. DEAE-Dextran Transfection

1. Tris-buffered saline (TBS): Prepare the following sterile solutions: Solution A: 80 g/L NaCl, 3.8 g/L KCl, 2 g/L Na_2HPO_4, 30 g/L Tris base. Adjust pH to 7.5.

Solution B: 15 g/L $CaCl_2$, 10 g/L $MgCl_2$. Filter sterilize both solutions and store at –20°C. To make a 100 mL working solution, add 10 mL of solution A to 89 mL of water and then add 1 mL of solution B. This solution should be filter sterilized and stored at 4°C.

2. Suspension Tris-buffered saline (STBS): 25 m*M* Tris-HCl, pH 7.4, 137 m*M* NaCl, 5 m*M* KCl, 0.6 m*M* Na_2HPO_4, 0.7 m*M* $CaCl_2$, 0.5 m*M* $MgCl_2$. Make in distilled H_2O and filter sterilize.
3. Phosphate-buffered saline (PBS): 137 m*M* NaCl, 2.7 m*M* KCl, 4.3 m*M* Na_2HPO_4. $7H_2O$, 1.4 m*M* KH_2PO_4, pH 7.3.
4. Dulbecco's modified Eagle's medium supplemented with 10% fetal bovine serum (or NuSerum; Collaborative Research, Bedford MA) (*see* Note 1).
5. DEAE-dextran: 10 mg/mL in TBS.
6. 10% DMSO.

2.2. Calcium Phosphate Coprecipitation

1. Dulbecco's modified Eagle's medium supplemented with 10% fetal calf serum.
2. CsCl-purified double banded DNA.
3. 2.5*M* $CaCl_2$ filter sterilized through a 0.45-µm filter.
4. 2X BES-buffered saline (BBS): 50 m*M* N,N-bis(2-hydroxyethyl)-2-aminoethane-sulfonic acid, 1.5 m*M* Na_2HPO_4, 280 m*M* NaCl. Adjust pH to 6.95 with 1*N* NaOH (*see* Note 4).
5. A 35°C 3% CO_2 humidified incubator.

3. Methods

3.1. DEAE-Dextran Methods

Two DEAE-dextran methods are commonly used. The first is the basic protocol which can be used on all anchorage-dependent cells. The second can be used on cells that normally grow in suspension or with anchorage-dependent cells that have been trypsinized and are in suspension. This procedure may increase transfection efficiency in some cells. For adherent cells, it is advisable to try the basic protocol first, then if transfection efficiency is low, try the suspension procedure.

3.1.1. Anchorage Dependent Cells

1. Plate 5×10^5 cells in a 10 cm tissue culture dish (*see* Note 2). Cells should be plated at least 24 h before transfection and should be no more than 40–60% confluent.
2. Ethanol precipitate 5 µg of DNA per plate in a 1.5-mL microcentrifuge tube and resuspend the pellet in 40 µL of TBS. If the same DNA is used for multiple plates, precipitate all the DNA in one tube. Ethanol precipitation sterilizes DNA.
3. Remove media from the cells and wash cells with 10 mL of PBS. After washing, add 4 mL (for a 10-cm dish) of DMEM supplemented with 10% NuSerum.
4. Add resuspended DNA to 80 µL of 37°C 10 mg/mL DEAE-dextran in TBS (80 µL/5 µg of DNA). Warm the DEAE-dextran in a 37°C water bath before use, and add DNA slowly while shaking the tube.

5. Add 120 µL of the DNA/DEAE-dextran mixture to the plate in a dropwise fashion using a 200 µL pipet tip. Swirl the plate after each drop is applied to ensure that the mixture is distributed evenly (*see* Note 3).
6. Incubate the plates for 4 h in a 37°C incubator with a 5% CO_2 atmosphere. This incubation time can be shortened for some cell types.
7. Remove the medium. At this point, the cells may look a little sick but this is normal.
8. Add 5 mL of 10% DMSO in PBS. Incubate for 1 min at room temperature. Remove the DMSO and wash with 5 mL of PBS. Replace the PBS with 10 mL of DMEM supplemented with 10% serum.

3.1.2. Cells in Suspension

1. Ethanol precipitate DNA and resuspend the pellet in 500 µL of STBS (*see* Section 2.). Use 10 µg of DNA per 2×10^7 cells. Cells can be either normally growing suspension cells, for example B-cells, or trypsinized anchorage-dependent cells.
2. Pellet cells in a 50-mL conical centrifuge tube.
3. Resuspend cells in 5 mL of STBS and repellet as in step 2.
4. Make a 2X solution of DEAE-dextran in STBS and add 500 µL of this solution to 500 µL of the DNA resuspended in 500 µL of STBS from step 1, mix well. Resuspend pelleted cells with this DEAE-dextran/DNA solution. Use a final concentration of 100–500 µg/mL of DEAE-dextran.
5. Incubate cells in a CO_2 incubator for 30–90 min. Tap cells occasionally to keep them from clumping. Incubation times vary and should be determined experimentally.
6. Add DMSO to cells dropwise to a final concentration of 10%; mix well while adding.
7. Incubate cell with DMSO for 2–3 min. Add 15 mL of STBS to cell.
8. Pellet cells, wash with 10 mL of STBS and pellet again. Wash cells in medium with serum and pellet. After this centrifugation, resuspend cells in complete medium. If cells are normally anchorage-dependent, replate on a tissue culture dish or flask. The onset of expression from transfected plasmids varies depending on cell type. Usually expression begins between 24–48 h posttransfection.

3.2. Calcium Phosphate Coprecipitation Method

Like DEAE-dextran transfection, two calcium phosphate transfection methods are routinely used, a HEPES-based method and a BES buffer method. Both are good for transient expression, but the BES-buffer procedure is much more efficient in making established constitutively expressing cell lines; in some cells 50% efficiency can be achieved. In addition, this procedure works better on a wider variety of cell types, is excellent for cotransfection and is easier to perform than the traditional HEPES-based method. Since the BES method is so versatile and offers all these advantages, this will be the method presented here. A representative HEPES-based method can be found in ref. *8*.

1. Plate approx 5×10^5 cells on a 10-cm tissue culture dish 24 h before transfection. Cell should be no more than 50% confluent for making established cell lines and about 70% confluent for transient expression. Smaller plates (example 6 cm) can be used and in some cases this is actually sufficient and easier.

2. Dilute the 2.5*M* $CaCl_2$ solution to 0.25*M* with sterile water.
3. Add 20–30 µg plasmid DNA per tube (Falcon # 2058) or 10–20 µg for a 6-cm plate (*see* Note 5). To the DNA, add 500 µL of 0.25*M* $CaCl_2$. Then add 500 µL of 2X BBS. Mix well and incubate at room temperature for 10–20 min (*see* Note 6). If 6-cm dishes are used, this mixture can be split into two and transfections can be done in duplicate. If duplicate transfections are not desired, then use half of the volumes of .25*M* $CaCl_2$ and 2X BBS and add the total volume to one 6-cm dish.
4. Add the calcium phosphate/DNA mixture to cells in a dropwise fashion, swirling the plate after each drop. Incubate the cells overnight in a 35°C 3% CO_2 incubator (*see* Note 7).
5. Wash cells twice with 5 mL of PBS, then add 10 mL of DME*M* with 10% FBS. Incubation of cells from this point on is done in a 5% CO_2 37°C incubator.
6. For transient expression, harvest cells 48 h posttransfection. For selection of stably integrated expression clones, split cells (1:10) 48 h posttransfection into selection medium.
7. For cotransfection *see* Note 8.
8. For cotransfection-replication assay *see* Note 9.

4. Notes

1. NuSerum is a special formulation of newborn calf serum and growth factors. Many investigators report greater transfection efficiencies when this serum is used in place of fetal calf serum. It also allows the cells to withstand the DEAE-dextran mixture for greater lengths of time.
2. Small dishes can be used. Higher densities of some cell types may be necessary to achieve good transfection efficiencies. If cell death is too high owing to the toxicity of DEAE, then try plating cells at a higher density.
3. It may be necessary to determine the optimal concentration of DEAE-dextran needed for good transfection efficiencies. Vary the amount of volume of TBS used to resuspend DNA and the amount of DEAE dextran. For example:

DNA in TBS, µL	DEAE-dextran, mg/mL
80	160
40	80
20	40

4. BBS pH is critical. Make three BBS solutions ranging in pH from 6.93–6.98. Usually a visual inspection of cells after an overnight transfection will indicate which BBS mixture and DNA concentration works best. A course and clumpy precipitate will form when DNA concentrations are too low; a fine, almost invisible precipitate will form when DNA concentrations are too high. An even granular precipitate forms when the concentration is just right. This usually correlates to the highest level of gene expression or formation of stably integrated cell lines.
5. Only high quality plasmid DNA will work. Use only double banded CsCl purified DNA. Carrier DNA is not necessary and actually will decrease efficiency. Also, linear DNA does not transfect well.

6. At this point no precipitate should form. Use three different concentrations of DNA to help identify the DNA concentration necessary for optimal transfection.
7. CO_2 level is critical. Measure the level with a Fyrite gas analyzer. Temperature is somewhat less critical. A 37°C incubator can be used.
8. When performing cotransfections, vary the amount of effector plasmid in relation to the other plasmids in the mix. The ratio of plasmids used in the mixture can be the difference between success and failure. We routinely find that a higher concentration of effector plasmid in the mix yields better results. We commonly use this cotransfection method to assay the level of promoter activity effected by certain vital transactivators.
9. Depending on the number of plasmids in the transfection mixture, vary the amount of transactivators and effector plasmid. As many as 12 plasmids can be transfected at one time. Each plasmid can contain one or many genes required for replication of a cloned origin.

References

1. Fujita, T., Shubiya, H., Ohashi, T., Yamanishi, K., and Taniguchi, T. (1986) Regulation of human interleukin-2 gene: Functional DNA sequences in the 5' flanking region for the gene expression in activated T lymphocytes. *Cell* **46,** 401–407.
2. Lopata, M. A., Cleveland, D. W., and Sollner-Webb, B. (1984) High-level expression of a chloramphenicol acetyltransferase gene by DEAE-dextran-mediated DNA transfection coupled with a dimethysulfoxide or glycerol shock treatment. *Nucleic Acids Res.* **12,** 5707–5711.
3. Reeves, R., Gorman, C., and Howard, B. (1985) Minichromosomes assembly of nonintegrated plasmid DNA transfected into mammalian cells. *Nucleic Acids Res.* **13,** 3599–3605.
4. Selden, R. F., Burke-Howie, K., Rowe, M. E., Goodman, H. M., and Moore, D. D. (1986) Human growth hormone as a reporter gene in regulation studies employing transient gene expression. *Mol. Cell. Biol.* **6,** 3173–3179.
5. Sussman, D. J. and Milman, G. (1984) Short-term, high-efficiency expression of transfected DNA. *Mol. Cell. Biol.* **4,** 1641–1646.
6. Pari, G. S. and Anders, D. G. (1993) Eleven loci encoding trans-acting factors are required for transient complementation of human cytomegalovirus oriLyt-dependent DNA replication. *J. Virol.* **67,** 6979–6988.
7. Pari, G. S., Kacica, M. A., and Anders, D. A. (1993) Open reading frames UL44, IRS1/TRS1, and UL36-38 are required for transient complementation of human cytomegalovirus oriLyt-dependent DNA synthesis. *J. Virol.* **67,** 2575–2582.
8. Graham, F. L. and Van der Eb, A. J. (1973) A new technique for the assay of infectivity of human adenovirus 5 DNA. *Virology* **52,** 456–460.
9. Wigler, M., Pellicer, A., Silverstein, S., and Axel, R. (1978) Biochemical transfer of single-copy eucaryotic genes using total cellular DNA as donor. *Cell* **14,** 725–731.
10. Chen, C. and Okayama, H. (1987) High-efficiency transformation of mammalian cells by plasmid DNA. *Mol. Cell. Biol.* **7,** 2745–2752.

24

Experimental Strategies in Efficient Transfection of Mammalian Cells

Electroporation

Donald C. Chang

1. Introduction

Electroporation is a method which utilizes an electric field to introduce DNA or other macromolecules into cells. (For review, *see* ref. *1*). When a cell is exposed to a pulse of high electric field, its cell membrane quickly becomes permeabilized. During this permeabilized state, macromolecules from the external medium can enter the cell. Following the removal of the external electric field, the cell membrane gradually repairs itself. In most cases, if the applied field is not excessively strong, the cell can recover.

Using this electroporation method, many biologically active substances, including DNA, proteins, drugs, antibodies, and metabolites, can be introduced into cells. At present, the mechanisms of molecular uptake by electroporation are not well understood. For small molecules, they may enter cells by simple diffusion during the period when the cell membrane is permeabilized by the electric field. For large macromolecules, such as DNA, the uptake mechanism could be much more complicated. The uptake process may involve DNA binding, electrophoretic movement and/or endocytosis *(1–4)*.

As a method of gene transfer, electroporation has been applied to many different cell types, including animal cells *(5–15)*, plant cells *(16)*, yeast cells *(17)*, and bacteria *(18)*. In comparison to other methods of gene transfer, such as the calcium phosphate method or the retrovirus method, electroporation has several advantages. First, it is simple. The gene transfer process takes only minutes, and no special treatment of DNA is required. Second, gene transfer efficiency using electroporation is usually significantly higher than that of the

From: *Methods in Molecular Biology, vol. 62: Recombinant Gene Expression Protocols*
Edited by: R. Tuan Humana Press Inc., Totowa, NJ

chemical methods *(18,19)*. Third, in comparison to transfection methods using chemicals or viruses, electroporation has very few biological or chemical side-effects. Fourth, since electroporation is a physical method, it is less cell-type dependent. Hence, electroporation is an efficient and highly versatile method of gene transfer.

2. Materials and Equipment

2.1. Pulse Generator

The basic equipment required for electroporation is a high-power pulse generator and a sample chamber for applying the electric field to the target cells. The major function of the pulse generator is to generate a high voltage pulse with large output current. When the cells are placed between a pair of electrodes with a gap distance *d*, the electric field is equal to

$$E = \mathrm{V}/d \tag{1}$$

where V is the voltage of the applied pulse. In most experiments, *d* may vary from 0.1–0.4 cm. Since the field strength required to electroporate animal cells is on the order of 1000–2000 V/cm, the pulse generator must be capable of supplying adequate voltage.

Because of the narrowness of the electrode gap and the fact that most cell suspension media contain a certain amount of salt, the effective resistance of the biological sample can be very low (e.g., on the order of 10 Ω). In order to supply a high voltage pulse, the pulse generator must be able to provide high output current. Hence, the pulse generator used for electroporation must be designed to give a high power output. This property makes the pulse generator a very hazardous equipment. It must be operated with great care!

Three different types of waveforms have been used to electroporate cells for gene transfer: rectangular pulse, exponential decay pulse, and radio-frequency pulse (*see* Fig.1). The exponential decay pulse is generated by discharging a large capacitor which is precharged at a high voltage; thus, it is also called the "capacitor-discharge" (CD) pulse. At present, most of the commercially available electroporation devices are generators of CD pulses. The major reason for the popularity of the CD pulse device is its cost effectiveness. The CD pulse device is relatively simple to design and thus can be manufactured at a lower cost. There is also evidence that, for some mammalian cells, the CD pulse can give higher transfection efficiency than the rectangular pulse *(7)*. Among the three different pulse types, the radio-frequency pulse is probably the most effective one for electroporation *(20,21)*. But, it is more complicated to use, and, at present, generators of this type are usually custom-built. Generators of CD pulses or rectangular pulses are currently available from a number of commercial sources. For example, Bio-Rad (Hercules, CA), BTX (San Diego),

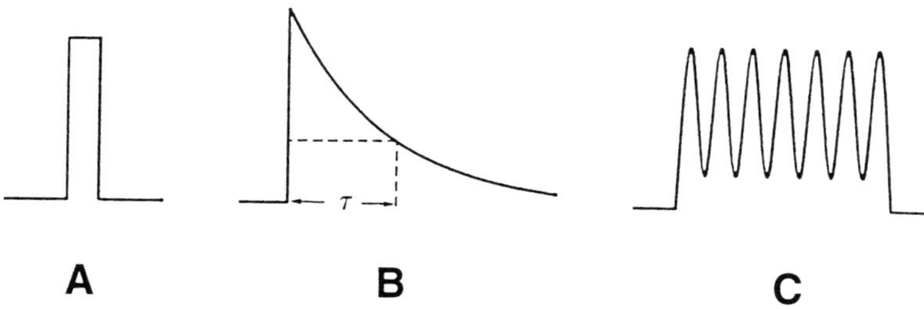

Fig. 1. Waveforms of electrical pulses commonly used in electroporation. **(A)** A rectangular pulse. **(B)** An exponential decade pulse, which is generated by discharging a capacitor and thus is also called a "capacitor discharge (CD) pulse." The "pulse width" is characterized by a time constant, τ, at which the field strength has fallen to 37% of the peak value. **(C)** A DC-shifted radio-frequency pulse (*see* ref. 20).

Hoefer (San Francisco), IBI-Kodak (New Haven, CT), Life Technologies (Gaithersburg, MD), and Shimadzu (Japan) are major supplier of such devices.

2.2. Sample Chamber

The chamber has two functions: to contain the cell mixture, and to provide a pair of electrodes for applying the external electric field. The main body of the chamber is made of transparent nonconducting material, such as polystyrene. The electrodes, of course, are made of metal.

There are two basic designs for the sample chamber. The first type is a plastic cuvet with two, parallel, built-in aluminum electrodes on the inside (Fig. 2). The cell mixture inside the cuvet is thus in direct contact with the electrodes, which are in turn connected to the output of the pulse generator. Cuvets of this type are commercially available and are usually sold as disposable items. Because they are presterilized and may be used only once, these disposable cuvets are most convenient to use for electroporating suspended cells.

In the second type of chamber design, the electrodes are separated from the cell container, which may be a plastic cuvet or a plastic cell culture plate. First, suspended cells are placed inside the container; then, a pair of metal electrodes is inserted into the container to make contact with the cell mixture. Such inserting electrodes are usually made of stainless steel or platinum. They are designed to be reusable; one can simply wash them and use them again and again. Electrodes of this type are also available commercially.

2.3. Solutions

The basic solutions needed for electroporation experiments are cell culture medium and poration medium. The cell culture medium is the same tissue culture medium normally used for culturing the specific cells under study. Its composition of course depends on the cell type used. For example, suppose the

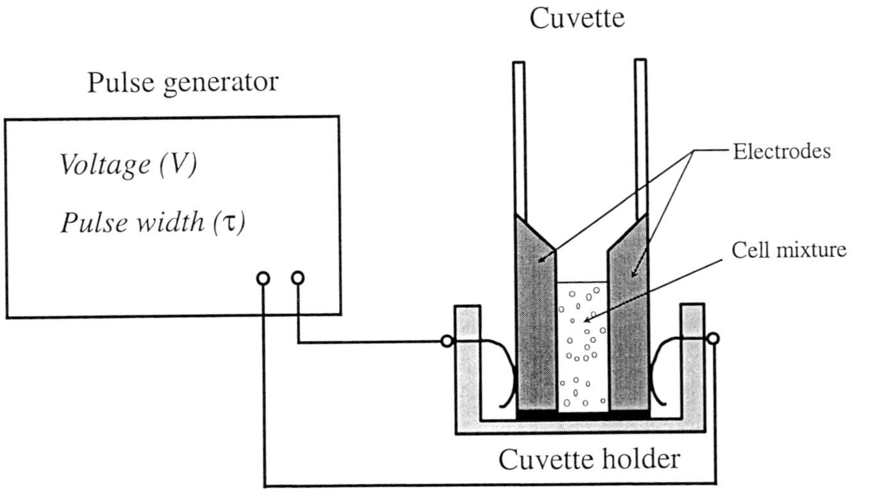

Fig. 2. A simplified diagram showing the basic components of the electroporation device and their arrangement. Here, the sample chamber is the type that has a pair of built-in electrodes inside the cuvet.

target cells for gene transfer are Swiss 3T3 cells; then the cell culture medium used will be DMEM supplemented with 10% fetal calf serum. The poration medium (PM) is basically a low salt isotonic buffer for mammalian cells. Its composition is:

1. 15 m*M* potassium phosphate.
2. 1 m*M* $MgCl_2$.
3. 250 m*M* sucrose (or mannitol).
4. 10 m*M* HEPES.
5. pH adjusted to 7.3.

The carbohydrate is needed for replacing part of the salts in the extracellular medium so that the conductance of the solution can be reduced although the osmotic pressure remains unchanged. The chemical composition is not very critical. The solution, however, should be made with triple distilled water and filtered in 0.2 µm filter for sterilization. The poration medium can be stored at 4°C and used over a period of several months.

Other solutions may be needed, according to the specific cells under study. For example, if the target cells used in the experiment are attached cells, they must first be detached using a trypsinizing treatment. In this case, one will also need the trypsinizing medium which is commonly used for detaching cultured cells from the flask.

2.4. DNA

The DNA used in the electroporation experiment does not require special treatment. It can be handled as is standard in most other molecular biological manipulations. The DNA can be used directly in its normal buffer, such as Tris buffer or distilled water. When plasmid DNA is used, some people may prefer to convert it to the linear form first. Such linearization may offer some advantage in stable transfection *(7)*. In transient transfection, however, the linearization step is unnecessary and may actually be undesirable *(7)*.

3. Methods

3.1. Selecting the Proper Electrical Parameters

In a gene transfer experiment using electroporation, there are three basic physical parameters that can greatly affect the outcome of the experiment. They are the strength of the applied electric field, the pulse width, and the number of applied pulses. In those studies using either rectangular pulse or radio-frequency pulse for gene transfer, multiple electrical pulses are normally applied during the electroporation process. In studies using CD pulse, however, the electroporation is usually accomplished with only one pulse. In fact, most commercially available pulse generators of the CD-pulse type are not capable of generating multiple pulses. Since the CD-pulse type generator is currently the most popular choice of electroporation devices for use in gene transfer, in the methodology discussed in the following, we will assume that the pulse generator is of such a type. Under this situation, the number of applied pulses is usually fixed at one.

3.1.1. Setting the Pulse Width

When electroporation was first used for gene transfer, the pulse width was always very short, typically set in the submillisecond range. Nowadays, short pulse width is still the standard setting for electroporation using rectangular pulses or in electroporation of prokaryotic cells. When CD pulse is used for electroporating mammalian cells, however, the gene transfer efficiency appears to be better with longer pulse width *(7)*. In the CD pulse device, the pulse width is determined by the time constant τ which is equal to

$$\tau = R\,C \qquad (2)$$

in which C is the capacitance of the pulse generator and R is the resistance of the cell sample. For mammalian cells, C is typically set at 500 µf or higher; R is related to the ionic strength of the cell medium, the volume of the cell sample, and the distance between the electrodes in the chamber. When a large amount of ions are contained in the medium (such as PBS), the resistance of the sample

Table 1
Parameters Used in Electroporation for Typical Mammalian Cultured Cells

Cell types	Field strength (kV/cm)	Pulse width (ms)/ Capacitance (μF)	References
Human fibroblasts HeLa cells CV-1 NIH3T3	0.5–0.7	7 ms	*7*
Human T-cells	0.2	700 ms	*8*
Mouse L Tk- cells	2	0.7 ms	*9*
Human fibroblast	1.5	20–25 ms	*10*
Murine BW lymphocytes	0.2	700 ms 1100 μF	*11*
Chinese Hamster Ovary (CHO) cells	2.875	0.4 ms	*12*
Primary human peripheral blood lymphocytes	0.63	45–50 ms 2000–2850 μF	*13*
Cos-M6 cells Human hematopoietic cells K-562	0.8–1.0	50 ms 500 μF	*14*
Fibroblast, 3T6 cells	0.8	50 ms 500 μF	*14*
Rat adipose cells	1.5	0.6 ms 25 μF (6 pulses)	*15*

is very low. On the other hand, if the cell medium contains very little salt, *R* can be very large. In some commercial devices, an internal shunt resistance is provided so that *R* can be determined by the circuit rather than by the cell medium of the sample. However, such arrangement is useful only when a low salt medium is used.

The starting value of the pulse width may be chosen based either on recommendation from the equipment manufacture or from values used in previous published works (*see* Table 1). If no recommended value of τ can be found, set the capacitance $C = 500$ μF and the shunt resistance $R = 100\ \Omega$, or less. In the case that the shunt resistance is not available and the cell medium may contain a large amount of salts, the capacitance C may be set to 800–1000 μF.

3.1.2. Setting the Field Strength

The parameter that can most greatly affect the gene transfer efficiency and the cell viability is the electric field strength, E. For mammalian cells, the workable range of E is on the order of one kV/cm. The optimal setting of E depends on the cell type. Sample values of E and τ for some commonly used mammalian cell types are listed in Table 1. More extensive lists of electroporation parameters based on published studies may also be available from companies that supply electroporation devices.

One cannot set the value of E directly in the pulse generator; one can only set the output voltage, V. These two parameters are related by the equation

$$E = \mathrm{V}/d \tag{3}$$

or

$$\mathrm{V} = E\,d \tag{4}$$

where d is the distance between the electrodes in the chamber. Hence, with the same cell type, the optimal setting of V may change when a new type of chamber is used.

The optimal setting of E differs between cell types. In theory, the value of E is inversely proportional to the diameter of the cell. If the optimal value is not previously known, it must be first determined by doing a voltage-dependent study to examine the effects of the different settings of electric field on gene transfer efficiency and cell viability. Thus, when one uses electroporation for the first time to transfect a given cell type, one must prepare multiple samples and electroporate each sample at a slightly different field strength. To determine the proper range of field strength for the initial experiment, one should first choose an assumed field strength, E_o, which may be estimated based on values used in previous electroporation studies of similar cell types (*see* Table 1). The experimental settings of field strength for the various samples then can be set at the following values: 0.5 E_o, 0.75 E_o, 1.0 E_o, 1.25 E_o, and 1.5 E_o (*see* Table 2). If no information is available to choose a proper E_o, one may assume E_o to be 1000 V/cm. Once the field strength is determined, the corresponding value of V can be calculated from equation 4.

3.2. Procedures for Gene Transfer by Electroporation

The following are common procedures for electroporation using a CD pulse device. Here, we will use five samples to examine the effects of field strength. Two additional samples are used as controls.

1. Grow cells in their normal culturing conditions. When cells reach the midlog growing phase, harvest them using a trypsin treatment. Suspend cells in culture medium and place in a 15-mL centrifuge tube.

Table 2
The Recommended Range of Voltage Settings in an Experiment Designed to Identify the Optimal Electroporation Parameter *(E)*

Sample #	E (V/cm)	V (If E_o is known)	V (If E_o is not known)
1	0.50 E_o	0.10 E_o	100
2	0.75 E_o	0.15 E_o	150
3	1.00 E_o	0.20 E_o	200
4	1.25 E_o	0.25 E_o	250
5	1.50 E_o	0.30 E_o	300

It is assumed here that a cuvet with a 0.2-cm electrode gap is used.

E_o (in unit of volt/cm) is the trial setting of E; it is usually estimated based on previous studies. If E_o is not known, it may be assumed to be 1000 V/cm.

2. Do a cell count using either a Coulter counter or a hemacytometer. To make up seven samples, about 5–10 million cells are needed. After counting, transfer the required amount of suspended cells to a new 15-mL centrifuge tube.
3. Concentrate the suspended cells by centrifugation at 200*g* for 3 min in a table top centrifuge. Then, remove supernatant and resuspend the pelleted cells in 1.05 mL of PM in a microcentrifuge tube. If the cells are not to be processed for electroporation right away, they may be stored on ice.
4. Turn on the power of the electroporation device. Let it warm up for about 5 min. Set the output voltage and the pulse width (or capacitance and resistance) according to the procedures outlined in Section 3.1. Make sure that the cuvet holder is properly connected to the output terminals of the pulse generator.
5. Transfer 150 µL of suspended cells to a cuvet. This will become the control sample without DNA.
6. Add DNA to the rest of the cell suspension contained in the centrifuge tube. The amount of DNA required for each sample may range from 4–20 µg, depending on the degree of optimization of the gene transfer process. (Remember that there are six samples here. The total amount of DNA added should equal the DNA needed for an individual sample times six). Mix the DNA thoroughly with the suspended cells by pumping the pipet in the cell mixture repeatedly. Then, aliquot 150 µL of cell mixture into six sterilized cuvets with built-in electrodes (with a electrode gap of 0.2 cm).
7. Place the first cuvet (with cell mixture inside) into the cuvet holder. Set the output voltage according to the value specified in Table 2. (Note: For the first sample, set the value to that specified for sample #1).
8. When the power supply indicates that the capacitor has been charged to the proper voltage, apply the high voltage pulse to the cell sample by pushing the "firing" button(s) on the electroporation device.
9. Remove the cuvet after the electrical pulse has been terminated.

10. After the electrical treatment, add 1 mL of normal culture medium to the cuvet, and transfer the cell mixture from the cuvet to a tissue culture container (Petri dish or culture plate). More culture medium is then added to give a proper final volume.
11. Place the next cuvet in the cuvet holder. Reset the output voltage to that of the next sample. Repeat steps 8–10.
12. The two control samples (one with DNA and the other without DNA) are treated in a similar manner as the experimental samples, except that no electrical pulse is applied to them.
13. Place the cell samples in the CO_2 incubator for normal culturing. Depending on the type of gene transfer experiment, the electroporated cells may be cultured for different lengths of time. For transient transfection, cells may be assayed for the expression of the transferred gene after culturing for 10–40 h. For stable transfection, the culturing period will be much longer. It can be several weeks, depending on the selection methods used.

The assay method depends on the nature of the introduced gene and thus will not be discussed here.

4. Notes

1. The optimal settings for the electroporation parameters are cell-type dependent. Thus, in order to produce the best results in gene transfer for a specific cell type, it is often necessary to conduct a few experiments to identify the optimal electroporation conditions. The most critical parameter is the applied electric field, E. The optimal setting of E, however, also depends on the chosen value of the pulse width, τ. In fact, E and τ can compensate for each other *(22,23)*. For example, a shorter τ will in general require a higher E, and inversely, if a longer τ is chosen, the optimal setting of E will be reduced. For electroporation using a CD pulse, the optimal setting of E is usually in a range in which approx one third of the electroporated cells can survive.
2. An important factor that can affect the electroporation efficiency is the growth state of the cultured cells. To achieve high efficiency, cells must be harvested during the midlog phase of active growth. If the cultured cells are grown for too long, the transfection efficiency will usually be low. Hence, it is recommended to replate the cultured cells within 2 d before they are used for electroporation.
3. The transfection efficiency can be improved by increasing the amount of DNA added. In fact, what really matters is not the total amount of DNA, but the concentration of DNA in the cell suspension. Thus, it is recommended to minimize the sample volume in using electroporation. In the electroporation process, it is quite common to concentrate cells to a density of 2×10^7 cells/mL. Thus, two million cells can easily fit into a volume of 0.1 mL. With a fixed amount of DNA, a smaller sample volume will have a higher DNA concentration and usually will result in higher transfection efficiency.
4. In some laboratories, carrier DNA, such as sonicated salmon sperm DNA, is also added to the cell mixture during electroporation *(7)*. The addition of carrier DNA

can improve the transfection efficiency to almost two-fold at optimal conditions. The improvement, however, depends on adding the proper amount of carrier DNA. If too much carrier DNA is added, there may be no improvement at all *(7)*.

5. In the early days when electroporation was first developed for gene transfer, it was frequently recommended to transfect cells at 0°C *(5)*. Later, it was found that transfecting cells at 20°C can actually give higher transfection efficiency *(7)*. Thus, for simplification, one may first carry out the entire electroporation procedures at room temperature. If the results are not satisfactory, one can then change the procedures by placing the cell sample in ice before and after electroporation.
6. The transfection efficiency can be enhanced by using a recovery treatment. After applying the electric pulse, cells may be first incubated in a recovery medium, which is the same as the normal culture medium except that Ca^{2+} ions are chelated with EGTA and substituted with a equal amount of Mg^{2+} ions. After incubation for 20–30 min, cells may be returned to their standard culture medium and cultured in normal culturing conditions.
7. In most commercially available electroporation cuvets, the built-in electrodes are made of aluminum. The surface resistance of these electrodes can build up significantly after a few usages, probably due to oxidation. In this situation, the effective electric field experienced by the suspended cells will be far less than the applied field, and could result in poor transfection efficiency. Thus, it is recommended not to use old cuvets with aluminum electrodes. If there is any doubt about the cuvet, use a new one.
8. Like most other molecular biology procedures, protocols for electroporation can vary significantly among different laboratories. For example, one can find many different compositions for the poration medium in the literature. Whereas some may use high ionic strength media, most others prefer low salt isotonic solutions. The optimal conditions, of course, also vary significantly depending on the specific cell type used. Thus, if the result of electroporation is less than satisfactory in spite of the efforts to optimize the voltage and pulse width, one may try to change the other experimental parameters, such as the poration medium, temperature, cell treatment, addition of carrier DNA, etc. In our experience, the recovery treatment is also very helpful.

References

1. Chang, D. C., Chassy, B. M., Saunders, J. A., and Sowers, A. E. (eds.) (1992) *Guide to Electroporation and Electrofusion*. Academic, San Diego.
2. Sukharev, S. I., Klenchin, C. A., Serov, S. M., Chernomordik, L. V., and Chizmadzhev, Y. A. (1992) Electroporation and electrophoretic DNA transfer into cells. The effect of DNA interaction with electropores. *Biophys. J.* **63,** 1320–1327.
3. Xie, T. D., Sun, L., and Tsong, T. Y. (1990) Study of mechanisms of electric field-induced DNA transfection I. DNA entry by surface binding and diffusion through membrane pores. *Biophys. J.* **58,** 13–19.
4. Chang, D. C. and Lu, P. (1993) Mechanisms of DNA-uptake during electroporation. *Mol. Biol. Cell* **4,** 223a.

5. Potter, H., Weir, L., and Leder, P. (1984) Enhancer-dependent expression of human κ immunoglobulin genes introduced into mouse pre-B lymphocytes by electroporation. *Proc. Natl. Acad. Sci. USA* **81**, 7161.
6. Smithies, O., Gregg, R. G., Boggs, S. S. Koralewski, M. A., and Kucherlapati, R. S. (1985) Insertion of DNA sequences into the human chromosmal β-globin locus by homologous recombination. *Nature* **317,** 230.
7. Chu, G., H. Hayakawa, and Berg, P. (1987) Electroporation for the efficient transfection of mammalian cells with DNA. *Nucleic Acids Res.* **15,** 1311–1326.
8. Hambor, J. E., Hauer, C. A., Shu, H. K., Groger, R. K., Kaplan, D. R., and Tykocinski, M. L. (1988) Use of an Epstein-Barr virus episomal replicon for antisense RNA-mediated gene inhibition in a human cytotoxic T-cell clone. *Proc. Natl. Acad. Sci. USA* **85,** 4010.
9. Murray, R., Hutchinson, C. A., III, and Frelinger, J. A. (1988) Saturation mutagenesis of a major histocompatibility complex protein domain: identification of a single conserved amino acid important for allorecognition. *Proc. Natl. Acad. Sci. USA* **85,** 3535.
10. Stevens, C. W., Brondyk, W. H., Burgess, J. A., Manoharan, T. H., Hane, B. G., and Fahl, W. E. (1988) Partially transformed, anchorage-independent human diploid fibroblasts result from overexpression of the c-sis oncogene: mitogenic activity of an apparent monomeric platelet-derived growth factor 2 species. *Mol. Cell. Biol.* **8,** 2089.
11. Tykocinski, M. L., Shu, H. K., Ayers, D. J., Walter, E. I., Getty, R. R., Groger, R. K., Hauer, C. A., and Medof, M. E. (1988) Glycolipid reanchoring of T-lymphocyte surface antigen CD8 using the 3' end sequence of decay-accelerating factor's mRNA. *Proc. Natl. Acad. Sci. USA* **85,** 3555.
12. Dunn, W. C., Tano, K., Horesovsky, G., Preston, R. J., and Mitra, S. (1991) The role of 06-alkylguanine in cell killing and mutagenesis in chinese hamster ovary cells. *Carcinogenesis* **12,** 83–89.
13. Cann, A. J., Koyanagi, Y., and Chen, I. S. Y. (1988) High-efficiency transfection of primary human-lymphocytes and studies of gene expression. *Oncogene* **3,** 123–128.
14. Chang, D. C. and Leung, D. (1992) Unpublished data.
15. Quon, M. J., Zarnowski, M. J., Guerre-Millo, M., Sierra, M. L., Taylor, S. I., and Cushman, S. W. (1993) Transfection of DNA into isolated rat adipose cells by electroporation: evaluation of promoter activity in transfected adipose cells which are highly responsive to insulin after one day in culture. *Biochem. Biophys. Res. Commun.* **194,** 338–346.
16. Fromm, M., Taylor, L. P., and Walbot, V. (1985) Expression of genes transferred into monocot and dicot plant cells by electroporation. *Proc. Natl. Acad. Sci. USA* **82,** 5824–5825.
17. Karube, I., Tamiya, E., and Matsuoka, H. (1985) Transformation of Saccharomyces cerevisiae spheroplasts by high electric pulse. *FEBS Lett.* **182,** 90.
18. Dower, W. J., Miller, J. F., and Ragsdale, C. W. (1988) High efficiency transformation of *E. coli* by high voltage electroporation. *Nucleic Acids Res.* **16,** 6127–6145.

19. Potter, H. (1988) Electroporation in biology: methods, applications and instrumentation. *Anal. Biochem.* **174,** 361–373.
20. Chang, D. C. (1989) Cell poration and cell fusion using an oscillating electric field. *Biophys. J.* **56,** 641–652.
21. Chang, D. C., Gao, P. Q., and Maxwell, B. L. (1991) High efficiency gene transfection by electroporation using a radio-frequency electric field. *Biophys. Biochim. Acta* **1992,** 153–160.
22. McNally, M. A., Lebkowski, J. S., Okarma, T. B., and Lerch, L. B. (1988) Optimizing electroporation parameters for a variety of human hematopoetic cell lines. *BioTechniques* **6,** 882–887.
23. Chang, D. C. (1992) Design of protocols for electroporation and electrofusion: selection of electrical parameters, in *Guide to Electroporation and Electrofusion* (Chang, D. C., Sowers, A. E., Chassy, B., and Saunders, J. A., eds.), Academic, San Diego, pp. 429–456).

25

Highly Effective Delivery of Foreign DNA to Adherent Cells via Polybrene/DMSO-Assisted Gene Transfer

Rémy A. Aubin, Michael Weinfeld, Marzieh Taghavi, Razmik Mirzayans, and Malcolm C. Paterson

1. Introduction

The ability to transfect cells with native genomic DNA and engineered gene/vector constructs has played a leading role in accelerating our understanding of gene regulation and function. Yet, despite significant diversification of gene delivery strategies, efforts to expand the repertory of cell types amenable to DNA-mediated gene transfer continue to stumble over a recurrent obstacle—namely, the sensitivity that recipient cells often manifest towards the gene transfer procedure itself *(1,2)*. To minimize the likelihood of cytopathic effects engendered by chemical and physical methods investigators, more and more, are adopting low toxicity synthetic polymers and ligand-mediated strategies to dispatch copious amounts of exogenous DNA to the surfaces of target cells *(3–10)*.

Polybrene (Fig. 1) is an amorphous polycation well known to retrovirologists for its ability to augment the infectivity of retroviral particles and nucleic acids in vitro *(11–13)*. The transfection-enhancing ability of polybrene is believed to reside in its propensity to establish electrostatic bridges between negatively charged viral particles (or nucleic acids) and anionic glycoprotein moieties protruding from the plasma membranes of recipient cells *(11)*. Inspired by the elegant simplicity of this adsorptive targeting process, Kawai and Nishizawa *(14)* were the first to report on the utility of combining polybrene with dimethyl sulfoxide (DMSO) to introduce biologically active (i.e., phenotypically selectable) Rous sarcoma viral genomic DNA into chicken embryo fibroblasts. Shortly thereafter, we *(9)* and others *(16–21)* demonstrated that the same basic principle could be applied to introduce plasmid, cosmid, or

From: *Methods in Molecular Biology, vol. 62: Recombinant Gene Expression Protocols*
Edited by: R. Tuan Humana Press Inc., Totowa, NJ

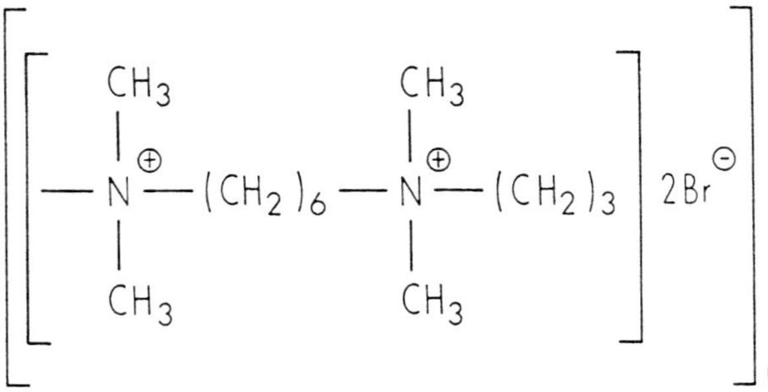

Fig. 1. Chemical structure of polybrene.

nonselectable genomic DNA segments in human, rodent or insect cells with equally high efficiency.

From a procedural perspective, polybrene/DMSO-assisted gene transfer is essentially a biphasic process consisting of adsorption followed by internalization. During the adsorption phase polybrene is used to coat target cells with exogenous DNA molecules. The latter are then internalized by the cells following a brief permeabilization period in the presence of DMSO. In a typical experiment, the gene delivery process begins by bathing adherent cells in a transfection cocktail composed of growth medium supplemented with polybrene and DNA. Monolayer cultures are then incubated under optimal growth conditions to allow polybrene/serum protein/DNA complexes to form and adhere to target cell membranes. The transfection cocktail is then removed and replaced with growth medium augmented with DMSO. A short exposure (typically 5 min) is generally sufficient to facilitate the uptake of exogenous DNA and initiate transit to the nucleus. Recipient cells are then rinsed to remove residual DMSO and allowed to recover from the ensuing osmotic stress. The transfected cultures can then be assayed either for "transient" expression of the newly acquired gene(s) or submitted to a dominant selection regime for the isolation of stable phenotypic transformants.

The method is simple to perform, lends itself readily to large-scale experiments and, once optimized for the cell type of interest, can be very effective, generating stable transfection frequencies reproducibly in the range 0.01–0.1% with only nanogram amounts of exogenous DNA *(15)*. The success of the approach is attributed in part to the low overall toxicity of the procedure which, in our hands, typically produces <15% reduction in clonogenic survival. Survival, in turn, does not appear to be under the direct influence of the polymer since most cell types can tolerate polybrene over a wide range of concentrations and for extended incubation periods without significant evidence of adverse effects. Cytotoxicity is, however, critically dependent on the DMSO permeabilization regime. Exposure to DMSO must be optimized independently for each cell line with respect to concentration, exposure time and incubation temperature. Also contributing to the success of the method is the ability of polybrene to protect DNA molecules against degradation by serum and cytoplasmic nucleases (*22,23* and Section 3.4. herein). In doing so, polybrene virtually ensures that multiple copies of unaltered DNA templates are delivered to cell nuclei. Investigators wishing to manipulate foreign gene dosage in stably transfected clones can do so easily by varying the amount of DNA in the transfection cocktails *(15)*.

The overall strategy is presented in five sections. In the first, the basic protocol is described using the mouse fibroblast line NIH 3T3 as an example. In the second, optimal conditions for the transfection of representative human fibroblast cultures are summarized. The third section provides a detailed guide for investigators wishing to customize the protocol for a particular cell type. Section four follows the fate of transfected plasmid DNA. The last section lists technical and cautionary notes. The advantages and limitations of the strategy are discussed throughout. The forerunner to this article has recently appeared in the inaugural issue of *Molecular Biotechnology (24)*.

2. Materials

1. Polybrene: Dissolve polybrene (1,5-dimethyl-1,5-undecamethylene polymethobromide; Aldrich, Milwaukee, WI, Cat. no. 10768-9) to 1 mg/mL in Ca^{++}/Mg^{++}-free Hanks balanced salt solution (HBSS) (GIBCO/BRL Life Technologies, Grand Island, NY, Cat. no. 310-4170AG). Sterilize by filtration (0.22 μ) and store 0.25 mL vol at –20°C in sterile microcentrifuge tubes. Once thawed, a tube of polybrene solution should not be refrozen for future use.
2. pSV2neo plasmid DNA (ATCC no. 37149; >95% form I, 10 ng/μL sterile nanopure water).
3. pBRSV plasmid DNA (ATCC no. 45019; >95% form I, 10 ng/μL sterile nanopure water).
4. G418: G418 sulfate (Geneticin™; GIBCO/BRL, Cat. no. 860-18 11IJ) is usually supplied at potencies ranging between 450 and 600 μg of active product per mg of powder. Prepare a 10 mg/mL stock solution at full potency by dissolving the necessary amount of powder in HBSS. Sterilize by filtration (0.22 μ) and store 5–10 mL vol at –20°C. G418 is acidic and the concentrated stock will assume a yellow color. The G418 solution will resist repeated freeze-thawing.

5. Dimethyl sulfoxide (DMSO): In principle, any source of DMSO certified for cell culture may be used. However, we have found that the product available from Fisher Scientific, Ottowa, Canada (Spectranalyzed DMSO, UV cutoff at 262 nm; Cat. No D-136) offers the most consistent performance.
6. Growth media: Murine NIH 3T3 fibroblasts are cultivated in Dulbecco's Modified Eagle Medium (DMEM; 1X liquid, 4500 mg/L D-glucose, with L-glutamine, without sodium pyruvate). Human fibroblasts are grown in Ham's F12 Nutrient Mixture (1X liquid, with L-glutamine). Growth media are supplemented with 10% or 15% (v/v) fetal calf serum (*see* Table 2 for details), 1 m*M* L-glutamine, 100 U/mL penicillin and 100% µg/mL streptomycin sulfate prior to use and stored at 4°C. Complete medium should be used within 14 d.
7. Trypsin (10X stock; 2.5% trypsin in 0.8 g/L NaCl): Prepare a 1X working solution by dilution in sterile phosphate buffered saline (PBS). Working solutions should be stored at –20°C. Repeated cycles of freezing and thawing will reduce the potency of the solution considerably.
8. Crystal violet staining solution [0.5% (w/v) crystal violet in 40% (v/v) methanol].

3. Methods

3.1. The Basic Protocol

A schematic representation of the basic gene delivery protocol, as optimized for murine NIH 3T3 fibroblasts *(15)*, is presented in Fig. 2.

1. NIH 3T3 fibroblasts are propagated in monolayer culture using Dulbecco's Modified Eagle Medium (DMEM; 4500 mg/mL glucose) supplemented with 10% fetal calf serum (FCS), 1 m*M* L-glutamine, 100 U/mL penicillin and 100 µg/mL streptomycin sulfate (henceforth referred to as "complete medium"). The cells are maintained in a 37°C incubator providing a humidified (75–85%) atmosphere of 5% CO_2 in 95% air and are fed complete medium twice a week. Cells destined for gene transfer should be routinely passaged in late log phase and never be allowed to remain in a confluent state for more than 24 h prior to seeding.
2. On the day preceding transfection, harvest log to late log cultures by brief exposure to dilute (0.25%) trypsin and seed NIH 3T3 cells at a density of 5×10^5 cells/60 mm diameter dish. Place the dishes in the incubator and allow the cells to attach and resume growth overnight.
3. The next day, initiate the process of DNA adsorption by replacing the growth medium in each dish with 2.0 mL of "transfection cocktail" consisting of 5.0 µg/mL polybrene and anywhere between 5 and 25 ng/mL form I pSV2neo plasmid DNA in prewarmed (37°C) complete medium. Prepare the cocktail immediately before use in a sterile culture tube by adding the medium first, the plasmid DNA second, and the polybrene last. Vortex the solution after the addition of plasmid DNA and polybrene. **Warning:** Polybrene should never be added directly to DNA in an empty vial. This will cause irreversible precipitation (*see* Note 4). Swirl the dishes gently to distribute the cocktail evenly over the cell monolayer

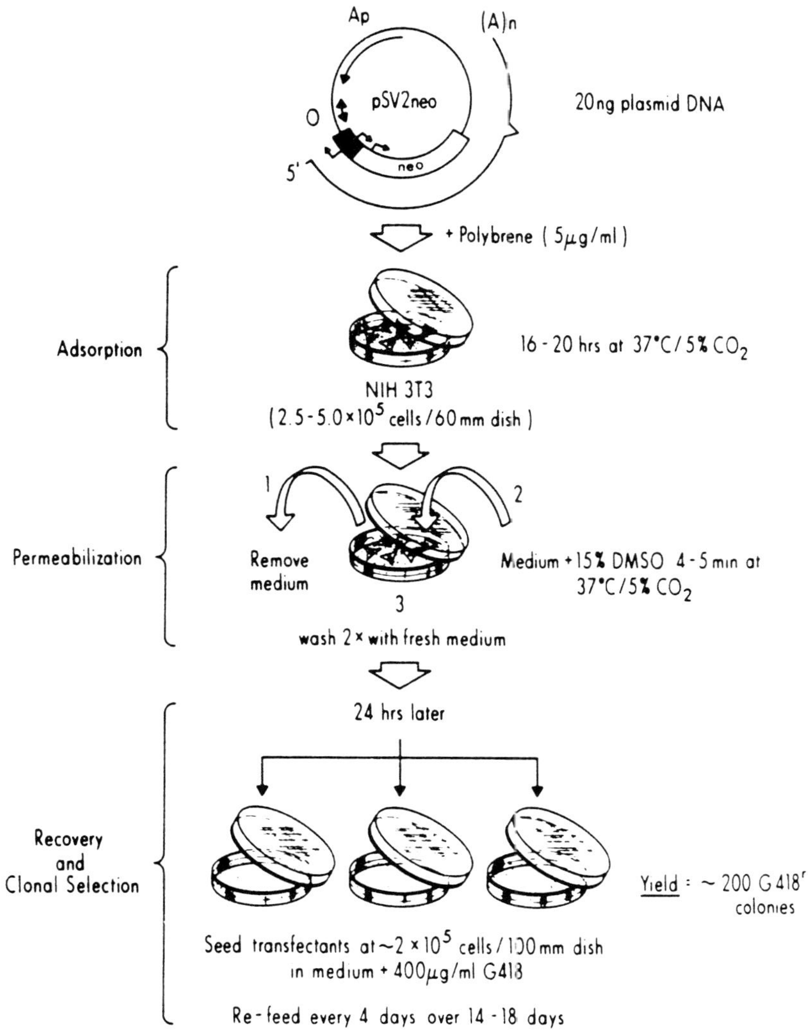

Fig. 2. Schematic outline of polybrene/DMSO-assisted gene transfer as optimized for NIH 3T3 fibroblasts.

and return the dishes to the incubator. Allow DNA-polybrene complexes to form and attach to the cells overnight (*see* Note 3).

4. Following adsorption, proceed to permeabilize the cells with DMSO. The permeabilization solution consists of complete medium augmented to 15% (v/v) DMSO. Prepare the solution immediately before use in a sterile glass bottle (or polypropylene centrifuge tube) and place in a water bath equilibrated to 37°C. Remove the transfection cocktail by aspiration and cover each monolayer with 4.0 mL of permeabilization medium. Distribute the solution evenly over the cells

by swirling each dish gently for 10–15 s. Transfer the dishes to the incubator and allow cell permeabilization to proceed for 4.5 min. Ensure adequate heat exchange by placing each dish in direct contact with the incubator shelf (i.e., do not stack the dishes). **Note:** Permeabilization medium should be kept at 37°C throughout the procedure. To minimize the toxicity of the treatment, the DMSO-augmented medium should be added from the side of each dish rather than directly over the monolayer.

5. Remove the DMSO solution by aspiration and rinse the cells twice with 4 mL of prewarmed complete medium. Add the medium from the side of each dish at a moderate rate and swirl the dishes slowly for 10–15 s to remove excess DMSO. It is very important to begin the rinsing schedule promptly after permeabilization and to maintain the rinse medium at 37°C. Treat the cells gently during these steps since they are submitted to extremes of osmotic pressure—especially during the washing regimen (Fig. 3.)
6. Cover the cells with 4 mL of growth medium following the last rinse and allow the transfectants to recover for 24 h in the incubator.
7. Re-seed the transfectants at densities of 2.0–2.5 × 10^5 cells using 100-mm diameter dishes containing 10 mL of warm complete medium supplemented with 400 μg/mL active G418. Replace the selection medium every fourth day over a 14–18 d period.
8. Determine the number of drug-resistant colonies by staining the dishes with crystal violet. To do this, decant the medium into a designated receptacle and rinse the dishes briefly under a gentle stream of tap water. Drain the dishes and pour approx 5 mL of crystal violet staining solution into each dish. Stain for 5 min. Decant the staining solution (which can be re-used) and rinse the dishes thoroughly but gently under tap water. Allow the dishes to dry in an inclined position. Score the colonies by eye.

The protocol will yield transfection frequencies varying between 0.02 and 0.1% over a dose range of 10–50 ng form I pSV2neo plasmid DNA (*see* Note 2).

3.2. Polybrene/DMSO-Assisted Gene Transfer of Human Fibroblasts

The principal features of the human fibroblast cultures considered in this section are outlined in Table 1. All cultures are available from the NIGMS Human Genetic Mutant Cell Repository (Camden, NJ). Conditions optimized for transfection of fibroblast strains and lines are listed in Table 2. Using the table as a guide, polybrene/DMSO-assisted gene transfer can be carried out by substituting the appropriate set of conditions into the basic transfection protocol described in Section 3.1.

SV40-transformed human fibroblast lines were adapted to Ham's F12 nutrient mixture supplemented with 15% (v/v) FCS in order to obtain uniform growth kinetics, high plating efficiency (>85%; defined as the proportion of

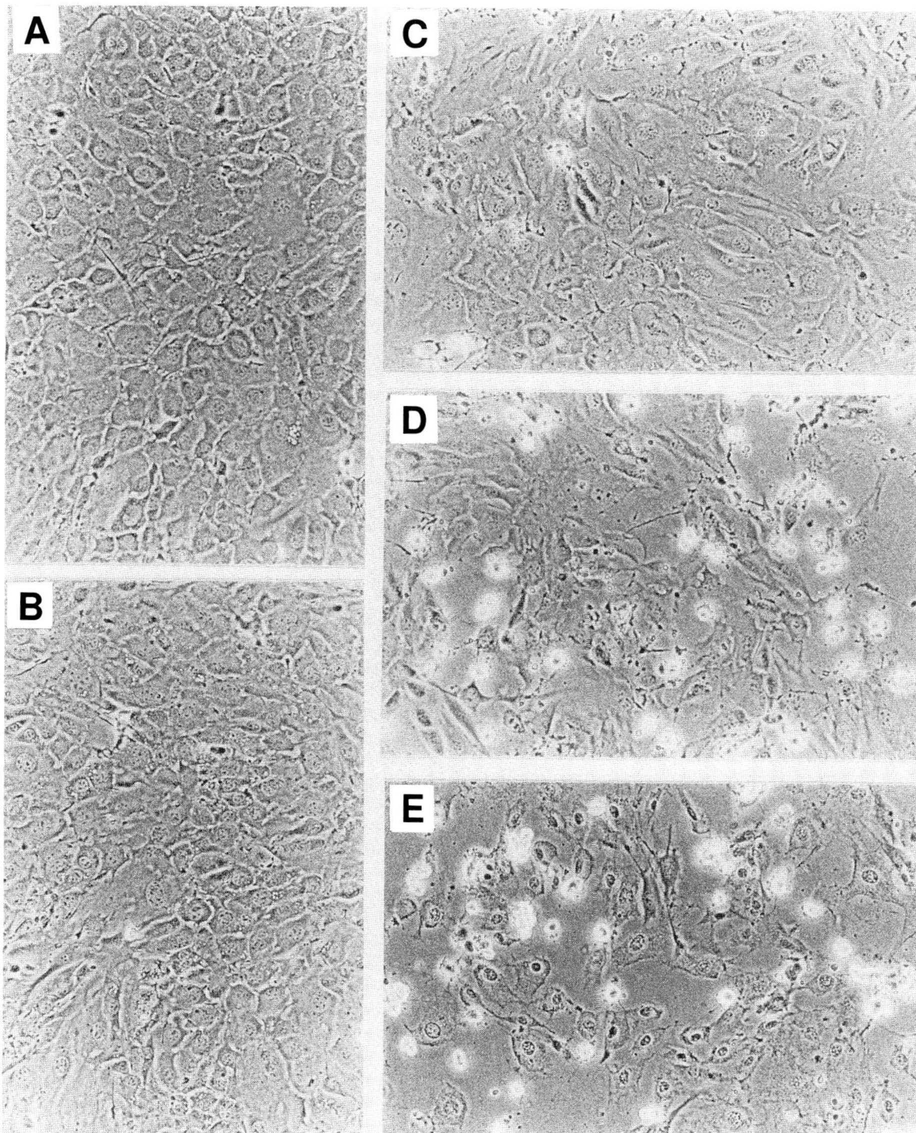

Fig. 3. Effect of DMSO permeabilization on NIH 3T3 cell morphology. **A**; NIH 3T3 cells exposed to transfection cocktail for 18 h. **B–E**; NIH 3T3 cells photographed at recovery 1 h following permeabilization with 10, 15, 20, and 25% DMSO, respectively. Note the swollen appearance of cells permeabilized with 15% DMSO and the cytotoxic effects at 20 and 25% DMSO. Magnification is 200× under phase contrast.

Table 1
Description of Human Fibroblast Cultures

Designation	Age	Sex	Comments
GM10	12 FW	M	Normal fetal strain
GM11	8 FW	M	Normal fetal strain
GM38	9	F	Normal fetal nonfetal strain
GM43	32	F	Normal adult strain
GM730	45	F	Normal adult strain
GM969	2	F	Normal nonfetal strain
GM4312A	7	F	SV40-transformed XP20S; xeroderma pigmentosum group A; U.V. sensitive
GM5849	18	M	SV40-transformed GM5823A (AT5BIVA); ataxia telangiectasia; radiosensitive
GM637A	18	F	SV40-transformed; normal

Abbreviations: F, female; FW, fetal weeks; M, male.

cells in the inoculum capable of attaching to the dish) and high cloning efficiency (typically between 40 and 50%; defined on the basis of colony forming ability in the absence of feeder cells). In our hands, cultures maintained in DMEM/10% FCS handled poorly and provided plating efficiencies of less than 40% and cloning efficiencies below 0.01%. Parameters for the transfection of fibroblast cultures were optimized using 100 ng form I pSV2neo plasmid DNA, a concentration which effectively saturates available DNA binding sites on the cell surfaces (*see* Note 6). These sets of conditions routinely provide transfection frequencies in the range of 0.01–0.05% for the nonimmortalized fibroblast strains and in the range of 0.05–0.1% for the SV40-transformed lines. For transfection experiments performed in 100-mm diameter dishes, SV40-transformed and nonimmortalized human fibroblasts should be seeded at 10^6 and 8.5×10^5 cells per dish, respectively. In addition, 4.5 mL of transfection cocktail, 6 mL of permeabilization medium, and 2×10 mL of rinse medium should be used.

3.3. Optimizing a Polybrene/DMSO-Assisted Gene Transfer Protocol

The success of a transfection experiment depends on a number of variables, not the least of which relate to the biology of the cultures under study. The first consideration, therefore, should be to ensure that the cells are growing under optimal in vitro conditions. Next, one should verify that the cells are neither sensitive to nor affected by exposure to polybrene or to DMSO. In our labora-

Table 2
Summary of Optimized Parameters for Polybrene/DMSO-assisted Gene Transfer of Human Fibroblast Cultures

Designation	Growth Medium[a]	Seeding Density[b]	Polybrene	DMSO Shock	G418
GM10	Ham's F12/10% FCS	3.5×10^5	6.25 μg/mL	30% 5 min	65 μg/mL
GM11	Ham's F12/10% FCS	3.5×10^5	6.25 μg/mL	30% 5 min	65 μg/mL
GM38	Ham's F12/10% FCS	3.5×10^5	2.50 μg/mL	30% 5 min 25% 18 min	50 μg/mL
GM43	Ham's F12/10% FCS	3.5×10^5	2.50 μg/mL	30% 5 min	50 μg/mL
GM730	Ham's F12/10% FCS	3.5×10^5	2.50 μg/mL	30% 5 min 25% 18 min	50 μg/mL
GM969	Ham's F12/10% FCS	3.5×10^5	2.50 μg/mL	30% 5 min 25% 18 min	75 μg/mL
GM4312A	Ham's F12/15% FCS	5.0×10^5	7.50 μg/mL	15% 5 min	75 μg/mL
GM5849	Ham's F12/15% FCS	5.0×10^5	5.00 μg/mL	15% 5 min	50 μg/mL
GMO637A	Ham's F12/15% FCS	5.0×10^5	7.50 μg/mL	15% 5 min	50 μg/mL

[a]The culture medium contains 1 m*M* L-glutamine, 100 U/mL penicillin and 100 μg/mL streptomycin sulfate.
[b]Expressed as the number of cells per 60 mm diameter dish.

tory, several tumor cell lines as well as fibroblasts and epithelial cells have been exposed to a wide range of polybrene concentrations for periods of up to 72 h without adverse effects. This, however, should be interpreted with caution not only because our survey was limited in scope but also because the polymer may affect cellular physiology in certain contexts. For example, polybrene has been reported to enhance retinoic acid-induced differentiation in embryonal carcinoma cell lines *(25)*. Polybrene also appears to stimulate replicative DNA synthesis in freshly isolated human, as well as murine, peripheral blood lymphocytes (unpublished observation) (*see* Note 1). Similar considerations apply to DMSO. DMSO is a well characterized inducer of differentiation in embryonal and other primordial cell systems *(25,26)*. The human prostatic carcinoma line LNCaP (CRL 1740; American Type Culture Collection, Rockville, MD) has also been observed to lose its adhesive properties upon brief (3 min) exposure to even low (10%) concentrations of DMSO (unpublished observation).

Having established tolerance, one must then determine the optima for polybrene and DMSO concentration. At first glance this may seem like a daunting exercise, especially if stable phenotypic transformation is used as the endpoint. However, both labor and tedium can be significantly reduced by streamlining the process in the following way. A matrix of 42 dishes is set up in which four basic parameters are kept constant. These are cell density (3.5×10^5 per dish or a sufficient number of cells to achieve 75% confluence), amount of input plasmid DNA (100 ng form I pSV2neo per dish), adsorption period (16–20 h), and permeabilization time (5 min). Only the polybrene and DMSO concentrations are varied in the assay. Cells are transfected according to the matrix strategy and stable phenotypic transformants are selected over a 14–21 d period (this time frame will vary depending on the growth rates of cultures) using an appropriate G418 dose *(27–29)*. Surviving G418^r colonies are fixed and stained with crystal violet and enumerated by eye. Polybrene and DMSO concentration optima are easily determined by arranging the stained dishes according to the original matrix pattern. From this basic, yet critical information, further refinements, such as varying the adsorption period, the permeabilization time or the initial cell density, can be incorporated if necessary.

3.3.1. Setting up the Transfection Matrix

Inoculate 60 mm diameter culture dishes with 3.5×10^5 log phase cells per dish according to the 42 dish matrix presented in Table 3. Allow the cells to attach and resume growth for 16–24 h in the CO_2 incubator.

3.3.2. Adsorption of pSV2neo Plasmid DNA

1. Prepare transfection cocktails in sterile 15-mL polystyrene centrifuge tubes. Each tube should correspond to a different polybrene concentration in μg/mL

Table 3
Arrangement of the Gene Transfer Optimization Matrix

DMSO (%)							
30	m	m	m	m	m	m	m
25	m	m	m	m	m	m	m
20	m	m	m	m	m	m	m
15	m	m	m	m	m	m	m
10	m	m	m	m	m	m	m
0	m	m	m	m	m	m	m
	0	1	2	4	6	8	10
	Polybrene (μg/mL)						

Table 4
Transfection Cocktail Formulations for the Optimization Matrix

	Polybrene, μg/mL							
Tube	0	0.5	1	2	4	6	8	10
Complete medium (mL)	11.94	11.93	11.93	11.92	11.89	11.87	11.84	11.82
pSV2neo (μL)	60	60	60	60	60	60	60	60
Polybrene (μL)	0	6	12	24	48	72	96	120
Final vol (mL)	12	12	12	12	12	12	12	12

(Table 4). Add the medium first, the DNA second (vortex gently but thoroughly) and the polybrene last (vortex again). The DNA concentration in each tube is kept constant at 50 ng/mL.

2. Remove the medium from the dishes in the "0" polybrene column (Table 3) and replace with 2 mL of transfection cocktail from the "0" polybrene tube (Table 4). Swirl the dishes gently to distribute the mixture evenly over the cells. Return the dishes to the CO_2 incubator and allow adsorption to proceed for 16–20 h. Repeat the process for the remaining columns of dishes, each corresponding to a different polybrene concentration series.

3.3.3. Permeabilization

1. Following adsorption, dishes in each of the polybrene series must be permeabilized with increasing concentrations of DMSO. Prepare the DMSO solutions in sterile glass bottles or polypropylene tubes (Table 5). Because the addition of DMSO to culture medium is exothermic, permeabilization solutions should be equilibrated for 20 min at 37°C before use. Keep the mixtures at 37°C and use within 1 h.
2. Take the first 4 dishes from the "0" DMSO row (corresponding to the 0, 1, 2, and 4 μg/mL polybrene dishes; Table 3) out of the incubator and set them on a clean

Table 5
Permeabilization Medium Formulations for the Optimization Matrix

	DMSO (%)					
Tube	0	10	15	20	25	30
Complete medium (mL)	25	22.5	21.25	20	18.75	17.5
DMSO (mL)	0	2.5	3.75	5	6.25	7.5
Final vol (mL)	25	25	25	25	25	25

paper towel inside the biological safety cabinet. The paper towel will insulate the dishes from the cold metal surface. Remove the transfection cocktail. Add 4 mL warm permeabilization medium from the "0" DMSO tube (Table 5) to each dish. Add the mixture slowly from the side of the dish. Swirl the dishes gently to distribute the solution evenly over the cell monolayer. Incubate at 37°C for no more than 5 min (use a timer and add permeabilization medium to the dishes at 60 s intervals).

3. Rinse the cells twice, gently but thoroughly, with 4 mL warm complete medium.
4. Add 4 mL fresh complete medium to each dish and let the cells recover in the incubator for 20 h before initiating the G418 selection regimen.
5. Permeabilize the remaining 3 dishes from the "0" DMSO row.
6. Proceed with the permeabilization of the remaining DMSO sets.

3.3.4. G418 Selection

1. Feed the cells fresh complete medium supplemented with G418. Replace the selection medium every 4 d.
2. Fix and stain the dishes after 14–21 d of selection. Position the stained dishes according to the transfection matrix. Determine the optimal polybrene and DMSO concentrations according to the number of visible G418^r colonies.

The results of a typical optimization matrix are presented in Fig. 4 using the human pancreatic adenocarcinoma cell line CRL1420 as an example.

3.3.5. Focus Formation Assay for the Optimization of Gene Transfer in Nonimmortalized Human Skin Fibroblasts

A convenient and cost-effective alternative to dominant selection with G418 for optimizing a polybrene/DMSO-assisted gene delivery scheme in nonimmortalized (i.e., primary) human skin fibroblasts is focus formation using plasmid pBRSV *(23)*. This hybrid construct contains the entire SV40 strain 776 genome subcloned in the Bam HI site of pBR322. In primate cells, pBRSV allows expression of the large T-antigen. Human cells expressing this protein become morphologically altered, require less serum to maintain their growth and acquire an extended lifespan in vitro *(30,31)*. Focus formation in human

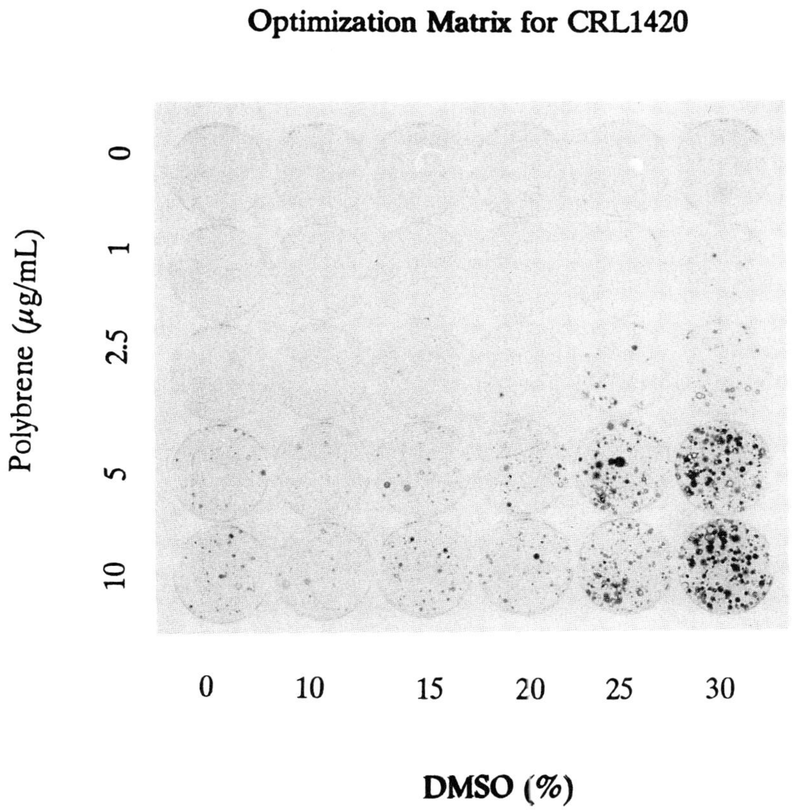

Fig. 4. Representative data generated from an optimization matrix experiment conducted on the human pancreatic adenocarcinoma cell line CRL1420. The optimal concentration for polybrene lies between 5 μg/mL and 10 μg/mL. Permeabilization appeared maximal in presence of 30% DMSO. No colonies were obtained for cells exposed to pSV2neo DNA in the absence of polybrene. For most cell lines permeabilization is also required. However, CRL1420 tumor cells appear to internalize adsorbed DNA in the absence of DMSO, albeit at low efficiency. Note the sharp threshold requirement for DMSO and how it contrasts with the wide range of efficiency for polybrene.

fibroblasts can also be induced by the large T antigen-carrying shuttle vector pSV3gpt *(32)*.

Optimization experiments are planned and executed as outlined in the previous section except for the following modifications—pBRSV (or pSV3gpt) plasmid DNA is substituted for pSV2neo, human fibroblasts are seeded in 100 mm diameter dishes at starting densities of 8.5×10^5 cells/dish, 4 mL transfection cocktail are added per dish for the adsorption phase, permeabilizations are carried out using 5 mL DMSO solution, and 5 mL growth medium are used for the

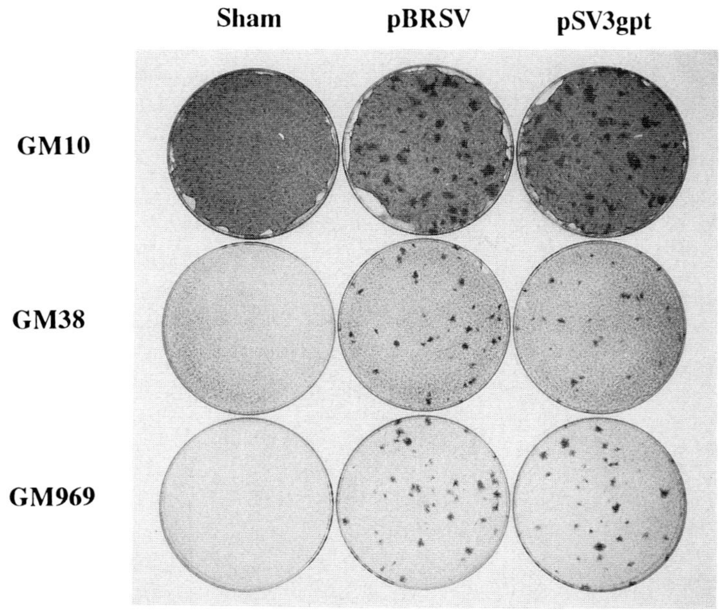

Fig. 5. Focus formation in GM10, GM38, and GM969 fibroblasts transfected with pBRSV and pSV3gpt plasmid DNAs (200 ng/ 8.5 × 10^5 cells in 100 mm diameter dish). Mock transfected dishes are aligned on the extreme left.

post-DMSO washes. Following permeabilization, cells are fed fresh complete medium (12 mL/dish) every 4 d and foci of morphologically transformed fibroblasts are usually scored 18 to 21 d later. Examples of focal growth induced by pBRSV and pSV3gpt transfection in human fibroblast strains GM10, GM38 and GM969 are presented in Fig. 5. Phase contrast micrographs of the accompanying change in cell morphology are shown in Fig. 6 for strain GM38.

3.3.6. Customizing the DMSO Permeabilization Step

Whereas polybrene can be introduced over a wide range of concentrations and for prolonged periods without adverse effects, permeabilization with DMSO, on the other hand, must be adjusted independently for each cell strain (or line) (*see* Note 5). Using stable phenotypic transformation as the endpoint, the DMSO concentration optimum for NIH3T3 cells was found to lie within a remarkably narrow range which could not be predicted on the basis of transient chloramphenicol acetyl transferase (CAT) reporter gene expression assays *(15)*. The lack of concordance between both endpoints probably reflects the fact that elevated concentrations of DMSO favor the influx of reporter plasmids while compromising the integrity of the plasma membrane to such an extent that cell viability and

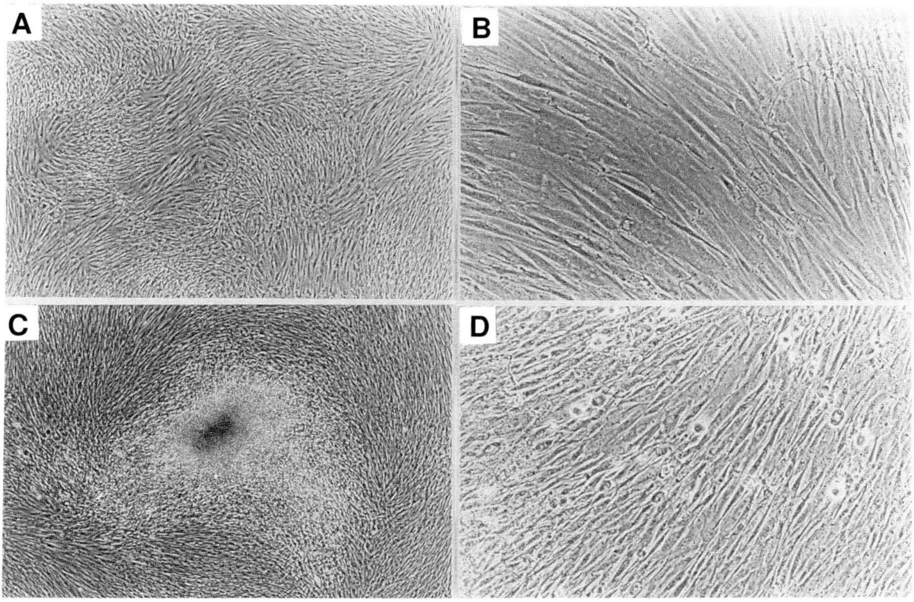

Fig. 6. Comparative assessment of cellular morphologies in parent and pBRSV-transformed GM38 fibroblasts. **A** and **B**; low (40×) and high (200×) magnification phase contrast appearance of a confluent GM 38 monolayer. **C**; pBRSV focus (40×). **D**; edge of the focus showing multiple cell layers and high mitotic activity (200×).

clonogenic survival are significantly reduced. The reporter gene assay, therefore, measures high levels of CAT activity in an essentially moribund cell population.

This phenomenon is also encountered in human fibroblasts. GM38 cells transfected with the reporter vector pSV2cat *(33)* and permeabilized over a 30 min period with a 25% DMSO solution show a rapid increase in CAT enzyme activity by 10 min (Fig. 7, top panel). CAT activity values then stabilize over the next 15 min of permeabilization. Shortly thereafter, by 30 min, CAT activity begins to decline. The overall profile of reporter gene activity suggests that the permeabilization process has reached maximum efficiency by 15 min. Scoring for G418^r colonies in a parallel series of transfections, however, revealed that the highest stable phenotypic transformation frequencies were obtained at ~20 min where cytopathic effects began to appear. High CAT activity values and maximal stable phenotypic transformation frequencies were also recorded when GM38 fibroblasts were permeabilized for 5 minutes with 30% DMSO (Fig. 7, middle panel). Values for both endpoints were reduced approx 10-fold when the cells were permeabilized for the same period of time with 25% DMSO. CAT activity values produced by permeabilization with 35%

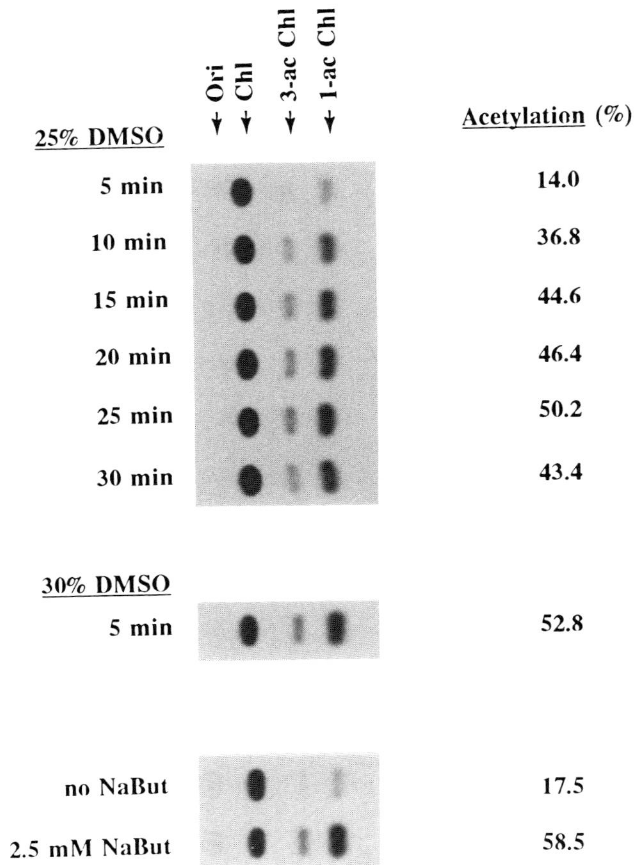

Fig. 7. Effects of permeabilization time, DMSO concentration and addition of sodium butyrate on transient CAT reporter gene activity in human fibroblasts (GM38; top and middle; XP12BE; bottom). *See* Section 3.3.6. for details.

DMSO approximated those encountered for 30% DMSO but the number of G418^r clones generated were reduced approx 20-fold. Optimal permeabilization conditions for generating stable transfectants appear, therefore, to reside at the threshold of toxicity. Substitution of glycerol or polyethylene glycol for DMSO was found to be ineffective.

3.4. Fate of Transfected Plasmid DNA

Under optimized conditions, polybrene/DMSO-assisted gene transfer is highly effective. Stable phenotypic transformants are generated at high frequencies with nanogram amounts of input plasmid DNA. Reporter gene activities are also easily recorded using 1 or 2 μg of construct (see Note 7). This

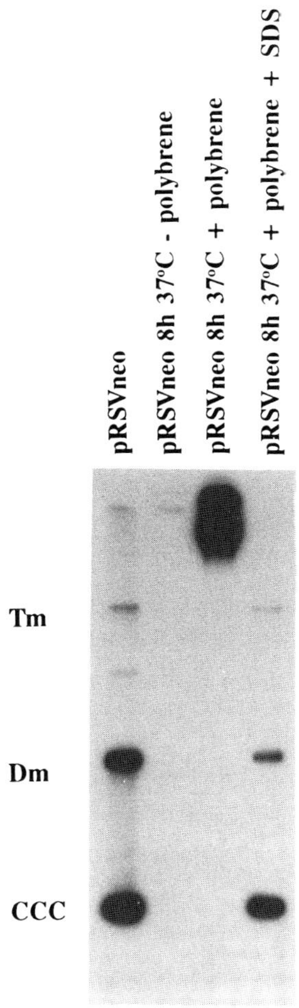

Fig. 8. Polybrene protects plasmid DNA against degradation by serum nucleases during the adsorption phase. One hundred nanograms of form I pRSVneo plasmid DNA were incubated for 8 h at 37°C in 2 mL growth medium with or without added polybrene (5 µg/mL). Ten picogram equivalents of plasmid DNA were retrieved, fractionated by agarose gel electrophoresis, transferred to a charge-modified nylon membrane and hybridized to a ^{32}P-labeled full length pRSVneo plasmid DNA probe. Untreated (control) parent pRSVneo DNA was run in the outer left lane. CCC; covalently closed circular (form I). Dm; dimer. Tm; trimer.

elevated success rate is attributed to several factors. First is the low overall toxicity of the process. Second is polybrene's ability to fully protect DNA molecules against degradation by the nucleases residing in fetal calf serum. As depicted in Fig. 8, pRSVneo form I DNA incubated at 37°C in transfection

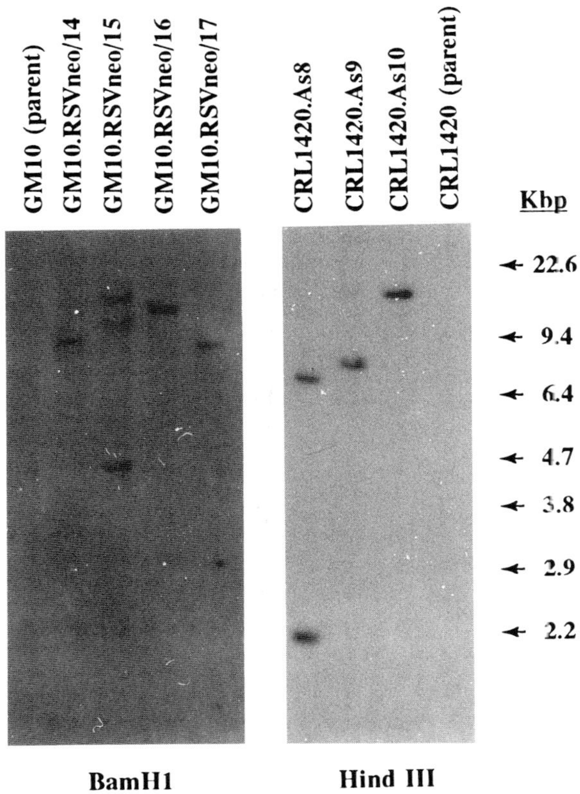

Fig. 11. Southern blot hybridization analysis of integrated *neo* gene sequences in transfected human GM10 fibroblasts and CRL1420 pancreatic adenocarcinoma cells.

2. The requirement for pure (i.e., free of contaminating bacterial chromosomal DNA and RNA) and structurally intact (i.e., >95% form I) plasmid DNA cannot be overstated. DNA prepared by several stringent methods is adequate *(37–40)*. Randomly linearized or nicked (open circular) plasmid molecules will decrease the efficiency of transient reporter gene assays significantly. However, controlled linearization with a suitable restriction enzyme will boost plasmid integration frequencies as well as transfection efficiencies two- to threefold. Exposure to chloroquine or sodium butyrate does not influence stable phenotypic transformation frequencies.
3. Uniform attachment of transfecting DNA to target cells can be assured by swirling the dishes periodically during the adsorption phase.
4. Large volumes (i.e., 100 mL) of adsorption cocktail can be prepared to accommodate experiments involving many dishes.
5. The most critical determinant of success is DMSO permeabilization. To ensure reproducibility between experiments and maximize both cell survival and

clonogenicity, the following points should be kept in mind. Prepare permeabilization solutions in glass bottles whenever possible. If polystyrene or polypropylene culture tubes are to be used, discard the solutions after 1 h. This is particularly important when solutions are augmented beyond 15% (v/v) DMSO. The solvent properties of DMSO release toxic plastisizers into the mixture. The addition of DMSO to culture medium is exothermic. Therefore, ensure that permeabilization solutions are equilibrated to 37°C before use. Maintain permeabilization solutions and rinse media at 37°C throughout the procedure. Certain cell types, like human fibroblasts, may be permeabilized as efficiently by lower DMSO concentrations using longer permeabilization times (*see* Table 2 and Section 3.3.6.). Under optimal conditions, the permeabilization scheme should produce no more than 15% cell killing as determined by colony forming ability.

6. Cell surface sites available for the binding of polybrene/DNA/protein complexes are saturable at relatively low DNA concentrations (i.e., 200 ng of pSV2neo DNA/5 × 10^5 fibroblasts). This should be taken into account when using genomic DNA or during cotransfection experiments where a nonselectable gene is likely to compete against a vector bearing a selectable marker for binding sites on cell membranes.
7. Transient reporter gene assays are routinely performed with 1 to 2 µg of plasmid. Exposure to sodium butyrate (2.5 to 10 m*M*) for 20 h following permeabilization will increase the signal considerably (Fig. 7, bottom panel). The peak of expression is usually found at 36–40 h postpermeabilization

Acknowledgments

The authors wish to acknowledge the financial support of the Medical Research Council of Canada, the National Cancer Institute of Canada, the Alberta Heritage Foundation for Medical Research, and the Alberta Cancer Board. The initial phase of this work was carried out under the auspices of the Atomic Energy of Canada Ltd. We are also indebted to M. Byron (Visual Communications, Health Canada) for artwork and photography.

References

1. Andreason, G. L., and Evans, G. A. (1988) Introduction and expression of DNA molecules in eukaryotic cells by electroporation. *BioTechniques* **6,** 650–660.
2. Watt, P. C., Sawicki, M. P., and Passaro, E. (1993) A review of gene transfer techniques. *Am. J. Surgery* **165,** 350–354.
3. McCutchan, J. H. and Pagano, J. S. (1968) Enhancement of the infectivity of simian virus 40 deoxyribonucleic acid with diethyl amino ethyl dextran. *J. Natl. Cancer Inst.* **41,** 351–357.
4. Gao, X. and Huang, L. (1996) Cationic liposome–mediated gene transfer. *Gene Therapy* **2,** 710–722.
5. Hara, T., Aramaki, Y., Takada, S., Koike, K., and Tsuchida, S. (1996) Receptor-mediated transfer of pSV2cat DNA to mouse liver cells using asialofetuin-labeled liposomes. *Gene Ther.* **2,** 784–788.

6. Cheng, P.-W. (1996) Receptor ligand–facilitated gene transfer, Enhancement of liposome–mediated gene transfer and expression by transferrin. *Hum. Gene Ther.* **7,** 275–282.
7. Erbacher, P., Bousser, M.-T., Raimond, J., Monsigny, M., Midoux, P., and Roche, A. C. (1996) Gene transfer by DNA/glycosylated polylysine complexes into human blood monocyte–derived macrophages. *Hum. Gene Ther.* **7,** 721–729.
8. Bond, V. C. and Wold, B. (1987) Poly-L-ornithine-mediated transformation of mammalian cells. *Mol. Cell. Biol.* **7,** 2286–2293.
9. Wagner, E., Zenke, M., Cotten, M., Beug, H., and Birnstiel, M. L. (1990) Transferrin-polycation conjugates as carriers for DNA uptake into cells. *Proc. Natl. Acad. Sci. USA* **87,** 3410–3414.
10. Midoux, P., Mendes, C., Legrand, A., Raimond, J., Mayer, R., Monsigny, M., and Roche, A. C. (1993) Specific gene transfer mediated by lactosylated poly-L-lysine into hepatoma cells. *Nucleic Acids Res.* **21,** 871–878.
11. Notter, M. F. D., Leary, J. F., and Balduzzi, P. C. (1982) Adsorption of Rous Sarcoma Virus to genetically susceptible and resistant chicken cells studied by laser flow cytometry. *J. Virol.* **41,** 958–964.
12. Toyoshima, K. and Vogt, P. K. (1969) Enhancement and inhibition of avian sarcoma viruses by polycations and polyanions. *Virology* **38,** 414–426.
13. Palmer, T. D., Hock, R. A., Osborne, W. R. A., and Miller, A. D. (1987) Efficient retrovirus-mediated transfer and expression of a human adenosine deaminase gene in diploid skin fibroblasts from an adenosine deaminase-deficient human. *Proc. Natl. Acad. Sci. USA* **84,** 1055–1059.
14. Kawai, S. and Nishizawa, M. (1984) New procedure for DNA transfection with polycation and dimethyl sulfoxide. *Mol. Cell. Biol.* **4,** 1172–1174.
15. Aubin, R., Weinfeld, M., and Paterson, M. C. (1988) Factors influencing the efficiency and reproducibility of polybrene-assisted gene transfer. *Somatic Cell Molec. Genet.* **14,** 155–167.
16. Morgan, T. L., Maher, V. M., and McCormick, J. J. (1986) Optimal parameters for the polybrene-induced DNA transfection of diploid human fibroblasts. *In Vitro Cell. Dev. Biol.* **22,** 317–319.
17. Chaney, W. G., Howard, D. R., Pollard, J. W., Sallustio, S., and Stanley, P. (1986) High-frequency transfection of CHO cells using polybrene. *Somatic Cell Molec. Genet.* **12,** 237–244.
18. Rhim, J. S., Park, J. B., and Jay, G. (1989) Neoplastic transformation of human keratinocytes by polybrene-induced DNA-mediated transfer of an activated oncogene. *Oncogene* **4,** 1403–1409.
19. Durbin, J. E. and Fallon, A. M. (1985) Transient expression of the chloramphenicol acetyltransferase gene in cultured mosquito cells. *Gene* **36,** 173–178.
20. Chisholm, O. and Symonds, G. (1988) Transfection of myeloid cell lines using polybrene/DMSO. *Nucleic Acids Res.* **16,** 2352.
21. Lee, M. S., Garkovenko, E., Yun, J. S., Weijerman, P. C., Peehl, D. M., Chen, L. S., and Rhim, J. S. (1994) Characterization of adult prostatic epithelial cells immortalized by polybrene-induced DNA transfection with a plasmid containing an origin-defective SV40 genome. *Int. J. Oncol.* **4,** 821–830.

22. Tidd, D. M., and Warenius, H. M. (1989) Partial protection of oncogene, anti-sense oligonucleotides against serum nuclease degradation using terminal methylphosphonate groups. *Br. J. Cancer* **60,** 343–350.
23. Mirzayans, R., Aubin, R. A., and Paterson, M. C. (1992) Differential expression and stability of foreign genes introduced into human fibroblasts by nuclear versus cytoplasmic microinjection. *Mut. Res.* **281,** 115–122.
24. Aubin, R. A., Weinfeld, M., Mirzayans, R., and Paterson, M. C. (1994) Polybrene/DMSO-assisted gene transfer: Generating stable transfectants with nanogram amounts of DNA. *Mol. Biotechnol.* **1,** 29–48.
25. Heath, J. K. (1987) Experimental analysis of teratocarcinoma cell multiplication and purification of embryonal carcinoma-derived growth factor, in *Terato-carcinomas and Embryonic Stem Cells: A practical Approach* (Robertson, E. J., ed.), IRL, Washington DC, Chapter 7, pp. 183–206.
26. McBurney, M. W. (1993) P19 embryonal carcinoma cells. *Int. J. Develop. Biol.* **37,** 135–140.
27. Santerre, R. B., Walls, J. D., and Grinnell, B. W. (1991) Use of vectors to confer resistance to the antibiotics G418 and hygromycin in stably transfected cell lines, in *Methods in Molecular Biology Vol. 7: Gene Transfer and Expression Protocols* (Murray, E. J., ed.), Humana Press, Clifton, NJ, Chapter 19, pp. 245–256.
28. Hanchett, L. A., Wang, S. J., Meegan, R. L., Baker, R. M., and Dolnick, B. J. (1992) Enhanced sensitivity to G418 of human KB cells adapted to certain media and sera. *BioTechniques* **12,** 482–486.
29. Sasaki, K., Mizusawa, H., Ishidate, M., and Tanaka, N. (1992) Regulation of G418 selection efficiency by cell-cell interaction in transfection. *Somatic Cell Mol. Genet.* **18,** 517–527.
30. Neufeld, D. S., Ripley, S., Henderson, A., and Ozer, H. L. (1987) Immortalization of human fibroblasts transformed by origin–defective Simian Virus 40. *Mol. Cell. Biol.* **7,** 2794–2802.
31. Shay, J. W., Wright, W. E., and Werbin, H. (1991) Defining the molecular mechanisms of human cell immortalization. *Biochim. Biophys. Acta* **1072,** 1–7.
32. Mulligan, R. and Berg, P. (1981) Selection for animal cells that express the *Escherichia coli* gene coding for xanthine–guanine phosphoribosyltransferase. *Proc. Natl. Acad. Sci. USA* **78,** 2072–2076.
33. Gorman, C. M., Moffat, L. F., and Howard, B. H. (1982) Recombinant genomes which express chloramphenicol acetyltransferase in mammalian cells. *Mol. Cell. Biol.* **2,** 1044–1051.
34. Saiki, R. K., Scharf, S., Faloona, F., Mullis, K. B., Horn, G. T., Erlich, H. A., and Arnheim, N. (1985) Enzymatic amplification of beta-globin genomic sequences and restriction analysis for diagnosis of sickle cell anemia. *Science* **230,** 1350–1354.
35. Takahashi, T., Carbone, D., Takahashi, T., Nau, M. M., Hida, T., Linnoila, I., Ueda, R., and Minna, J. D. (1992) Wild-type but not mutant p53 suppresses the growth of human lung cancer cells bearing multiple genetic lesions. *Cancer Res.* **52,** 2340–2343.

36. Frégeau, C. J., Aubin, R. A., Elliott, J. C., Gill, S. S., and Fourney, R. M. (1995) Characterization of human lymphoid cell lines GM9947 and GM9948 as intra- and interlaboratory reference standards for DNA typing. *Genomics* **28,** 184–197.
37. Aubin, R., Weinfeld, M., and Paterson, M. C. (1991) Preparation of recombinant plasmid DNA for DNA-mediated gene transfer, in *Methods in Molecular Biology Vol. 7: Gene Transfer and Expression Protocols* (Murray, E. J., ed.), Humana Press, Clifton, NJ, Chapter 1, pp. 3–13.
38. Alberti, S. and Fornaro, M. (1990) Higher transfection efficiency of genomic DNA purified with a guanidinium thiocyanate-based procedure. *Nucleic Acids Res.* **18,** 351–353.
39. Ehlert, F., Bierbaum, J., and Schorr, J. (1993) Importance of DNA quality for transfection efficiency. *BioTechniques* **14,** 546.
40. Goldberg, G. S. and Lau, A. F. (1993) Transfection of mammalian cells with PEG-purified plasmid DNA. *BioTechniques* **14,** 548.

26

Selection of Transfected Cells

Magnetic Affinity Cell Sorting

Raji Padmanabhan, Snorri S. Thorgeirsson, and R. Padmanabhan

1. Introduction

DNA-mediated gene transfer techniques have revolutionalized molecular biology and are used extensively to study the function and regulation of eukaryotic genes in a variety of cell types. In general, expression of genes in mammalian cells can be studied by either stable transformation or by transient expression. Stable transformation of cells expressing the gene of interest can be achieved by cotransfection of the gene cloned in an appropriate expression vector under the control of a constitutive or inducible eukaryotic cellular or viral promoter and a vector that carries a dominant selectable marker such as Eco-gpt *(1)* or neo *(2)*. Since the DNA-mediated gene transfer methods target only a fraction of cells for gene expression, the isolation of stably transfected cells with a selectable marker gives rise to a cell population expressing the gene of interest free of untransfected cells under conditions of drug selection. However, this method of selection is a long-term process and may have adverse effects on host cell functions such as cell growth or chromosomal rearrangements due to integration of a single or multiple copies of the selectable marker gene. Transient expression without a dominant selectable marker allows functional analysis of the transfected gene within 24–72 h after transfection but suffers from the drawback that the presence of a large fraction of untransfected cells in the milieu of cells expressing the gene of interest may give rise to problems of interference due to high background. Therefore, it was necessary to develop a method for isolation of transiently transfected cells free of untransfected cells within 24–72 h after transfection. With this overall goal in mind, the previously developed "panning" methodology was modified to

From: *Methods in Molecular Biology, vol. 62: Recombinant Gene Expression Protocols*
Edited by: R. Tuan Humana Press Inc., Totowa, NJ

isolate transiently transfected cells expressing the gene of interest together with a cotransfected cell surface marker gene using the magnetic affinity cell sorting (MACS) technology (*3*; for reviews, *see* refs. *4–8*). The MACS methodology allows the separation of cells expressing a surface protein away from those lacking the marker. The cell surface marker could be either introduced into cells by DNA-mediated gene transfer techniques or be an endogenously-expressing protein on the surface of selective cell type. In either case, the antibodies against the surface protein attached to a magnetic matrix are used to selectively "pull out" cells expressing that surface marker with the application of a magnetic field under appropriate experimental conditions whereas the cells lacking the marker remain unaffected. Expression of any cell surface protein for which a suitable antibody is available in a variety of cell types using DNA-mediated gene transfer methods allows this methodology to be useful in a wide range of biological applications *(6,8,9)*. In the initial stages of development of this application of MACS methodology to transfected cells, readily assayable reporter gene product such as chloramphenicol acetyltransferase (CAT) was used as the gene of interest in conjunction with the cell surface markers such as the vesicular stomatitis virus glycoprotein (VSV-G) and the Tac subunit of interleukin 2 receptor (IL-2R) for transient expression in mammalian cells by DNA-mediated transfection techniques *(3,10,11)*. Subsequently, it was shown that MACS could be successfully used to select a rare population of cells expressing the P-glycoprotein, the product of multiple drug resistant *(mdr)* gene *(6,8,9)* among human lymphomas as well as for selection of virus-infected cells expressing a surface protein (using dengue virus as an example) *(6)*. The experimental conditions for MACS methodology have undergone some improvement over the original protocol published *(3)*, and the modified procedure is described in this chapter.

2. Methods

2.1. Overview

One of the simplest applications of MACS methodology is the estimation of the efficiency of transient transfection by DNA-mediated gene transfer techniques. A surface protein marker, for which a suitable antibody (polyclonal or monoclonal) is readily available, and a reporter gene product that can be readily assayed are coexpressed. For reporter gene products, CAT *(12)*, firefly luciferase *(13)*, or *Escherichia coli* β-galactosidase *(14)* can be used. In the initial studies *(3)* the antibody against the surface protein was covalently linked to the magnetic beads (*see* Note 1), and in subsequent studies *(5,6,8,9,15)*, the primary antibody against the surface antigen was noncovalently attached to the beads through a secondary antibody. Both methods are described below.

2.2. Preparation of Magnetic Particles Attached to an Antibody

2.2.1. Materials

1. Pyridinium HCl (*see* Section 2.2.2.) coupling buffer (10 m*M*).
2. Phosphate coupling buffer (10 m*M* K_2HPO_4).
3. Glycine quenching solution (1*M*).
4. Buffer A (0.01*M* K_2HPO_4/0.15*M* NaCl/0.1% BSA/0.1% sodium azide. Disolve 1.74 g K_2HPO_4, 8.7 g NaCl, 1.0 g BSA, 1.0 g sodium azide in distilled H_2O; adjust pH to 7.4, and bring to 1 L).
5. Glutaraldehyde, 50% solution (EM Sciences or equivalent) in sealed ampules.
6. Bovine serum albumin (BSA) (Sigma, St. Louis, MO).
7. Phosphate-buffered saline (PBS), pH 7.4 (0.15*M* NaCl in 10 m*M* sodium phosphate) without Ca^{2+}/Mg^{2+} (Gibco-BRL, Gaithersburg, MD, and Quality Biologicals, Gaithersburg, MD).
8. PBS/0.1% BSA solution.
9. PBS/0.1% BSA/0.02% sodium azide solution.
10. HEPES buffer stock solution (1M) (Gibco-BRL).
11. Fetal calf serum (FCS) (Gibco-BRL).
12. Dulbecco modified Eagle medium (DMEM) (Gibco-BRL and Quality Biologicals).
13. Modified Eagle medium (MEM) (Quality Biologicals).
14. Magenetic beads (BioMag 4100 from Advanced Magnetics, Cambridge, MA or Dynal Corp.; distributed by P & S Biochemical, Gaithersburg, MD). Immunocolloids or ferrofluids are purchased from Immunicon, Huntington Valley, PA).
15. Magnetic slabs (Advanced Magnetics).
16. pCMV-IL-2R (Tac subunit of IL-2 receptor gene under the control of cytomegalovirus [CMV] early promoter).
17. pCMV-βGal (gene encoding *E. coli* β-galactosidase under the control of CMV promoter.
18. The MAb against IL-2R *(16)*.
19. Monkey kidney (CV-1) cells.

2.2.2. Method I

1. To prepare pyridinium-HCl coupling buffer, pipet 0.4 mL pyridine in fume hood (with sash lowered and with gloves on) into 400 mL distilled water.
2. Adjust pH to 6.0 with 10 m*M* HCl. Bring volume to 500 mL and filter sterilize.
3. Prepare the glycine quenching solution by dissolving 75 g of glycine in water and adjusting the pH to 7.0 with HCl.
4. Dilute BioMag M4100 particles (0.2 mL; low settling particle lot from Advanced Magnetics Inc.) into 1 mL pyridine coupling buffer in 1.5 mL Eppendorf tubes.
5. Separate the particles magnetically in 3–4 min, and aspirate the supernatant.
6. Wash the particles with 1 mL pyridine coupling buffer by vigorous vortexing and repeat 2X.
7. Prepare a solution of glutaraldehyde (5%) by mixing 1 mL of 50% solution with 9 mL pyridine coupling buffer. Add 0.4 mL of 5% glutaraldehyde solution to the particles from step 6.

8. Mix and incubate the contents in a 1.5 mL Eppendorf tube at room temperature for 3 h.
9. Sonicate the particles in a cuphorn at high power for 10 s. Wash the particles with 1 mL pyridine coupling buffer by mixing and vortexing and separate magnetically.
10. Aspirate the supernatant. Repeat the wash and magnetic separation cycles 5X.
11. Dilute the purified MAb raised against IL-2R *(16)* (200 μg–200 μL) using the phosphate coupling buffer containing 8 mg/mL crystalline BSA (e.g., antibody in 50 μL in 150 μL coupling buffer; *see* Note 2).
12. Mix the magnetically separated particles and the antibody solution, and incubate the tube in a rotary mixer and the antibody coupling reaction was allowed to proceed at room temperature overnight.
13. Add the glycine quenching solution to a final volume of 1 mL, and mix the contents for 10 min on a rotary mixer. Separate the particles magnetically, and aspirate the supernatant.
14. Wash the particles with 1 mL buffer A 2X, each time separating the particles magnetically and discarding the supernatant.
15. Store the particles at 4°C (short-term) or at –20°C (long-term) in 1 mL storage buffer containing glycerol (50% v/v)/sodium azide buffer (0.02%, w/v).

2.2.3. Method II

The method described in this section is useful if the commercial magnetic particles (Dynabeads from Dynal Corp.) are already precoated with an anti-IgG antibody such as goat antimouse IgG for attachment of an MAb or beads precoated with goat antirabbit IgG for attachment of a polyclonal rabbit antibody. The Dynabeads have reactive OH groups on their surface to which the anti-IgG is covalently attached *(17)*.

1. Wash the Dynabeads with PBS/0.1% BSA solution 4X in a 1.5 mL Eppendorf tube. Add an aliquot of an MAb (to Dynabeads attached to goat antimouse IgG) or a rabbit polyclonal antibody (to Dynabeads attached to goat antirabbit IgG) containing 100–200 μg IgG purified from a hybridoma culture supernatant, ascites fluid, or rabbit immune serum.
2. Mix the contents of the tube and incubate in a rotary mixer for about 24 h.
3. Wash the beads 5X with PBS/0.1% BSA magnetically separating the beads after each wash to remove any unbound antibody.
4. Resuspend the beads in PBS/0.1% BSA/0.02% sodium azide solution to give rise to a particle concentration of 1×10^8 particles/mL.

2.3. Magnetic Affinity Cell Sorting Protocol

For the application of MACS methodology, a cell surface marker and the gene of interest (X) are coexpressed in mammalian cells. In addition, as an internal control, a reporter gene is often included to monitor the efficiency of transfection and MACS procedures (*see* Note 3). An aliquot of cells isolated by MACS could be used for monitoring the efficiency as described in this section. The remainder of the cells could be used for the functional analysis of a gene X. For simplicity, only the reporter gene and the gene encoding the Tac subunit of IL-2R are included in the MACS protocol described here. For the functional analysis of the gene X, an expression plasmid encoding the gene X should be included in the transfection protocol. Moreover, in the original protocol published *(3)* DNA-mediated transfections were carried out on monolayers of cells and after 48–72 h posttransfection, and the transfected cells were dislodged from monolayers and resuspended in a sorting medium containing components to prevent aggregation of cells and to reduce nonspecific binding of antibody-coated magnetic beads (*see* Note 4). However, this original protocol is simplified as follows:

1. Plate CV-1 or any other cell type (3×10^5/25-cm^2 flask) cells in DMEM + 10% FCS and incubate overnight at 37°C in a CO_2 incubator.
2. Feed the cells with 5 mL fresh DMEM + 10% FCS 3 h prior to transfection.
3. Transfect using a protocol appropriate for the cell type. For liposome-mediated transfection, dilute 5 µg of pCMV-β-Gal and 2 µg of pCMV-IL-2R with 100 µL of DMEM w/o serum in an Eppendorf tube (1.5 mL capacity), and a 30 µL aliquot of lipofectin with 100 µL DMEM w/o serum in a 15-mL polystyrene tube. Transfer the diluted plasmid DNA to the polystyrene tube and gently mix the contents. Incubate at room temperature for 15 min.
4. Aspirate medium from a monolayer of CV-1 cells (about 50% confluency) and wash the cells 1X with 5 mL of DMEM w/o serum. Aspirate the medium again and add the DNA-lipofectin mixture to the cell monolayer. Spread the solution evenly to cover the cells and incubate at 37°C in a CO_2 incubator for 3–5 h. Refeed cells with the complete medium.
5. For transfection by $Ca{\cdot}PO_4$-DNA precipitation method, follow the protocol as described *(18)*. Prepare the $Ca{\cdot}PO_4$-DNA precipitate with 4 µg of pCMV-β-Gal + 1 µg of pCMV-IL-2R. Add the precipitated DNA to cells containing medium (0.5 mL $Ca{\cdot}PO_4$-DNA precipitate to 5 mL medium), and incubate for at least 4 h. Remove the precipitate by aspiration. Wash cells 2X with medium w/o serum. Subject cells to glycerol shock for 2–3 min as appropriate for different cell types. Wash twice with medium w/o serum and refeed cells with fresh DMEM + 10% FCS.
6. Forty-eight–seventy-two h after transfection, wash 100 µL suspension of beads (1×10^7 beads) with PBS/0.1% BSA 3X to remove sodium azide, and resuspend in 1 mL medium w/o serum.
7. Take 0.1 mL suspension (containing 1×10^6 beads) and mix with 2.5 mL of fresh medium w/o serum. Gently vortex to mix the beads.
8. Remove medium from cells and gently wash 1× with PBS to remove any floating cells. Add the 2.5 mL suspension of beads to cells (magnetic beads to cell ratio of 1:1) and gently spread over the monolayer of cells. Incubate at 37°C in a CO_2 incubator to allow beads to attach for 15–30 min.

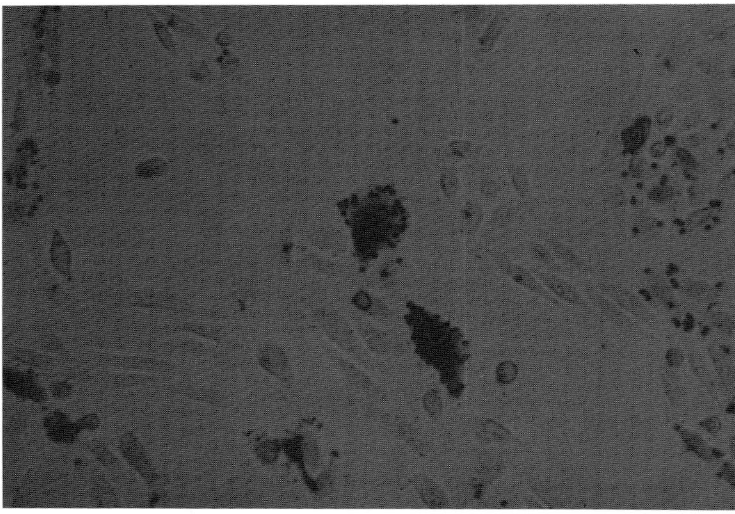

Fig. 1. Coexpression of *E. coli* β-galactosidase and IL-2R in CV-1 cells. CV-1 cells were transfected with pCMV-IL-2R *(15)* and pCMV-β-Gal expression plasmids, and the transfected cells were treated with anti-Tac antibody-coated magnetic beads. Subsequently, cells were stained for *E. coli* β-galactosidase activity as described in the text (Section 2.4.). Note that the anti-Tac antibody-coated magnetic beads bind specifically to cells coexpressing the IL-2R and the *E. coli* β-galactosidase.

9. Remove the excess beads not attached to cells. Wash once with medium w/o serum.
10. Stain for expression of *E. coli* β-galactosidase while the cells are still attached to magnetic beads as described below (Fig. 1). Alternatively, gently add a prewarmed solution of trypsin/EDTA (0.25% trypsin- 1 m*M* EDTA) and incubate at 37°C to dislodge the cells from the flask and monitor under the microscope (*see* Note 5). Magnetically sort the cells and plate them for further analysis.

2.4. Staining of Mammalian Cells for E. coli β-Galactosidase

2.4.1. Materials

1. Phosphate buffered glutaraldehyde solution (2 mL of 50% glutaraldehyde solution in 100 mL of 0.1*M* sodium phosphate, pH 7.4).
2. Staining solution (200 mL containing 0.1*M* sodium phosphate buffer, pH 7.4/2 m*M* $MgCl_2$/0.02%, v/v NP40 [40 μL]/0.01%, w/v, sodium deoxycholate [20 mg]/ 5 m*M* potassium ferricyanate [330 mg]/5 m*M* potassium ferrocyanide [442 mg]). To this solution, add X-Gal stock in dimethylformamide freshly before use to a final concentration of 1 mg/mL.
3. Dimethyl sulfoxide (DMSO; 3% solution) in 0.1*M* sodium phosphate buffer, pH 7.4.

2.4.2. Method

1. Wash the cells with PBS. Fix cells for 15 min in 1% glutaraldehyde (or 3.7% formaldehyde) buffered with 0.1*M* sodium phosphate pH 7.4. Wash once with 0.1*M* sodium phosphate pH 7.4. Stain the fixed cells for β-galactosidase (*see* Note 6).
2. Cover the cell layer with the staining solution (4 mL) and incubate at 4°C for 12–24 h. Wash the cell layer 1X with 3% DMSO solution and score for blue colonies. Calculate the transfection efficiency from the number of blue-stained cells (Fig. 1).

2.5. MACS Protocol to Isolate Cells Endogenously Expressing a Surface Protein: Application to Isolate Tumor Cells Expressing the MDR1 Gene Product, P-Glycoprotein

2.5.1. Overview

The MACS protocol can be successfully applied to isolate a population of cells that transiently express a surface protein during normal development or in response to external as well as internal stimuli. As one example, resistance to cancer chemotherapeutic drugs such as doxorubicin, vincristine, vinblastine, and etoposide may result from the overexpression of the MDR1 gene product, the 170-kDa surface protein which functions as a drug efflux pump called the multidrug transporter or P-glycoprotein. The MDR1 gene expression has previously been monitored using cDNA probes to detect the mRNA levels or using specific MAbs such as MRK-16 to detect protein levels (*19*; for reviews *see* refs. *20,21*). Such studies revealed various levels of MDR1 gene expression in normal tissues such as kidney, liver, intestine, adrenal, pancreas, brain, and endothelial cells, as well as in cancers *(22–24)*. The expression of P-glycoprotein in normal tissues is believed to reflect normal transport function of P-glycoprotein to excrete toxic natural products present in the diet or to excrete endogenous metabolites. Hence a drug therapy which inactivates the P-glycoprotein (such as by treatment with verapamil or tiapamil, which are inhibitors of P-glycoprotein-mediated transport and thereby reverse the multidrug resistance) may have deleterious side-effects on normal tissues. Moreover, it is important to monitor the levels of MDR1 expression in human cancers prior to and during the course of chemotherapy in order to accurately predict the outcome of a particular therapeutic regimen. Human lymphomas were chosen to study the feasibility of applying MACS for developing a sensitive method to detect cells expressing the MDR1 gene prior to chemotherapy.

2.5.2. Materials

1. Human tumor samples. (In the experiments described here, patient samples were provided by Drs. W. Wilson, A. Fojo, and S. Bates of the National Cancer Institute. Samples were labeled as 1, 2, 3, and 4).

Table 1
MACS of Tumor Cells Isolated from Lymphoma Patients

Patient sample	Total No. of cells	Total No. of sorted cells	% sorted
1	2.5×10^7	2.8×10^5	1.2
2	1.8×10^7	1.5×10^6	8.7
3	2.5×10^7	5.0×10^6	20
4	4.8×10^6	5.0×10^4	1.1

2. KB3-1 (drug-sensitive parental human adenocarcinoma cell line).
3. KB-8-5 cell line (*see* Note 7).
4. The MAb MRK-16 against MDR1 protein (from Dr. T. Tsuruo, Institute of Applied Microbiology, Univ. of Tokyo, Japan) *(19)*.
5. Dynabeads (M450) coated with goat antimouse IgG.
6. RPMI medium (Gibco-BRL).
7. PBS.
8. Fetal calf serum.
9. Trypsin-EDTA.

2.5.3. Methods

1. Couple the MRK-16 MAb to the Dynabeads as described in Section 2.2.3.
2. Suspend patients' samples (from 5–25 × 10^6 cells/mL) in RPMI medium supplemented with 2% FCS. Save an aliquot from each sample as total cell population.
3. Add the antibody-coupled magnetic bead suspension to the remainder of cell suspension at a bead-to-cell ratio of 1:4.
4. Incubate at 4°C with gentle mixing in a rotary mixer for 30 min.
5. Magnetically sort the cells using a magnet designed for tubes rather than magnetic slabs which are designed for cell culture flasks. Save the bound fraction of cells as well as the supernatant containing the unbound cells.
6. Wash the bound cells with PBS (2X) and repeat the magnetic sorting.
7. Count the number of cells attached to beads (Table 1; *see* Notes 8–10). Freeze cells at –70°C for mRNA analysis by Northern hybridization or by reverse transcriptase-polymerase chain reaction (RT-PCR) protocol.

2.6. Cell Fusion and Isolation of Somatic Cell Hybrids by MACS

2.6.1. Overview

Cell fusion methodology in which two different cell types are fused giving rise to a hybrid cell is widely used in somatic cell genetics and in production of MAbs. Using this approach it is possible to determine whether a particular phenotype is dominant or recessive in the cell hybrid, and to analyze several mutants of a given genetic trait and classify them into different complementation groups. One frequently used approach for isolation of cell hybrids is that one parental cell line lacking the thymidine kinase (TK^-) gene, and thus unable to use exogenous thymidine, is fused with another cell line that is deficient in hypoxanthine-guanine phosphoribosyl transferase ($HGPRT^-$) and therefore unable to utilize the purine hypoxanthine. Hybrid cells are selected on the basis of their ability to grow in a medium containing high levels of hypoxanthine, aminopterine, and thymidine (HAT) due to complementation of mutant phenotypes *(25,26)*. One disadvantage of this selection method is that if the selective pressure is not maintained, the hybrid cells usually revert to a parental wild-type. It became apparent that the DNA-mediated gene transfer techniques followed by the MACS protocol could be successfully used for isolation of somatic cell hybrids between any two cell types. The basic principle of this approach is that one of the parental cell types is transfected with a plasmid encoding Neo^r (producing the enzyme aminoglycoside phosphotransferase which confers resistance to the analog of neomycin (G418) in the growth medium), and the other with the plasmid encoding IL-2R and selected by MACS using the anti-IL-2R antibody-coated magnetic beads as described in Section 2.3. The two transfected cell types are then fused and subjected to G418 selection followed by MACS to isolate the cell fusion hybrids (Fig. 2). Only the heterokaryons formed between $IL\text{-}2R^+$ and Neo^r cell types will survive in the G418 medium. The cells expressing only the IL-2R will not survive and the cells expressing only Neo^r would not be selected by MACS. Moreover, the isolation of hybrid cells can be accomplished within 8–10 d and, after selection, the cells can be grown in the absence of G418 in the medium. The protocols for cell fusion between two parental cell types (cell type I and cell type II) and the MACS are described below.

2.6.2. Materials

1. Polyethylene glycol (PEG; MW 4000) (American Type Culture Collection).
2. DMEM or appropriate growth media for the two cell types (Gibco-BRL).
3. Fetal calf serum (FCS).
4. Gentamycin and glutamine solutions (2 m*M* each).
5. G418 medium (G418 was obtained from Gibco-BRL).
6. Cell types I and II.
7. Plasmids: pCMV-IL-2R, pSV2-Neo^r.
8. PBS (w/o Ca^{2+}/Mg^{2+}).

2.6.3. Methods

1. Plate cell types I and II in 25-cm^2 flasks at a density of 3–3.5 × 10^5/flask in DMEM (or an appropriate growth medium) containing 10% FCS and 2 m*M* each gentamycin (or penicillin/streptomycin) and glutamine.
2. Add fresh medium 4 h prior to transfection.

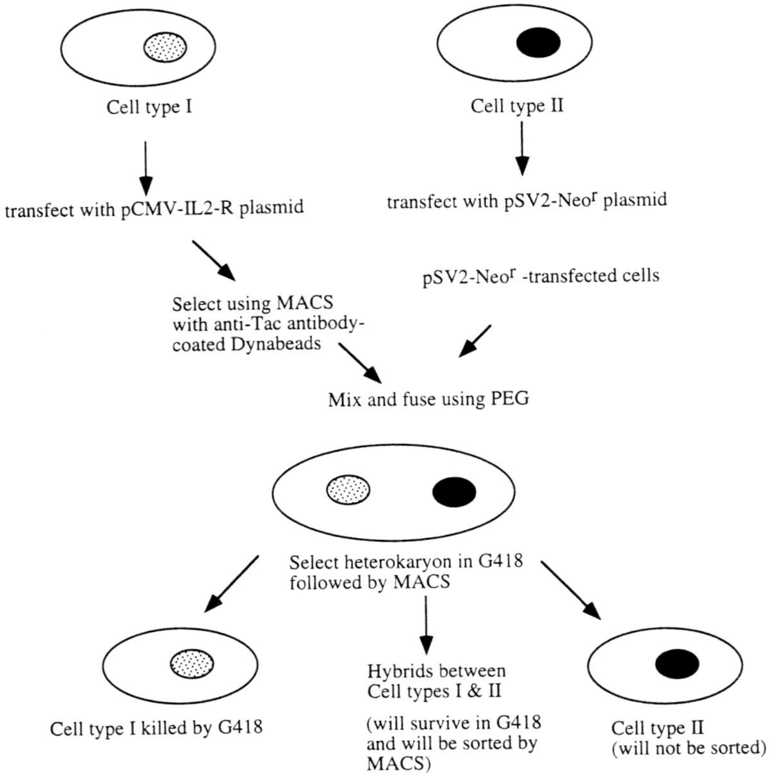

Fig. 2. Strategy for isolation of somatic cell hybrid cells. Rat liver epithelial (RLE) cells and the spontaneously transformed (C4T) cells *(28)* were individually transfected with pCMV-IL-2R and pSV2-Neo^r plasmids (5 μg each), respectively. The IL-2R-transfected cells were sorted using MACS and then fused with the pSV2-Neo^r-transfected cells using the PEG medium as described in the text. Fused heterokaryons were allowed to grow in G418 selection medium for 6–7 d until all the parental RLE and RLE:RLE hybrid cells were no longer viable in the culture dishes. The cells after 7 d in G418 medium were subjected to a second round of the MACS protocol. By this procedure only the heterokaryons formed between RLE-IL-2R($^+$) and C4T(Neo^r) were enriched.

3. Transfect the cell type I with pCMV-IL-2R and the cell type II with pSV2-Neo^r plasmids (5 μg each) using either Ca.PO_4-DNA precipitation method or Lipofectin (described in Section 2.3).
4. Select the cells expressing IL-2R by MACS procedure 48 h posttransfection. Calculate percentage of cells transfected from the number of sorted cells and the total number of cells used.
5. Melt PEG (2 g) in a 45°C water bath for 15 min or until it goes into solution; cool to a temperature range from 35°C to room temperature.

6. Dilute PEG to 50% with DMEM w/o serum, mix well, and centrifuge to remove air bubbles (*see* Note 11).
7. Mix the sorted cells from step 4 with the pSV2-Neo^r-transfected cells (1 × 10^6 cells from each), centrifuge, and wash 1X with PBS (w/o Ca^{2+}/Mg^{2+}) and resuspend cells in 100 μL PBS.
8. Add one mL of 50% PEG solution to cell suspension with constant stirring with the tip of a pipet for about 90 s.
9. Slowly add 10 mL of DMEM (w/o serum) with constant mixing over 3–5 min and then bring the final volume to 20 mL.
10. Centrifuge the PEG cell suspension at 800*g* for 10 min in a clinical centrifuge and aspirate the supernatant. Resuspend cells in 10 mL growth medium containing 10% FCS and plate in 50-cm^2 cell culture dishes for G418 selection.
11. Add G418 solution (300 μg/mL) to the growth medium containing cells 6 h after they attach to the cell culture plates.
12. Eight days after G418 selection, select the surviving colonies by MACS protocol with anti-Tac (IL-2R) antibody-coated magnetic beads as described in Section 2.3. to remove cells expressing only the Neo^r gene (Fig. 2).
14. Analyze the cell hybrids for the phenotypic characteristics of either of the parental cell types or a new phenotype. Stain for IL-2R expression by indirect immunofluorescence (Fig. 3).

2.7. Use of Immunomagnetic Colloids in MACS Technology

2.7.1. Overview

Since the development of Dynabeads, new magnetic matrices called ferrofluids have been designed for cell separations *(7,27)*. Ferrofluids are in essence supermagnetic colloids containing crystalline magnetic cores in the size range of 5–60 nm with a surface which could be coated with any polymer macromolecule (DNA, RNA, protein, etc.). Ferrofluids have no net magnetization in the absence of an external magnetic field. Ferrofluids coated with goat antimouse IgG are commercially available from Immunicon Corp. (Huntington Valley, PA) with an average size of 60 nm with an iron concentration of 1 mg/mL, and a particle concentration of 1 × 10^{12} particles/mL. The iron particles are completely covered with the antibody which reduces the electrostatic charge interactions between the particles, and thus, nonspecific binding of cells to magnetic particles. Ferrofluids allow easy recovery of cells after separation in a viable condition much more readily within minutes rather than the larger Dynabeads which require a longer incubation periods (between 12–24 h). These particles are supplied in sterile-filtered form and can be resterilized using a 0.2 μm sterile filter.

2.7.1. Materials

1. Ferrofluids (Immunicon Corp. Huntington Valley, PA).
2. PBS/0.5% BSA/ gelatin.

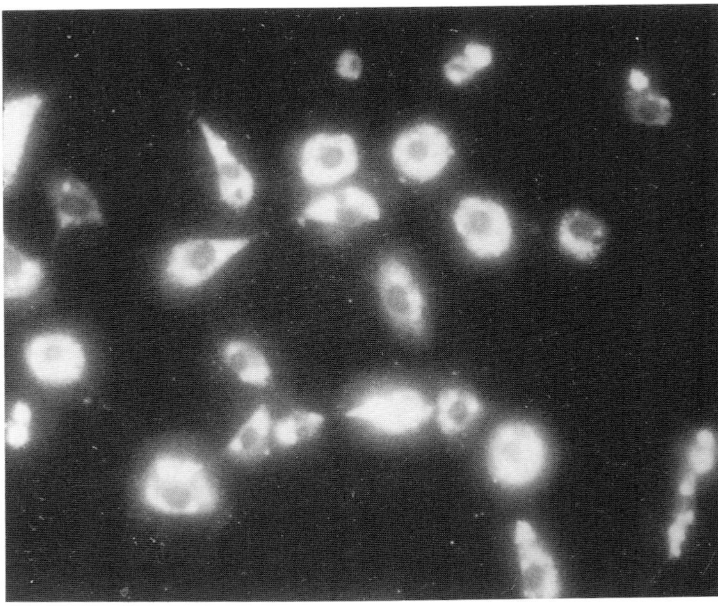

Fig. 3. Immunofluorescence of hybrid clones. Hybrid cells which survived in G418 medium (as described in Fig. 2) were fixed with paraformaldehyde and incubated with the anti-Tac antibody used as the primary antibody. The cells were then reacted with the rhodamine-labeled goat antimouse IgG used as the secondary antibody. The surface expression of IL-2R and the nuclei of the heterokaryons are shown here.

3. HeLa-S3 cells.
4. DMEM and FCS.
5. pCMV-IL-2R plasmid.
6. Immunicon high gradient magnetic cell separator device.
7. FITC-tagged IL-2R (Beckton Dickinson, Rutherford, NJ).

2.7.2. Methods

1. Grow HeLa-S3 cells in two T-25 flasks in DMEM containing 10% FCS to about 50% confluency. Transfect HeLa-S3 cells growing in one T-25 flask using 5 µg of pCMV-IL-2R plasmid using either Ca·PO4-DNA precipitation method or the lipofectin method as described in Section 2.3. and use the other flask for untransfected cell control.
2. After 48 h, trypsinize the control and the transfected cells using milder conditions (0.005% trypsin, 0.01 m*M* EDTA). Resuspend in 200 µL of PBS/0.5% BSA buffer.
3. Add 2 µg of anti-Tac (IL-2R) antibody to both cells and incubate in ice for 30 min (*see* Note 12).
4. Dilute ferrofluid solution 1:20 with PBS/0.5% BSA.

5. At the end of incubation period, add 200 μL of diluted ferrofluid to achieve a final dilution of 1:40. Incubate for 15 min with occasional mixing by gentle tapping on the tube.
6. Separate the cells expressing IL-2R using the cell separator device. Collect cells in a medium. Carry out further analysis on the sorted cells.

3. Notes

1. This general procedure for covalent attachment of an antibody to the magnetic particles may be used for other purposes such as attachment of not only MAbs but also a wide variety of macromolecules which include lectins, enzymes, and both single- and double-stranded DNA. Attachment of avidin to these beads would allow binding of a biotinylated moiety of DNA or RNA. The limitation of this procedure is that glutaraldehyde treatment may alter the antibody-antigen interaction in some cases.
2. High titer ascites fluid can also be used in some cases after removing particulate debris from ascites fluid by spinning in a microfuge for 1.5 min.
3. The reporter gene product *E. coli* β-galactosidase is often used as an internal control to monitor the efficiency of DNA-mediated transfection techniques. It is especially useful as an internal marker in transfected and sorted cells in which the expression of a gene X is studied. By inclusion of a set of control flasks for transfection, sorting, and β-gal expression in the presence and absence of the gene X, it is possible to estimate the transfection efficiency as well as to verify whether or not the gene X has any effect on cell growth.
4. In these original experiments, trypsin was avoided due to the assumption that it has the potential to digest the cell surface marker, thus decreasing its ability to bind to the antibody-coated magnetic beads. Therefore, cell monolayers after transfection was treated with the MS medium (spinner MEM containing 40 m*M* EDTA, 10 m*M* HEPES, pH 7.3) to dislodge the cells. Then the medium C (spinner MEM containing 10% v/v FCS, 20 m*M* HEPES, pH 7.3, and 100 μg/mL chondroitin sulfate) was used to complete the detachment of cells and to prevent cell to cell adhesion. The detached cells were pelleted by centrifugation and resuspended in the sorting medium, PBS containing 4 m*M* EGTA/1 m*M* $MgCl_2$/10 m*M* HEPES (pH 8.0), 100 μg/mL chondroitin/1 mg/mL gelatin, 8 mg/mL nonfat dry milk, and 10 μg/mL BSA prior to the addition of antibody-coated magnetic beads at a bead-to-cell ratio of 10:1 *(3)*. The finding that cell monolayers could be treated with the antibody-coated magnetic beads and that subsequent treatment with trypsin dislodged the cells without affecting their attachment to the beads in a number of cases made the MACS protocol simple (*5,6,8,9,15*; unpublished results). If trypsin is found to disrupt cell attachment to the antibody-coated magnetic beads, the original protocol may prove useful.
5. Most of the cells will remain attached to beads under these conditions. In our experience, this attachment of antibody-coated beads to a cell surface protein is resistant to trypsin in a number of cases tested so far. At this point, it is unknown

17. Kemshead, J. T., Heath, L., Gibson, F. M., Katz, F., Richmond, F., Treleaven, J., and Ugelstad, J. (1986) Magnetic microspheres and monoclonal antibodies for the depletion of neuroblastoma cells from bone marrow: experiences, improvements, and observations. *Br. J. Cancer* **54,** 771–778.
18. Gorman, C. M., Padmanabhan, R., and Howard, B. H. (1983) High efficiency DNA-mediated transformation of primate cells. *Science* **221,** 551–553.
19. Hamada, H. and Tsuruo, T. (1986) Functional role for the 170 to 180 KD glycoprotein specific to drug resistant tumor cells as revealed by monoclonal antibodies. *Proc. Natl. Acad. Sci. USA* **83,** 7785–7789.
20. Gottesman, M. M. and Pastan, I. (1993) Biochemistry of multidrug resistance mediated by the multidrug transporter. *Annu. Rev. Biochem.* **62,** 385–427.
21. Germann, U. A., Pastan, I. and Gottesman, M. M. (1993) P-glycoproteins: mediators of multidrug resistance. *Semin. Cell. Biol.* **4,** 63–76.
22. Fojo, A. T., Ueda, K., Slamon, D. J., Poplack, D. G., Gottesman, M. M., and Pastan, I. (1987) Expression of a multidrug resistant gene in human tumors and tissues. *Proc. Natl Acad. Sci. USA* **84,** 265–269.
23. Goldstein, L. J., Fojo, A. T., Ueda, K., Crist, W., Green, A., Brodeur, G., Pastan, I., and Gottesman, M. M. (1990) Expression of the multidrug resistance, MDR1, gene in neuroblastomas. *J. Clin. Oncol.* **8,** 128–136.
24. Thiebaut, F., Tsuruo, T., Hamada, H., Gottesman, M. M., Pastan, I., and Willingham, M. C. (1987) Cellular localization of the multidrug-resistance gene product P-glycoprotein in normal human tissues. *Proc. Natl. Acad. Sci USA* **84,** 7735–7738.
25. Szybalska, E. H. and Szybalski, W. (1962) Genetics of human cell lines. IV. DNA mediated heritable transformation of a biochemical trait. *Proc. Natl. Acad. Sci. USA* **48,** 2026.
26. Littlefield, J. W. (1964) Selection of hybrids from matings of fibroblasts in vitro and their presumed recombinants. *Science* **145,** 709.
27. Hancock, J. P. and Kemshead, J. T. (1993) A rapid and highly selective approach to cell separations using an immunomagnetic colloid. *J. Immunol. Meth.* **164,** 51–60.
28. Huggett, A. C., Ellis, P. A., Ford, C. P., Hampton, L. L., Rimoldi, D., and Thorgeirsson, S. S. (1991) Development of resistance to the growth inhibitory effects of transforming growth factor β_1 during the spontaneous tranformation of rat liver epithelial cells. *Cancer Res.* **51,** 5929–5936.

27

Selection of Transfected Cells and Coamplification of Transfected Genes

Susan E. Kane

1. Introduction

There are many applications for gene transfer and gene expression technology in mammalian somatic cell biology. Whether for functional analysis of a newly cloned gene or for purifying large amounts of a protein for its subsequent biochemical characterization, it is often important to overexpress foreign genes in the "native" environment of a mammalian cell. Furthermore, rapidly developing gene therapy technology relies on the ability to express foreign sequences at significant levels in a stable, long-term fashion.

Expression of foreign genes in mammalian cells requires some method of gene transfer into the appropriate host cell. Most gene transfer (transfection) methods are very inefficient, thus some type of selection is required to isolate those cells that take up the gene of interest. However, most genes do not themselves confer a selectable growth advantage on their host cells. In that case, the foreign cDNA can be transfected along with a drug-selectable marker and transfected cells can be isolated by an appropriate drug selection *(1,2)*. Furthermore, with amplifiable selectable markers, the transfected cells can be grown in increasing concentrations of drug. This imposes increasingly stringent selective pressure on the transfected cells and requires overexpression of the marker gene for them to survive. The foreign cDNA that was transfected along with the marker is concomitantly overexpressed, and very high-level expression of the gene can often be achieved.

The success of this approach depends very much on which drug-selectable marker is used for isolating and amplifying expression of the transfected genes. Common selectable markers confer resistance to antibiotics, such as G418 (neomycin) or hygromycin, or encode genes involved in nucleotide biosynthe-

From: *Methods in Molecular Biology, vol. 62: Recombinant Gene Expression Protocols*
Edited by: R. Tuan Humana Press Inc., Totowa, NJ

sis (thymidine kinase, hypoxanthine guanine phosphoribosyltransferase, dihydrofolate reductase). These markers are limited in their usefulness, however, either because they are not amplifiable or because they are recessive markers that can only be used with mutant cells lacking the corresponding normal gene activity *(1,2)*.

One dominant and amplifiable selectable marker that has proven to be useful for a variety of gene expression applications is the human multidrug resistance (*MDR*1) gene *(3)*. The *MDR*1 gene encodes a 170-kDa membrane glycoprotein termed P-glycoprotein (Pgp). It functions as a drug transporter to prevent the intracellular accumulation of cytotoxic agents and thereby confers resistance to those agents. Pgp transports a broad range of drugs, but the most common one used for in vitro gene transfer procedures is colchicine. Expression of an *MDR*1 cDNA is sufficient for conferring colchicine resistance to otherwise sensitive cell lines. Expression and copy number of transfected *MDR*1 can be amplified by growing transfected cells in increasing concentrations of colchicine. Foreign genes that are transferred into cells along with *MDR*1 are concomitantly amplified and overexpressed *(4,5)*.

This chapter will discuss the use of *MDR*1 as a selectable marker for gene transfection, drug selection, and gene amplification. As a special case, we will discuss viral transduction as a method for transferring *MDR*1 plus foreign genes into appropriate target cells. The general method of *MDR*1-mediated gene transfer, selection, and amplification has been used by a number of investigators to obtain high-level expression of foreign genes and of a fusion gene encoding a bifunctional Pgp-adenosine deaminase protein *(4–8)*. The technique is also described in ref. *9*.

2. Materials

2.1. Transfection and Selection

1. 2.5 M $CaCl_2$.
2. Sterile H_2O.
3. 2X HBS: 50 m*M* N-2-hydroxyethylpiperazine-N'-2-ethanesulfonic acid (HEPES), pH 7.1, 280 m*M* NaCl, 1.5 m*M* Na_2HPO_4.
4. Filter sterilize reagents 1–3 and store at 4°C. The pH of the 2X HBS is critical and should be checked periodically and adjusted to pH 7.05–7.1 if necessary.
5. Phosphate-buffered saline (PBS) containing Ca^{2+} and Mg^{2+} (Gibco Life Technologies, Gaithersburg, MD).
6. Colchicine (Sigma, St. Louis, MO) dissolved in dimethylsulfoxide. Do not filter. Stock solutions are 10 mg/mL and 1 mg/mL. Keep frozen at –20°C, wrapped in foil.
7. Methylene blue solution: 0.5% methylene blue (w/v), 50% ethanol (v/v).

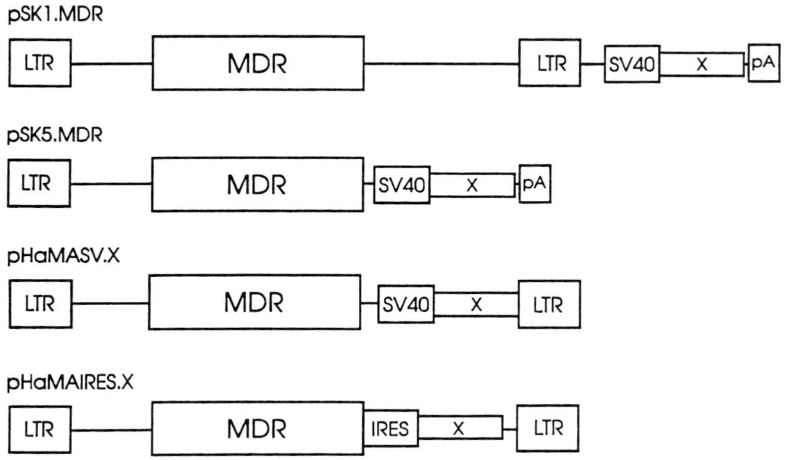

Fig. 1. Schematic drawings of *MDR*1 expression vectors. Four different configurations of expression vectors are shown. Each uses the Harvey murine sarcoma virus long terminal repeat (LTR) to control expression of *MDR*1 *(12)*. The foreign gene of interest (X) is controlled by the internal SV40 early promoter or by an internal ribosome entry site (IRES) *(14)*. Transcription is terminated either in the downstream LTR or in a SV40 polyadenylation signal (pA). Only pHaMASV.X and pHaMAIRES.X can be used for the production of retrovirus carrying both *MDR*1 and gene X. Vector pSK1.MDR is described in ref. *5*. Vectors pHaMASV.X and PhaMAIRES.X are described in ref. *17*. All vectors are available from the author. Drawings are not to scale.

2.2. Retrovirus Packaging and Viral Transduction

1. Retroviral packaging cell lines, GP+E86 *(10)* and GP+envAm12 *(11)*.
2. Polybrene (Sigma), 400 µg/mL in H_2O, filter sterilized.

2.3. Expression Vectors

Figure 1 shows a number of expression vectors that use *MDR*1 as the selectable marker for expressing foreign genes in mammalian cells. The features of these vectors vary slightly and the choice of which vector to use will depend on the particular needs of the experiment. All of the vectors shown in Fig. 1 use the Harvey murine sarcoma virus to drive expression of *MDR*1 *(12)*. The foreign gene is inserted downstream of *MDR*1 and is expressed under the control of either an internal SV40 promoter *(13)* or an internal ribosome entry site (IRES) from encephalomyocarditis virus *(14)*. With the internal promoter, two mRNAs are obtained, one (full-length) encoding *MDR*1 and the other (from the SV40 promoter) encoding the foreign gene of interest. In the case of the IRES element, a single, bicistronic message is expressed; the downstream gene product is translated by virtue of an internal site for ribosome loading and translation initiation *(14)*.

Vectors pSK1.MDR *(5)* and pSK5.MDR can only be used in gene transfection procedures in which the plasmids are introduced into target cells as naked DNA. Vectors pHaMAIRES.X and pHaMASV.X can be used directly for gene transfection or can be used to package retroviruses carrying both *MDR*1 and the second gene (X). The retroviruses can then be used to virally transduce the genes into target cells. Both gene transfection and viral transduction methods will be described below. The cloning methods for inserting foreign cDNA into any of these vectors are beyond the scope of this chapter. The vectors are available upon request from the author and detailed cloning instructions will be provided at that time.

3. Methods

3.1. Gene Transfection and Selection with Colchicine (see Fig. 2)

1. On d 0, plate 5×10^5 cells/10-cm tissue culture dish. Set up one dish per DNA to be transfected plus a negative control dish for a mock transfection without DNA.
2. On d 1, change the medium of the cells 3 h before the transfection. Use the same medium in which the cells normally grow.
3. For the transfection, set up two 15-mL, conical tubes for each transfection: tube 1: 250 µL 2X HBS; tube 2: DNA (5–10 µg), 25 µL 2.5*M* $CaCl_2$, H_2O to 250 µL.
4. Using a 1-mL pipet, bubble air into the solution of 2X HBS; at the same time, dropwise add DNA solution to the HBS. Continue bubbling air through the solution for 5 s after all the DNA is added. The bubbling of air will allow a fine $CaPO_4$-DNA coprecipitate to form. Let the mixture sit at room temperature for 30 min to allow precipitate to form completely.
5. Add DNA/HBS mix to cells in complete medium, dropwise with gentle swirling. Incubate overnight.
6. On d 2, wash the cells three times with PBS and add fresh complete medium overnight.
7. On d 3, split the cells 1:4 into selective medium (*see* Note 5) and incubate for 10–14 d.
8. After colonies have formed, stain one dish (*see* Note 5) to determine the transfection efficiency. With the other plates, either pick individual colonies or trypsinize the entire plate of cells and replate in a T75 tissue culture flask. (**Note:** Colchicine-resistant colonies will appear as distinct clusters of cells visible by eye when the dish is held up to the light. These can be circled with a marker on the bottom of the dish and examined microscopically to determine whether the cells are healthy.)

3.2. Retrovirus Packaging and Viral Transduction

1. Transfect retroviral expression vector into GP+E86 or GP+envAm12 packaging cell lines using the method described above. The concentration of colchicine to use for selection is 20 ng/mL.

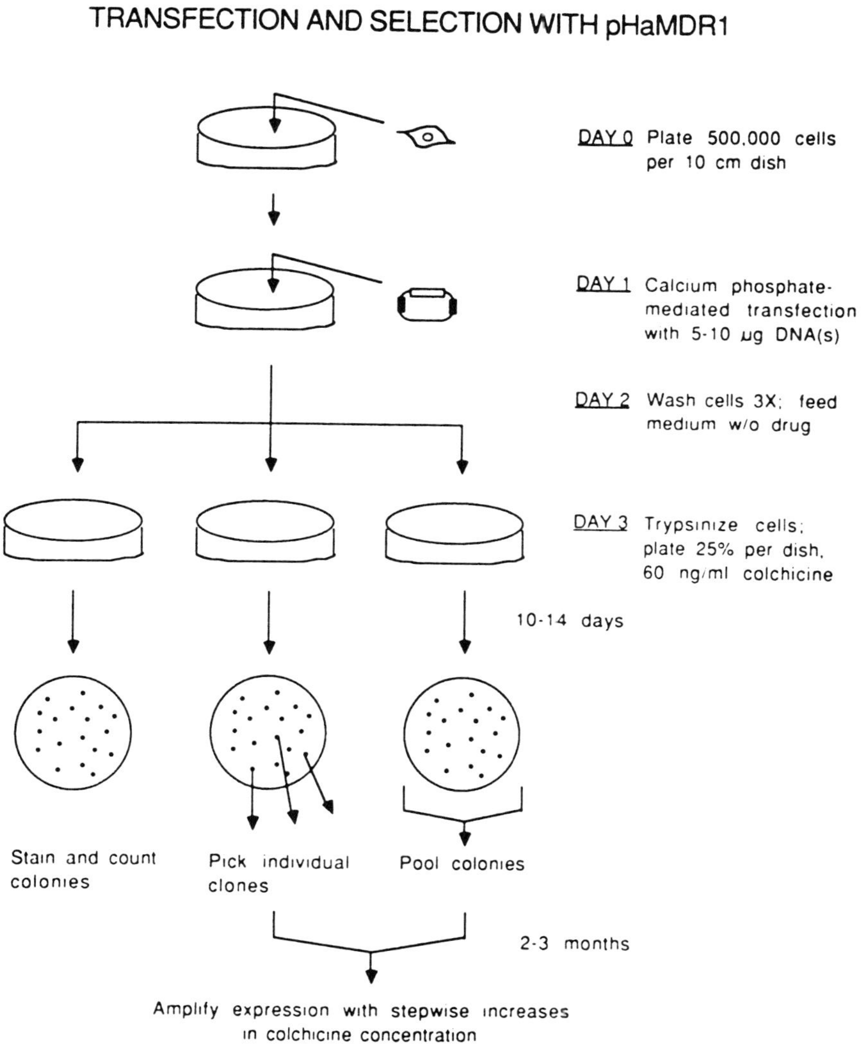

Fig. 2. Schematic flow diagram of gene transfection. Cells are plated on d 0 and allowed to attach. DNA-mediated transfection with appropriate plasmid occurs on d 1, followed 2 d later by plating of cells in selective medium. After 10–14 d, colonies are stained and counted to determine transfection efficiency, individual colonies are isolated, or colonies are pooled into populations of transfected cells. Colonies or populations are then exposed to increasing concentrations of drug to obtain amplified expression of transfected genes.

2. Grow and expand transfected packaging cells (producer lines), maintaining in colchicine, until they have undergone a number of passages in T75 tissue culture flasks. Both of the packaging cell lines need to be maintained at relatively high density for optimal growth (at least 40% confluence).

3. To collect retroviruses carrying *MDR*1 plus the foreign gene of interest, grow producer lines until 80% confluent in a T75 or T150 flask. Change the medium to complete growth medium lacking any colchicine (10 mL on a T75, 20 mL on a T175) and culture 16–24 h. Collect the medium (virus supernatant) and put immediately on ice. Spin out dead cells for 5 min at 5000*g*, 4°C. Virus supernatant can also be filtered through a 0.45- or 0.2-µm filter, with some loss of virus titer. Aliquot virus and store at –80°C. Minimize freezing and thawing and always keep virus on ice when thawed.
4. To determine the titer (concentration) of infectious particles in the virus supernatant, plate 5×10^4 NIH3T3cells in each of five 60-mm dishes on d 0. On d 1, add 0.1 mL of 10-fold serial dilutions (undiluted to 10^{-4} diluted) of virus to the cells in fresh medium containing 8 µg/mL polybrene. On d 3, add fresh medium containing 60 ng/mL colchicine and incubate for 8–10 d. Stain colchicine-resistant colonies with methylene blue, as in Note 5. Count colonies and determine colony forming units per mL (cfu/mL) of the original virus supernatant by correcting for the dilution factor.
5. To transduce target cells with recombinant retrovirus, plate cells at 1-2 $\times 10^4$ cells per 35-mm or 60-mm dish on d 0. On d 1, add virus to cells at a multiplicity of infection (total cfu/cell) of 1–10, in fresh growth medium containing 4–8 µg/mL polybrene. Incubate for 2 d and add fresh medium containing colchicine at the appropriate concentration (*see* Section 3.1., step 7, and Notes 4 and 5). Transduction efficiency can be from 10–100% depending on the target cell. Colchicine-resistant cells should appear in 5–10 d and can be treated as above and below for transfected cells.

3.3. Amplification

1. From a confluent T75 of transfected or transduced cells, with cells growing at the initial selecting concentration of drug, trypsinize the cells and replate in a twofold higher drug concentration. Cells should be replated at no less than 25% confluence at any time during the amplification process.
2. Maintain cells at a given drug concentration for at least two passages.
3. Repeat the amplification in twofold increments of drug concentration for at least four rounds of amplification (*see* Note 9).
4. Cells at low, moderate, and high levels of drug resistance can be analyzed for DNA, RNA, and protein levels corresponding to *MDR*1 (*see* Chapter 58 of this volume) and the foreign gene of interest.

4. Notes

1. The choice of cells will depend on the experiment, but the $CaPO_4$-mediated method described here *(15,16)* works best for cells that grow as monolayers. Suspension cells and cell lines that are not efficiently transfected by this method can often be transfected by electroporation or lipid-mediated techniques *(1,2)*. Some cell lines that are refractory to transfection will require viral transduction, also described here.

2. DNA for transfections should be of high quality and purity, but it is not necessary to use CsCl-purified DNA. Commercially available plasmid preparation kits are generally adequate for this purpose, as long as supercoiled DNA free of RNA is obtained. Dilute the DNA to 250 ng/μL and run 1 μL of this on an agarose gel to check the integrity and quantity of the DNA. Use 20–40 μL of this dilution in the transfection.
3. Do not incubate cells with the $CaPO_4$-DNA precipitate for longer than 16 h.
4. When plating the cells on d 3 of the transfection, the density of cells plated will vary from cell type to cell type. NIH3T3 cells that are split 1:4 on d 3 will result in 100–300 colonies per 10-cm dish by 10–14 d after addition of colchicine. Dishes should not be disturbed during the selection process. It is sometimes necessary to change the medium once, after about 5 d, taking care not to disrupt colonies.
5. Colchicine is one drug of choice for selections with *MDR*1, but other *MDR*1-related drugs can also be used, such as vinblastine, etoposide, or doxorubicin *(3)*. The concentration of drug to use must be determined for each cell line. To determine the appropriate drug concentration, plate cells at about 5×10^5 cells/10-cm dish in multiple dishes and allow them to attach. Remove the medium and replace with medium containing increasing concentrations of drug (colchicine). For murine cell lines, 20–80 ng of colchicine/mL is a good range of concentrations to test. For human cell lines, test 4–8 ng/mL. Incubate in the presence of drug for 10 d, then remove the medium and add 5 mL of methylene blue solution for 5 min. Wash dishes to visualize stained colonies. Use the concentration of drug at which no cells grow to form colonies. Be aware that colchicine prevents cells from dividing but does not kill cells outright. The result is that colchicine-sensitive cells appear as large, multinucleated cells that give a background blue staining with methylene blue.
6. Retroviral packaging cell lines express all the *gag*, *pol*, and *env* viral gene products required for producing functional virus particles *(10,11)*. RNAs expressed from retroviral expression vectors contain *cis*-acting nucleotide sequences required for packaging into virus particles. The GP+E86 and GP+envAm12 cell lines are recommended because of their low probability of producing recombinant, replication-competent retroviruses.
7. The GP+E86 cells produce ecotropic retroviruses capable of infecting only mouse cells. The GP+envAm12 cells produce amphotropic retroviruses with a broad-species host range. Titers should generally be on the order of 10^4–10^6 cfu/mL, with ecotropic titers being 5- to 10-fold higher than amphotropic titers. To obtain higher titers of amphotropic virus, first transfect retroviral vector DNA into GP+E86 cells, select for colchicine resistance, collect virus supernatant, and use this to transduce GP+envAm12 cells, as described.
8. During the transduction process, a fresh aliquot of virus can be added to the cells (in fresh growth medium containing polybrene) after 24 h and incubated for another 24 h before adding colchicine to select transduced cells.
9. Before beginning the amplification process, it is critical to have healthy, drug-resistant cells growing at a low concentration of drug (the same concentration

used to select transfected or transduced cells initially). Grow individual colonies or populations of cells for 3–4 passages before freezing or proceeding to the amplification step. This will eliminate multinucleated, colchicine-sensitive cells that tend to survive the first few platings.

10. Aliquots of cells should be frozen at each drug concentration during the amplification process. Cells growing at some of these concentrations can be stably maintained in culture for further analysis.
11. The maximum drug concentration to which cells can be amplified will vary. With NIH3T3 cells, transfected or transduced DNA is stable for months in culture in about 1 μg/mL colchicine. Amplification of Chinese hamster ovary cells has been reported up to 12.8 μg/mL colchicine *(6)*.

References

1. Kane, S. E. (1991) High-level expression of foreign genes in mammalian cells, in *Genetic Engineering* (Setlow, J. K., ed.) Plenum, New York, pp. 167–182.
2. Articles within a series (1990) *Methods Enzymol.* **185,** 485–611.
3. Gottesman, M. M. and Pastan, I. (1993) Biochemistry of multidrug resistance mediated by the multidrug transporter. *Ann. Rev. Biochem.* **62,** 385–427.
4. Kane, S. E., Troen, B. R., Gal, S., Ueda, K., Pastan, I., and Gottesman, M. M. (1988) Use of a cloned multidrug resistance gene for coamplification and overproduction of major excreted protein, a transformation-regulated secreted acid protease. *Mol. Cell. Biol.* **8,** 3316–3321.
5. Kane, S. E., Reinhard, D. H., Fordis, C. M., Pastan, I., and Gottesman, M. M. (1989) A new vector using the human multidrug resistance gene as a selectable marker enables overexpression of foreign genes in eukaryotic cells. *Gene* **84,** 439–446.
6. Konig, R., Ashwell, G., and Hanover, J. A. (1989) Overexpression and biosynthesis of CD4 in Chinese hamster ovary cells: coamplification using the multiple drug resistance gene. *Proc. Natl. Acad. Sci. USA* **86,** 9188–9192.
7. Germann, U. A., Gottesman, M. M., and Pastan, I. (1989) Expression of a multidrug resistance-adenosine deaminase fusion gene. *J. Biol. Chem.* **264,** 7418–7424.
8. Noiman, S., Yaniv, A., Tsach, T., Miki, T., Tronick, S. R., and Gazit, A. (1991) The Tat protein of equine infectious anemia virus is encoded by at least three types of transcripts. *Virology* **184,** 521–530.
9. Kane, S. E. and Gottesman, M. M. (1993) Use of multidrug resistance gene in mammalian expression vectors. *Methods Enzymol.* **217,** 34–47.
10. Markowitz, D., Goff, S., and Bank, A. (1988) A safe packaging line for gene transfer: Separating viral genes on two different plasmids. *J. Virol.* **62,** 1120–1124.
11. Markowitz, D., Goff, S., and Bank, A. (1988) Construction and use of a safe and efficient amphotropic packaging cell line. *Virology* **167,** 400–406.
12. Velu, T. J., Vass, W. C., Lowy, D. R., and Tambourin, P. E. (1989) Harvey murine sarcoma virus: Influences of coding and noncoding sequences on cell transformation in vitro and oncogenicity in vivo. *J. Virol.* **63,** 1384–1392.

13. Gorman, C. M., Moffat, L. F., and Howard, B. H. (1982) Recombinant genomes which express chloramphenicol acetyltransferase in mammalian cells. *Mol. Cell. Biol.* **2,** 1044–1051.
14. Ghattas, I. R., Sanes, J. R., and Majors, J. E. (1991) The encephalomyocarditis virus internal ribosome entry site allows efficient coexpression of two genes from a recombinant provirus in cultured cells and in embryos. *Mol. Cell. Biol.* **11,** 5848–5859.
15. Graham, F. and van der Eb, A. (1973) A new technique for the assay of infectivity of human adenovirus 5 DNA. *Virology* **52,** 456–457.
16. Gorman, C. M., Merlino, G. T., Willingham, M. C., Pastan, I., and Howard, B. H. (1982) The Rous sarcoma virus long terminal repeat is a strong promoter when introduced into a variety of eukaryotic cells by DNA-mediated transfection. *Proc. Natl. Acad. Sci. USA* **79,** 6777–6781.
17. Metz, M. Z., Matsumoto, L., Winters, K. A., Doroshow, J. H., and Kane, S. E. (1996) Bicistronic and two-gene vectors for using *MDR*1 as a selectable marker and a therapeutic gene. *Virology* **217,** 230–241.

28

Optimization of Growth, Viability, and Specific Productivity for Expression of Recombinant Proteins in Mammalian Cells

Jennie P. Mather, Alison Moore, and Robert Shawley

1. Introduction

Although optimization of recombinant protein production is an important part of expression, it is difficult to provide "cookbook" techniques. We will instead outline general approaches to optimization with specific methods described where appropriate. Optimal titers are reached by designing an environment where growth, viability, and specific productivity (protein/cell/time) are balanced so as to give maximum titers (protein/volume of medium). The strategy used will depend on the cell line and the characteristics of the protein being produced. Other considerations are the amount of protein needed, how much time is available for developing the process, and whether production of the protein will need to be repeated. Additionally, purification is much more efficient if the starting material contains a significant proportion (10–80%) of the protein of interest.

The remarks below are directed toward optimizing production of protein from a stably integrated gene in which a moderate initial effort in optimizing titer will pay off in simplified purification and future savings of time and money *(1,2)*. These methods are for cloning of cells, suspension adaptation, adapting cells to reduced serum, optimizing medium formulation, and measurements of cell viability. They can be used together to obtain cells that produce optimal titers and are easily scaled to the desired production size from a few hundred milliliters to thousands of liters *(3)*.

1.1. Cloning

Stable transfection and selection using a selectable marker results in a population of cells with a wide variety of growth characteristics and specific pro-

From: *Methods in Molecular Biology, vol. 62: Recombinant Gene Expression Protocols*
Edited by: R. Tuan Humana Press Inc., Totowa, NJ

duction rates. The composition of this population changes over time, usually with the low producers predominating after multiple passages. Obtaining a clonal cell population is important for maintaining stability of expression in a line. The purpose of cloning is to assure that all cells in the culture are descended from a single cell, i.e., they are genetically identical. This prevents the rapid and unpredictable changes in culture phenotype that may occur in mixed cell populations when conditions change to favor one cell type over another (e.g., a nonproducer over one producing a recombinant product).

Additionally, cloning allows the screening of a large number of cell lineages and the selection of the cell line with the highest product yield. Since most transfection techniques yield cells with widely differing numbers of gene copies, specific productivities, and growth rates, cloning allows one to choose a cell strain with the optimal properties *(4)*. There can, however, be considerable change with time even in cloned populations. These changes can be genetic, and therefore irreversible, or a phenotypic response to changing or marginal culture conditions that can be controlled or reversed.

1.2. Suspension Adaptation

The purpose of suspension adaptation is to obtain a cell line that will grow as single cells unattached to a substrate. There are several approaches to obtaining this end: alter the medium so that the ability of cells to attach is eliminated or diminished; select for cells that will not attach in the standard medium conditions; or select for cells that will grow in suspension in the standard media when the surfaces available have been treated to prevent attachment (these cells may still attach to surfaces treated for tissue culture) *(5)*. The first approach has the disadvantage in that the media devised to promote cell detachment (generally with much reduced magnesium and calcium, e.g., Joclik's medium) are frequently suboptimal for supporting high titers of desired proteins. The second approach is adequate for production cell lines but is more difficult than the third, and the resulting cell lines are less flexible. The third approach is generally (but by no means always) rapid, and results in a line that can be grown in an attached state for further manipulation such as cloning or transfection.

The method described in this chapter (Section 3.2.) is designed to suspension adapt cells in this third sense with as little alteration in other cell properties as possible. The one exception to this rule is that we have sometimes chosen to suspension adapt in a reduced serum, hormone-supplemented medium in order to obtain a line that will grow continuously in these conditions. In at least one case, this strategy also improved our ability to suspension adapt the cells and obtain a stable phenotype.

1.3. Medium Selection and Optimization

The first step involved in identifying media from which to narrow down the selections involves the cell line itself and the tissue of origin. This alone can help in the selection of media composition and requirements for growth factors. As an example, cell lines derived from liver often have a unique requirement for taurine and other bile salts. Many commercially available media have components that are required for some but not all cell types. These commercial media are, however, still good starting points for selecting a medium for optimizing cell growth and production. Individual optimization of all media components, although the only way to derive the ideal medium for any cell line, is a lengthy process that will usually not be undertaken without a strong scientific or commercial payoff *(1,6)*.

Many producers of cell culture media and components have developed their own proprietary cell culture media specific for those cell types widely used for production of recombinant products. These media may offer shortcuts when proteins must be rapidly produced; however, sometimes there is not complete disclosure of a medium's components.

Many of the commercially available media were originally designed to support low density or clonal growth of cell lines, a difficult task since an accurate balance of nutrients is required to allow growth while still avoiding toxicity. Thus, many of these media are well suited for growth at clonal densities involved in selection of cell lines. Often these same media are used for high-density cultivation of cell lines, which, in concert with high serum concentrations, appears to work satisfactorily. When serum-free culturing is desired, the shortfalls of these media become apparent.

While most mammalian cell culture media contain similar components, the appropriate balance of nutrients is critical to optimal cell growth. This optimal balance is different for different cell types and for cells grown at low vs high densities *(7,8)*. Accurate control of pH is also critical in cell culture, especially under serum-free conditions. Excessively high pH will cause a culture to lag, eventually leading to cell death. Low pH, whether owing to improper starting pH or owing to production of CO_2 and lactic acid by the cells, will also prevent the culture from reaching higher densities and can be as toxic as high pH. Buffers in the medium along with CO_2 provided by the incubator atmosphere are necessary to control these pH fluctuations.

1.4. Eliminating Serum

If the cell line to be used for production is growing in the usual 5–10% serum containing medium, one may wish to eliminate or reduce the serum level. This makes the medium less expensive in many cases, simplifies purification by eliminating serum proteins, and often improves product quality and/or cell growth *(9–11)*. To optimize recombinant protein expression and recovery, the goal is sometimes not to maximize growth rate, but rather to optimize viability and productivity. Likewise, whereas serum reduction is, almost always, useful in meeting this goal, complete elimination might be too time consuming or expensive to be desirable

1.5. Measuring Cell Viability and Death in Recombinant Cell Lines

Accurate measurement of cell viability and/or characterization of cell death is key to the optimization of any production cell line. Viability can be readily measured by a variety of well-known techniques; however, it has more recently been established that many cell lines grown under batch culture conditions undergo cell death via apoptosis *(12,13)*. Apoptosis is an active programmed cell death that exhibits specific morphological traits, such as nuclear condensation and fragmentation, maintenance of organelle and membrane integrity, and gross cell shrinkage *(14,15)*. This contrasts with necrotic cell death, which is passive in nature and recognized by the degradation of cellular membranes and organelles, and cell swelling *(16)*. These different forms of cell death are distinct in morphology, mechanisms, and in the way in which they can be identified. An understanding of such is required for accurate reciprocal viability quantification, since the observation of simple loss of membrane integrity may be insufficient to account for all cell death owing to apoptosis. This may also be important in the accurate calculation of specific productivity.

Viability can be assessed most simply by the exclusion of charged dyes such as trypan blue or propidium iodide (PI), which move freely into dead cells and exhibit minimal uptake into live cells *(17)*. The extent of cell death can then be quantified by microscopy, or in the case of PI, by flow cytometry. Loss of plasma membrane integrity also results in the release of intracellular contents into the media, which may then be quantified, such as the enzyme lactate dehydrogenase (LDH) *(18)*. The LDH content can be assayed in both the cellular and supernatant fractions, allowing relative quantification of cell death.

Apoptotic morphology specifically can be readily identified by staining fixed cells with a fluorescent dye such as acridine orange, which fluoresces green when bound to double-stranded nucleic acids *(19)*. The classic apoptotic morphology of condensed and/or fragmented nuclear material can then be observed. An additional characteristic of apoptotic cell death that can be utilized for quantification is that of DNA strand breaks, which are introduced at the internucleosomal linker sections of DNA in apoptotic cells *(20,21)*. The ultimate loss of short DNA fragments coupled with reduced accessibility to intercalating fluorochromes such as PI because of the condensation of chromatin, often result in the congregation of apoptotic cells within a fixed cell population to a "hypodiploid peak," when analyzed by flow cytometry *(22)*. The percentage of cells localized to this peak can be quantified if not obscured by excessive amounts of cellular debris.

The free 3' hydroxyl groups at the ends of the apoptotic DNA strand breaks may also be labeled with a fluorescent nucleotide using terminal deoxynucleotidyl transferase *(23,24)*. This technique allows the fluoresceination of apoptotic cells with only low level or absent labeling of nonapoptotic cells. Cells labeled in such a way may be counterstained with PI allowing concurrent cell cycle analysis, so that exit from the cell cycle can be tracked by flow cytometry.

2. Materials

2.1. Cloning

1. Medium: standard serum-free F12/DME (or medium of choice). Supplement with: 2–10% fetal bovine serum (*see* method 4, Section 3.4.); insulin, 5 µg/mL; and HEPES buffer, 15 m*M*.
2. Culture dishes: 96-well, flat bottom tissue culture dish (Corning or Falcon).
3. Trypsin. EDTA solution (Gibco-BRL, Gaithersburg, MD).

2.2. Suspension Adaptation

1. Medium: standard serum-free F12/DME (or medium of choice). Supplement with: 2–10% fetal bovine serum (*see* method 4, Section 3.4., for serum reduction or elimination); insulin, 5 µg/mL; pluronic F-68 (Gibco-BRL #24040-016), 0.1% *(5)*; and HEPES buffer, 15 m*M*.
2. Spinners: Use well siliconized (*see* Note 2) spinners. We prefer 250-mL spinners (Bellco Glass, Vineland, NJ).

2.3. Medium Selection and Optimization

A selection of medium can be obtained in liquid or powdered form from Life Technologies (Bethesda, MD) or Gibco-BRL. Special purpose media can also be obtained: e.g., CHO-S-SFM, (#12050, Gibco-BRL) or Hybridoma-SFM, (#12045, Gibco-BRL), as well as products from other companies such as JRH Scientific (Lenexa, KS), and Sigma (St. Louis, MO).

2.4. Reducing or Eliminating Serum

Growth factors can be obtained from the following suppliers: Life Technologies; Gibco-BRL; Upstate Biotechnology (Lake Placid, NY); Collaborative Research (Bedford, MA); ICN, Cell Biology Catalog (Costa Mesa, CA); and Sigma. Lipid supplement suppliers are: bovine lipoprotein (Pentex Ex-Cyte, Miles Laboratories [Rexdale, Ontario, Canada] *[11]*);

LipoMAX, Gibco-BRL #23000), or lipid-rich bovine serum albumin (AlbuMAX, Gibco-BRL #11020).

2.5. Measuring Viability and Cell Death

2.5.1. Reagents for Dye Exclusion

1. Phosphate buffered saline (PBS).
2. Trypan blue (Sigma), or propidium iodide (Molecular Probes, Eugene, OR).
3. Light microscope (TB) or fluorescence microscope (PI).

2.5.2. Reagents for Lactate Dehydrogenase Meaurement

1. Saponin (100 mg/L).
2. LDH standard photometric method (Procedure #228-UV, Sigma).

2.5.3. Reagents for the Identification of Apoptosis by Nuclear Morphology

1. Methanol.
2. Acridine orange (Sigma) solution: 6 μg/mL in PBS plus 0.1% NaEDTA.
3. Fluorescence microscope.

2.5.4. Reagents for the Quantification of Cell Death (DNA Content) by Flow Cytometry

1. Methanol.
2. 0.5% iso-osmotic paraformaldehyde: 5% solution of formaldehyde (10%, EM grade, Polysciences, Warrington, PA) in 0.05*M* PBS.
3. Propidium Iodide, 10–20 μg/mL in PBS.
4. 180 U/mL DNase-free RNase A (Boehringer Mannheim, Mannheim, Germany).
5. Flow cytometer.

2.5.5. Reagents for the TdT Assay

1. 0.5% iso-osmotic paraformaldehyde: 5% solution of formaldehyde (10%, EM grade, Polysciences) in 0.05*M* PBS.
2. Reaction buffer: 0.2*M* potassium cacodylate, 25 m*M* Tris-HCl, 0.25 mg/mL BSA.
3. 2.5 m*M* $CoCl_2$.
4. 0.25*M* NaEDTA, pH 8.0.
5. PBS.
6. Terminal deoxynucleotidyl transferase (TdT) (Boehringer Mannheim).
7. FITC-ddUTP (Boehringer Mannheim).
8. Flow cytometer, and/or fluorescence microscope.

3. Methods

3.1. Cloning

The method outlined below, cloning by limiting dilution, can be used with suspension or attached cells. It is the only method that should be used with suspension cells or with cells that are very mobile when attached. Because the cells are plated at a low cell/medium volume ratio, conditioning of the medium (*see* Note 1) is frequently necessary to get good growth.

1. Treat the plate of cells to be cloned with trypsin to obtain a single cell suspension. Visually inspect to ensure that most cells in the suspension are separate. Wash cells twice with medium by centrifuging at low speed. Take cells up in a complete growth medium to a final density of 10–100 cells/mL. Plate at 100 µL/well in a 96-well plate. The density used will depend on the plating efficiency of the cell line used. To maximize the chances of having a single cell/well, approximately half of the wells should be empty.
2. Check colonies visually 12–24 h after plating and mark wells or colonies that arise from a single cell. If two cells are seen in a well, it should be discarded.
3. To increase the statistical probability of obtaining a single cell clone, cloning should be repeated 2–3 times in succession.
4. The medium used should be the normal maintenance medium (e.g., F12/DME + 7.5% FBS, *see* method 4, Section 3.4. for medium selection). The addition of 5 µg/mL insulin during the low-density growth during cloning and during the first subsequent passage is helpful for many cell types. Conditioned medium can be mixed with normal growth medium (1:1 v/v) if the cells are difficult to grow at low density. After the cells grow to a higher density, the selective agent can be returned to the medium. This is less stressful to the cells and allows a higher plating efficiency.
5. Clones can be initially screened for protein production directly from the 96-well dish using whatever assay is available for the protein of interest. If an antibody is available for a Western blot, then the clones can be screened by transferring an aliquot of medium directly onto a nitrocellulose filter in a 96-well manifold. The filter can then be handled as for a normal Western blot. Cells from the wells that give the strongest reaction can be chosen for further expansion.
6. When clones are picked, they should be passaged first into a well of a 24-well dish, then to 35-mm, and 100-mm. Conditioned medium may also be used in the first passage after cloning, if necessary.

3.2. Suspension Adaptation

1. We prefer 250 mL spinners with 50–100 mL volume of medium. Spinners are run at 50–80 rpm. Set up two spinners in parallel. Carry both with subculturing on different days. If the cells cake around the spinner shaft and on the sides of the spinner at the medium surface, the spinner is not properly siliconized.
2. Remove cells from the starting culture with trypsin and neutralize the trypsin with serum (or soybean trypsin inhibitor if using serum-free culture).
3. Set up the spinners at 3–5 × 10^5 cells/mL. Check cells daily for growth and viability. More sodium bicarbonate may be added if the pH drops below 6.8 (*see* Section 3.3.3.).
4. After the first day or so, the caps on the spinners should be loosened to allow for increased oxygen and carbon dioxide exchange.

5. On d 3 or 4, the cells should be counted and passaged. Initially, cells should be centrifuged and fresh medium added to bring the cell number back to 3–5 × 10^5/mL. As the cells start growing logrithmically to densities over 1 × 10^6/mL, they may be passaged by dilution of the suspended cells with fresh medium. The cell density should be such as to allow at least a 1:5 split. If cells clump, excessively large clumps should be allowed to settle and not be passaged.
6. After 2–10 wk, the cells should be capable of logarithmic growth in suspension to reach densities of >10^6 cells/mL when inoculated at densities of 5–10 × 10^4. Cell viability should remain at >90% throughout the growth period. At this point, the cells are termed "suspension adapted" even though they will still attach to tissue culture plastic, especially in medium containing serum or attachment factors. The cells may exhibit some clumping during prolonged growth in suspension, but this is not necessarily a disadvantage.

3.3. Medium Selection and Optimization

3.3.1. Identify a Selection of Media to Be Tested

1. The first step involved in identifying media from which to narrow down the selections involves the cell line itself and the tissue of origin. Search the literature for media used for the cell line or cell type you wish to grow.
2. Many producers of cell culture media and components have developed their own proprietary cell culture media for specific cell types used in production of recombinant products (e.g., CHO-S-SFM, #12050 or Hybridoma-SFM, #12045, Gibco-BRL, as well as products from other companies such as JRH, Lenexa, KS, and Sigma, St. Louis, MO). These media may offer a shortcut when proteins must be produced in short order; however, incomplete disclosure of the medium components can be a cause for concern.

3.3.2. Comparing Media and Subsequent Experimentation

After identifying and purchasing a selection of commercially available media that suit the cell line of interest, apply a small scale preliminary screening assay.

1. Using 60-mm tissue culture plates, inoculate 3–10 × 10^5 cells/mL or more (3–4 mL/plate) for production optimization. The media selected are inoculated identically and screened for optimal growth or colony formation (for clonal density screening) or both growth and product accumulation (for productivity screening) at two time points, e.g., 4 and 7 d.
2. The best candidates from the initial screening are then blended at differing ratios to identify any significant growth and/or productivity benefits. The widely used 50:50 blend of Ham's F-12 nutrient mixture and Dulbecco's modified eagle medium has proved useful for a variety of cell lines. We have observed that, in some cases, tailoring the ratio of these two media has offered some improvement. One common requirement for prolonging cell viability at high densities is the addition of glucose to levels above that found in the normal media formulations.

3.3.3. Buffering and Osmolality

Buffering in most all cell culture media is based primarily on bicarbonate (carbonic acid/bicarbonate base), although phosphate (monobasic/dibasic) and some amino acids (e.g., glycine) in the medium contribute some buffering.

1. The addition of some amount of HEPES (10–50 m*M*) can provide additional buffering capacity and may be especially important at low cell densities.
2. Spinners filled to the recommended volume frequently cannot vent CO_2 or acquire oxygen by diffusion from the head space at a rate sufficient to support high cell densities. We therefore usually run these at 1/3 volume to allow sufficient CO_2 and oxygen exchange with the air space.
3. High-density cell culture presents added difficulties even when cultured in a controlled CO_2 environment, since lactate and CO_2 produced by the cells tend to accumulate, driving the pH down. This can be compensated for by the addition of sterile sodium carbonate or sodium hydroxide solutions to spinners and bioreactors as the pH falls later in the culture period. However, these additions will increase the osmolality of the medium, which can have a negative impact (*see* Note 3).

3.4. Reducing or Eliminating Serum

3.4.1. Weaning Cells from Serum

The general practice is to wean the cells by reducing the serum concentration over several passages. This method often works best, but is exceptionally time consuming.

1. Set up spinners of suspension adapted cells at 1–5 × 10^5 cells/mL at the serum concentration they are normally grown in (e.g., 10%).
2. Reduce serum through a series of passages. Serum can be reduced in small steps, e.g., by adding 10, 5, 2, 1, 0.5, and 0.1%, and finally serum-free over successive passages. Allow two or more passages at each concentration if a reduction in growth rate is observed. Keep the cell number in the range of 2–10 × 10^6 during these passages (*see* Note 4).

3.4.2. Deriving a Hormone Supplement to Replace Serum

The elimination or reduction of serum to below 1% necessitates the addition of hormones, growth factors, trace elements, and lipids *(6,9–11)*. Because of the low concentrations of these components, the absence of binding and carrier proteins provided by serum, and the instability of the components in culture medium, these must be added to the media after filtration and just prior to use.

1. Check the literature to see if there is a description of a hormone supplemented, serum-free formula for the cell type you wish to grow. If not, follow steps 2–9, Section 3.4.2.

2. Keep the cells at a density of 2–5 × 10^5 or higher. This allows the accumulation of autocrine factors and eliminates the need for the addition of components the cells can make.
3. Experimentally determine the lowest concentration of serum in which cells will survive, but not necessarily grow (this may be 0–2%).
4. Add the most commonly required additives, which are insulin (1–10 μg/mL), transferrin (1–100 μg/mL), and selenium (10–30 n*M*) (*see* Note 5).
5. Some cell lines have an added requirement for lipids. In addition to the insulin, transferrin, and selenium, test for a lipid response by adding commercially available lipid supplements such as bovine lipoprotein (Pentex Ex-Cyte, Miles Laboratories *[11]*; LipoMAX, Gibco-BRL #23000) or lipid-rich bovine serum albumin (AlbuMAX, Gibco-BRL #11020). Chose the most beneficial of these and add it to the mix.
6. Some cell lines have more complex requirements for growth factors such as epidermal growth factor (EGF), steroid hormones, or other small molecules. A more complete description of deriving the appropriate hormone supplement in order to eliminate serum is given in Barnes and Sato *(10)* and Hewlett *(11)*.
7. For cell lines that are particularly fragile, or if the intended goal is suspension adaptation of the cell line, Pluronic F-68 *(5)* and Pluronic F-127 have each been found to be extremely effective in preventing shear-associated cell lysis. However, it should be noted that the addition of F-68 can often prevent cells from attaching, especially in medium devoid of serum.
8. If the media in step 3, Section 3.4.2. did not start with 0% serum, eliminate the serum in the presence of all of the additives determined to be useful. If the growth is not satisfactory, repeat steps 4–7 in the presence of the growth promoting factors determined in the first round.
9. Once the serum concentration has been reduced to the lowest possible concentration supporting adequate growth, a large population of the cells should be frozen down as a stock supply before further screening. Addition of 0.1% carboxymethylcellulose and 5–10% dimethyl-sulfoxide to the medium generally provides a satisfactory freezing medium. It is preferable to freeze the cells in serum-free medium rather than re-expose the cells to serum.

3.5. Measuring Viability and Death

The following methods range from simple microscopy through to more sophisticated flow cytometric techniques. These have been used successfully to assess viability and cell death in production cell lines in our laboratory; however, this is in no way a conclusive list of viability measurement options.

3.5.1. Dye Exclusion

1. Incubate unfixed cells with either trypan blue solution or PI for 1 min at room temperature.
2. View by light or fluorescence microscopy.

3.5.2. Lactate Dehydrogenase Measurement

1. Samples of culture fluid plus cells, and culture fluid alone are treated with 100 mg/L saponin and vortexed for 2 min.
2. Centrifuge at 4800*g* at room temperature for 1 min.
3. The concentration of LDH in the supernatant is measured spectrophotometrically at 340 nm following the oxidation of NADH following the addition of pyruvate.

3.5.3. Identification of Apoptosis by Nuclear Morphology

1. Prepare slides of fresh cells using a cytocentrifuge, or by smearing.
2. Air dry briefly and fix in ice-cold methanol for 5 min.
3. Stain for approx 1 min in acridine orange solution, and rinse well in PBS.
4. View cells using fluorescence microscopy (*see* Note 6).

3.5.4. Quantification of Cell Death (DNA Content) by Flow Cytometry

1. Fix cells (approx 10^6 cells/sample) in ice-cold absolute methanol and store at –20°C.
2. Alternatively, fix in iso-osmotic paraformaldehyde for 15–30 min on ice. Spin out of fixative, resuspend in PBS, and store at 4°C.
3. Permeablize cells by washing in 0.1% Triton-X solution.
4. Incubate cells in labeling solution containing PI and RNase in PBS for 1 h at 37°C.
5. Remove samples to 4°C in the dark overnight.
6. Analyze by flow cytometry measuring red cell fluorescence (>620 nm).

3.5.5. The TdT Assay

1. Fix approx 10^6 cells/sample in 0.5% iso-osmotic paraformaldehyde for 15–30 min on ice.
2. Wash cells in PBS and resuspend in a 50-µL volume of reaction buffer, 2.5 m*M* $CoCl_2$, 25 U terminal transferase, and 100 pmol FITC-ddUTP.
3. Appropriate controls should be included, prepared in the absence of terminal transferase.
4. Incubate cells for 30 min at 37°C in the dark, and terminate the reaction by the addition of 5 µL 0.25*M* NaEDTA.
5. Wash cells twice in PBS.
6. Cells may be viewed by fluorescence microscopy, or analyzed by flow cytometry in the presence or absence of propidium iodide if concurrent cell cycle analysis is desired. Green (apoptotic labeling; 530 ± 20 nm) and red (DNA content; >620 nm) fluorescence can be measured using 488 nm excitation.

4 Notes

1. Conditioned medium is prepared by allowing the cells to grow to near confluency. The medium is changed and fresh growth medium placed on the cells for 24–48 h. This medium is collected, filtered through a 0.22-µm filter to remove any floating cells and used for conditioned medium described. It can generally be stored at 6°C for 1–2 wk.

2. Siliconization of spinners is most essential during suspension adaptation. Cells also require lower concentrations of some hormones, such as insulin, in siliconized vessels when grown in serum-free defined medium. The following method works well.
   a. Acid wash and dry spinners.
   b. Dip into a 20:1 mixture of dimethyldichlorosilane:1,1,1 trichloroethane (the trichloroethane reduces the flashpoint of the mixture).
   c. Rinse 5 times in alcohol.
   d. Rinse 3 times in water.
   e. Reassemble and autoclave.

   An alternative method is to use Dow Corning "Dowcoat," which is GMP approved. However, this must be baked on and releases noxious fumes when heated.
3. The osmolality of most cell culture media is in the range 270–310 mosM, in which optimal growth of mammalian cells is observed. Unless additions to the media are adjusted to the same osmolality as that of the medium, they will contribute additional osmolality.
4. A potential problem is an increase in the amount of carry-over of spent medium if cells being cultured in suspension grow too slowly with serum reduction. This increased proportion of medium carryover, along with toxic metabolites produced by the cells, and toxic breakdown products of the medium itself (e.g., lactate, ammonia from glutamine breakdown, and vitamin breakdown products) results in sequentially poorer growth. If growth slows appreciably at a passage, pellet the cells by gentle centrifugation between passages and replace all of the medium with fresh medium to ameliorate this problem.
5. Completely prepared supplement solutions containing insulin, transferrin, and selenium are available (GMS-S, GMS-A, GMS-G, and GMS-X supplements, Gibco-BRL).
6. If the existence of apoptotic cell death within a population is in doubt, it may be confirmed by the hallmark laddering pattern of extracted DNA when electrophoresed in agarose gel. The extent of DNA degradation owing to apoptosis may also be quantified according to the method described by Tilly and Hsueh *(25)*.
7. Apoptotic cells are cleared in vivo very rapidly by phagocytosis; however, in vitro, apoptotic cells and fragments may persist much longer, eventually losing membrane integrity and becoming secondarily necrotic. Additionally the existence of cell death arising from primary necrosis within a culture should also be investigated if the TdT assay is to be used. The random degradation of DNA during necrotic cell death results in low level labeling in the TdT assay, the extent of which may vary depending on the cell line.
8. The TdT assay may also be used to identify apoptotic cells adhered to a tissue culture plate, or also in tissue sections, using evaluation by microscopy instead of flow cytometry.

References

1. Mather, J. and Tsao, M. T. (1990) Expression of cloned proteins in mammalian cells: Regulation of cell-associated parameters, in *Large Scale Mammalian Cell Culture Technology* (Lubiniecki, A., ed.), Marcel Dekker, New York, pp. 161–177.

2. Mather, J. P. (1990) Optimizing the cell and culture environment for the production of recombinant proteins, in *Gene Expression Technology* (Goeddel, D., ed.), Methods in Enzymology, vol. 185, Academic, New York, pp. 157–167.
3. Arathoon, W. R. and Birch, J. R. (1986) Large-scale cell culture in biotechnology. *Science* **232,** 1390–1395.
4. Griffiths, J. B. and Racher, A. J. (1994) Cultural and physiological factors affecting expression of recombinant proteins. *Cytotechnology* **15,** 3–9.
5. Swim, H. E. and Parker, R. F. (1960) Effect of pluronic F68 on growth of fibroblasts in suspension on a rotary shaker. *Proc. Sci. Exp. Biol. Med.* **103,** 252–254.
6. Ham, R. G. and McKeehan, W. L. (1979) Media and growth requirements, in *Methods in Enzymology*, vol. 58, Academic, New York, pp. 44–93.
7. McKeehan, W. L., McKeenan, K. A., and Ham, R. G. (1977) Improved medium for clonal growth of human diploid fibroblasts at low concentrations of serum protein. *In Vitro* **13,** 399–416.
8. Zetterberg, A. and Engström, W. (1981) Effects of alkaline pH and glutamine on cell growth and multiplication, in *The Biology of Normal Human Growth* (Ritzén, M., ed.), Raven, New York, pp. 47–57.
9. Bottenstein, J., Hayashi, I., Hutching, S. H., Masui, H., Mather, J., McClure, D. B., Okasa, S., Rizzino, A., Sato, G., Serroro, G., Wolfe, R., and Wu, R. (1979) The growth of cells in serum-free hormone-supplemented media, in *Methods in Enzymology*, vol. 58, Academic, New York, pp. 94–109.
10. Barnes, D. and Sato, G. H. (1980) Methods for the growth of cells in serum free medium. *Anal. Biochem.* **102,** 255–270.
11. Hewlett, G. (1991) Strategies for optimizing serum-free media. *Cytotechnology* **5,** 3–14.
12. Moore, A., Donahue, C. J., Hooley, J., Stocks, D. L., Bauer, K. D., and Mather, J. P. (1995) Apoptosis in CHO cell batch cultures: examination by flow cytometry. *Cytotechnology* **17,** 1–11.
13. Franek, F., Vomastek, T., and Dolníková, J. (1992) Fragmented DNA and apoptotic bodies document the programmed way of cell death in hybridoma cultures. *Cytotechnology* **9,** 117–123.
14. Kerr, J. F. K., Wyllie, A. H., and Currie, A. H. (1972) Apoptosis, a basic biological phenomenon with wider implications in tissue kinetics. *Brit. J. Cancer* **26,** 239–245.
15. Arends, M. J. and Wyllie, A. H. (1991) Apoptosis: mechanisms and roles in pathology. *Int. Rev. Exp. Path.* **32,** 223–254.
16. Moore, J . V. (1987) Death of cells and necrosis of tumours, in *Perspectives in Mammalian Cell Death* (Potten, C. S., ed.), Oxford University Press, Oxford, UK, pp. 295–325.
17. Horan, P. K. and Kappler, J. W. (1977) Automated flourescent analysis for cyrotoxicity assays. *J. Immunol. Meth.* **18,** 309–316.
18. Goergen, J. L., Marc, A., and Engasser, J. M. (1993) Determination of cell lysis and death kinetics in continuous hybridoma cultures from the measurement of lactate dehydrogenase release. *Cytotechnology* **11,** 189–195.

19. Dive, C. and Wyllie, A. H. (1993) Apoptosis and cancer chemotherapy, in *Frontiers in Pharmacology: Cancer Chemotherapy* (Hickman, J. A and Tritton, T. R., eds.), Oxford University Press, Oxford, UK, pp. 21–56.
20. Wyllie, A. H. (1980) Glucocorticoid induced thymocyte apoptosis is associated with endogenous endonuclease activation. *Nature* **284,** 555,556.
21. Duke, R. C., Chervenak, R., and Cohen, J. J. (1983) Endogenous endonuclease-induced DNA fragmentation: an early event in cell-mediated cytolysis. *Proc. Natl. Acad. Sci. USA* **80,** 6361–6365.
22. Telford, W. G., King, L. E., and Fraker, P. J. (1992) Comparative evaluation of several DNA binding dyes in the detection of apoptosis-associated chromatin degradation by flow cytometry. *Cytometry* **13,** 137–143.
23. Gorczyca, W., Bruno, S., Darzynkiewicz, R. J., Gong, J., and Darzynkiewicz, Z. (1992) DNA strand breaks occurring during apoptosis: their early in situ detection by the terminal deoxynucleotidyl transferase and nick translation assays and prevention by serine protease inhibitors. *Int. J. Oncol.* **1,** 639–648.
24. Gorczyca, W , Bigman, K., Mittleman, A., Ahmed, T., Gong, J., Melamed, M. R., and Darzynkiewicz, Z. (1993) Induction of DNA strand breaks associated with apoptosis during treatment of leukemias. *Leukemia* **7,** 659–670.
25. Tilly, J. L. and Hsueh, A. J. W. (1993) Microscale autoradiographic method for the qualitative and quantitative analysis of apoptotic DNA fragmentation. *J. Cell. Physiol.* **154,** 519–526.

VII

Expression in Transgenic Organisms

29

Promoters to Express Cloned Genes Uniformly in *Drosophila*

Jean-Paul Vincent and Charles Girdham

1. Introduction

Forcing ubiquitous expression of a given gene in transgenic *Drosophila* has become a powerful tool of molecular genetics. Gene misexpression usually causes a dominant phenotype and one may deduce potential functions for the gene from an analysis of this phenotype. Evidently, when a dominant phenotype arises, uniform expression must be inducible such that the ubiquitously expressing construct can be maintained in a silent state in the transgenic stock. Even more informative than uniform expression throughout the animal, is expression in a large patch of tissue, especially when the gene considered acts in a non-cell-autonomous fashion. Here we review the methods available to achieve uniform expression of any cloned gene in *Drosophila*. We then go on to discuss how expression can be made conditional and/or in large patches of tissue.

2. Constitutive Promoters

Finding a promoter that is ubiquitously active at all stages of development is not as trivial as it seems. The hsp70 promoter has been a workhorse of gene misexpression in the past several years (*see* ref. *1*). The heat shock response seems to occur in all cells at almost all stages of development. Several vectors are available that allow cloning of any cDNA downstream of the heat shock promoter. The most useful one is pCaSpeR-hs made by Carl Thummel and colleagues (Fig. 1). The hsp70 promoter has the advantage of being inducible such that the time of expression is under experimental control (although basal expression is nonzero). The disadvantages are that expression is only transient and that repeated or prolonged heat shocks suppress expression of other genes and can be deleterious (heat shock is actually lethal when performed before

From: *Methods in Molecular Biology, vol. 62: Recombinant Gene Expression Protocols*
Edited by: R. Tuan Humana Press Inc., Totowa, NJ

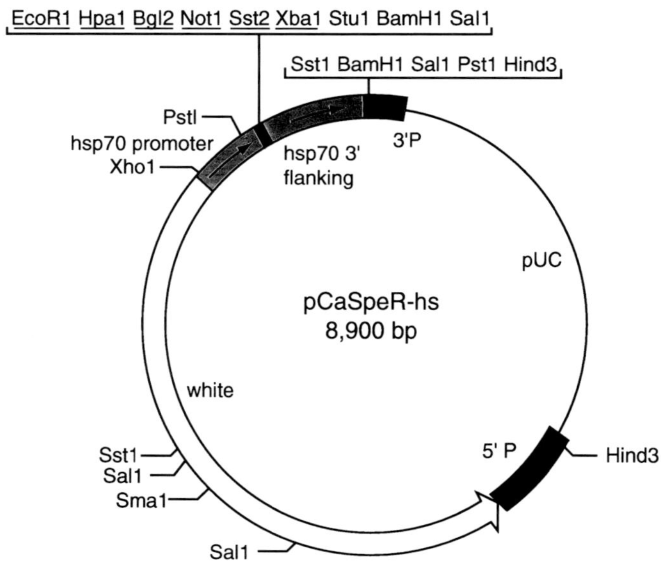

Fig. 1. Map of pCaSpeR-hs, a P-element vector allowing cloniong of a cDNA downstream of the hsp70 promoter. This figure is redrawn from the map obtained by the authors from Carl Thummel. Unique cloning sites are underlined. The white gene is used to identify transformants.

nuclear cycle 14, the time of cellularization). Here, we focus on truly constitutive (not inducible) promoters.

The simplest test of a candidate constitutive promoter is to link it to the Lac Z gene and verify that β-galactosidase is uniformly produced, either using antibodies or X-gal histochemistry. This way, dominant phenotypes do not interfere since β-galactosidase has not been observed to have toxic effects, even at high levels. Obvious candidates for constitutive promoters are those of genes presumably required in all cells, e.g., those encoding cytoskeletal components. For example, both the nonmuscle actin and tubulin promoters have been used. These have been shown to be uniformly active in imaginal disks *(2,3)*. The actin and tubulin promoters have two shortcomings. First, expression levels are relatively low: for example, expression of the *wingless* cDNA under actin control is several-fold lower than expression of *wingless* under its normal promoter *(2)*. Second, expression in embryos commences relatively late and is initially patchy (our own observation). In the case of the tubulin promoter, activity is never completely uniform in early embryos *(4)*. Much patterning of the larva occurs during this period and an ideal ubiquitous promoter should be active and uniform then. Variation in expression levels could complicate interpretation of the phenotype.

Another, less obvious, candidate uniform promoter is that of the segment polarity gene, *armadillo (5)*. This promoter was chosen to overcome the problem of the weak and spotty activity of the two cytoskeletal promoters during early embryogenesis. Transgenic embryos carrying an *armadillo* promoter-Lac Z fusion do produce β-galactosidase uniformly immediately after cellularization *(6)*. Expression appears uniform in imaginal disks as well. This promoter is therefore a good choice although not all tissues have been tested for expression (only embryos and imaginal disks have been thoroughly analyzed). Also, the level of expression has not been directly compared to that of the actin and tubulin promoter. Our impression is that all three (actin, tubulin, and *armadillo*) have comparable levels of activity, with *armadillo* being possibly more active. The *armadillo* promoter has been inserted into a bluescript vector (available from either author) and can be conveniently subcloned with a variety of restriction enzymes (Fig. 2). Alternatively, it can be recloned as a PCR fragment using the sequences indicated in the legend of Fig. 2 as primers.

One of us (C.H.G.) has recently investigated another ubiquitous promoter, that of the polyubiquitin gene *(7)*. This promoter may turn out to be the best choice, as it is expressed in early embryos (Fig. 3) as well as in imaginal disks and its level of activity appears higher than that of the three promoters mentioned above. The polyubiquitin gene inserted in pUC19 (pPUb from ref. *7*) is also available from one of us (C.H.G.). The promoter can be excised as a Bgl2 fragment, in which case a protein fusion must be made since this fragment incorporates the first three codons of the polyubiquitin open reading frame. Alternatively, the polyubiquitin promoter (without the initiation ATG) can be obtained by PCR using the sequence information given in the legend of Fig. 2. In conclusion, we think that either the *armadillo* or the polyubiquitin promoters are useful for a variety of applications where a uniform promoter is required.

3. Making a Constitutive Promoter Conditional

If misexpression of a gene is lethal, expression from a ubiquitous promoter-cDNA construct must be kept silent until transgenic stocks are established. One way to achieve this is to insert a transcription termination signal (denoted PolyA here) between the promoter and the cDNA. If this signal is flanked by FRTs (Flp recombination targets, where Flp is a site-specific recombinase from yeast), it can be excised at the appropriate time by producing Flp recombinase (usually under heat shock control) *(2,8,9)*. Thus transgenic flies containing the transgene [polyubiquitin > PolyA > cDNA] (where > denotes an FRT site) are crossed to flies carrying a hsp70-Flp construct (these flies are generally available within the *Drosophila* community). The

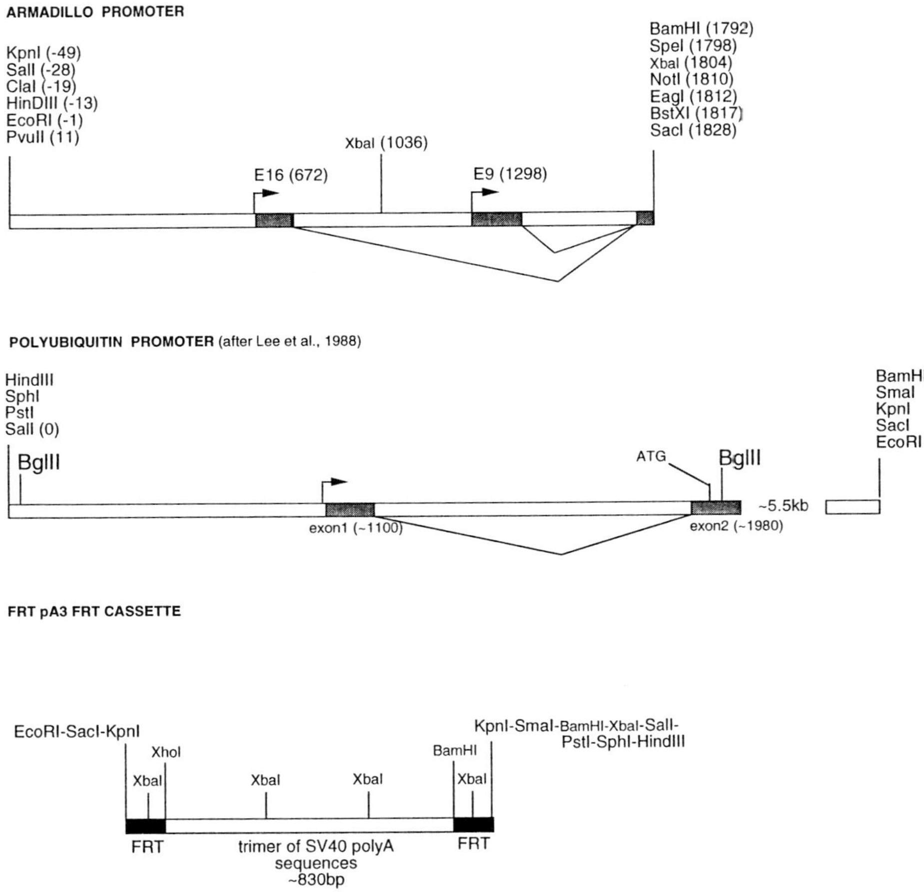

Fig. 2. Schematic maps of the *armadillo* and polyubiquitin promoters and of the FRT.PolyA.FRT cassette. In *armadillo*, all sites indicated can be used except for Xba1. The sequence flanking the two promoters is as follows:

5' TCCGCCGCCAGCTGCTGTGAC.../*armadillo* promoter/...TCTCTTTCTT-GCAGGTGTGGT 3'

5' AACAGCTATGACCATGATTACGCCAAGCTTGCATGCC.../polyubiquitin promoter/...CCCGCAGAATAATCCAAA**ATG**CAGATCT 3'

DNA fragments amplified with primers hybridizing to these sequences will contain the complete promoter, including the transcription start (and, in the case of polyubiquitin, the translation start if desired). For polyubiquitin, the 5' sequence indicated is that in pUC19, in which the polyubiquitin genomic region has been inserted (to our knowledge, the actual 5'end of the polyubiquitin promoter fragment has not been sequenced). The ATG is indicated in bold. The last six nucleotides at the 3' end constitute the *Bgl*II site.

(continued)

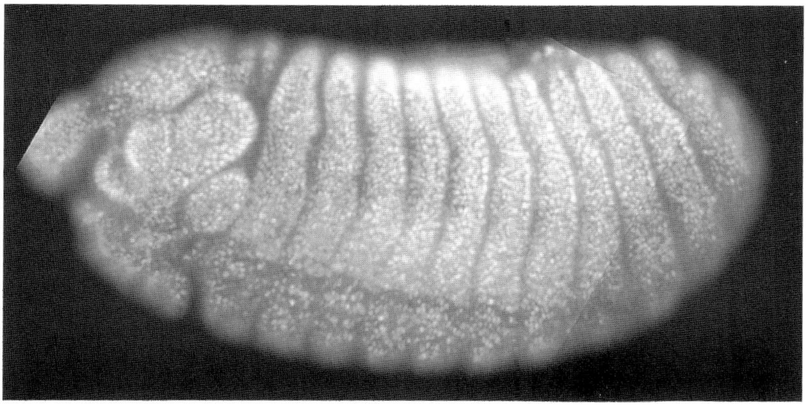

Fig. 3. Stage 13 *Drosophila* embryo expressing nuclear targeted β-galactosidase under the control of the polyubiquitin promoter.

progeny is heat shocked at the appropriate time, the termination sequence is thus excised in most or all cells, and uniform expression of the cDNA commences. Again, a DNA fragment containing a transcription termination sequence flanked by FRTs (> PolyA >) inserted in pUC19 is available (from C.H.G.; Fig. 2).

An alternative way to make constitutive expression conditional would be to utilize the Gal4/UAS system *(10)*. This system is based on two transgenes which are brought together in the same animal by a genetic cross. The first transgene contains a cDNA under the control of the yeast transcription factor Gal4. (It carries a Gal4 responsive element called upstream activating sequence [UAS] within its promoter.) The second transgene would produce Gal4 constitutively, e.g., under the control of the *armadillo* or polyubiquitin promoter. (One might choose to provide this transgene from the male if one wanted to avoid the maternal contribution of these promoters.) Transformants bearing the latter transgene are now available upon request to J.P.V. The advantage of this approach is its versatility. The disadvantage, however, is that the experimenter would have no control on the time of onset of constitutive expression, as is the case with the hsp70-Flp/FRT system.

Fig. 2. *(continued)* A trimerized repeat of an SV40 polyadenylation sequence was inserted between direct repeats of 48bp minimal FRT sites in pUC19. The SV40 fragment has polyadenylation signal sequences on both strands and the cassette can therefore function in both orientations to terminate transcription.

4. Constitutive Expression in a Tissue Patch

The Gal4/UAS system has become a popular method to express a given cDNA in various patterns. In this case, Gal4 is not produced uniformly as described in the previous paragraph. Rather, it is expressed in a defined pattern, usually obtained by enhancer trapping *(10)*. The Gal4/UAS system allows expression of any cloned gene in many patterns and is widely useful except maybe when the boundary of the patch needs to be precisely known (low level expression of Gal4 outside the main domain of expression may induce the expression of the cDNA at undetectable, yet functional levels).

There are other ways besides the Gal4/UAS system to make patches of tissue expressing a given cDNA constitutively. We discuss two ways here, both of which allow a precise determination of the patch boundary. One way is to make mosaic animals by nuclear transplantation. Details on the methodology can be found in refs. *11–13*. Donor nuclei would contain the following three transgenic constructs, [polyubiquitin > PolyA > cDNA], [hsp70-Flp, and [polyubiquitin-lacZ] whereas host embryos are wild-type. The latter transgene allows identification of donor tissue. Because of the random colonization by donor nuclei, such tissue occurs in random patches in the host animal. Expression of the cDNA would be induced at the desired time (by heat shock) throughout most of the donor patch.

A second method to create patches of tissue expressing a given cDNA would use animals containing the polyubiquitin > PolyA > cDNA and hsp70-Flp transgenes in the same manner as described in ref. *2*. Namely, a mild heat shock early in development induces Flp-mediated recombination in a few randomly located cells. These cells will begin to express the cDNA and, as proliferation proceeds, will give rise to a clonal patch expressing this cDNA. One drawback is that generation of a large patch of tissue through cell proliferation takes time. Therefore, this system is not suitable to study the temporal effects of inducing expression of the cDNA in a large patch since expression begins as soon as the founder cell of the patch is singled out. To do that, one has to resort to nuclear transplantations, or else, consider the following hypothetical scheme.

5. Inducing Constitutive Gene Expression in a Defined Large Patch of Tissue

What follows here is a suggested method to induce expression of a given gene in a large patch of tissue at a given time. This method makes use of both the Gal4-UAS and the Flp-FRT system. Suppose that a defined enhancer is active in the region where one wishes to express the cDNA at a particular time. For example, there is an enhancer of the *distalless* gene which is active in the center region of each imaginal disk. Fusion of this enhancer to a minimal promoter and the Flp recombinase coding sequence would make a [dll-Flp gene]. Consider now the following construct, [UAS.hsp70-Gal4], in which Gal4 is produced either by heat shock or when Gal4 is present. A single heat shock will trigger a positive feedback loop and Gal4 expression will be stably maintained. If a > PolyA > cassette is introduced downstream of the transcription start in this construct, to make [UAS.hsp70 > PolyA > Gal4], the feedback loop will only operate in cells which have excised the cassette. Thus, in flies carrying [dll-Flp], [UAS-cDNA], and [UAS.hsp70 > PolyA > Gal4], a heat pulse will initiate Gal4 expression in the center of disks (in which the PolyA sequence has been excised from the Gal4 construct by Flp recombinase expressed under the control of the dll enhancer). The feedback mechanism of [UAS.hsp70-Gal4] will ensure continuous expression of the cDNA in this domain. We are currently testing whether such a scheme is feasible (one potential problem is that overproduction of Gal4 may be lethal). As described, this system would induce expression only in a defined patch (in the example described, in the dll expression domain) but we think that it could be modified to produce random patches.

References

1. Heemskerk, J., DiNardo, S., Kostriken, R., and O'Farrell, P. H. (1991) Multiple modes of *engrailed* regulation in the progression towards cell fate determination. Nature, **352,** 404–10.
2. Struhl, G. and Basler, K. (1993) Organizing activity of Wingless protein in *Drosophila. Cell* **72,** 527–540.
3. Basler, K. and Struhl, G. (1994) Compartment boundaries and the control of Drosophila limb pattern by hedgehog protein. *Nature* **368,** 208–14.
4. Harrison, D. A. and Perrimon, N. (1993) Simple and efficient generation of marked clones in *Drosophila. Curr. Biol.* **3,** 424–433.
5. Riggleman, B., Wieschaus, E., and Schedl, P. (1989) Molecular analysis of the *armadillo* locus: uniformly distributed transcripts and a protein with novel internal repeats are associated with a *Drosophila* segment polarity gene. *Genes Dev.* **3,** 96–113.
6. Vincent, J.-P., Girdham, C. H., and O'Farrell, P. H. (1994) A cell-autonomous, ubiquitous marker for the analysis of *Drosophila* genetic mosaics. *Dev. Biol.* **164,** 328–331.
7. Lee, H., Simon, J. A., and Lis, J. T. (1988) Structure and expression of ubiquitin genes of *Drosophila melanogaster. Mol. Cell. Biol.* **8,** 4727–4735.
8. Golic, K. G. and Lindquist, S. (1989) The FLP recombinase of yeast catalyzes site-specific recombination in the *Drosophila* genome. *Cell* **59,** 499–509.
9. Golic, K. G. (1991) Site-specific recombination between homologous chromosomes in *Drosophila. Science* **252,** 958–961.
10. Brand, A. H. and Perrimon, N. (1993) Targeted gene expression as means of altering cell fates and generating dominant phenotypes. *Development* **118,** 401–415.

11. Lawrence, P. A. and Johnston, P. (1984) On the role of the *engrailed*$^+$ gene in the internal organs of *Drosophila*. *EMBO J.* **3,** 2839–2844.
12. Lawrence, P. A. and Johnston, P. (1989) Analysis of function of the pair-rule genes *hairy*, *even-skipped* and *fushi tarazu* in mosaic *Drosophila* embryos. *Development* **107,** 847–853.
13. Vincent, J.-P. and Lawrence, P. A. (1994) *Drosophila wingless* sustains *engrailed* expression only in adjoining cells: evidence from mosaic embryos. *Cell* **77,** 909–915.

30

Application of Micromechanical Piercing Structures for Genetic Transformation of Nematodes

Sarwar Hashmi and Randy Gaugler

1. Introduction

Several methods are available to accomplish gene transfer into mammalian cells, plant cells, yeast, and other organisms. Three of the most promising DNA delivery systems are electroporation *(1,2)*, DNA-coated microprojectiles *(3)*, and microinjection (4,5). Electroporation involves the formation of transient pores in biological membranes by the discharge of electrical current. DNA diffuses into cells through the pores created by this process. The microprojectile gene transfer technique shoots particles coated with genetic materials into target cells. In the heavily studied free-living nematode *Caenorhabditis elegans*, the functional characterization of cloned genes has been greatly facilitated by the availability of an efficient DNA transformation system. The only procedure for transformation of *C. elegans* has been microinjection of DNA into the gonad of individual nematodes *(4,5)*. Microinjection has also been used successfully for transforming the insect parasitic nematode *Heterorhabditis bacteriophora (6)*. Although the available techniques have been successful in genetic transformation of various organisms, these techniques are expensive, laborious, and require considerable expertise.

A new micromechanical device to inject DNA-coated microprobe through the nematode *(H. bacteriophora)* cuticle has been developed *(7)*. Microprobes are micromachined needles made using silicon fabrication techniques. This technique allows arrays of very sharp pyramidal points to be etched on a silicon substrate. The tip of the microprobe is very sharp, with a radius of curvature of less than 1 µ. The penetration depth of the microprobes can be precisely controlled, and the height of the probe can vary from less than ten to several hundred micrometers. The arrays of extremely sharp tips project up from the surface of chip.

From: *Methods in Molecular Biology, vol. 62: Recombinant Gene Expression Protocols*
Edited by: R. Tuan Humana Press Inc., Totowa, NJ

Microprobe arrays etched on a silicon wafer were used for successful genetic transformation of *H. bacteriophora*. The nematodes were placed on the surface of a microprobe array coated with the desirable plasmid DNA. When nematodes crawled on these DNA-coated tips, the points penetrated through the nematode cuticle, allowing DNA to enter the nematode. Some of the tips usually pierce the nematode gonad. Central to the success of introducing foreign materials into cells directly is clean penetration without causing any significant damage to a cell or organism. Apparently, microprobes do not cause any significant damage to nematode.

Microprobes have also been successfully used to inject DNA into tobacco (Chee-Kok Chin, personal communication). In this experiment, the microprobe array attached to a plastic handle coated with DNA was applied manually to the surface of tobacco leaves, where the cells were exposed by removing the epidermal layer. Transformants evaluated by GUS assay showed transient expression in tobacco cells.

In this chapter, we describe methods to use the microprobe array for genetic transformation of nematodes. The microprobe-mediated DNA transformation into nematode is performed under sterile conditions and involves several important steps: (1) preparation of nematode strain and culture; (2) purification of plasmid DNA; and (3) DNA transformation.

2. Materials

1. Microprobe array (chip). The size of a microprobe array is usually 5 × 8 mm. Researchers interested in using the microprobe array should contact Dr. Ahmad Hakim Elahi (Office of Corporate Liaison and Technology Transfer, Rutgers University, New Brunswick, NJ 08903).
2. Dissecting microscope.
3. Sterile forceps.
4. M9 buffer (3 g KH_2PO_4, 6 g Na_2HPO_4, 5 g NaCl, 0.25 g 1*M* $MgSO_4$, per liter H_2O) autoclaved and stored at room temperature.
5. 95% Ethanol.
6. 1X TE buffer (10 m*M* Tris, 1 m*M* EDTA, pH 8.0) stored at room temperature.
7. 2.5*M* $CaCl_2$ stored at –20°C.
8. 0.1*M* spermidine-free base (Sigma, St. Louis, MO) (used for DNA aggregation and coating on the microprobe) stored at 4°C.
9. Purified plasmid DNA (stored at –20°C for short-term storage or at –70°C for long-term storage). Plasmid DNA used for transformation should contain a selectable marker for identification of transformants (e.g., the *E. coli lacZ* encoded enzyme, β-galactosidase).
10. Agar plate pre-seeded with bacteria (For example, Photorhabdus luniniscence bacteria for *H. bacteriophora*; *E. coli* for *C. elegans*).
11. Incubator (25 C for *H. bacteriophora*; 20°C for *C. elegans*).

12. Autoclaved ultrapure (distilled deionized) water.
13. Worm brush (eye-brow hair glued to a needle).
14. Micropipeter (various sizes).
15. Nematodes to be injected.

3. Methods

3.1. Nematode Strain and Culture

1. Culture nematodes in several petri plates (60 × 15 mm) with nutrient agar and bacteria for nematode food, maintain at appropriate incubation temperature (20°C or 25°C depending on nematode species). Nematodes must be free from contamination.
2. Collect adult hermaphrodites nematodes containing 4–6 eggs from in vitro plate by pouring approx 20–25 mL M9 buffer onto the plate (amount of M9 buffer may vary depending on the size of culture plate). Transfer the nematode suspension to a sterile centrifuge tube.
3. Centrifuge at 1500*g* for 6–7 min. Remove the supernatant and add 20–25 mL fresh M9 buffer, centrifuge again. Repeat this step 3–4 times.
4. After about four wash/spin, decant the supernatant from the tube. Leave just enough M9 buffer in the tube so that nematodes do not desiccate. These nematodes are to be used for transformation. M9 buffer is used to reduce the osmotic stress on nematodes.

3.2. Purification of Plasmid DNA

1. DNA to be injected can be prepared from rapid minipreps. Passage of miniprep DNA over a Sephacryl-S400 (Pharmacia) column, or a Quiagen column is recommended to remove tRNA.
2. Measure DNA concentration using spectrophotometer, and dilute DNA to a desirable concentration.

3.3. DNA Transformation

1. All procedures should be performed in a sterile environment.
2. Obtain several microprobe array and soak them in 95% ethanol in a 60 × 15 mm sterile petri plate for about 30 min. Dry them in air by transferring in another sterilized Petri plate using sterilized forceps.
3. Transfer a single microprobe array using a sterilized forceps onto a nutrient agar plate preseeded with bacteria.
4. Make DNA mixture in a 1.5-mL eppendorf tube as follows: add 2.5 μL of plasmid DNA (2 μg/mL in TE buffer, pH 8.0), 12.5 mL of $CaCl_2$ (from 2.5*M* solution), and 5 μL of spermidine-free base (0.1*M* solution) to make 20 μL DNA mixture.
5. Transfer 10 μL of DNA mixture onto the microprobe array. Make sure the DNA covers the microprobes, because the adsorption of the DNA on the tips of the microprobes is a very important step.

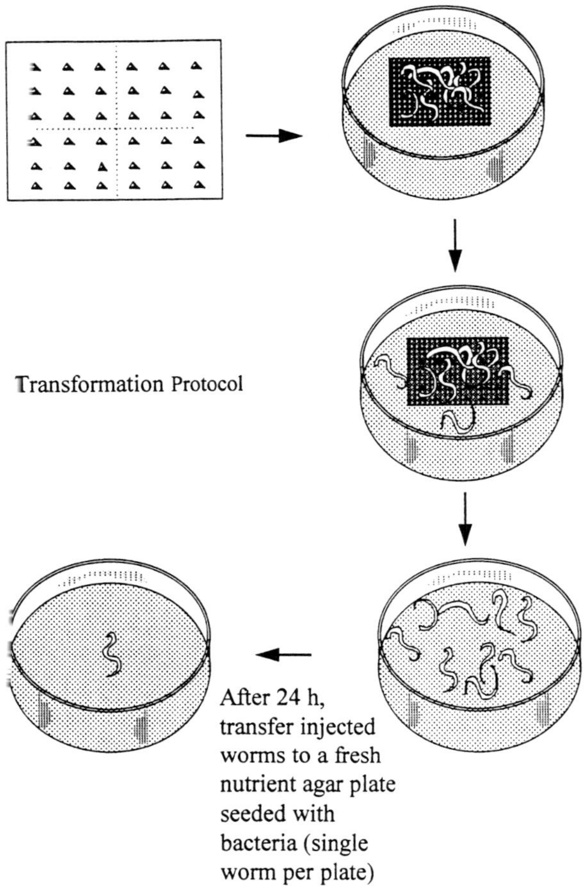

Fig. 1. A schematic drawing of microprobe-mediated gene transfer method. There are 200–250 microprobes on a 5 × 8 mm substrate. The nematodes were placed on a DNA-coated microprobe array.

6. After 10 min, pipet 5–10 µL of a concentrated nematode suspension (approx 200–250 nematodes) onto the microprobe array and leave at room temperature for 8–10 min (Fig. 1).
7. After 10 min, most nematodes will crawl through the tips of the microprobes onto the bacterial lawn. If some nematodes remain on the tips of the microprobes, use a worm brush to transfer them onto the bacterial lawn.
8. Remove microprobe array from the plate using a sterile forceps, seal the Petri plate with parafilm, and incubate plate at a temperature suitable for nematode growth and reproduction (e.g., 25°C for *H. bacteriophora*; 20°C for *C. elegans*).
9. After 24 h, transfer injected nematode from the original bacterial plate to another fresh bacterial plate (i.e., one injected nematode per plate).

10. Incubate these plates for 5–6 d or until injected nematodes produce progeny.
11. Examine progeny for the expression of the transformation marker in the first and subsequent generation to determine if the introduced DNA passed onto the progeny. We found that an average of 8% of transformants segregated the marker to their progeny.

4. Notes

Microprobe-mediated DNA transformation is a highly efficient genetic transformation system for nematodes. The method is quick, at least 100 times faster than nematode microinjection *(4–6)*. In microinjection, individual nematodes are injected, whereas in microprobe-mediated DNA transformation more than 100 nematodes can be injected at one time. Injection by microprobe however, is not site-directed. Apparently, the tips of the microprobe act as injection needles. Because of the presence of many of the microprobes in a small area, many injections are possible though they are random. We have observed that a tip of the microprobe is capable of penetrating through cuticle without causing apparent damage to the nematode. Therefore, this approach appears suitable for nematodes. The microprobe's simplicity of use and high degree of efficiency makes it an extremely useful tool that speeds up studies involving transgenic plants and animals. In addition, microprobe array is less expensive than other available techniques for DNA transformation *(1–5)*, as it does not require sophisticated equipment and specialized training.

Gene transfer efficiency is influenced by the age of the nematode and the DNA concentration. The ideal nematodes are those containing 4–6 eggs at the time of injection, because this is the stage at which egg formation has just begun. As oocytes mature, the membranes encapsulate individual germ nuclei along with the portions of the core cytoplasm as well as the introduced DNA.

Another important factor to consider in genetic transformation is DNA concentrations. In our protocol, we used a DNA concentration ranging from 100 to 200 µg/mL. We found that at concentrations ≥100 µg/mL, the frequency of transformation remained the same. Mello et al. (1991) *(5)* used the microinjection procedure for introducing plasmid pRF4 into *C. elegans* and reported that the F1 progeny obtained at 50 µg/mL of DNA concentration were far less likely to transmit the transformant phenotype to their progeny than were F1 phenotype at 100 µg/mL. Therefore, we suggest a DNA concentration of 100 µg/mL or higher in any nematode transformation experiment using microprobe array. Although we used 10 µL of DNA mixture to load onto the microprobe, this amount may be changed depending on the size of the microprobe array. The concentration of $CaCl_2$ and spermidine-free base used to coagulate DNA on the tips of the microprobes can also be adjusted according to the requirement.

The transfer of nematodes onto the DNA-coated microprobes is also an important step: those nematodes should be distributed evenly over the microprobe array. The amount of solution on the microprobes should be adjusted so that nematodes do not float in the solution, but rather crawl onto the tip of the microprobes. It is important not to touch the microprobe array on the top surface; it may damage the tips. Therefore, a sterilized forceps should be use to handle the microprobe array.

Although the microprobe-mediated DNA delivery system has been used for nematodes, it has broad potential for use with other popular systems, such as *Drosophila* and plant cells. Many variables can be further optimized to increase the efficiency of DNA delivery by the microprobes, such as using a higher density probe, or control on microprobe penetration depth in the target cell. The higher density probe, along with control on the penetration depth that impacts the target cells, may increase the number of cells that are penetrated. Microprobes with an average tip size of 1 μm proved effective in nematode transformation, yet tip size may need to be optimized for specific cell types.

Acknowledgment

We thank Ghazala Hashmi for her help and advice in preparation of this manuscript.

References

1. Potter, H., Weir, L., and Leder, P. (1984) Enhancer-dependent expression of human *k* immunoglobulin genes introduced into mouse pre-B lymphocytes by electroporation. *Proc. Natl. Acad. Sci. USA* **81,** 7161–7165.
2. Fromm, M., Taylor, L. P., and Walbot, V. (1985) Expression of genes transferred into monocot and dicot plant cells by electroporation. *Proc. Natl. Acad. Sci. USA* **82,** 5824–5828.
3. Klein, T. M., Wolf, E. D., Wu, R., and Sanford, J. C. (1987) High-velocity microprojectiles for delivering nucleic acids into living cells. *Nature* **327,** 70–73.
4. Fire, A. (1986) Integrative transformation of *Caenorhabditis elegans*. *EMBO J.* **5,** 2673–2680.
5. Mello, C. C., Kramer, J. M., Stinchcomb, D., and Ambros, V. (1991) Efficient gene transfer in *C. elegans*: Extrachromosomal maintenance and integration of transferring sequences. *EMBO J.* **10,** 3959–3970.
6. Hashmi, S., Hashmi, G., and Gaugler, R. (1995) Genetic transformation of an entomopathogenic nematode by microinjection. *J. Invertebr. Pathol.* **66,** 293–296.
7. Hashmi, S., Ling, P., Hashmi, G., Reed, M., Gaugler, R., and Trimmer, W. (1995) Genetic transformation of nematodes using arrays of micromechanical piercing structures. *BioTechniques* **19,** 766–770.

31

Strategies in Generating Transgenic Mammals

Olena Jacenko

1. Introduction

The ability to manipulate genes in mammals is providing insights into most aspects of modern biology, including the regulation and function of genes, the mechanisms of developmental and pathological processes, and the generation of animal models for human disorders. Furthermore, the development of gene transfer techniques is stimulating efforts to treat human diseases with gene-based therapies, and is establishing a new area for biotechnology in which transgenesis can be used for the improvement of domestic animals and plants, as well as for the production of rare products. The focus of this chapter will be to provide an overview of the strategies that can be used to alter the mammalian genome through gene transfer. Advantages and disadvantages of each approach will be discussed, and specific examples of how each strategy can be applied to address problems in mammalian biology will be provided in order to illustrate the potential scope of transgenesis.

A transgenic animal is defined here as one whose genome contains DNA of exogenous origin that has been introduced through experimental manipulation. By this definition, all animals with an *experimentally* altered genome resulting either from microinjection of recombinant DNA, infection with recombinant retroviruses, replacement of pre-existing genes with inactivated or mutated variants by gene targeting, or introduction of altered multipotent stern cells (e.g., hematopoietic, liver), are transgenic. Likewise, a transgene is the exogenous DNA introduced through experimental manipulation into the animal's genome, and includes recombinant DNA or retroviral constructs used for microinjection/infection, as well as replacement and insertional vectors used for gene targeting.

From: *Methods in Molecular Biology, vol. 62: Recombinant Gene Expression Protocols*
Edited by: R. Tuan Humana Press Inc., Totowa, NJ

2. Establishment of Methods for Manipulating Genes in Mammals

Historically, the mouse has been the mammal of choice for genetic analysis because of its size, short gestation period, relatively large litters, availability of inbred strains, and its numerous spontaneous mutations mimicking human genetic diseases. Furthermore, mouse embryology and immunology have been extensively studied. For these reasons, it is not surprising that the development of transgenic and embryonic stem cell technology was first achieved in the mouse, rather than in less complex organisms such as flies, worms, or fish. The first transgenic mouse was generated by Jaenisch and Mintz in 1974 *(1)*, when the simian virus 40 (SV40) DNA was microinjected into the blastocyst cavity of mouse embryos. Subsequently, germ line transmission of retroviral DNA was demonstrated following the exposure of early mouse embryos to a solution containing infectious retroviruses *(2,3)*. Infection of mouse embryos with recombinant retroviruses however, constitutes only one of at least six methods of transgene transfer (Fig. 1). The most commonly used technique to date involves direct microinjection of recombinant DNA into the pronucleus of a fertilized egg *(4)*. Although considerable work occurred in the 1960s and 1970s to provide a foundation for the development of transgenic mammals (*see* ref. *5* for review), the first report describing the presence of microinjected sequences in newborn mice appeared in 1980 *(6)*. The detailed protocol for gene transfer through microinjection remains virtually unchanged to date *(4)*, and represents the method through which the majority of transgenic mice are produced. Immediately following this initial report, five laboratories demonstrated stable integration of microinjected DNA into the host chromosome, and the expression of these genes in embryos and mice *(7–11)*. The accelerated growth of mice carrying a metallothionein-growth hormone fusion gene provided the most dramatic demonstration that the integrated genes were expressed and functional *(12)*. The genetically-engineered mice were termed "transgenic" (*see* ref. *5,13–18* for background).

The almost parallel development of a complementary approach involving gene targeting, the process of homologous recombination between an introduced altered gene and an endogenous chromosomal allele, has greatly facilitated the studies of mammalian development. The transfer of DNA into totipotent embryonic stem (ES) cells, which are capable of contributing to the germ line when reintroduced into the host, has been widely used to overexpress or inactivate genes both in cell culture and in vivo *(19–24)*. The basis for this technique stemmed from experiments performed as early as the 1960s with teratocarcinoma cells *(25)*, embryonal carcinoma cells *(26)*, as well as hematopoietic multipotent stem cells *(5,16,17)*, and was greatly enhanced by the development of transgenic techniques during the 1980s. The current strategy for germ line modification through homologous recombination, as well as a

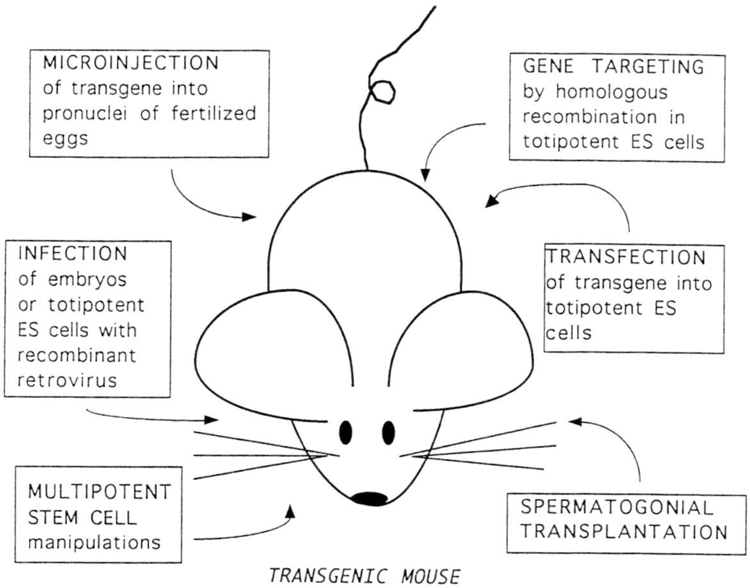

Fig. 1. Strategies for introducing transgenes into mice.

method for enrichment of ES cells in which the desired targeting event has occurred, are outlined by Capecchi and coworkers *(20,27)*. Thus, within one decade, significant advances were made in the ability to study gene regulation and function in the context of a whole animal; the technology became available for producing an animal with a variety of desired genotypes by experimental means. Adding to this progress, Brinster and coworkers recently described a novel method involving spermatogonial transplantation *(28–30)*, which offers the potential for transgenesis. The current strategies for generating transgenic mammals:

1. Microinjection of transgene into fertilized eggs;
2. Infection with recombinant retroviruses;
3. Gene targeting in ES cells;
4. Manipulation of multipotent stem cells; and
5. Spermatogonia transplantations include the following and are summarized (Fig. 1).

3. Transgenesis by Pronuclear Microinjection of Recombinant DNA

3.1. Methodology

The technique described by Gordon et al. *(4)* remains the method of choice for dissecting the intricate regulatory elements governing gene regulation, and

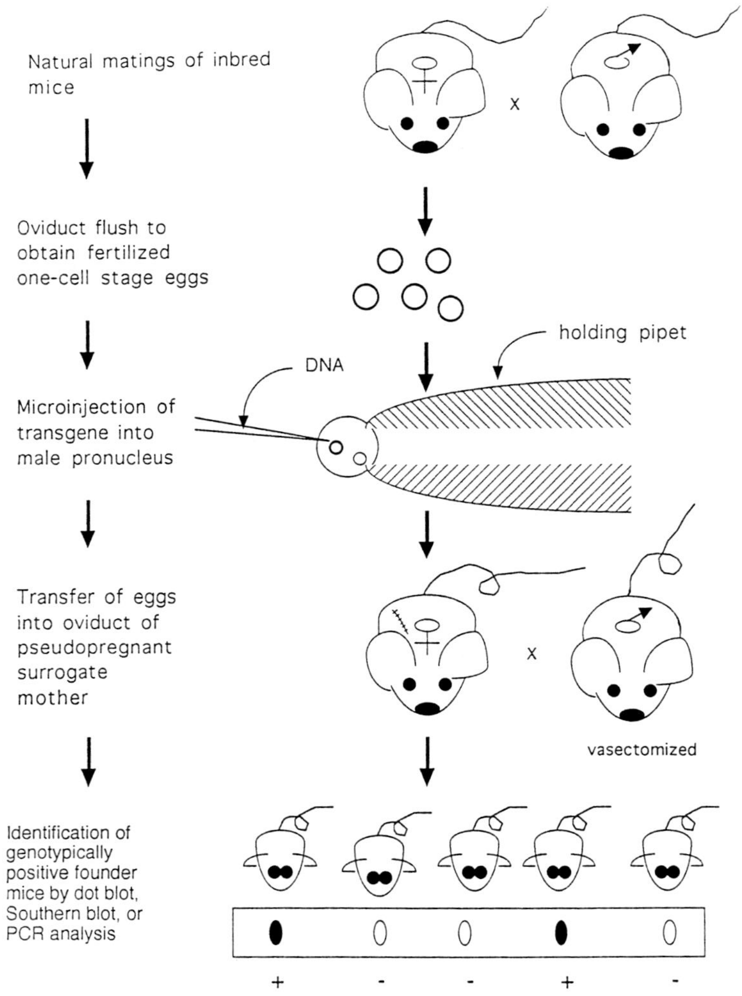

Fig. 2. Generation of transgenic mice through pronuclear microinjection of transgene. *See* text for details.

for expressing a given gene in almost any tissue (Fig. 2). Briefly, freshly isolated fertilized mouse eggs at the one-cell stage are cultured for 1–2 h until the pronuclei become visible. A solution containing the transgene of interest (present in linearized form with minimal vector sequences) is then microinjected via a glass micropipet into the male pronucleus of a fertilized egg that is restrained; a successful injection is evidenced by pronuclear swelling. The injected eggs are then surgically transferred into the oviduct of a pseudopregnant surrogate mother, who has been previously mated with a vasectomized male. The resultant pups (f_0) are analyzed for the presence of the transgene by either genomic Southern blotting, dot blotting, or PCR, using DNA obtained from tail biopsies. Typically, approx 10–20% of the pups born carry the transgene *(14,31,32)*. Each animal positive for the transgene is referred to as a founder, and represents the result of an independent transgene microinjection and integration event. The founders are bred with wild-type mice to obtain offspring (f_1) that also carry the transgene, thereby establishing unique families of mice, or transgenic lines. If germ line transmission is achieved, the interbreeding of f_1 hemizygotes (carrying the transgene on one of the two chromosomal alleles) gives rise to a portion of mice homozygous for the transgene. Homozygotes are identified by the intensity of transgene hybridization signals on genomic Southern blots; furthermore, they are "proven" homozygotic by backcrossing with wild-type mice, which should generate 100% hemizygous offspring. Genotyping and expression analysis of these mice are essential for determining if and where the transgene is expressed, and whether the transgene segregates with the observed phenotype. Furthermore, the integration site of the transgene in the chromosome also may influence the pattern and level of expression, necessitating the analysis of at least two transgenic lines per transgene construct. Although transgene expression is often stable over a number of generations, rearrangements and deletions may occur *(7,33)*, and should therefore be screened for. These critical issues are discussed in Section 3.3.4.

Transgene integration will usually occur at the one-cell stage; therefore the germ cells and all somatic cells of the founder will contain the foreign DNA. However, if integration occurs at a later point, not all cells may have the transgene. In such a case, the mouse is a mosaic for the transgene *(31)*. Furthermore, transgene integration is random; therefore the DNA may insert anywhere in the genome, and by doing so, may disrupt endogenous gene function, leading to a phenotype. Approximately 10% of the random integration events result in insertional mutagenesis, which most commonly manifests as a recessive phenotype *(4,34)*. This event represents one unexpected and major benefit of gene transfer through microinjection, as well as through retroviral infection; transgenes can act as insertional mutagens that can inactivate and thereby identify endogenous genes involved in specific developmental processes (*see* Section 3.3.5.) *(34,35)*.

It is also possible to microinject a transgene construct that will express a protein that will contribute to a phenotype independent of the integration site (Fig. 3). To result in transgene expression, the transgene construct must include a promoter. This promoter may be constitutive, inducible, cell-specific, viral, or that of a housekeeping gene. Likewise, the transgene whose expression is driven by this promoter may either be a reporter gene whose activity can be

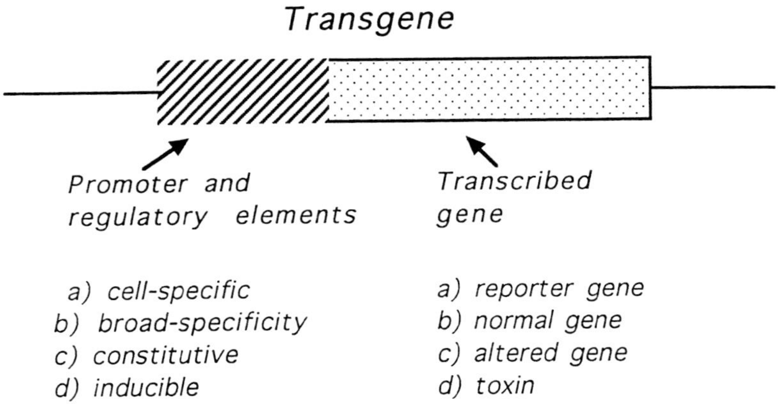

Fig. 3. Transgene construct for pronuclear microinjection into fertilized eggs. A hybrid construct can be designed by combining different types of promoters, which will drive the expression of a variety of genes to address gene regulation or function.

monitored histochemically or enzymatically (such as β-galactosidase, luciferase, chloramphenicol transferase), a normal or an altered mouse gene, a gene from a different species, or even a synthetic gene. This type of a "hybrid" transgene construct can be designed to address issues concerning either gene regulation or function.

3.2. Gene Regulation Studies

Many early studies using transgenic mice generated by pronuclear microinjections were designed to address the control and tissue-specificity of gene regulation *(36–40)*. By altering the nature and extent of the transgene promoter and by monitoring the expression of a reporter gene, regions within the promoter that are required for temporal and cell-specific expression could be mapped. These reporter genes are often, but not always, of prokaryotic origin and usually encode proteins that are not typically expressed in most eukaryotic cells. The expression pattern of the selected reporter gene, and thus the specificity of the promoter, can be determined either histochemically, by *in situ* hybridization, or biochemically in tissue homogenates *(40)*. For example, among the most commonly-used reporter genes is *lacZ*, encoding for bacterial β-galactosidase. *LacZ* activity can be successfully localized in mouse embryos by incubating whole embryos with X-gal, a substrate for β-galactosidase that is converted to a deep blue product *(38–41)*. However, *lacZ* staining in postnatal mice often yields unreliable results owing to nonspecific staining, in which case alternate reporter constructs may be designed. The activity of reporter genes such as firefly luciferase or bacterial chloramphendicol transferase (CAT) could be detected in tissue extracts in the presence of the appropriate substrate either spectroscopically (for luciferase), or biochemically (for CAT) *(38,40)*.

It is important to note that results obtained from such in vivo analyses in transgenic mice do not always mimic those obtained from in vitro assays. Since most data on gene regulation is generated through transient transfection expression analyses, it suffers from the general shortcoming of such an approach, namely, the absence of proper chromatin structure for controlled expression of the transfected DNA. Furthermore, many transfection studies are being carried out in cells that have no endogenous expression of the gene of interest (owing to the difficulty in isolating and culturing certain cell types), or in cells from different species, making the data difficult to interpret *(42)*. The availability of an in vivo approach has been essential for testing the in vitro observations, and has confirmed the identification of transcription regulatory regions as major determinants of tissue-specific gene expression in the whole organism. To date, transgenesis by pronuclear microinjection of promoter-reporter constructs remains the most successful strategy for mapping regulatory elements in genes. The importance of identifying regulatory elements and knowing how genes are controlled is a prerequisite for targeting genes to specific sites, and is of utmost relevance for gene therapy.

3.3. Strategies to Study Gene Function

The ability to express genes in selected cells and tissues has led to even more profound possibilities: transgene products could interfere with specific gene functions or protein interactions in complex systems, and consequences of these alterations could be monitored *(15,16,43,44)*. Through design of a hybrid transgene construct comprised of a tissue-specific promoter linked to a normal or altered gene of interest, the deregulated transgene expression may yield insights into gene and/or tissue function. Several strategies outlined can be used for analysis of gene function using transgenic mice:

3.3.1. Antisense RNA

This strategy involves blocking the expression of an endogenous gene by preventing translation of sense transcripts. In principle, the antisense approach is possible, and has had some success in vitro *(45,46)*, as well as in the generation of transgenic flies *(47)* and frogs *(48)*. In practice, its use has been limited in mammals, although some success has been documented *(49,50)*.

3.3.2. Dominant Interference; Dominant Negative

A more powerful approach involves generating a dominant interference phenotype in transgenic mice, by blocking the function of a gene at the protein

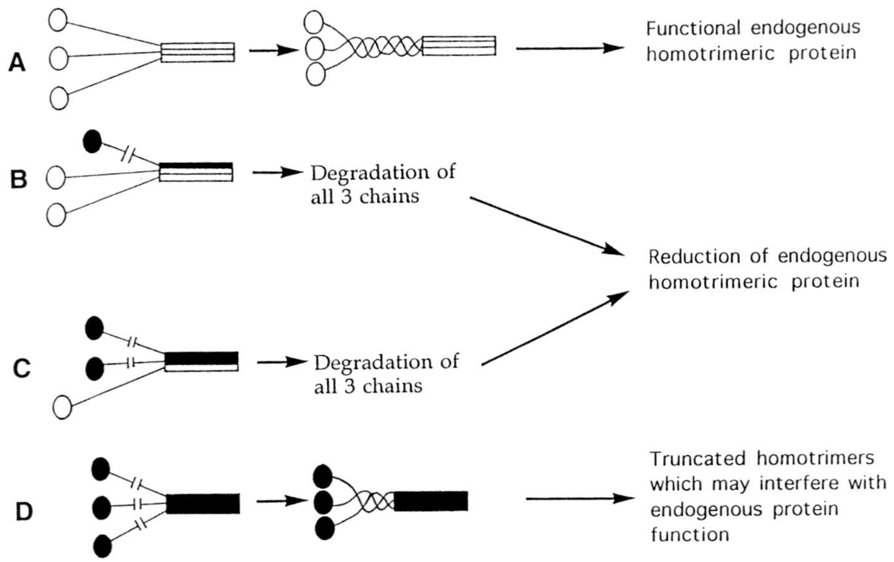

Fig. 4. Schematic representation of dominant interference. Expression of a transgene encoding truncated polypeptides (shaded molecules) results in a competition with the endogenous polypeptides (clear molecules) for binding, followed by the inability of hybrid molecules to form stable trimers **(B** and **C)** unlike endogenous molecules **(A)**. Such a scenario will likely lead to degradation of the hybrid chains through a protein suicide mechanism, leading to a partial or complete loss of function of the endogenous gene. Truncated homotrimers may also persist and interfere with the function of wild-type trimers **(D)**. This may contribute to a loss of function phenotype, but may also result in gain of function. *See* text for discussion. (This diagram is adapted from ref. *70*.)

level through expression of an inhibitory variant of the same protein *(43)*. This approach is particularly effective for multimeric proteins (e.g., collagens), proteins with multiple functional domains or subunits (e.g., gene-regulatory proteins), or enzymes whose activity is limited by substrate availability (Fig. 4). The resultant phenotype is considered dominant, because even very low levels of the inhibitor will have an effect (usually disruptive) on the normal function of the endogenous protein. A dominant negative phenotype may result from a partial or complete loss of function of the endogenous gene product. An excellent example of this approach is provided by work on collagens (*see* references within refs. *51,52*), where the first dominant negative mutation was introduced within the Col1a1 gene *(53)*. Type I collagen, the most abundant structural extracellular matrix protein predominantly found in dense connective tissues, has been associated with the inherited disease osteogenesis imperfecta in humans *(54)*. Transgenic mice bearing single residue substitutions within one type I collagen gene developed a dominant phenotype characteristic of the human disease, and demonstrated that as little as 10% of mutant gene expression was needed to disrupt folding of collagen chains into functional trimers *(53)*.

It is important to realize, however, that although the dominant interference approach is designed to disrupt endogenous gene function at the protein level, overproduction of an inactive or a modified protein can have an opposite effect leading to a new phenotype or a gain of function. It is also noteworthy that many naturally occurring mutations may function by dominant interference, resulting in loss and/or gain of function phenotypes.

3.3.3. Overexpression

The second successful approach involves overexpressing a transgene product in appropriate or inappropriate cells to create an imbalance in the concentration of the correct gene product. Often, this will create a competition situation (as in dominant interference) between the transgene product and the endogenous protein, leading to a dominant negative, and a loss of function phenotype. However, there are also examples where this approach has led to a new phenotype as a result of a gain of function. One example of this approach is provided by the deregulated expression of the proto-oncogene *c-fos* by Wagner and coworkers *(32)*. Expression of *c-fos* in a number of transgenic mouse tissues has resulted in an effect only in bone and the thymic epithelium, identifying the cells within these tissues as the targets for *c-fos* action. Such results from *c-fos* overexpression studies are currently enabling the investigators to unravel the complex pathways leading to oncogenic transformation. An additional dramatic example involves the use of a broad specificity inducible promoter to express the human growth hormone gene in mice *(12)*. The resultant "big mice" demonstrated the role of the growth factor in organismal growth, which has been subsequently used to correct growth deficiency in dwarfed mice *(55)*.

3.3.4. Ablation of Cells

The strategy involving the selective destruction of cell types and tissues is summarized in the review by Hanahan *(15)*. Briefly, the transgene construct is designed to consist of a cell or tissue-specific promoter linked to a toxin gene such as diptheria toxin A or ricin *(56–58)*. Such an approach has potential for addressing questions concerning cell function, lineage, and interactions during development *(59)*. However, the use of toxins in transgenic mice has revealed a problem with penetrance of the transgene, resulting in limited cell death *(56,57)*. A variation of this approach involved the design of a suicide vector containing a "drug-conditional" promoter, whose expression resulted in cell death only when a drug was administered *(60)*. However, one shortcoming of this method is realized when a herpes simplex virus thymidine kinase (HSV-TK) promoter is used; in such a case, cell death occurs only in proliferating cells through the incorporation of the drug gancyclovir, a nucleoside analog, into replicating DNA. Perhaps the recent cloning and characterization of apoptosis genes *(61,62)* may provide the necessary tools for cell- or tissue-specific ablation without the aforementioned limitations.

3.3.5. Insertional Mutagenesis

Insertional mutations arising from the random integration of defined DNA sequences are especially valuable for identifying genes with developmental roles. In this approach, the integrated DNA serves two purposes. First, the transgene disrupts the endogenous gene, leading to a mutant phenotype; second, it acts as a molecular "tag" marking the integration locus. By recloning the integrated sequences, the disrupted endogenous sequences can be recovered. Through random insertional mutagenesis, a number of genes with developmental effects have been cloned *(34,63)*. One of the first examples involves the limb deformity mutation characterized by Leder and coworkers *(64)*. While investigating the role of *c-myc*, several transgenic lines were generated and bred to homozygosity with respect to the inserted gene. In one of these lines, a recessive mutation has resulted in severe dysmorphism in limbs, and has thus provided a link with the control of pattern formation in the developing mammalian embryo. The first mutation arising from retroviral insertional mutagenesis involved the *Mov-13* transgenic line generated by Jaenisch and coworkers *(65,66)*, which likely represents the most thoroughly characterized transgenic mice to date. In these mice, retroviral infection of postimplantation embryos resulted in a single proviral insertion into the first intron of the $\alpha 1(I)$ collagen gene, causing a recessive perinatal lethal phenotype. These mice are continuing to provide data on the role(s) of fibrillar collagens during embryogenesis and postnatal life, on collagen gene regulation, on fibroblast, osteoblast, and odontoblast cell lineages, and on angiogenesis.

A prerequisite to the insertional mutagenesis approach is the demonstration that the mutant phenotype results from transgene disruption of an endogenous gene (*see* Section 3.3.6.), as well as the isolation of the altered gene. One complicating factor in cloning the disrupted gene when insertional mutagenesis results from the pronuclear injection of the transgene, stems from transgene integration in tandemly repeated copies, as well as potential rearrangements of the endogenous gene near the integration sites *(7,14,33)*. To facilitate the screening and isolation of genes involved in morphogenesis, new strategies involving "enhancer traps" and "gene traps" were developed for "tagging" the mutated genes of interest *(63,67–69)*. The transgene constructs used in these strategies both carry a gene for β-galactosidase, and are electroporated into pluripotent ES cells, which are subsequently reintroduced into the blastocyst (*see* Section 5.). As mentioned previously, *lacZ* expression is observed histochemically by staining early embryos with X-gal, a substrate for β-galalactosidase that is converted to a deep blue product. In "enhancer traps," *lacZ* is linked to a weak promoter, and expression is dependent on the vector's integration near an enhancer. In "gene traps," the *lacZ* gene lacks regulatory sequences except for a splice acceptor site; expression is only achieved if the vector integrates into an intron of a cellular gene, and if splicing results in a chimeric mRNA that produces a functional fusion protein. Thus in both cases, the *lacZ* transgene constructs are used to rapidly screen many integration events, and to identify and clone regions of the mouse genome that are active in a temporal and spatial pattern during development *(67)*.

3.3.6. Analysis of Transgenic Mice

The basic requirement in the analysis of transgenic mice is to establish the involvement of the transgene in any newly identified alteration in phenotype. First, to exclude a spontaneous mutation arising coincidentally in the transgenic strain as the cause of the phenotype, the observed murine phenotype and the transgene must be demonstrated as genetically inseparable. This is accomplished by standard genetic crosses and genomic DNA analysis by Southern blotting and/or PCR to identify genotypically positive pups, and to monitor transgene cosegregation with the observed phenotype. Second, to establish whether the phenotype results from insertional mutagenesis or from transgene expression, mice from several transgenic lines (carrying the same transgene) need to be compared based on genotype, phenotype, and transgene expression. To rule out insertional mutagenesis, a minimum of two lines (representing at least two independent microinjection events and transgene insertion sites) must show the same phenotype and express the transgene message/product in a similar temporal pattern. Southern blot analysis of genomic DNA obtained from tail biopsies from mice in these lines should reveal different insertion sites, evidenced by differences in migrations of specific DNA fragments following digestion with the same restriction enzyme (e.g., ref. *70*; Fig. 2A). Southern analysis will also reveal transgene dosage, as well as head-to-tail arrangements. It is important to realize that transgene expression does not necessarily correspond to transgene copy number, but is influenced by the insertion site microenvironment. Furthermore, transgene deletions and rearrangements can occur over a number of generations, and thus may influence transgene expression and the resultant phenotype. These points further underline the importance of initially analyzing mice from several transgenic lines, and then maintaining two-to-three of these lines for characterization.

If more than one transgenic line is not available, or if the phenotype appears to result from insertional mutagenesis, characterization of the transgene insertion site becomes necessary. This is accomplished by first cloning the genomic DNA flanking both sides of the inserted transgene, and later using these clones to isolate the intact gene. As mentioned previously, a significant drawback of gene transfer through pronuclear microinjection stems from transgene integration in multiple head-to-tail concatenated arrangements; these long blocks of tandemly repeated copies often are larger than the capacity of standard cloning vectors, making it difficult to clone the junction sites. Furthermore, complex rearrangements are also known to occur at the integration point, complicating gene isolation. To circumvent this potential problem, a cosmid library and cloning vector may need to be generated. Subsequent mapping of the integration site to a chromosome would permit comparison of this locus to that of other known genes, or mapped mutations. Identification of similar mutant phenotypes near this locus may lead to genetic complementation tests *(64)*, and may provide conclusive proof for insertional mutagenesis.

3.3.7. Summary

3.3.7.1. Advantages of Generating Transgenic Mice Through Pronuclear Microinjection

1. This represents the most successful approach to date for mapping regulatory elements in genes.
2. The design of hybrid promoter-transgene constructs permits expression of exogenous DNA in virtually any site.
3. The dominant interference strategy may provide insights into specific protein action and mechanisms of pathogenesis, and requires only the expression of a mutant gene product rather than the inactivation of an endogenous gene. Furthermore, such mutations may be representative of many heritable disorders.
4. Random insertional mutagenesis may lead to the identification of developmentally active genes.
5. There is no restriction with respect to size or type of DNA, which is microinjected (unlike the case with retroviral vectors and gene targeting through homologous recombination).
6. This represents the only means to date of generating transgenic domestic animals *(18)*.

3.3.7.2. Disadvantages

1. The random introduction of exogenous DNA into the genome may result in an unexpected (and often complex and difficult to interpret) phenotype, independent of the desired effect.
2. Gene integration is not targeted; therefore, it is difficult to predict tissue-specific expression; this is a significant concern for gene therapy.

3. Data relating to gene function may be difficult to interpret; although the transgene construct may be designed to disrupt endogenous gene function, the overproduction of an inactive product may have the opposite effect.
4. The integration site is difficult to clone owing to transgene tandem arrays and gene rearrangements (this is circumvented by using retroviral vectors).
5. Pronuclear microinjections require the availability of expensive microinjection facilities, as well as technical experience unavailable to most laboratories. Thus, government subsidized programs (such as DNX, Princeton, NJ) and university core facilities are beginning to extend these services. Nevertheless, the establishment, characterization, and maintenance of transgenic lines in virus-free animal facilities remains a labor-intensive and expensive process.

4. Transgenesis by Retroviral Infection

4.1. Methodology

One advantage of using viruses (most frequently recombinant retroviral vectors) for transgene introduction either directly into embryos, into ES cells that can then be used to form chimeras, or into multipotent stem cells that can replace an endogenous tissue, is the technical simplicity of the protocol. Briefly, stem cells or embryos at various stages of development are infected at an efficiency approaching 100% by their coculture with cells producing the virus. The efficient infection and expression in a wide variety of cells represent the fundamental advantage of using retroviruses. Furthermore, stable and accurate integration of a single viral transgene copy into the host DNA facilitates the identification of the insertion site. For this reason, viruses are superior agents when genetic tagging of chromosomal loci by insertional mutagenesis or when marking cell lineages in stem cell differentiation and during embryo development is desired *(35,71)*. The disadvantages include size constraint of the retroviral vectors, with the insert size being no larger than ~8 kb of DNA (and in some cases significantly smaller), precluding the expression of many genomic DNAs. Furthermore, infection by retroviruses requires cell replication; therefore, nondividing cells cannot be targets. Relatively low virus titers and low expression of the inserted genes are also problematic, as well as the instability of the retroviral vector structure. Finally, proper functioning of regulatory elements of the inserted DNA may be affected when positioned close to the viral long terminal repeats, altering cell-specific expression. Nevertheless, retroviral infection remains the only means at present through which DNA can be stably introduced into many somatic tissues and multipotent stem cells (*see* Section 6.) *(14,16,18,35,71–74)*, and thus holds great promise as a strategy for gene therapy.

4.2. Summary

4.2.1. Advantages of Generating Transgenic Mice Through Retroviral Infection

1. Simplicity of protocol; no need for embryo microinjection or extensive ES cell screening in culture.
2. Most proliferating cells can be infected.
3. Integration of a single transgene copy occurs in transcriptionally active regions of the genome, and does not include chromosomal rearrangements. This facilitates the identification of the insertion site, and is ideal for tagging chromosomal loci and marking cell lineages.
4. Retroviral infection represents the only means of introducing a transgene into certain somatic cells.

4.2.2. Disadvantages

1. Exogenous DNA is randomly introduced into the genome and may complicate data interpretation.
2. Retroviral vectors have an insert size constraint.
3. Expression of the transgene is often poor.
4. Retroviral vectors are often unstable.
5. The juxtaposition of viral regulatory elements with those of the insert is problematic.

5. Transgenesis by Gene Targeting Through Homologous Recombination in ES Cells

Gene modification through targeting allows the derivation of mice with a predesigned genetic composition. This strategy has a major advantage over other transgenic approaches for the analysis of gene function, because rather than introducing exogenous DNA into the genome, the endogenous gene is replaced with a modified cloned gene during homologous recombination (HR) in cultured ES cells. In principle, any gene can be modified in a defined manner in any species from which ES cells can be obtained, and upon their reintroduction into the animal, the specific effects of the introduced modification can be observed *(20,24,75)*.

The finding that ES cells can link genetic manipulations in vitro to analysis of function in vivo was pivotal in the establishment of gene targeting through HR *(5,20)*. The ES cells are derived from the mouse blastocysts (3 d postcoitus), specifically from the inner cell mass from which the embryo develops. These pluripotent cells can be explanted and maintained as stable diploid cell lines under well-defined conditions *(76)*. A targeting vector containing the desired gene mutation and two selectable markers, the bacterial neomycin-resistance gene (neo) and the Herpes Simplex virus (HSV) thymidine kinase (TK) gene (Fig. 5), can be introduced into ES cells by electroporation or microinjection.

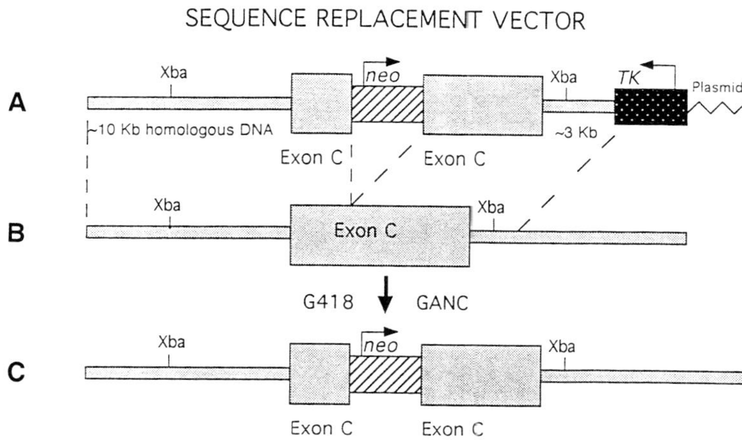

Fig. 5. Homologous recombination through use of a replacement vector. **(A)** Sequence replacement vector containing ~10 kb of homology with the endogenous locus, and ~3 kb 3' of the *neo* insertion, which interrupts the coding sequence within exon C. The genomic sequence is flanked on the 3' end by *TK* sequences. Arrows indicate transcriptional orientations of the *neo* and *TK* promoters; dotted lines indicate the regions of homology within which recombination may occur. **(B)** The endogenous wild-type locus in the region homologous to the replacement vector sequences. **(C)** The predicted structure of the altered endogenous allele following homologous recombination with the replacement vector shown in (A). Through this process, the endogenous sequences are replaced by the vector sequences containing *neo*.

In most cells, this vector will insert randomly into the ES genome; in a few cells however, the introduced DNA will pair with the cognate chromosomal DNA sequence and transfer the mutation into the genome through HR. The frequency of this double crossover event is very low (at best one out of ~3×10^4 ES cells electroporated), and appears to depend on the extent of homology between the exogenous and chromosomal sequences in the cells (for DNA with 2–4 kb of homology, the frequency is one out of ~5×10^7–5×10^6 ES cells electroporated) *(19,20,27)*.

Sequence replacement or insertion targeting vectors can be designed. For gene inactivation, a replacement vector has been more commonly used (Fig. 5). This vector should ideally contain approx 10 kb of DNA homologous to the target gene to increase targeting frequency. A neo gene is inserted, along with its promoter, into an exon of the target gene sequence, thereby serving as a mutagen as well as a selectable marker. A TK gene is also cloned into the vector adjacent to the 5' or 3' region of homology. The vector is linearized outside the region of homology, and HR results in the replacement of the

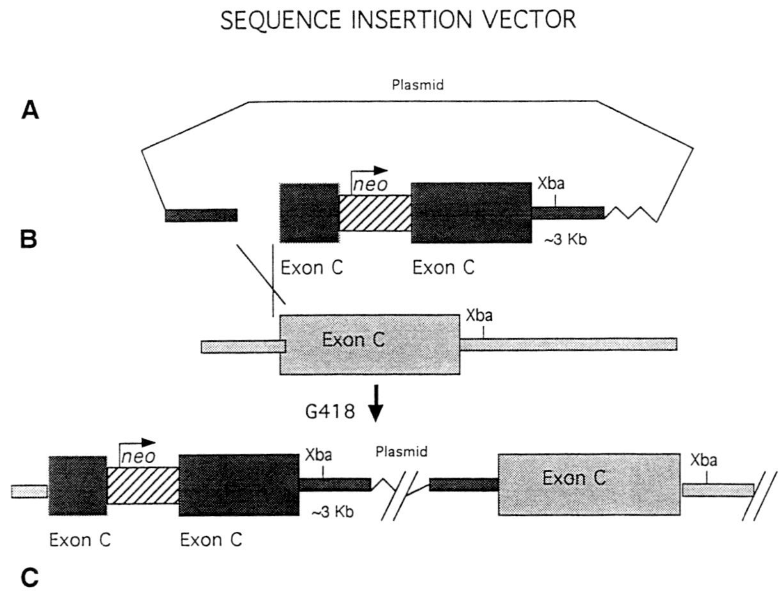

Fig. 6. Homologous recombination through use of an insertional targeting vector. **(A)** The sequence insertion vector, containing the recombinant DNA homologous to the endogenous locus (dark gray), with a *neo* insertion interrupting the coding sequence within exon C. Prior to electroporation, the vector is linearized within the region of homology, and the 5' and 3' ends lie adjacent to one another. **(B)** The endogenous wild-type locus in the region homologous to the insertion vector sequences. **(C)** The predicted structure of the altered endogenous allele following homologous recombination with the insertion vector shown in A. Upon pairing of homologous sequences and recombination, the entire vector is inserted into the endogenous gene. This procedure produces a duplication of the endogenous gene represented in the vector.

endogenous gene with the neo-containing genomic sequence; transfer of TK does not occur, because it lies distal to the region of homology. The insertion vector (Fig. 6) has been used both for gene inactivation and introduction of subtle site-specific mutations into the gene of interest *(77,78)*. Prior to electroporation, the vector is linearized within the region of homology; HR results in the entire vector being incorporated into the endogenous gene, producing a partial duplication of the target sequence. Similar gene targeting frequencies have been reported for both types of vectors *(27)*.

The rare ES cells carrying the targeted mutation are enriched by a positive/negative selection (PNS) procedure in culture *(20)* (Fig. 7). Briefly, ES cells are positively selected in G418-containing medium for clones with insertion of neo in their genome by either homologous or random integration. Negative

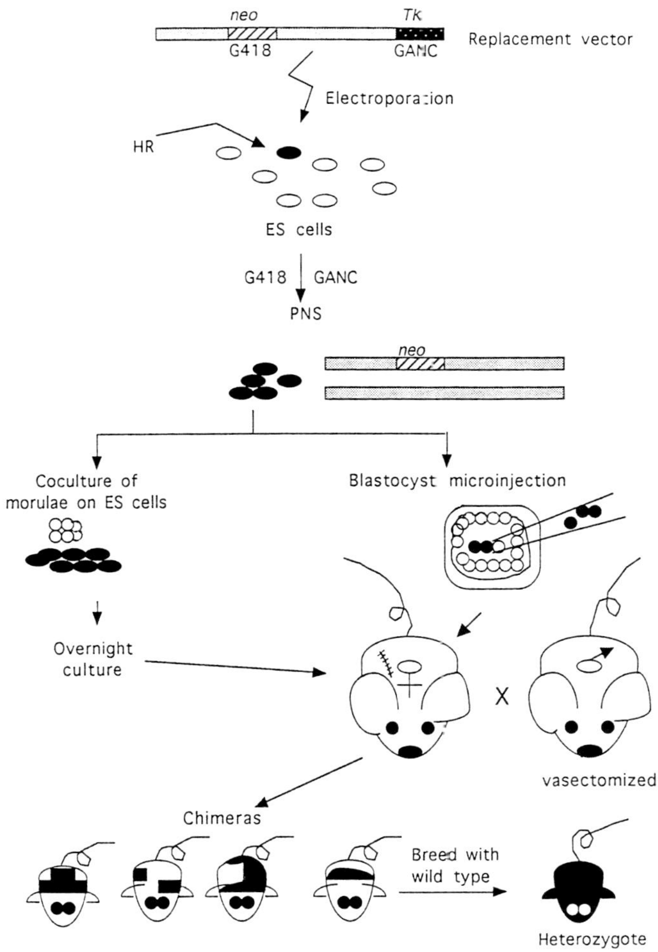

Fig. 7. Generation of germ line chimeras from embryo-derived totipotent stem cells carrying a targeted gene disruption. *See* text for details.

selection in the presence of gancyclovir (GANC) or FIAU selects against all clones with random integrations containing the TK product, thus enriching for cells containing the targeted mutation. The G418/GANC resistant clones are expanded, and screened by PCR and/or Southern blot analysis for the targeted DNA. The appropriate cells are then cloned and maintained as a pure population. Ideally, a minimum of two clones carrying independent HR events are selected for transfer into embryos. This is accomplished either by microinjection into blastocysts, or by mixing with morulas. Blastocyst microinjection *(76)*, which has been the predominant method used to date, requires the availability of microinjection equipment, as well as experience with the technique. On the other hand, the recently described aggregation protocol *(79)* involves coculture of ES cells with morulae, which readily adhere. Using this technique, high success rates were reported (nearly 96% chimeric embryos) for ES cell incorporation in the inner cell mass, where they participate in somatic and germ cell development. The degree of embryo manipulation is also reduced, and the requirement for microinjection equipment is bypassed.

The ES cell-containing embryos are then surgically implanted into the uterus of a surrogate mother, where development proceeds to term. The resultant animal is a chimera, being composed of cells derived from both the donor ES cells and the host blastocyst. This is usually manifested by coat color, since chimeras have patches of both the color of the blastocyst strain, as well as the color of the host. Breeding of chimeras with wild-type mice tests for germ line transmission, and establishes heterozygotes (mice containing one allele with an altered gene, and the other with the wild-type gene). Interbreeding of such heterozygotes generates animals homozygous for the mutation (Fig. 7). Screening and analysis of animals is as described for transgenic mice; however, rather than testing for the presence and expression of the transgene, the absence of wild-type gene expression and function is assayed.

5.1. Transgenesis by Pronuclear Microinjection vs Gene Targeting by Homologous Recombination

Transgenesis (via pronuclear microinjection) and gene targeting are often directed toward different ends. As described previously, when studying function, the former can be used to introduce a foreign gene and to observe either its specific dominant effect on the endogenous gene, or the result of its deregulated expression. The latter strategy involving gene targeting can also be used to generate a gain-of-function scenario through ectopic overexpression; bypassing HR and PNS of the transgene *(32)*. However, its most powerful use involves loss-of-function through inactivation of an endogenous gene. Among the many successful examples, the targeted disruptions of *int-1 (80,81)* and *c-fos (82)* proto-oncogenes demonstrate the power of this technique. The *int-1* is temporally and spatially restricted during CNS development; its inactivation in mice resulted in the inability of portions of the brain to develop, implicating the role of *int-1* in the induction of the mesencephalon and cerebellum. The inactivation of *c-fos* has surprisingly resulted in osteopetrosis in mice. Through use of intricate hematopoietic cell culture and marrow transplantation assays, it was established that this proto-oncogene regulates the osteoclast-macrophage lineage determination, and thus affects bone remodeling *(84,85)*. Among the many successes, however, there are also disappointments. The lack of an identifiable phenotype after gene inactivation has underlined the potential redundancy in gene function in nature. However, it is difficult to accept that certain genes are completely compensated for. It is more likely that we are unable to perform the proper diagnostic experiments (e.g., measure mouse intelligence, adaptability, and so on) to reveal the phenotype in all successful transgenic animal experiments.

5.2. Summary

5.2.1. Advantages of Gene Targeting Through HR

1. The genome is altered through the replacement of the endogenous gene with an altered one, rather than by random insertion of exogenous DNA; thus the precise consequences of mutations can be analyzed.
2. Although this method involves labor-intensive cell culture work, it is technically simpler than pronuclear microinjections.

5.2.2. Disadvantages

1. For increased targeting frequency, ~10 kb of homologous DNA are recommended, preferably from the same strain as the ES cells.
2. Gene inactivation occasionally results in either no detectable phenotype, or a very mild one, and is often considered not informative; caution should be taken in the interpretation of such data.

6. Transgenesis by Manipulation of Multipotent Stem Cells

6.1. Methodology

Strategies used for generating transgenic mice are not limited to experiments directed only at germ line modifications. In specific cases, it is advantageous to modify only certain somatic tissues of the organism via stem cells (e.g., hematopoietic, liver, epithelial, lung, etc.). As cell culture methods develop, it is becoming possible to use the available gene targeting protocols to correct defective genes in the appropriate tissues, leading to somatic gene therapy.

Two multipotent stem cell systems have been shown to be capable of repopulating their specific organs: the hematopoietic *(16,17,62)* and the liver *(5,86)* stem cells. Hematopoietic stem cells are multipotent in that they could regenerate cells of all the lymphoid and myeloid blood cell lineages, as well as produce more stem cells through self renewal. In vivo assays have been established where the regeneration capacity of these stem cells and their ability to maintain a functional hematopoietic system can be tested. These assays involve the removal of bone marrow (which represents the major site of blood production in the adult) from a donor animal, and its transplantation into a recipient host with a compromised or destroyed (by irradiation) hematopoietic system. The donor stem cells could also be modified through retroviral transgene transfer, enabling one to investigate specifically the effects of individual genes on

subsequent expression in adult mice and their offspring. *Proc. Natl. Acad. Sci. USA* **78,** 6376–6380.

12. Palmiter, R. D., Brinster, R. L., Hammer, R. E., Trumbauer, M. E., Rosenfeld, M. J., Birnberg, N. C., and Evans, R. M. (1982) Dramatic growth of mice that develop from eggs microinjected with metallothionein-growth hormone fusion genes. *Nature* **300,** 611–615.
13. Palmiter, R. D. and Brinster, R. L. (1986) Germ-line transformation of mice. *Annu. Rev. Genet.* **20,** 465–499.
14. Jaenisch, R. (1988) Transgenic animals. *Science* **240,** 1468–1474.
15. Hanahan, D. (1989) Transgenic mice as probes into complex systems. *Science* **246,** 1265–1275.
16. Wagner, E. F. and Keller, G. (1992) The introduction of genes into mouse embryos and stem cells, in *Development: The Molecular Genetic Approach* (Russo, V. E. A., Brody, S., Cove, D., and Ottolenghi, S., eds.), Springer-Verlag, Berlin and Heidelberg, pp. 440–458.
17. Wagner, E. F. (1990) On transferring genes into stem cells and mice. *EMBO J.* **9,** 3025–3032.
18. Cameron, E. R., Harvey, M. J. A., and Onions, D. E. (1994) Transgenic science. *Br. Vet. J.* **150,** 9–24.
19. Mansour, S. L., Thomas, K. R., and Capecchi, M. R. (1988) Disruption of the protooncogene int-2 in mouse embryo-derived stem cells: a general strategy for targeting mutations to non-selectable genes. *Nature* **336,** 348–352.
20. Capecchi, M. R. (1989) Altering the genome by homologous recombination. *Science* **244,** 1288–1292.
21. Zijlstra, M. R., Li, E., Sajjadi, F., Subramani, S., and Jaenisch, R. (1989) Germ-line transmission of a disrupted B2 microglobulin gene produced by homologous recombination in embryonic stem cells. *Nature* **342,** 435–438.
22. DeChiara, T. M., Efstratiadis, A., and Robertson, E. J. (1990) A growth-deficiency phenotype in heterozygous mice carrying an insulin-like growth factor II gene disrupted by targeting. *Nature* **345,** 78–80.
23. Koller, B. H., Marruck, P., Kappler, J. W., and Smithies, O. (1990) Normal development of mice deficient in B2M MHC class I proteins, and CD8+ T cells. *Science* **248,** 1227–1230.
24. Fassler, R., Martin, K., Forsberg, E., Litzenburger, T., and Iglesias, A. (1995) Knockout mice: how to make them and why. The immunological approach. *Int. Arch. Allergy Immunol.* **106,** 323–334.
25. Pierce, G. B., Jr. (1975) Teratocarcinoma: introduction and perspectives, in *Teratomas and Differentiation* (Sherman, M. I. and Softer, D., eds.), Academic, New York, pp. 3–12.
26. Dewey, M. J., Martin, D. W., Jr., Martin, G. R., and Mintz, B. (1977) Mosaic mice with teratocarcinoma-derived mutant cells deficient in hypoxanthin phosphoribosyl transferase. *Proc. Natl. Acad. Sci. USA* **74,** 5564–5568.
27. Thomas, K. R. and Capecchi, M. R. (1987) Site-directed mutagenesis by gene targeting in mouse embryo-derived stem cells. *Cell* **51,** 503–512.

28. Brinster, R. L. and Zimmerman, J. W. (1994) Spermatogenesis following male germ cell transplantation. *Proc. Natl. Acad. Sci. USA* **91,** 11,298–11,302.
29. Brinster, R. L. and Avarbock, M. R. (1994) Germline transmission of donor haplotype following spermatogonial transplantation. *Proc. Natl. Acad. Sci. USA* **91,** 11,303–11,307.
30. Dym, M. (1994) Spermatogonial stem cells of the testes. *Proc. Natl. Acad. Sci. USA* **91,** 11,287–11,289.
31. Brinster, R. L., Chen, H. Y., Trumbauer, M. E., Yagle, M. K., and Palmiter, R. D. (1985) Factors affecting the efficiency of introducing foreign DNA into mice by microinjecting eggs. *Proc. Natl. Acad. Sci. USA* **52,** 4438–4442.
32. Grigoriadis, A. E., Wang, Z. Q., and Wagner, E. F. (1993) c-fos oncogene expression in cartilage and bone tissues of transgenic and chimeric mice, in *Cellular and Molecular Biology of Bone* (Noda, M., ed.), Academic, London, pp. 497–537.
33. Palmiter, R. D., Chen, H. Y., and Brinster, R. L. (1982) Differential regulation of metallothionein-thymidine kinase fusion genes in transgenic mice and their offspring. *Cell* **29,** 701–710.
34. Kuehn, M. R. and Stoye, J. P. (1992) Insertional mutagenesis and mouse development, in *Development: The Molecular Genetic Approach* (Russo, V. E. A., Brody, S., Cove, D., and Ottolenghi, S., eds.), Springer-Verlag, Berlin and Heidelberg, pp. 420–439.
35. Soriano, P. and Jaenisch, R. (1986) Retroviruses as probes for mammalian development: allocation of cells to the somatic and germ cell lineages. *Cell* **46,** 19–29.
36. Brinster, R. L., Chen, H. Y., Warren, R., Sarthy, A., and Palmiter, R. D. (1982) Regulation of metallothionein-thymidine kinase fusion plasmids injected into mouse eggs. *Nature* **296,** 39–42.
37. Brinster, R. L. and Palmiter, R. D. (1986) Introduction of genes into the germ line of animals. *Harvey Lect.* **80,** 1–38.
38. Niederreither, K., D'Souza, R. N., and De Crombrugghe, B. (1992) Minimal DNA sequences that control the cell lineage-specific expression of the proα2(1) collagen promoter in transgenic mice. *J. Cell Biol.* **119,** 1361–1370.
39. D'Souza, R. N., Niederreither, K., and De Crombrugghe, B. (1993) Osteoblast- specific expression of the alpha 2(I) collagen promoter in transgenic mice: correlation with the distribution of TGF-beta 1. *J. Bone Min. Res.* **8,** 1127–1136.
40. Cui, C., Wani, M. A., Wight, D., Kopchick, J., and Stambrook, P. J. (1994) Reporter genes in transgenic mice. *Transgenic Res.* **3,** 182–194.
41. Bonnerot, C. and Nicolas, J.-F. (1993) Application of lac Z gene fusions to postimplantation development. *Methods Enzymol.* **225,** 451–472.
42. Jacenko, O., Olsen, B. R., and LuValle, P. (1991) Organization and regulation of collagen genes. *Eukaryotic Gene Expr.* **1,** 327–353.
43. Herskowitz, I. (1987) Functional inactivation of genes by dominant negative mutations. *Nature* **329,** 219–222.
44. Westphal, H. and Gruss, P. (1989) Molecular genetics of development studied in the transgenic mouse. *Ann. Rev. Cell Biol.* **5,** 181–196.

45. Izant, J. G. and Weintraub, H. (1984) Inhibition of thymidine kinase gene expression by anti-sense RNA: a molecular approach to gene analysis. *Cell* **36,** 1007–1015.
46. Izant, J. G. and Weintraub, H. (1985) Constitutive and conditional suppression of exogenous and endogenous genes by anti-sense RNA. *Science* **229,** 345–352.
47. Rosenberg, U. B., Preiss, A., Seifert, H. E., Jackle, H., and Knipple, D. C. (1985) Production of phenocopies by kruppel anti-sense RNA injection into drosophila embryos. *Nature* **313,** 703–706.
48. Melton, D. A. (1985) Injected anti-sense RNAs specifically block messenger RNA translation in vivo. *Proc. Natl. Acad. Sci. USA* **82,** 144–148.
49. Katsuki, M., Sato, M., Kimura, M., Yokoyama, M., Kobayashi, K., and Nomura, T. (1988) Conversion of myelin basic protein antisense cDNA in transgenic mice. *Science* **241,** 593–595.
50. Simons, M., Edelman, E. R., Dekeyser, J. L., Langer, R., and Rosenberg, R. D. (1992) Antisense c-myb oligonucleotides inhibit intimal arterial smooth muscle cell accumulation *in vivo*. *Nature* **359,** 67–70.
51. Jacenko, O., Olsen, B. R., and Warman, M. L. (1994) Of mice and men: heritable skeletal disorders. *Am. J. Hum. Genet.* **54,** 163–168.
52. Jacenko, O. and Olsen, B. R. (1995) Transgenic mouse models in studies of skeletal disorders. *J. Rheumatol.* **22,** 39–41.
53. Stacy, A., Bateman, J., Choi, T., Mascara, T., Cole, W., and Jaenisch, R. (1988) Perinatal lethal osteogenesis imperfecta in transgenic mice bearing an engineered mutant pro-α1 (I) collagen gene. *Nature* **332,** 131–135.
54. Prockop, D. J., Baldwin, C. T., and Constantinou, C. D. (1990) Mutations in type I procollagen genes that cause osteogenesis imperfecta. *Adv. Hum. Genet.* **19,** 105–132.
55. Hammer, R. E., Brinster, R. L., and Rosenfield, M. G. (1985) Expression of human growth hormone-releasing factor in transgenic mice results in increased somatic growth. *Nature* **315,** 413–416.
56. Palmiter, R. D., Behringer, R. R., Quaife, C. J., Maxwell, F., Maxwell, I. H., and Brinster, R. L. (1987) Cell lineage ablation in transgenic mice by cell-specific expression of a toxin gene. *Cell* **50,** 435–443.
57. Behringer, R. R., Mathews, R. D., Palmiter, R. D., and Brinster, R. L. (1988) Dwarf mice produced by genetic ablation of growth hormone-expressing cells. *Genes Dev.* **2,** 453–544.
58. Bruggeman, L. A., Xie, H. X., Brown, K. S., and Yamada, Y. Y. (1991) Developmental regulation for collagen II gene expression in transgenic mice. *Teratol.* **44,** 203–208.
59. Camper, S. A., Saunders, T. L., Kendall, S. K., Keri, R. A., Seasholtz, A. F., Gordon, D. F., Birkmeier, T. S., Keegan, C. E., Karolyi, I. J., Roller, M. L., Burrows, H. L., Samuelson, L. C. (1995) Implementing transgenic and embryonic stem cell technology to study gene expression, cell-cell interactions and gene function. *Biol. Reprod.* **52,** 246–257.
60. Borrelli, E., Heyman, R. A., Arias, C., Sawchenko, P. E., and Evans, R. M. (1989) Transgenic mice with inducible dwarfism. *Nature* **339,** 538–544.

61. Vaux, D. L. (1993) Toward an understanding of the molecular mechanisms of physiological cell death. *Proc. Natl. Acad. Sci. USA* **90,** 786–789.
62. Weissman, I. L. (1994) Developmental switches in the immune system. *Cell* **76,** 207–218.
63. Rossant, J. and Hopkins, N. (1992) Of fin and fur: mutational analysis of vertebrate embryonic development. *Genes Dev.* **6,** 1–13.
64. Woychik, R. B., Stewart, T. A., Davis, L. G., D'Eustachio, P., and Leder, P. (1985) An inherited limb deformity created by insertional mutagenesis in a transgenic mouse. *Nature* **318,** 36–40.
65. Schnieke, A., Harbers, K., and Jaenisch, R. (1984) Insertion of retrovirus into the first intron of αl(I) collagen gene. *Nature* **304,** 315–320.
66. Hartung, S., Jaenisch, R., and Breindl, M. (1986) Retrovirus insertion inactivates mouse αl(I) collagen gene by blocking initiation of transcription. *Nature* **320,** 365,366.
67. Gossler, A., Joyner, A. L., Rossant, J., and Skarnes, W. C. (1989) Mouse embryonic stem cells and reporter constructs to detect developmentally regulated genes. *Science* **244,** 463–465.
68. Rijkers, T., Peetz, A., and Ruther, U. (1994) Insertional mutagenesis in transgenic mice. *Transgenic Res.* **3,** 203–215.
69. Gossen, J. A., de Leeuw, W. J., and Vijg, J. (1994) LacZ transgenic mouse models: their application in genetic toxicology. *Mutation Res.* **307,** 451–459.
70. Jacenko, O., LuValle, P., and Olsen, B. R. (1993) Spondylometaphyseal dysplasia in mice carrying a dominant negative mutation in a matrix protein specific for cartilage-to-bone transition. *Nature* **365,** 56–61.
71. Keller, G., Paige, C., Gilboa, E., and Wagner, E. F. (1985) Expression of a foreign gene in myeloid and lymphoid cells derived from multipotent haematopoietic precursors. *Nature* **318,** 149–154.
72. van der Putten, H., Botteri, F. M., Miller, A. D., Rosenfeld, M. G., Fan, H., Evans, R. M., and Verma, I. M. (1985) Efficient insertion of genes into the mouse germ line via retroviral vectors. *Proc. Natl. Acad. Sci. USA* **82,** 6148–6152.
73. Stewart, C. L., Ruther, U., Garber, C., Vanek, M., and Wagner, E. F. (1986) The expression of retroviral vectors in murine stem cells and transgenic mice. *J. Embryol. Exp. Morphol.* **97 Suppl.,** 263–275.
74. Temin, H. M. (1989) Retrovirus vectors: promise and reality. *Science* **246,** 983.
75. Capecchi, M. (1990) Tapping the cellular phone. *Nature* **344,** 105.
76. Robertson, E. J. (1987) Embryo-derived stem cell llines, in *Teratocarcinomas and Embryonic Stem Cells: A Practical Approach* (Robertson, E. J., ed.), IRL, Oxford, pp. 71–112.
77. Valancius, V. and Smithies, O. (1991) Testing an "in-out" target procedure for making subtle genomic modifications in mouse embryonic stem cells. *Mol. Cell. Biol.* **11,** 1402–1408.
78. Hasty, P., Ramirez-Solis, R., Krumlauf, R., and Bradley, A. (1991) Introduction of a subtle mutation into the Hox-2.6 locus in embryonic stem cells. *Nature* **350,** 243–246.

79. Wood, S. A., Allen, N. D., Rossant, J., Auerbach, A., and Nagy, A. (1993) Non-injection methods for the production of embryonic stem cell-embryo chimeras. *Nature* **365,** 87–89.
80. Thomas, K. R. and Capecchi, M. R. (1990) Targeted disruption of the murine *int-1* proto-oncogene resulting in severe abnormalities in midbrain and cerebellar development. *Nature* **346,** 847–850.
81. McMahon, A. P. and Bradley, A. (1990) The Wnt-1 (int-1) proto-oncogene is required for development of a large region of the mouse brain. *Cell* **62,** 1073–1085.
82. Wang, Z. Q., Ovitt, C., Grigoriadis, A. E., Steinlein, U. M., Ruther, U., and Wagner, E. F. (1992) Bone and hematopoietic defects in mice lacking c-fos. *Nature* **360,** 742–745.
83. Johnson, R. S., Spiegelman, B. M., and Papaloannou, V. (1992) Pleiotropic effects of a null mutation in the c-fos proto-oncogene. *Cell* **71,** 577–586.
84. Grigoriadis, A. E., Wang, Z. Q., Cecchini, M. G., Hofstetter, W., Felix, R., Fleisch, H. A., and Wagner, E. F. (1994) c-Fos: a key regulator of osteoclast-macrophage lineage determination and bone remodeling. *Science* **266,** 443–448.
85. Jacenko, O. (1995) c-fos and bone loss: a regulator of osteoclast lineage determination. *BioEssays* **17,** 277–281.
86. Rhim, J. A., Sandgren, E. P., Degen, J. L., Palmiter, D. R. D., and Brinster, R. L. (1994) Replacement of diseased mouse liver by hepatic cell transplantation. *Science* **263,** 1149–1151.
87. Heckel, J. L., Sandgren, E. P., Degen, J. L., Palmiter, R. D., and Brinster, R. L. (1990) Neonatal bleeding in transgenic mice expressing urokinase-type plasminogen activator. *Cell* **62,** 447–456.

32

Use of Gonadal Primordial Germ Cells (PGCs) as Tools for Gene Transfer in Chickens

Nathalie Allioli, Gérard Verdier, and Catherine Legras

1. Introduction

As a result of morphological and developmental characteristics of the early avian embryo, several approaches, taking place at early stages, have been investigated to introduce foreign DNA into the avian genome. They include:

1. Insemination with sperm cells to which foreign DNA was coated *(1,2)*;
2. Transfection of DNA into follicles before fertilization *(3)*;
3. Direct injection of naked DNA into newly fertilized ova or into germinal disk *(4–7)*;
4. Microinjection of replication-defective retroviral vectors *(8–10)* into unincubated embryo blastoderm; and
5. Retroviral infection of primordial germ cells, PGCs *(11,12)*.

Chicken PGCs are located in the germinal crescent from the primitive streak stage. At the first heartbeats they are transported by the bloodstream to the germinal ridge where they settle and rapidly proliferate. Consequently, chicken PGCs can be isolated from the germinal crescent of a 1-d-old embryo *(12,13)*, from embryonic blood of 2.5-d-old embryos *(12,14)*, or from undifferentiated gonads of 5- to 7-d-old embryos *(15,16)*.

Use of gonadal PGCs as a tool for gene transfer offers several advantages over the two other sources:

1. The number and proportion of PGCs are higher in the gonad (5%) than in the germinal crescent (1.5%) or in blood (0.003%); and
2. Gonadal PGCs actively proliferate in vivo once they have reached the gonads; thus, we can expect them to retain this proliferation capacity in vitro.

In this chapter, we describe a protocol for using gonadal PGCs as a tool for gene transfer. This protocol (Fig. 1) involves:

From: *Methods in Molecular Biology, vol. 62: Recombinant Gene Expression Protocols*
Edited by: R. Tuan Humana Press Inc., Totowa, NJ

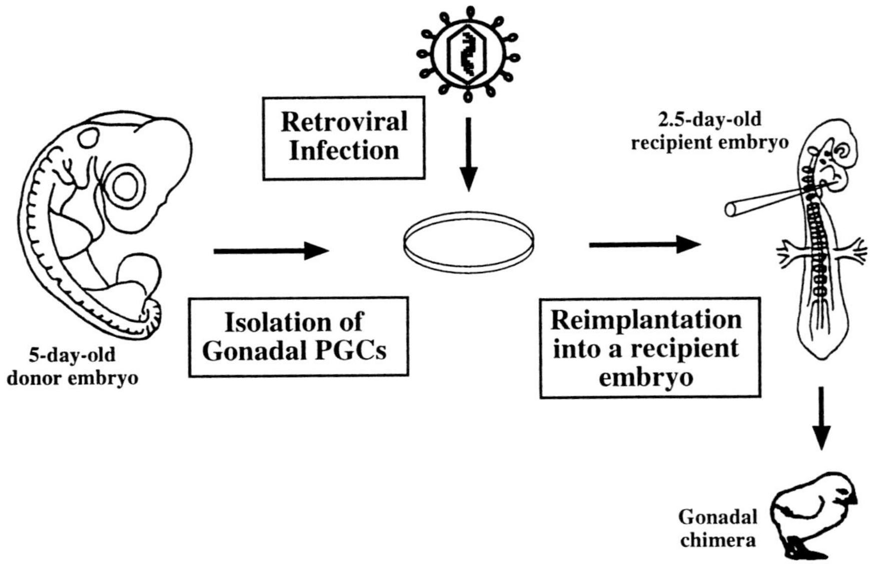

Fig. 1. Strategy of use of gonadal PGCs for chicken gene transfer.

1. Isolation of gonadal PGCs from gonads of 5-d-old embryo;
2. Their proliferation in vitro;
3. The introduction of exogenous DNA information using retroviral infection; and
4. The reintroduction of these genetically modified gonadal PGCs into 2.5-d-old recipient embryos, by intracardiac injection.

2. Materials

2.1. Collection and Culture of PGCs

1. Embryos can be obtained from Brown Leghorn eggs (C/E), from the Station de Pathologie Aviaire INRA, Centre de Recherche de Tours-Nouzilly, France.
2. 10X Phosphate buffered saline (PBS) (Gibco-BRL, Paisley, Scotland) without $Ca^{2+}Mg^{2+}$.
3. 10X Trypsin solution (0.25%; Gibco-BRL).
4. 10X solution stock of Erythrosin B: dissolve 3.6 mg of Erythrosin B in 10X PBS and store at –20°C.
5. Schiff's reagent (Sigma, St. Louis, MO): pararosaniline HCl 1%, sodium bisulfate 4% in hydrochloric acid 0.25 M. Store at 4°C.
6. Periodic acid (Sigma). Store at 4°C.
7. Sodium metabisulfite (Sigma). Store at 4°C.
8. 37% Paraformaldehyde solution (Sigma). Store at room temperature.
9. PGC culture medium: DMEM/F12 (1:1) medium (Gibco-BRL) containing 10% fetal calf serum, 1% penicillin/streptomycin (5000 IU/mL, 5000 UG/mL,

Gibco-BRL) and 1% glutamine (200 m*M*, Gibco-BRL). Can be stored at 4°C for 1 mo, in dark.

10. All instruments, fine microsurgery forceps and dissecting Pascheff wolff scissors, are sterilized for 90 min at 200°C.
11. Tissue culture chambers: Lab-tek Chamber slides (Nunc, Naperville, IL).

2.2. Retroviral Vector Infection and Detection of PGCs Expressing the LacZ Marker Gene

1. Viral stock production: Replication-defective retroviral vector used for infection is the NLB vector carrying both the *Neo* selectable and *Escherichia coli LacZ* marker genes driven by *cis*-acting regulatory sequences from the Rous Associated Virus type 2 *(17)*. This vector is produced as helper-free avian leukosis virus particles from SEnta packaging cell line expressing E (NLB/E) subgroup specificity *(18)*.
2. Packaging cell line culture medium: SEnta packaging cell line is grown in F10 medium composed as follow: 10% Ham's F10 10 X (Gibco-BRL), 10% tryptose phosphate broth 29 g/L (Difco, Detroit, MI), 5% newborn calf serum (Gibco-BRL), 1% chicken serum (Gibco-BRL), 0.18% sodium bicarbonate (Gibco-BRL), 1% fungizone 2.5 mg/mL (Gibco-BRL), 1% penicillin 5000 IU/mL-streptomycine 5000 UG/mL (Gibco-BRL). SEnta packaging cell line is grown under appropriate selective conditions: 50 µg/mL of hygromycin B (Boehringer Mannheim, Mannheim, Germany) and Phleomycin (Cayla), 200 µg/mL of G418 (Gibco-BRL). These three drugs are dissolved in PBS and sterilized by filtration through 0.22-µm-pore-size filters.
3. X-Gal solution: 2.5 m*M* $MgCl_2$, 5 m*M* potassium hexacyanoferrate III, 5 m*M* potassium hexacyanoferrate II, 3% dimethylsulfoxyde (DMSO), 1 mg/mL in DMSO of X-Gal (4-chloro-5-bromo-3 indolyl-β-D-galactoside, Boehringer Mannheim), 1X PBS. Make fresh. Light sensitive.
4. Collagen stock solution: 0.6% of collagen (calf skin acid soluble collagen; Sigma) in acetic acid 100 m*M*. Make fresh. Coat slides by immersing them into the stock solution of collagen diluted to 1/100^e in PGC medium, for a night. Then rinse slides in ultrapure water.
5. Paraformaldehyde solution: 4% in 1X PBS. Make fresh.

2.3. Injection of Gonadal PGC into the Blood Circulation of Recipient Embryos

1. The glass microcapillaries used for injection were 20–30 mm inner diameter at the tip. They were produced with a microcapillary puller (Sutter Instrument Co.). To avoid the loss of PGCs by adherence to the inner wall of the microcapillary, the inner surface of the glass microcapillaries was coated with a solution of silicon (dimethyl dichlorosilane, Sigma), heated for 1 min in a microwave oven, and rinsed with ultrapure water. For injection, microcapillary was plugged into a 10-mL hypodermic syringe.
2. 10X PBS (Gibco-BRL) with Ca^{2+} and Mg^{2+}.

3. Methods

3.1. Collection of PGCs

1. Incubate donor eggs at 37.5°C in an incubator (60% hygrometry) for 5 d to obtain embryos of stages 27–28, staging by Hamburger and Hamilton *(19)*.
2. Swab the outer surface of the shell of each donor egg with 70% ethanol and cut out a round window (approx 1.5 cm in diameter) in sterile conditions.
3. Dissect five donor embryos to recover the gonads. Gonads from stages 27–28 embryos *(19)* appear as a slight swelling of the peritoneal epithelium located between the mesonephros and the dorsal mesentery.
4. Collect 10 gonads from such donor embryos in PGC medium and then rinse the gonads in PBS, before dissociation.
5. For dissociation, transfer the gonads into 0.02% EDTA solution in PBS for 15 min at 37°C. Centrifuge the resulting cell suspension at 1500 rpm for 5 min and resuspend the pellet in 1 mL of fresh PGC medium (*see* Note 1). In these conditions, a cell suspension in which PGCs are released as single cells and as small aggregates (5–10 cells) can be obtained. These small aggregates could be broken up by gentle pipeting. This gonadal cell suspension contains approximately 8000 PGCs (about 800 PGCs released per embryo) that represent 5% of the total cell suspension.

3.2. PGC Identification and Viability

1. To determine the number of PGCs, air-dry an aliquot of this gonadal cell suspension on a slide and fix the cells in a 4% paraformaldehyde solution in PBS for 10 min at 4°C. Then, immerse the slides for 10 min in periodic acid, rinse in distilled water for 10 min, and stain in Schiff's reagent for 10 min. Wash slides in three successive 1 min baths of sodium metabisulfite M/20 and finally in running tap water for 5 min. After staining, PGCs are clearly recognizable from gonadal somatic cells by their large size (20 µm), their voluminous and eccentrically located nucleus, and their cytoplasmic magenta staining owing to the abundant cytoplasmic deposits of glycogen.
2. To determine the viability of PGCs, stain an aliquot of the gonadal cell suspension using Erythrosin B, at a final concentration of 0.36 mg/mL, as an exclusion test. Score viable PGCs with a Malassez hemocytometer. The percentage of viable PGCs is always greater than 90%.

3.3. PGC Culture and Determination of the Propitious Time for Retroviral Infection

1. Plate 250 mL of this gonadal cell suspension at 5×10^2–10^3 PGCs/tissue culture chamber. Incubate cell cultures at 37°C in 95% air and 5% CO_2. Somatic gonadal cells adhere on plastic, spread, and form mainly a layer of fibroblasts, whereas PGCs do not adhere to the plastic culture chamber. PGCs appear as small colonies by aggregation of 5–10 PGCs, which adhere on top of this fibroblast feeder layer.

2. To determine the propitious time for retroviral infection, i.e., when PGCs actively proliferate, the number of PGCs might be counted each day (*see* Note 2). In our culture conditions, a drastic decrease in the number of PGCs to only 10% of the initial number is observed, after 1 d in culture. At d 2, a fourfold increase in the number of PGCs is observed. By contrast, after 2 d in culture, PGCs do not proliferate *(15)*. Therefore, the propitious time for allowing retroviral infection corresponds to d 2, and 200–400 PGCs can be obtained at this period.

3.4. Retroviral Infection

1. Packaging cell line culture conditions: Maintain the SEnta packaging cell line at 37°C in 5% CO_2. When cells are subconfluent, rinse cultures in PBS and incubate cells in 0,025% trypsin solution in PBS for 3 min at 37°C. Add cold F10 medium to stop the trypsin reaction and centrifuge the cell suspension at 460*g* for 5 min. Resuspend the pellet in F10 medium and plate cells at a density of 5×10^6 cells per 100-mm dish.
2. Viral stock collection: Recover NLB helper-free virus stocks from subconfluent SEnta packaging cell line in 5 mL of fresh Optimem I medium (Gibco-BRL) after 12–16 h. Remove cell debris by centrifugation (10 min, 1850*g*, 4°C). Concentrate the supernatant containing the virus stock by ultracentrifugation, for 30 min at 140,000*g* and 4°C, in 50.2 TI Beckman rotors. Resuspend sedimented viral particles in 1/100 of initial volume in Optimen I medium. Store the resulting concentrated helper-free virus stocks at –80°C before use. Titers of these concentrated helper-free NLB/E vector viruses are ranging from 10^5 and 10^6 FFU^{Lac}/mL.
3. Gonadal cell suspensions are infected, at d 2 (i.e., during the proliferation period) by helper-free NLB/E retroviral vector with a multiplicity of 1–10 particles/cell. For this purpose, add 100 µL of NLB/E viral stocks (containing 10^4 to 10^5 particles), 24 h after seeding, to gonadal cell suspensions (containing 200–400 PGCs and 10^4 somatic gonadal cells).
4. To check the efficiency of infection, recover an aliquot of this infected cell suspension, 24 h after infection and stain this sample by X-Gal. For this, cytocentrifuge 100 µL of gonadal cell culture at 200*g* for 5 min on a collagen-coated slide. Fix the cells in a solution of 4% paraformaldehyde in PBS for 10 min at 4°C, and rinse in PBS for 10 min at 4°C. Then, incubate slides in X-Gal solution at 37°C for 5 h and rinse in distilled water. After X-Gal staining, cells expressing the *LacZ* gene present a cytoplasmic blue coloration. We generally obtain, 50–80% of PGCs express the *LacZ* gene (*see* Note 3).

3.5. Injection of Gonadal PGCs into the Blood Circulation of Recipient Embryos

1. Incubate recipient eggs at 37.5°C for 2.5 d, reaching stages 17–19 *(19)*.
2. Swab the surface of the shell of each recipient egg with 70% ethanol.
3. Expose the recipient embryos by removing a small piece of shell and determine their stage of development, according to the shape of the wing-buds. Remove the vitelline membranes and the amnion from the injection site.

4. Four hours after retroviral infection, recover the infected gonadal cell suspension, by successive pipeting, and centrifuge (460*g*, 5 min) to pellet cells. Resuspend this pellet in DMEM/F12 medium to obtain a final concentration of 100 PGCs/μL.
5. Inject 1 μL of this gonadal cell suspension into the heart of recipient embryos. Seal the window in the shell with transparent adhesive tape and subsequently incubate the operated embryos normally.

4. Notes

1. A good PGC suspension can be also obtained by incubating gonads with trypsin (0.025% in PBS) for 10 min at 37°C.
2. To estimate the proportion of dividing PGCs in culture, an immunocytochemical BrdU labeling test could be performed on an aliquot of the PGC cultures. For that, incubate this aliquot with BrdU (10 m*M* in fresh PGC medium), for 1 h at 37°C in 5% CO_2. Wash cells twice in PBS, resuspend cells in PBS-5% BSA and centrifuge cells onto a collagen-coated glass slide with a cytocentrifuge (200*g*, 5 min). Realize the following steps as recommended by the supplier (Boehringer Mannheim Biochemical). PGCs which have incorporated BrdU into DNA could be visualized by light microscopy. Under our culture conditions, we found that approximately, 10–20% of PGCs were dividing during the second day of culture. **Note:** *BrdU is classified as highly toxic. It is a mutational compound and there is danger of cumulative effects. Avoid contact with skin, do not swallow, and follow good laboratory practices.*
3. We describe in this protocol a procedure of infection using subgroup E viral particles (NLB/E). However, we have investigated the PGC susceptibility to other viral envelope subgroups. For this, the NLB vector was packaged as subgroup A, using Isolde packaging cell line *(20)*, and as subgroups B and C, using Haïdee-PhE**B** and Haïdee-PhE**C** packaging cell lines, respectively *(18)*. PGCs are found susceptible to these three other ALV subgroups. Comparatively to the efficiency of infection of subgroup E, we found a similar efficiency of infection with subgroup C, but a lower one with subgroup A, and intermediate values with subgroup B. Therefore, we prefer to use the subgroup E rather the subgroup C because the infection is restricted to PGCs, owing to the resistance of Brown Leghorn chicken embryo fibroblasts to subgroup E.

Acknowledgments

The authors would like to thank their colleagues, Corinne Ronfort for helpful discussions, and in particular Jean-Luc Thomas for technical comments. They also thank Thierry Jaffredo and Françoise Dieterlen-Lievre (from Laboratoire d'Embryologie Moléculaire et Cellulaire, Nogent sur Marne, France) for the demonstration of the intracardiac injection method. This work was supported by research grants from the Commission of European Communities, the Centre National de la Recherche Scientifique, the Institut National de la Recherche Agronomique. We also thank the Ministère de la Recherche et de l'Espace for fellowship (Nathalie Allioli).

References

1. Gavora, J. S., Benkel, B., Sasada, H., Cantwell, W. J., Fiser, P., Teather, R. M., Nagai, J., and Sabour, M. P. (1991) An attempt at sperm-mediated gene transfer in mice and chickens. *Can. J. Anim. Sci.* **71,** 287–291.
2. Rottman, O. J., Antes, R., Höfer, P., and Maierhofer, G. (1992) Liposome mediated gene transfer via spermatozoa into avian egg cells. *J. Anim. Breed. Genet.* **109,** 64–70.
3. Perry, M., Morrice, D., Hettle, S., and Sang, H. (1991) Expression of exogenous DNA during the early development of the chick embryo. *Roux's Arch.* **179,** 85–110.
4. Naito, M., Agata, K., Otsuka, K., Kino, K., Ohta, M., Hirose, K., Perry, M. M., and Eguchi, G (1991) Embryonic expression of the β-actin-lacZ hybrid gene injected into the fertilized ovum of the domestic fowl. *Int. J. Dev. Biol.* **35,** 69–75.
5. Sang, H. and Perry, M. M. (1989) Episomal replication of cloned DNA injected into the fertilized ovum of the hen, gallus domesticus. *Mol. Rep. Dev.* **1,** 98–106.
6. Shuman, R. M. (1986) Gene transfer in avian species using retrovirus vectors. 7th European Poultry Conference, (Paris, Ed. World's Poultry Science Association) **1,** 24–37.
7. Love, J., Gribbin, C., Mather, C., and Sang, H. (1994) Transgenic birds by DNA microinjection. *Biotechnology* **12,** 60–63.
8. Bosselman, R. A., Hsu, R. Y., Briskin, M. J., Boggs, T., Hu, S., Nicolson, M., Souza, L. M., Schultz, J. A., Rishell, W., and Stewart, R. G. (1990) Transmission of exogenous genes into the chicken. *J. Reprod. Fert.* **41,** 183–195.
9. Bosselman, A. R., Hu, R. Y., Boggs, T., Hu, S., Bruszewski J., Ou, S., Kozar, L., Martin, F., Green, C., Jacobsen, F., Nicolson, M., Schultz, J.A., Semon, K. M., Rishell, W., and Stewart, R. G. (1989) Germline transmission of exogenous genes in chicken. *Science* **243,** 533–535.
10. Thoraval, P., Afanassieff, M., Cosset, F. L., Lasserre, F., Verdier, G., and Dambrine, G. (1995) Germ line transmission of exogenous genes in chicken by use of helper free ecotropic avian leukosis viruses based vectors. *Transgenic Research* **4,** 369–376.
11. Simkiss, K., Rowlett, K., Bumstead, N., and Freeman, B. M. (1989) Transfer of primordial germ cell DNA between embryos. *Protoplasma* **151,** 164–166.
12. Vick, L., Li, Y., and Simkiss, K. (1993) Transgenic birds from transformed primordial germ cells. *Proc. R. Soc. Lond. B.* **251,** 179–182.
13. Wentworth, B. C., Tsai, F., Hallett, J. H., Gonzales, D. S., and Rajcic-Spasojevic, G. (1989) Manipulation of avian primordial germ cells and gonadal differentiation. *Poultry Sc.* **68,** 999–1010.
14. Kuwana, T. and Fujimoto, T. (1984) Locomotion and scanning electron microscopic observations of primordial germ cells from the embryonic chick blood in vitro. *Anat. Rec.* **209,** 337–343.

15. Allioli, N., Thomas, J. L., Chebloune, Y., Nigon, V. M., Verdier, G., and Legras, C. (1994) Use of retroviral vectors to introduce and express the beta galactosidase marker gene in cultured chicken primordial germ cells. *Devel. Biol.* **165,** 30–37.
16. Guido, T. C., Abott, U. K., and Carrey, J. R. (1991) The interspecific transfer of avian primordial germ cells. Manipulation of the avian genome. Keystone symposia on Molecular and Cellular Biology. *J. Cell Biochem.* **supp15E,** 204.
17. Cosset, F. L., Legras, C., Thomas, J. L., Molina, R. M., Chebloune, Y., Faure, C., Nigon, V. M., and Verdier, G. (1991) Improvement of avian leukosis virus (ALV)-based retroviruses vectors by using different cis-acting sequences from ALVs. *J. Virol.* **65,** 3388–3394.
18. Cosset, F. L., Ronfort, C., Molina, R. M., Flamant, F., Drynda, A., Benchaïbi, M., Valsesia, S., Nigon, V. M., and Verdier, G. (1992) Packaging cells for avian leukosis virus-based vectors with various host ranges. *J. Virol.* **66,** 5671–5676.
19. Hamburger, V. and Hamilton, H. L. (1951) A series of normal stages in the development of the chick embryo. *J. Morphol.* **88,** 49–92
20. Cosset, F. L., Legras, C., Chebloune, Y., Savatier, P., Thoraval, P., Thomas, J. L., Samarut, J., Nigon, V. M., and Verdier, G. (1990) A new avian leukosis virus (ALV)-based packaging cell line using two separate complementing helper genomes. *J. Virol.* **64,** 1070–1078.

33

Strategies for the Production of Transgenic Chickens

Robert J. Etches and Ann M. Verrinder Gibbins

1. Introduction

Insertion of foreign DNA into the germline requires access to the chromosomes of cells that give rise to sperm or eggs. From a theoretical point of view, the newly fertilized zygote is the most appropriate cell for genetic alteration because any modification of the genome will be inherited by every cell that subsequently develops within the organism. In mammals, random insertion of genes via injection of DNA into the pronucleus is the most cost-effective route for modification of the genome (*see* Chapter 31). If specific, targeted modifications to the genome are desired, however, an elaborate culture system is required that simultaneously supports proliferation of pluripotential embryonic stem cells and facilitates selection of cells that bear the specific genetic change (*see* Chapter 31). In the chicken, injection into the pronuclei of the newly fertilized egg is not feasible for two reasons. Firstly, the oocyte is located on the surface of the large and fragile yolk and techniques to identify, manipulate, and inject the female pronucleus have not yet been developed. Secondly, fertilization in avian species is associated with the entry of several sperm into the egg *(1)* and it is not yet possible to distinguish individual male pronuclei or to determine which male pronucleus will unite with the female pronucleus. Strategies for gaining access to the germline of chickens, therefore, have arisen through an improved understanding of the morphological and physiological events that occur between fertilization and differentiation of the primordial germ cell lineage.

2. Early Embryonic Development and Derivation of the Germline

Fertilization of the oocyte occurs in the infundibulum of the oviduct within 15 min after ovulation (Figs. 1 and 2). Several sperm usually enter the oocyte

From: *Methods in Molecular Biology, vol. 62: Recombinant Gene Expression Protocols*
Edited by: R. Tuan Humana Press Inc., Totowa, NJ

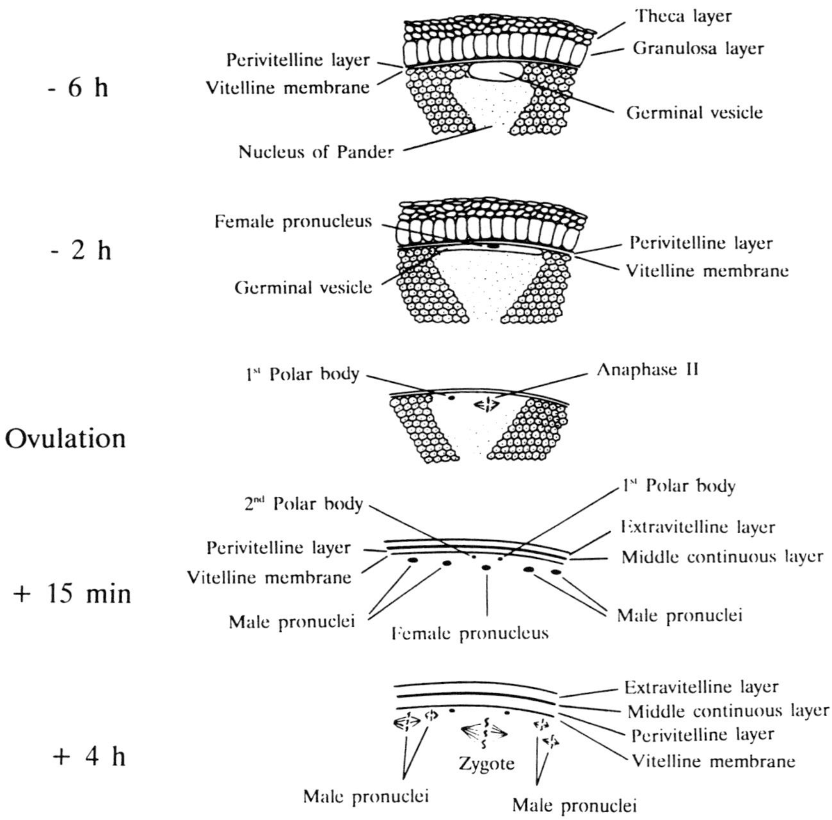

Fig. 1. The nuclear changes that take place in the ovum of the hen during the 6 h before and the 4 h after ovulation. The germinal vesicle begins to disintegrate following the initiation of the preovulatory surge of luteinizing hormone 6 h prior to ovulation. At ovulation, the first polar body is extruded and the second meiotic division begins. Within 15 min, fertilization occurs, the second meiotic division is completed, and several sperm enter the egg. As the egg traverses the infundibulum, two additional layers are added to the yolk membrane. During the fourth hour after ovulation, the zygote enters the first mitotic division, and the remaining male pronuclei undergo a synchronous mitotic division. Reprinted with permission from ref. *45*.

before the secretions of the tubular glands and the epithelial cells in the infundibulum cover the sperm receptors located on the apex of protrusions of the vitelline membrane through the perivitelline layer *(2)*. During the next 4 h, the egg traverses the magnum, which secretes albumen onto the yolk; and the isthmus, which secretes the egg shell membranes onto the albumen. As the egg leaves the isthmus and enters the shell gland, one male pronucleus and the

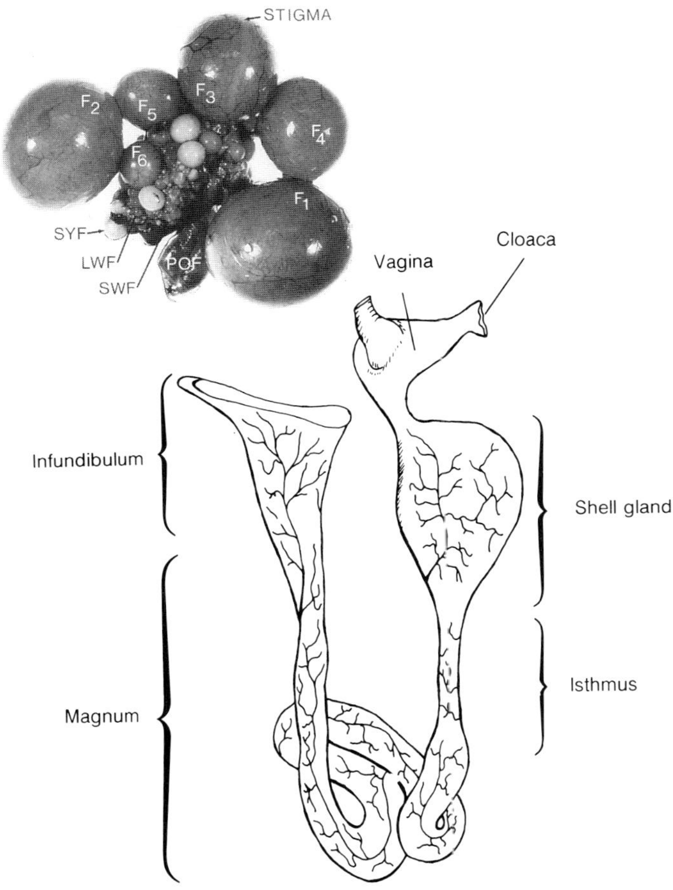

Fig. 2. The anatomy of the reproductive system of the hen illustrating the ovary, infundibulum, magnum, isthmus, shell gland, and vagina. Ova are released from the largest (F_1) ovarian follicle into the body cavity or the infundibulum directly and are subsequently coated with layers of membrane, albumen and shell as the ovum is propelled down the tract. Reprinted with permission from ref. *45*.

female pronucleus fuse and the first mitotic division is initiated. The egg is retained for 18–22 h in the shell gland as crystals of calcium carbonate are deposited to form the shell. When the egg is expelled from the shell gland at the time of oviposition, the embryo contains approx 30,000–60,000 cells. Assuming that the rate of cell division in the embryo is constant during shell deposition, the cell cycle is completed 16× or approximately once every 75 min between fertilization and oviposition. The morphology of the embryo during this period has been described by Eyal-Giladi and Kochav *(3)* and Watt et

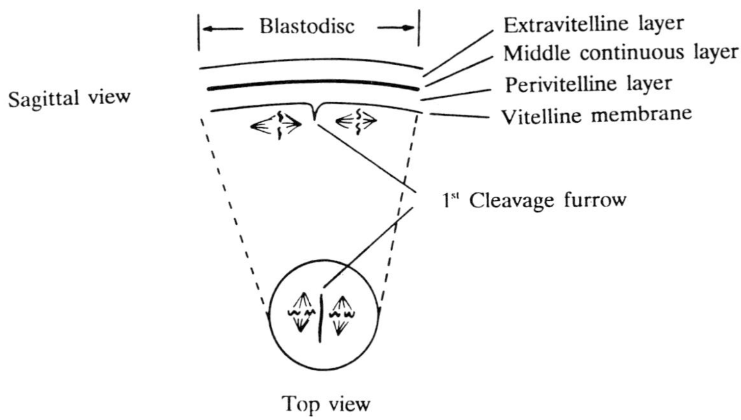

Fig. 3. The formation of the first cleavage furrow in the embryo of the domestic hen viewed from above and in cross-section. Reprinted with permission from ref. *45*.

al. *(4)*. The first cleavage planes in the vitelline membrane are perpendicular to the surface and penetrate 10–20 μm into the ovum (Fig. 3). The blastomeres that are formed at each of the early cell divisions are open to the underlying and peripheral yolk. After six or seven cell divisions, a cleavage plane forms that is parallel to the surface of the ovum and the first blastomeres that are completely surrounded by a cell membrane are formed. When several thousand cells are present in the embryo, the apical surface of the embryo that faces the albumen begins to pump fluid to the dorsal surface that faces the underlying yolk. As fluid is pumped through the dorsal surface of the embryo, a subgerminal cavity begins to form. During the last few hours before oviposition, the embryo expands into a radially symmetric dome of cells (blastomeres) overlying the subgerminal cavity (Fig. 4) and, at the time of lay, the embryo has been designated as a stage X embryo by Eyal-Giladi and Kochav *(4)*.

The blastomeres in a stage X (E-G & K) embryo at oviposition are morphologically similar (Fig. 5). Morphological similarity, however, does not necessarily imply a lack of differentiation and it is not yet resolved if some or all of these cells are pluripotential, multipotential, or committed to germline or somatic fates. It is clear, however, that there is only one somatic lineage, since the primitive streak has not formed and gastrulation has not occurred at this stage of development.

After oviposition, embryonic development ceases until the egg is incubated. After incubation for 18 h at 37.5°C and 50% relative humidity, the primitive streak has formed and the primordial germ cells are located in extraembryonic tissue anterior to the embryo (Fig. 6). At this time, the primordial germ cells

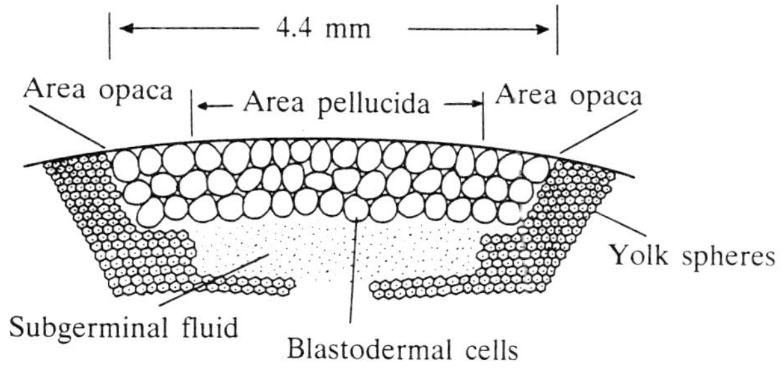

Fig. 4. The structure of an embryo at the time of laying illustrated as a cross-section through the embryo perpendicular to the surface of the yolk. At this time, the embryo contains 30,000–60,000 cells and is referred to as a stage X (E-G & K) embryo. Reprinted with permission from ref. *45*.

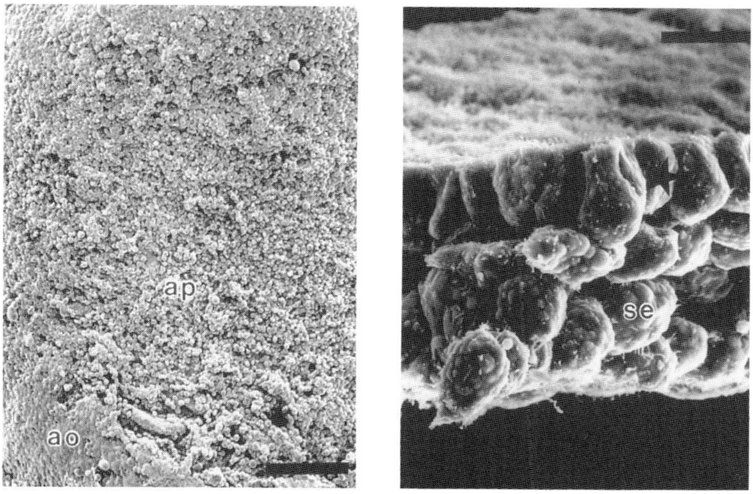

Fig. 5. Left panel: A scanning electron micrograph of the blastomeres of a stage X embryo viewed from the dorsal surface showing the area opaca (AO) and the area pellucida (AP); scale bar = 300 μm. Right panel: a transverse view of the central region of a stage X embryo revealing an upper layer of epithelial-like cells (star) and cells in a subepithelial position (se); scale bar = 20 μm. Reprinted with permission from ref. *4*.

can be recognized as large, periodic-acid Schiff's reagent-positive cells *(5)*, but there is little understanding of the morphology and movements of the primordial germ cells as they diverge from the undifferentiated dome of cells that develop while the egg is in the reproductive tract into the highly specialized cells committed to the germline. Subsequently, the primordial germ cells enter

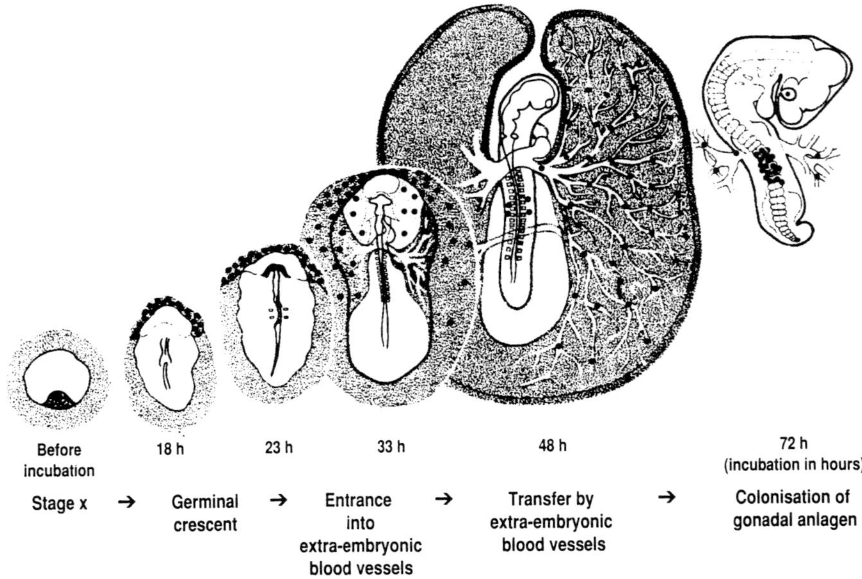

Fig. 6. A summary of the migration of primordial germ cells (PGC) in the chick embryo. These cells may begin to differentiate in the stage X (E-G & K) embryo but do not have their characteristic morphology until after the primitive streak has formed. At this time, the cells are recognized by their location anterior to the embryo in the germinal crescent, their large size, and their characteristic glycogen-laden cytoplasm. During the next 18 h, the PGC will move into the lateral extraembryonic membranes and enter the developing vascular system. By 48 h of incubation, the PGC have started to migrate through the vascular system and take up residence in the future gonad. By the end of the third day of incubation, the PGC have concluded their migration and continue to develop in the ovary and testis in association with the derivatives of the embryonic mesoderm. Reprinted with permission from ref. *46*.

the vascular system and are transported to the germinal ridge where they colonize the developing gonad. In male chickens, the primordial germ cells differentiate into spermatogonia within the testis, whereas in female chickens, the primordial germ cells differentiate into oogonia within the ovary.

3. Insertion of Genetic Information into the Newly Fertilized Zygote

The newly fertilized zygote recovered from an egg in the magnum contains several male pronuclei and a female pronucleus (Fig. 1). The newly fertilized zygote recovered at this stage can be incubated to term in three successive surrogate systems *(6–9)* (Fig. 7). During the first 24 h, the ovum is submerged

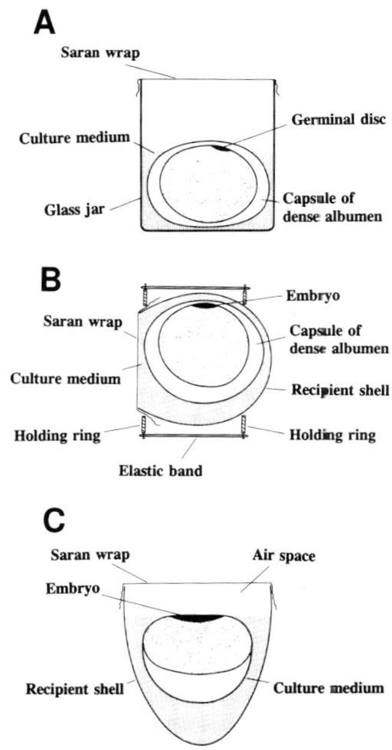

Fig. 7. Three surrogate culture systems are used to provide the appropriate environment to support development of a chicken embryo from syngamy to "hatching." **(A)** The culture system used for the first 24 h after removal of the newly fertilized ovum from the reproductive tract. The ova were placed in 60-mL glass jars and medium was added to a level in line with the germinal disc, but below the upper surface of the albumen capsule. **(B)** During the next 3 d, the embryos were incubated in surrogate shells prepared by removing the sharp end of the shell. Air was excluded from the culture system and the open end of the egg was sealed with cling film. **(C)** Throughout the remainder of development, the embryos were incubated in surrogate shells prepared by removing the blunt end of an egg that was approx 15–18 mL larger than the preceding surrogate eggshell. The aperture was covered with cling film, which was removed at the end of development as the chick "hatched." Reprinted with permission from ref. *6*.

in fluid that is exposed to atmospheric gas exchange. The embryo is then transferred to a surrogate shell, which was prepared by removing the pointed end of the egg and discarding the contents. The open end of surrogate shell is sealed with a plastic film after carefully excluding air bubbles by completely filling the surrogate shell with thin albumen. The embryo develops throughout the next 4 d in the surrogate shell which restricts gaseous transfer between the embryo and ambient air. Between the fourth and sixth day, the embryo is transferred to a larger surrogate shell from which the blunt end has been removed.

After the embryo is inserted into the second surrogate shell, the open end is sealed with cling film. On d 22 of incubation, the cling film is removed and the chick emerges from the shell.

DNA has been injected into newly fertilized zygotes which have subsequently been nurtured in the surrogate systems described above. This approach has produced a transgenic chicken *(10)*, but several similar attempts have failed to yield birds in which the injected DNA is integrated into the somatic or germline lineages *(11–14)*. Embryos and chicks produced by direct injection of DNA are mosaics comprised of genetically modified and nonmodified cells and the incorporation of the genetic modification in the germline is a rare event *(10)*. Direct injection of DNA into the cytoplasm of the newly fertilized egg, therefore, is a route through which heritable modifications can be introduced into the genome, but the yield of transgenic chickens is low. In part, transgenic birds are produced infrequently by this approach because injected DNA is incorporated only infrequently into the chromosomes of the newly fertilized egg and the blastomeres of the early embryo. The technology is also constrained by the expense of obtaining newly fertilized eggs from donor hens, by the technical skill required to manipulate the embryo through the three surrogate culture systems, and by mortality of embryos during development in vitro.

4. Insertion of Genetic Information into Primordial Germ Cells

Introduction of genetic information that can be inherited in a Mendelian fashion can theoretically be accomplished by directing genetic modifications to the primordial germ cell lineage at any point between differentiation from the somatic tissues and the onset of meiosis. At about 18 h of incubation, primordial germ cells can be recognized in the germinal crescent as large, periodic-acid, Schiff's reagent positive cells. During the next few days, the primordial germ cells migrate through the vascular system and settle in the developing gonad (Fig. 6). In females, the primordial germ cells undergo several mitotic divisions during embryonic development to produce oogonia. Within a few days before or after hatching, the oogonia enter meiotic prophase and are arrested for months or years at this stage until meiosis is initiated by the preovulatory luteinizing hormone (LH) surge 6 h before ovulation. In males, the primordial germ cells differentiate into spermatogonia, which serve as a diploid stem cell population for spermatocytes throughout adult life.

Primordial germ cells have been isolated from blood collected from stage 13–15 *(15)* embryos and transferred to the vasculature of recipient embryos at a slightly later stage of development. Donor primordial germ cells colonized the germline of the recipients, yielding germline chimeras that produced offspring derived from both the donor and recipient embryos *(16,17)*. Similarly, primordial germ cells have been collected from the germinal crescent of stage 5–11 (H & H) embryos and injected into the vasculature of stage 15 (H & H) recipients to yield germline chimeras that produced fertile gametes front both the donor-derived and recipient-derived lineages within the germline *(18)*. These experiments clearly demonstrated that if genetic modifications were introduced into donor primordial germ cells, they could be transmitted to the next generation.

Genetic modifications have been introduced into the genome of primordial germ cells by retroviral infection of a vector (NLB) containing the long terminal repeat of the Rous associated virus (RAV-2), a gene encoding neomycin resistance, and *lacZ* which encodes the bacterial enzyme, β-galactosidase. Transfer of primordial germ cells that were exposed to NLB into recipient embryos produced two roosters that transmitted the vector to the next generation *(19)* demonstrating that genetic modification of primordial germ cells provides a route to the germline. Nonretroviral constructs containing the gene encoding β-galactosidase have also been introduced into primordial germ cells by injecting a mixture of the lipofection reagent, DOTAP™, and *lacZ* into the vasculature of stage 10–15 (H & H) embryos *(20)*. Approximately 0.1–1.4% of the primordial germ cells that were migrating in the vasculature expressed *lacZ* by stage 24–27 (H & H) indicating that transfection of primordial germ cells using lipofection reagents is possible. The low yield of genetically modified cells, however, is likely to restrict application of this approach. Genetic modifications have also been introduced into primordial germ cells in the germinal crescent by firing tungsten particles coated with DNA from a ballistic device *(21)*. When DNA was extracted from sperm produced by roosters derived from embryos that had been shot with DNA-coated tungsten particles, the foreign sequences were preset The genetic modification was inherited by the offspring of these roosters, but subsequently disappeared. It is possible, therefore, that the foreign sequences persisted episomally in the embryos that were treated and in their offspring *(21)*.

5. Insertion of Genetic Information into Blastodermal Cells

Access to the germline can be achieved in stage X (E-G & K) embryos contained in newly laid eggs because the precursors of the primordial germ cells are among the 30,000–60,000 cells at this stage of development. Since a precise description of the fate of individual cells within the early embryo is not available, it is not known if the precursors of the primordial germ cells in a stage X (E-G & K) embryo are pluripotential stem cells or if they are already committed to the germline. In the absence of information regarding the pluripotentiality, commitments and fate of individual cells within stage X (E-G & K) embryos, the number of cells that can carry a genetic modification to the next generation is also unknown. Despite the lack of a precise understanding of the mechanism through which cells in the stage X (E-G & K) embryo give rise to the germline, genetic modifications have been produced in blastoderms in newly laid eggs using retroviral vectors and lipofection protocols that introduce the genetic modification into an array of blastodermal cells including at least some cells destined to enter the germline.

Stage X (E-G & K) embryos have been infected with both replication competent *(22–25)* and replication defective retroviruses *(26–28)* to produce transgenic birds that transmitted the genetic modification to their offspring. To date, retroviral infection remains the best documented method for making transgenic chickens, but the technology is fraught with disadvantages. First, the proportion of embryos from virus-infected eggs that transmit the genetic modification to their offspring is relatively low. Second, hundreds or thousands of eggs must usually be inoculated and a similar number of offspring must be examined for the presence of the transgene to identify a genetically modified chicken. Third, replication competent vectors introduce chronic viremia, whereas replication defective viruses are difficult to propagate at high titer. Fourth, the size of the gene inserted into the viral vector is limited to approx 2000 base-pairs so that many genes of interest cannot be accommodated. Retroviral vectors remain very attractive, however, because they ensure that a single copy of the foreign DNA is transferred immediately into the chicken genome, whereas all other forms of transfection introduce DNA into the cytoplasm. Translocation of DNA from the cytoplasm into the nucleus occurs frequently and yields transient expression of the plasmid DNA, but stable integration of the genetic modification into the genome occurs in only a small proportion of the cells receiving the construct *(29)*. Some of the problems associated with retroviral infection may be circumvented by cotransfecting two plasmid sequences, one of which encodes the gene of interest but lacks some of the endogenous viral genes required for replication, whereas the other encodes the missing retroviral genes. This procedure, which is termed virofection, has the potential to yield a high proportion of genetically modified blastodermal cells, which will give rise to genetically modified cells within the germline without producing infectious viruses *(30)*.

If the cells within a stage X (E-G & K) embryo that are destined for the germline could be identified and cultured in vitro, an ideal scenario for introducing genetic modifications into the germline could be achieved. Under these conditions, cells containing the genetic modification would be selected in vitro by incorporating genes conferring antibiotic resistance and/or the ability to grow in selective media before the cells were introduced into a recipient embryo in which they would colonize the germ line (Fig. 8). Although all of the technologies that are required to support this ideal system have not been realized, the major components have been developed and preliminary data indicate that the remaining components can be achieved.

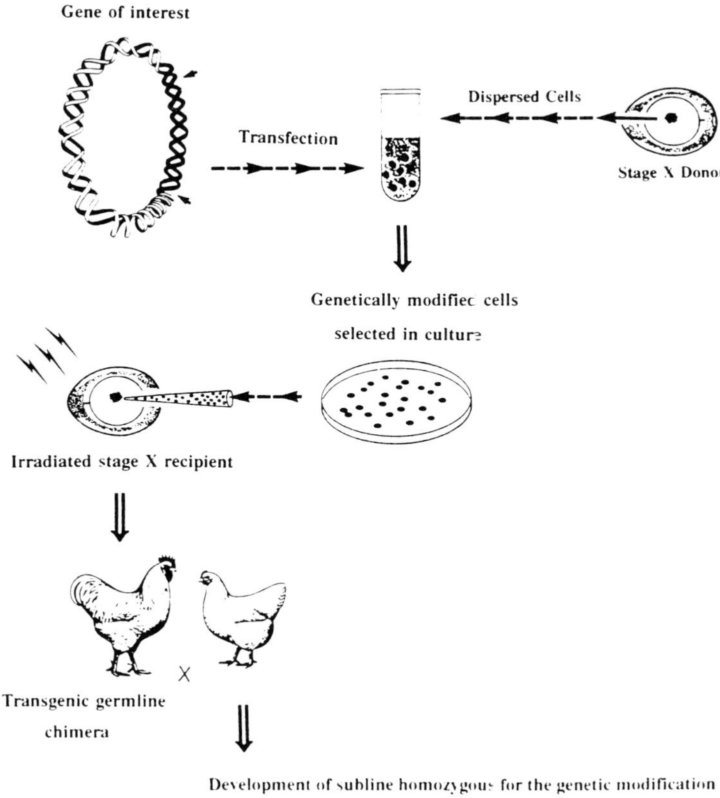

Fig. 8. A strategy to introduce genetic material into the genome of chickens. Cells from a stage X (E-G & K) embryo are removed and dispersed. The dispersed cells are transfected with a gene of interest and genetically modified cells are selected in culture. The genetically modified cells are injected into a recipient embryo that is at the same stage of development as the donor embryo and a chimera is produced. Development of the recipient embryos is compromised by exposure to irradiation prior to injection of donor cells. If genetically modified cells are incorporated into the gonad of the recipient, the genetic modification will be passed to the next generation in a classical Mendelian pattern, and a new line of chickens bearing the trait can be produced. Reprinted with permission from ref. *37*.

The ability of blastodermal cells from stage X (E-G & K) embryos to enter the germ line of recipient embryos when injected at the same stage of development was first demonstrated by Petitte et al. *(31)*. Although these initial experiments demonstrated that somatic and germ line chimeras were formed under these conditions, the rate of colonization of donor-derived cells in the germline of chimeras was very low. In part, the low frequency of germ line chimerism was the consequence of injecting several hundred donor cells into an uncompromised recipient embryo containing several thousand cells. The ratio of donor:recipient cells was increased by exposing the recipient embryo to sufficient γ-irradiation to arrest, but not stop, embryonic development before the donor cells were introduced *(32)*. Exposure to approx 600 rad of γ-irradiation delays development of the recipient embryo or chimera for about 24 h while the nonirradiated donor cells continue to proliferate, presumably at a rate that reflects the very short cell cycle of blastodermal cells up to stage X (E-G & K) *(see above)*. By compromising development of the recipient embryo, the rate of germline chimerism was increased from a rare event to approx 50% of the somatic chimeras that were raised to sexual maturity *(32,33)*.

One of the advantages of using blastodermal cells as a route to the germline is that the contributions of donor cells to somatic tissues can be used as an indicator of the potential of a chimera to transmit the genetic modification to its offspring before the germline can produce functional gametes. For example, the likelihood that White Leghorn embryos into which Barred Plymouth Rock donor cells have been injected will contain Barred Rock-derived cells can be assessed at hatch by the color of the down. The absence of black down is a good indicator (but not a guarantee) that the chick will not contain donor-derived cells in the germline. Conversely, about 50–60% of embryos that exhibit somatic chimerism in plumage color will produce donor-derived offspring. However, the extent of feather pigmentation, which can range from a few feathers to a completely Barred Rock plumage pattern, has no relationship to the extent of germline chimerism, which can range from a fraction of 1–100% (Fig. 9).

Development of chimeras during the first 4 d after donor cells are introduced proceeds normally, but subsequent development of the chorioallantois around the hole in the shell through which the donor cells were introduced into the recipient embryo is impaired. Although this impairment does not induce gross abnormalities in the extraembryonic membranes, only 5–15% of the potentially chimeric embryos completed development and hatched *(32)*. To increase the proportion of embryos injected with donor cells that survive development and hatch, the chimeras are routinely transferred into surrogate shells (*see* Fig. 7) on the fourth day of development *(34)*. Whereas the physiological and developmental processes that are ameliorated by removing the embryo from its original shell are not completely understood, the survival of injected embryos to hatching has consistently ranged from 30–40% when the last 17 d of development are completed in a surrogate shell.

The utilization of chimeras as intermediates in the production of transgenic chickens requires the ability to introduce foreign DNA into stage X (E-G & K) blastodermal cells, and still retain the ability of the cells to form germline chimeras. The *lacZ* gene has been introduced into chicken blastodermal cells using lipofection reagents and β-galactosidase has been expressed by these cells

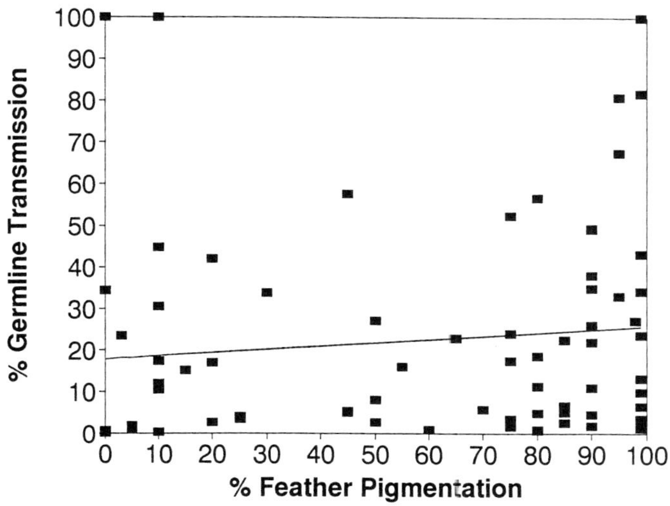

Fig. 9. The relationship between donor-derived feather pigmentation and the production of donor-derived gametes in chimeras made by injecting blastodermal cells from Barred Plymouth Rocks into irradiated White Leghorn recipients. The correlation coefficient between somatic and germline chimerism was 0.12, which was not statistically significant. Note that three germline chimeras have been observed that produced 0.2, 35, and 100% donor-derived offspring and exhibited no evidence of somatic chimerism.

(29). When cells transfected with *lac*Z have been introduced into recipient embryos, β-galactosidase has been revealed in extraembryonic tissues, and in ectoderm, mesoderm, and endoderm of the resulting chimera *(29,35)*. Amplification of DNA from semen produced by a germline chimeric male injected with donor cells that were transfected with *lacZ* has indicated that the exogenous sequence is contained within the germline (unpublished data) and the ability of this male to transmit *lac*Z to his offspring is currently being evaluated.

Exogenous *lac*Z sequences have also been identified in blood DNA from chimeras produced by injecting FACS-sorted cells that were transfected with *lac*Z (36).

A strategic advantage of using blastodermal cells for gaining access to the germline is the potential for deriving either embryonic stem cells or manipulating primary cultures of cells that retain the ability to enter the germline. If in vitro conditions could be developed that supported the proliferation of genetically modified cells, the population of cells that is injected into recipient embryos could be selected to facilitate the growth of cells in which targeted mutations have been introduced. Preliminary reports suggest that it is possible to obtain germline transmission from donor-derived blastodermal cells that have been maintained in culture for 48 h *(37)*, but these procedures require refinement before they can be applied routinely. Recent results have demonstrated that specific modifications to the genome can be induced by homologous recombination of blastodermal cells *(38)*. If these data can be extended to produce targeted genetic modification in cells that can enter the germline, it will be possible to make specific modifications to the genome of chickens.

6. Insertion of DNA into Spermatogonia and Spermatozoa

Accessing the germline after sexual differentiation has received very little attention in any species, but recent evidence has demonstrated that this is a feasible approach in mice *(34,35)*. The seminiferous tubules contain a population of spermatogonia, which proliferate to repopulate the periphery of the tubule with spermatogonial stem cells and produce functional spermatozoa through spermatogenesis. When donor spermatogonia are transferred to recipient testes, donor stem cells colonize the recipient seminiferous tubules and give rise to donor-derived offspring *(39,40)*. If methods for the culture of spermatogonia can be developed, this approach has the potential to provide an alternative route through which DNA can be introduced into the germline. In mice, the transfer of spermatogonia can be accomplished between lines that are immunotolerant, but this strategy is not universally available when applied to other species. General applicability of this approach, therefore, may rely on the ability to control immunological reactions against donor-derived antigens.

Since the sperm cell enters the ovum at fertilization, several attempts have been made to introduce DNA into the genome by attaching marker genes to sperm cells. In most cases these attempts were unsuccessful *(41–43)*. A proportion of the second and third generation descendants of embryos fertilized with sperm that had been incubated with a very low density lipoprotein-chloramphenicol acetyl transferase (VLDLCAT) fusion gene contained sequences that could be amplified by the polymerase chain reaction *(44)*. However, only portions of the VLDLCAT gene could be recovered in genomic DNA from the offspring and the ratio of offspring that inherited the putative VLDLCAT sequence was not consistent with the expectations of Mendelian inheritance. Additional data are required, therefore, to demonstrate the effectiveness of this approach.

7. Anticipated Utility of Transgenic Chickens

The application of transgenic technology to poultry will probably be exercised in commercial poultry breeding, in biomedical research, and in the production of pharmaceutically important proteins. In poultry breeding, the technology is likely to find application in the development of strains of chickens that possess resistance to some of the diseases that are currently controlled by vaccination and/or medication. If the production of transgenic birds can be made into a routine procedure, the chicken may become the preferred model for the acquisition of a molecular understanding of development. In the future it may be possible to combine the detailed morphological descriptions of avian development with a genetic and molecular analysis by creating precise experimental models via transgenic technology. In the production of pharmaceuticals, the chicken may find application because novel products could be deposited in, and subsequently harvested from the egg. For example, sequences encoding the production of novel immunoglobulins could be transferred to chimeras that would serve as founders of strains depositing the product in eggs. Since approx 100 mg of immunoglobulins are contained within each egg, the potential utility of transgenic chickens for immunoglobulin production is significant.

References

1. Perry, M. M. (1987). Nuclear events from fertilization to the early cleavage stages in the domestic fowl *(Gallus domesticus)*. *J. Anat.* **150,** 99–109.
2. Bakst, M. R. (1978). Scanning electron microscopy of the vitelline membrane of the hen ovum. *J. Reprod. Fert.* **52,** 361–364.
3. Eyal-Giladi, H. and Kochav, S. (1976) From cleavage to primitive streak formation: a complementary normal table and a new look at the first stage of the development of the chick. I. General morphology. *Dev. Biol.* **49,** 321–337.
4. Watt, J. M., Pettite, J. M., and Etches, R. J. (1993) Early development of the chick embryo. *J. Morph.* **214,** 1–18.
5. Meyer, D. B. (1964). The migration of primordial germ cells in the chicken. *Devel. Biol.* **10,** 154–190.
6. Perry, M. M. (1988) A complete culture system for the chick embryo. *Nature* **331,** 70–72.
7. Naito, M. and Perry, M. M. (1989) Development in culture of the chick embryo from cleavage to hatch. *Brit. Poultry Sci.* **30,** 251–256.
8. Naito, M., Nirasawa, K., and Oishi, T. (1990) Development in culture of the chick embryo from fertilized ovum to hatching. *J. Exp. Zool.* **254,** 322–326.
9. Naito, M., Nirasawa, K., and Oishi, T. (1995) An *in vitro* culture method for chick embryos obtained from the anterior portion of the magnum of the oviduct. *Brit. Poultry Sci.* **36,** 161–164.
10. Love, J. Gribbin, C., Mather, C., and Sang, H. (1994) Transgenic birds by DNA microinjection. *Biotechnology* **12,** 60–63.
11. Sang, H. H. and Perry, M. M. (1989) Episomal replication of cloned DNA injected into the fertilized ovum of the hen, *Gallus domesticus*. *Mol. Reprod. Devel.* **1,** 98–106.

12. Perry, M. M., Morrice, D., Hettle, S., and Sang, H. (1991) Expression of exogenous DNA during the early development of the chick embryo. *Roux's Arch. Devel. Biol.* **200,** 312–319.
13. Naito, M., Agata, K., Otsuka, K., Kino, K., Ohta, M., Hirose, K., Perry, M. M., and Eguchi, G. (1991) Embryonic expression of β-actin-lacZ hybrid gene injected into the fertilized ovum of the domestic fowl. *Int. J. Dev. Biol.* **35,** 69–75.
14. Naito, M., Sasaki, E., Ohtaki, M., and Sakurai, M. (1994a) Introduction of exogenous DNA into somatic and germ cells of chickens by microinjection into the germinal disc of fertilized ova. *Mol. Reprod. Devel.* **37,** 167–171.
15. Hamburger, V. and Hamilton, H. L. (1951). A series of normal stages of development of the chick embryo. *J. Morph.* **88,** 49–92.
16. Tajima, A., Naito, M., Yasuda, Y., and Kuwana, T. (1993) Production of germ line chimera by transfer of primordial germ cells in the domestic chicken *(Gallus domesticus)*. *Theriogenology* **40,** 509–519.
17. Naito, M., Tajima., A., Yasuda, Y., and Kuwana, T. (1994b) Production of germline chimeric chickens with high transmission rate of donor-derived gametes, produced by transfer of primordial germ cells. *Mol. Reprod. Devel.* **39,** 153–161.
18. Vick, L., Luke, G., and Simkiss, K. (1993a) Germ-line chimeras can produce both strains of fowl with high efficiency after partial sterilization. *J. Reprod. Fert.* **98,** 637–641.
19. Vick, L., Li, Y., and Simkiss, K. (1993b) Transgenic birds from transformed primordial germ cells. *Proc. Roy. Soc., Lond., B.* **251,** 179–182.
20. Watanabe, M., Naito, M., Sasaki, E., Sakuai, M., Kuwana, T., and Oishi, T. (1994) Liposome mediated DNA transfer into chicken primordial germ cells in vivo. *Mol. Reprod. Devel.* **38,** 268–274.
21. Li, Y., Benham, J., and Simkiss, K. (1995) Ballistic transfection of avian primordial germ cell *in ovo*. *Transgenic Res.* **4,** 26–29.
22. Salter, D. W., Smith, E. J., Hughes, S. H., Wright, S. E., Fadly, A. M., Witter, R. L., and Crittenden, L. B. (1986) Gene insertion into the chicken germ line by retroviruses. *Poultry Sci.* **65,** 1445–1458.
23. Salter, D. W., Smith, E. J., Hughes, S. H., Wright, S. E., and Crittenden, L. B. (1987) Transgenic chickens: Insertions of retroviral genes into the chicken germ line. *Virology* **157,** 236–240.
24. Chen, H. Y., Garber, E. A., Mills, E., Smith, J., Kopchick, J. J., DiLella, A. G., and Smith, R. G. (1990) Vectors, promoters, and expression of genes in chick embryos. *J. Reprod. Fert.* **Suppl. 41,** 173–182.
25. Petropoulos, C. J., Payne, W., Salter, D. W., and Hughes, S. H. (1992) Using avian retroviral vectors for gene transfer. *J. Virol.* **66,** 3391–3397.
26. Bosselman, R. A., Hsu, R., Boggs, T., Hu, S., Bruszewski, J., Ou, S., Kozar, L., Martin, F., Green, C., Jacobsen, F., Nicholson, M., Schulz, J. A., Semon, K. M., Richell, W., and Stewart R. G. (1989a) Germline transmission of exogenous genes in the chicken. *Science* **243,** 533–535.
27. Bosselman, R. A., Hsu, R., Boggs, T., Hu, S., Bmszewski, J., Ou, S., Souza, L., Kozar, L., Martin, F., Nicholson, M., Rishell, W., Schulz, J. A., Semon, K. M.,

and Stewart, R. G. (1989b) Replication defective vectors of reticuloendotheliosis virus transduce exogenous genes into somatic stem cells of the unincubated chicken embryo. *J. Virol.* **63,** 2680–2689.
28. Briskin, M. J., Hsu, R.-Y., Boggs, T., Schultz, J. A., Rishell, W., and Bosselman, R. A. (1991) Heritable retroviral transgenes are highly expressed in the chicken. *Proc. Natl. Acad. Sci. USA* **88,** 1736–1740.
29. Brazolot, C. L., Petitte, J. N., Etches, R. J., and Verrinder Gibbins, A. M. (1991) Efficient transfection of chicken cells by lipofection and introduction of transfected blastodermal cells into the embryo. *Mol. Reprod. Dev.* **30,** 304–312.
30. Flamant, F., Demeneix, B., Benoist, C., Markossian-Belin, S., and Samarut, J. (1994) Virofection: a new procedure to achieve stable expression of genes transferred into early embryos. *Int. J. Devel. Biol.* **38,** 751–757.
31. Petitte, J. N., Clark, M. E., Liu, G., Verrinder Gibbins, A. M., and Etches, R. J. (1990) Production of somatic and germline chimeras in the chicken by transfer of early blastodermal cells. *Development* **108,** 185–189.
32. Carsience, R. S., Clark, M. E., Verrinder Gibbins, A. M., and Etches, R. J. (1993) Germline chimeric chickens from dispersed donor blastodermal cells and compromised recipient embryos. *Development* **117,** 669–675.
33. Thoraval, P., Lasserre, F., Coudert, F., and Dambrine, G. (1994) Production of germline chimeras obtained from Brown and White leghorns by transfer of early blastodermal cells. *Poultry Sci.* **73,** 1897–1905.
34. Etches, R. J., Clark, M. E., Verrinder Gibbins, A. M., and Cochran, M. (1995) Development of chimeric chickens in surrogate shells. *Poultry Sci.* **74 Suppl. 1,** 26.
35. Fraser, R. A., Carsience, R. S., Clark, M. E., Etches, R. J., and Verrinder Gibbins., A. M. (1993) Efficient incorporation of transfected blastodermal cells into chimeric chick embryos. *Int. J. Devel. Biol.* **37,** 381–385.
36. Speksnijder, G. J., Etches, R. J., and Verrinder Gibbins, A. M. G. (1995). The construction of chimeric chicken embryos using transfected blastodermal cells sorted by FACS. *Poultry Sci.* **74 Suppl. 1,** 11.
37. Etches, R. J., Carsience, R. S., Fraser, R. M., Clark, M. E., Toner, A., and Verrinder Gibbins, A. M. (1993) Avian chimeras and their use in manipulation of the avian genome. *Poultry Sci.* **72,** 882–889.
38. Liu, G., Etches, R. J., and Verrinder Gibbins, A. M. (1995) Targeted insertion of foreign DNA into the vitellogenin II (VTGII) gene in cultured chicken blastodermal cells, unpublished data.
39. Brinster, R. L. and Zimmerman, J. W. (1994) Spermatogenesis following male germ-cell transplantation. *Proc. Natl. Acad Sci. USA* **91,** 11,298–11,302.
40. Brinster, R. L. and Avarbock, M. R. (1994) Germline transmission of donor haplotype following spermatogonial transplantation. *Proc. Natl. Acad Sci. USA* **91,** 11,303–11,307.
41. Freeman, B. M. and Bumstead, N. (1987). Transgenic poultry: theory and practice. *World's Poultry Sci. J.* **43,** 180–189.

42. Gavora, J. S., Benkel, B., Sasada, H., Cantwell, W. J., Fiser, P., Teather, R. M., Nagai, J., and Sabour, M. P. (1991) An attempt at sperm mediated gene transfer in mice and chickens. *Cancer J. Anim Sci.* **71,** 287–291.
43. Rottman, O. J., Antes, R., Hofer, P. and Maierhofer, G. (1992) Liposome mediated gene transfer via spermatozoa into avian egg cells. *J. Anim. Breed. Genet.* **109,** 64–70.
44. Squires, E. J. and Drake, D. (1994) Transgenic chickens by liposome-mediated gene transfer. *Proc. 5th World Cong. Genet. App. Livestock Prod.* **21,** 350–353.
45. Etches, R. J. (1995) *Reproduction in poultry.* CAB, Wallingford, England.
46. Nieuwkoop, P. D. and Sutasurya, L. A. (1979) *Primordial germ cells of the chordates.* Cambridge University Press, Cambridge.

VIII

Expression in Plants

34

Plant Cell Transfection by Electroporation

William B. Terzaghi and Anthony R. Cashmore

1. Introduction

A crucial step in the characterization of recombinant genes is their introduction into a suitable host. Numerous techniques have been developed for introducing recombinant genes into plant cells (see other chapters in this volume, or vol. 6 in this series) *(1)*; one of the most versatile and generally applicable is direct DNA transfer by electroporation. Gene transfer by electroporation was first demonstrated in cultured animal cells *(2,3)*, and was soon adapted for use in plant protoplasts *(4,5)*. It has been used both for studying transient gene expression *(5,6)* and for generating stably transformed plant cells *(4,6,7)*. Electroporation can also be used for introducing other molecules such as RNA, proteins, or dyes *(8,9)*.

The principle underlying electroporation is that cells exposed to a high-voltage electric field pulse become transiently permeable to molecules including DNA as large as 150 kb *(8,9)*. This is thought to occur because temporary pores are created by electrical breakdown of the plasma membrane, which then reseal in a process which can be prolonged by keeping the cells at 0°C (for discussion of the biophysics of pore formation and molecular transfer, *see* refs. *8,10*). Effectiveness of electroporation is determined by the voltage and duration of the pulse. High voltages require short pulses and low voltages require longer pulses for effective transfection. Pulses which are too weak will not permeabilize the cells, whereas pulses which are too strong destroy them. Optimal settings vary with cell type and need to be determined for each application, but as a general rule, there is an inverse relationship between cell size and voltage needed to induce pore formation, and settings which result in about 50% survivorship are a good starting point for optimization *(9,11,12)*. Two basic types of pulses have been used for electroporation: rectangular and expo-

From: *Methods in Molecular Biology, vol. 62: Recombinant Gene Expression Protocols*
Edited by: R. Tuan Humana Press Inc., Totowa, NJ

2. Resuspend pellets in 2 vol KN2M, then add 1 vol protoplasting enzyme solution (final concentration 0.5% cellulysin, 0.1% macerase).
3. Place 30 mL in 150 x 25 mm Petri dish, seal with parafilm, then leave shaking at 40 rpm for 16 h at 25°C (checking periodically under the phase contrast microscope for conversion to protoplasts).
4. Filter the protoplast solution through 250- and 53-μm mesh nylon. Rinse the plates with 1/2 vol EP and use this to wash the filters, then centrifuge the filtrate for 10 min at 40*g*.
5. Resuspend the pellet in 9 mL 50% Percoll solution and dilute to 15 mL with EP (30% Percoll final). Carefully layer on 10 mL EP, then centrifuge for 10 min at 40*g*.
6. Remove the interface band with a wide-bore pipet and dilute to 50 mL with EP. Take an aliquot for counting in a hemocytometer, then centrifuge for 10 min at 40*g*.
7. Count the aliquot in the hemocytometer while the protoplasts are pelleting. Determine viability by mixing 5 μL protoplasts with 5 μL Evan's Blue solution (live protoplasts exclude this dye, whereas dead cells become stained).
8. If proceeding directly to electroporation resuspend pellet in 50 mL W5; otherwise resuspend in KN2M at 1–5 × 10^6/mL, transfer to Petri dish of suitable capacity, seal dish with parafilm and store in dark at 25°C until ready to electroporate.

3.3. Electroporation

1. Harvest protoplasts, resuspend pellet in 50 mL W5 (if cultured in KN2M), and leave at least 45' at 25° C.
2. Centrifuge for 10 min at 40*g*.
3. Resuspend pellet in EP at 5 × 10^6/mL.
4. Heat shock 5 min at 42°C.
5. Place cells on ice for 1 min, then mix 1 mL protoplasts with 10–100 μg plasmid, 20–40 μg internal control (optional, *see* notes), and carrier salmon sperm DNA to 300 μg total DNA. Total volume of DNA should not exceed 100 μL, and be sure to mix well.
6. Transfer 2 × 500 μL aliquots of protoplasts to electroporation cuvets, then place on ice once cells have been mixed with DNA for 5 min.
7. Once protoplasts have been mixed with DNA for 5 min place cuvets on ice and leave for 10 min.
8. Flick cuvets to resuspend settled cells, then place cuvets in prechilled sample holder and administer one pulse of 150 V, 960 μF.
9. Leave 10 min on ice.
10. Use a sterile Pasteur pipet stuffed with cotton wool to transfer protoplasts to 10 mL chilled KN2M in 15-mL tubes and centrifuge for 10 min at 40*g*.
11. Resuspend pellet in 1 mL KN2M and transfer to a 60-mm Petrie dish. Set aside an aliquot to determine viability.
12. Seal Petri dish with parafilm and culture in dark or under dim (<20 μmol · $m^{-2} \cdot s^{-1}$) light for 2 d at 25°C.

13. Harvest cells with a wide-bore pipet and rinse plate with KN2M. Set aside an aliquot to determine viability then centrifuge 10' at 40*g*.
14. Resuspend in suitable extraction buffer and lyse by sonication or by freezing and rapidly thawing three times.
15. Perform the appropriate reporter assay(s).

3.4. Selecting Stable Transformants

1. Electroporate as described in Section 3.3. (using the same settings), except use plasmid DNA linearized with a restriction enzyme which cuts outside the transgene.
2. After resuspending cells in 1 mL KN2M (*see* Section 3.3., step 11) mix protoplasts with an equal volume of 2.8% sodium alginate in 400 m*M* mannitol.
3. Take up protoplasts with a sterile Pasteur pipet plugged with cotton wool and add drop-wise to a Petri dish containing 50 m*M* $CaCl_2$ in 400 m*M* mannitol pH 5.6. Each drop should form a solid bead within a few minutes.
4. After 1 h at room temperature, replace the 50 m*M* $CaCl_2$ in 400 m*M* mannitol with KN2M.
5. After 1 h at room temperature, replace the KN2M with fresh KN2M, then seal the Petri dishes with parafilm and culture under dim fluorescent light at 25°C.
6. Periodically check the cells under the inverted microscope and once over 10% of the cells have divided (which should occur within 1 wk), replace medium with a 1:1 mix of KN2M and KN2, each supplemented with 50 mg/L kanamycin sulfate.
7. One week later, replace medium with a 3:1 mix of KN2 and KN2M, each supplemented with 50 mg/L kanamycin sulfate.
8. One week later, and at weekly intervals thereafter, replace medium with KN2 supplemented with 50 mg/L kanamycin sulfate.
9. After 8–10 wk kanamycin-resistant calli should be apparent. Entire beads may be transferred to Petri dishes containing KN2 supplemented with 50 mg/L kanamycin sulfate and solidified with 0.8% purified agar. Alternatively, calli may be released from the alginate beads by treatment with 100 m*M* sodium citrate or 100 m*M* EDTA. After depolymerization, the calli are mixed with KN2, centrifuged for 5 min at 100*g*, then resuspended in KN2 supplemented with 50 mg/L kanamycin sulfate and spread on KN2 plates supplemented with 50 mg/L kanamycin sulfate.
10. Subculture kanamycin-resistant calli to fresh KN2 supplemented with 50 mg/L kanamycin sulfate plates every 2 wk, and analyze for the presence of the transgene once adequate amounts of biomass are available.

4. Notes

1. Plant cell growth conditions and media will need to be optimized for your particular application. As a general rule, protoplasts are most easily prepared from rapidly growing tissues, and one should optimize cell growth then test cultures for protoplast yield at various times after subculture if difficulty in obtaining protoplasts is experienced (we rarely get more than 30% conversion to proto-

plasts, even with rapidly growing SB-P cells; whereas other cell lines which we use are much easier to convert to protoplasts). We grow our cultures under lights to keep them photosynthetically active, but for other cultures this may be unnecessary or even undesirable. If using intact plant tissue, it is convenient to start with axenic plants grown in sterile containers; however, you may also start with plants grown under nonsterile conditions provided the material is surface-sterilized with 70% ethanol or a 50% dilution of commercial bleach and thoroughly washed with sterile water before adding to protoplasting enzymes.

2. Clean, healthy protoplasts are essential. Poor quality protoplasts are unlikely to survive long enough to assay, and even if they do will express the transgenes poorly. One element affecting protoplast quality is the incubation in enzymes. For SB-P cells, prolonged incubation at relatively low enzyme concentrations works best, as was also found for tobacco *(19)* and Arabidopsis *(20)* mesophyll cells, but for other materials 1–3 h incubations at higher enzyme concentrations work better *(5,19)*. Therefore, the preparations should be checked regularly under the microscope, and harvested when protoplasting appears to have gone to completion or when lysis begins to occur. We use protoplasting enzymes from Calbiochem (La Jolla, CA), but other manufacturers provide equally suitable products, and some cell types may respond better to enzymes from different manufacturers. The important thing is to supply a mixture of cellulases, hemicellulases, and pectinases. A second element is thorough cleaning, and most successful protocols include flotation through Percoll or other density gradients to purify the protoplasts followed by washing to remove residual enzyme. Many protocols include two or more washes; we prefer to increase the dilution and minimize centrifugations as protoplasts are lost or destroyed at each centrifugation.
3. Evans' Blue provides a simple but imperfect assay for viability. If fluorescence optics are available, a better assay is staining with fluorescein diacetate (FDA). FDA is nonfluorescent and freely permeable, but once inside cells is cleaved to form the impermeable fluorescent molecule fluorescein and only living cells fluoresce upon excitation with UV light.
4. Transgenes driven by a constitutive promoter such as the CaMV 35S promoter are expressed well in freshly isolated protoplasts. However, we detect little expression of transgenes driven by light-regulated promoters until the cells begin to divide, and we get better results if we culture the protoplasts for 2 d in KN2M before electroporation. This also gives us a better indication of the protoplast quality: we do not continue until cells begin to regenerate their cell walls and divide.
5. The incubation in W5 followed by the heat shock is not strictly necessary, but improves the transfection efficiency in our hands.
6. Preincubating the protoplasts with DNA is also unnecessary; the minimum handling time will be adequate. However, it is a good idea to ensure that all samples spend the same amount of time at each stage to minimize variation (arising, for example, from DNA degradation by nucleases released from the protoplasts).

7. Amounts of transient expression are directly related to amounts of plasmid DNA up to at least 200 µg/mL cells. Carrier DNA will increase expression from a given amount of plasmid DNA, although amounts of expression from, for example, 50 µg of plasmid and 150 µg of carrier are not as great as from 200 µg of plasmid *(21)*. Carrier DNA is also reported to enhance stable transformation *(19)*.
8. If the experiment involves comparing amounts of expression from different promoters or when the protoplasts are subjected to different treatments, it is a good idea to include a second plasmid encoding a second reporter driven by a constitutive promoter to normalize for sample to sample variation. Results are then expressed as the ratios of reporter and internal control activities. We routinely use luciferase driven by various light-regulated promoters as our reporter and GUS driven by the CaMV 35S promoter as our constitutive internal control.
9. We split our cells into two aliquots to reduce variation due to individual samples and to recover some data in the event that one cuvet is lost (e.g., by leaking, losing its cap, etc.).
10. Some workers report that more transient expression is observed if electroporation is conducted at room temperature *(9,12,19)*. However, the transient permeabilization is prolonged if cells are kept at 0°C; prechilling and then placing the cells back on ice after electroporation also negates heating from the electrical pulse. In general, pulsing chilled cells is a good idea when using low voltage pulses of long duration, but is unnecessary for short, high voltage pulses *(9,19)*.
11. Used electroporation chambers can be washed and resterilized by a variety of techniques, although the electrodes corrode with use and transfection efficiency is lower when recycled cuvets are used. We recycle our Bio-Rad (Hercules, CA) chambers by thoroughly rinsing with water, then autoclaving (but do not autoclave the lids!). Test a few samples from other manufacturers before committing to this approach! We resterilize our lids by soaking for 10 min with 70% ethanol and drying in a laminar flow hood. Also, 70% ethanol can be used to recycle chambers, but it will not destroy leftover DNA. Treatment for 10 min in 50% bleach followed by several washes with sterile water will sterilize the chambers and destroy residual DNA, but it will also reduce efficiency and shorten the lifespan of the cuvets. Cuvets recycled by any means should be checked for cracks and badly corroded electrodes. Use new cuvets for critical experiments.
12. Optimal settings vary with plant material and must be determined empirically. Successful transfection has been reported for both high voltage/short duration *(4,19)* and low voltage/long duration pulses *(5,21)*. In general, it is best to start with low voltage and high capacitance when working with capacitance discharge devices; settings which give about 50% survivorship tend to be optimal *(9,11,12)*. For rectangular wave generators, most workers have used high voltage pulses of short duration *(3,11)*. When using low voltage and high capacitance, a single pulse generally works best, whereas workers using high voltage pulses frequently administer multiple pulses *(3,11,19)*.
13. Electroporation may be performed in either low ionic strength medium such as 400 m*M* mannitol containing $MgCl_2$ or $CaCl_2$ to stabilize the protoplasts *(19)*, or

in higher ionic strength medium such as PBS supplemented with enough mannitol to osmotically stabilize the protoplasts *(5)*. Optimal medium varies with cell type and should be determined empirically. Some cell types such as maize mesophyll protoplasts are damaged by the high salt in PBS *(22)*, whereas transient expression in others such as maize suspension cells was optimal when electroporation was performed in medium containing 90 m*M* NaCl *(5)*. There is some debate regarding the influence of divalent cations, especially $MgCl_2$. $MgCl_2$ was reported to reduce electroporation efficiency in mammalian cells *(3,9)*, but was found to be essential for electroporation of carrot protoplasts *(23)*. $CaCl_2$ is also beneficial because it helps stabilize protoplasts and hence improves survivorship *(21)*. Intuitively, we anticipate that medium which approximates the cytoplasmic contents will be the best choice as it reduces the deleterious effects of losing electrolytes while the cells are permeabilized. Some workers have included reducing agents in their electroporation medium *(22)*.

14. Addition of PEG to the electroporation medium just before electroporation has been reported to enhance transfection *(19,23)*. In our experience there is some improvement, but the PEG solutions are viscous and difficult to administer accurately and mix thoroughly. Therefore, we do not find that the increased transfection is worth the trouble (and increased source of variation between samples).
15. For transient assays there is little difference between supercoiled and linear DNA, but for stable transformation, linear DNA gives better results. For transient expression most workers use 40–100 µg/mL DNA; the addition of calf thymus or salmon sperm carrier DNA is reported to be beneficial *(21)*. Conditions which give optimal transient expression also generally give optimal stable transformation *(9,11)*. There is also a recent report that propagating the plasmid DNA in methylation-deficient hosts such as GM2163 gives improved regulation of introduced promoters *(24)*.
16. Embedding protoplasts in solid medium improves survivorship and cell division. For transient assays, it is rarely worth the additional trouble, but it greatly improves recovery of stable transformants. It also simplifies the various media changes during the selection process. Low-melting-temperature agarose has also been used as the embedding medium by many workers *(19)*; we favor sodium alginate because of the ease with which embedded cells can be released by dissolving the matrix.

References

1. Pollard, J. W. and Walker, J. M., eds. (1990) *Methods in Molecular Biology: Plant Cell and Tissue Culture*, vol. 6, Humana, Clifton, NJ.
2. Wong, T. K. and Neumann, E. (1982) Electric field mediated gene transfer. *Biochem. Biophys. Res. Commun.* **107,** 584–587.
3. Neumann, E., Schaeffer-Ridder, M., Wang, Y., and Hofschneider, P. H. (1982) Gene transfer into mouse lymphoma cells by electroporation in high electric fields. *EMBO J.* **1,** 841–845.

4. Shillito, R. D., Saul, M. W., Paszkowski, J., Müller, M., and Potrykus, I. (1985) High efficiency direct gene transfer to plants. *Bio/Technology* **3,** 1099–1103.
5. Fromm, M. E., Taylor, L. P., and Walbot, V. (1985) Expression of genes transferred into monocot and dicot plant cells by electroporation. *Proc. Natl. Acad. Sci. USA* **82,** 5824–5828.
6. Ow, D. W., Wood, K. V., DeLuca, M., de Wet, J. R., Helinski, D. R., and Howell, S. H. (1986) Transient and stable expression of the firefly luciferase gene in plant cells and transgeneic plants. *Science* **234,** 856–859.
7. Fromm, M. E., Taylor, L. P., and Walbot, V. (1986) Stable transformation of maize after gene transfer by electroporation. *Nature* **319,** 791–793.
8. Chang, D. C., Chassy, B., Saunders, J. A., and Sowers, A. E., eds. (1992) *Guide to electroporation and electrofusion.* Academic, New York.
9. Potter, H. (1993) Application of electroporation in recombinant DNA technology. *Methods Enzymol.* **217,** 461–478.
10. Neumann, E., Sowers, A. E., and Jordan, C. A., eds. (1989) Electroporation and electrofusion in cell biology. Plenum, New York.
11. Chang, D. C. (1992) Design of protocols for electroporation and electrofusion: selection of electrical parameters, in *Guide to Electroporation and Electrofusion* (Chang, D. C., Chassy, B., Saunders, J. W., and Sowers, A. E., eds.), Academic, San Diego, CA, pp. 429–455.
12. Saunders, J. W. and Bates, G. W. (1992) Genetic manipulation of plant cells by means of electroporation and electrofusion, in *Guide to Electroporation and Electrofusion* (Chang, D. C., Chassy, B., Saunders, J. W., and Sowers, A. E., eds.), Academic, San Diego, CA, pp. 471–483.
13. Lindsey, K. and Jones, M. G. K. (1987) Transient gene expression in electroporated protoplasts and intact cells of sugar beet. *Plant Mol. Biol.* **10,** 43–52.
14. Matthews, B. F., Abdul-Baki, A. A., and Saunders, J. A. (1990) Expression of a foreign gene in electroporated pollen grains of tobacco. *Sex. Plant Reprod.* **3,** 147–151.
15. Dekeyser, R. A., Claes, B., De Rycke, R. M. U., Habets, M. E., Van Montagu, M. C., and Caplan, A. B. (1990) Transient expression in intact and organized rice tissues. *Plant Cell.* **2,** 591–602.
16. Rogers, S. M. D., Ogren, W. L., and Widholm, J. M. (1987) Photosynthetic characteristics of a photoautotrophic cell suspension culture of soybean. *Plant Physiol.* **84,** 1451–1456.
17. Murashige, T. and Skoog, F. (1962) A revised medium for rapid growth and bioassay with tobacco tissue culture. *Physiol. Plant.* **15,** 473–476.
18. Menczel, L., Nagy, F., Kiss, Z., and Maliga, P. (1981) Streptomycin resistant and sensitive somatic hybrids of *N. tabacum* + *N. knitiana* : correlation of resistance to *N. tabacum* plasmids. *Theor. Appl. Genet.* **59,** 191–195.
19. Shillito, R. D. and Potrykus, I. (1987) Direct gene transfer to protoplasts of dicotyledenous and monocotyledenous plants by a number of methods, including electroporation. *Methods Enzymol.* **153,** 313–336.

20. Damm, B. and Willmitzer, L. (1988) Regeneration of fertile plants from protoplasts of different *Arabidopsis thaliana* genotypes. *Mol. Gen. Genet.* **213,** 15–20.
21. Fromm, M., Callis, J., Taylor, L. P., and Walbot, V. (1987) Electroporation of DNA and RNA into plant protoplasts. *Methods Enzymol.* **153,** 351–366.
22. Sheen, J. (1990) Metabolic repression of transcription in higher plants. *Plant Cell.* **2,** 1027–1038.
23. Langridge, W. H. R., Li, B. J., and Szalay, A. A. (1987) Uptake of DNA and RNA into cells mediated by electroporation. *Methods Enzymol.* **153,** 336–350.
24. Tovar Torres, J., Block, A., Hahlbrock, K. and Somssich, I. E. (1993) Influence of bacterial strain genotype on transient expression of plasmid DNA in plant protoplasts. *Plant J.* **4,** 587–592.

35

Transformation and Foreign Gene Expression in Plants Mediated by Microprojectile Bombardment

Henry Daniell

1. Introduction

Most common strategies for gene transfer in plants derive from the fact that *Agrobacterium tumefaciens* contains Ti plasmids which facilitate transfer of T-DNA from the plasmid genome into plant cells with attendant integration of the T-DNA region containing foreign genes into the plant nuclear genome *(1)*. Alternatively, DNA may be introduced into protoplasts using various chemical fusagens *(2)*, by electroporation *(3)*, or by microinjection *(4)*. In all these approaches foreign DNA is introduced into the nuclear genome and mostly single genes driven by independent promoters flanked by polyA-specifying terminators have been expressed. Despite several advantages of the *Agrobacterium*-mediated transformation, major limitations still remain; the limited host range should be extended to major crop species. Monocots, especially cereals, are highly recalcitrant to *Agrobacterium* infection. Since regeneration from cultured tissues is a prerequisite, problems of somaclonal variation resulting from tissue culture induced mutations should also be addressed. Direct DNA transfer by electroporation or using chemical fusagens require the use of protoplasts; unfortunately, most crops can not be regenerated from single cells *(5)*.

Sanford and coworkers *(6)* invented a transformation technique that relies on the bombardment of recipient cells, tissues or organs with high velocity metal particles coated with foreign DNA. Using this biolistic delivery system (the "gene gun"), several groups have demonstrated stable transformation of foreign genes into the nuclear genomes of higher plants, mitochondrial genomes of yeast, and chloroplast genomes of *Chlamydomonas* and higher plants *(5,6)*. In addition to plants, transgenic animals (e.g., Rainbow trout and Zebra fish), insects (e.g., fruitfly), and several species of algae, fungi, and bac-

From: *Methods in Molecular Biology, vol. 62: Recombinant Gene Expression Protocols*
Edited by: R. Tuan Humana Press Inc., Totowa, NJ

teria have been obtained using various biolistic devices. Biolistic devices are also employed in human gene therapy *(7)*. The gene gun use competes favorably with other transformation methods because of the simple vectors employed (e.g., no T-DNA borders required), unlimited host range, and truly genotype-independent transformation; this method also bypasses some of the difficulties encountered in other transformation methods such as protoplast or callus culture, culture induced mutations, somaclonal variation, recovery of infertile regenerants, somatic embryogenesis, etc. DNA delivery by the gene gun into cultured tissues or the scutellum of immature embryos of several cereal crops is generally followed by selection for transformed tissues. Regenerated transgenic plants have been obtained from maize, rice, wheat, oat, barley, sorghum, sugarcane, and numerous dicots, including tree species like poplar and pine. For an updated list of transgenic plants obtained through microprojectile bombardment see Table 1 which demonstrates the broad applicability of the gene gun. The gene gun is also most commonly used in studies of transient foreign gene expression in order to optimize conditions for stable transformation and to study gene regulation, tissue specificity, and promoter functions in plant and animal cells. Even *Escherichia coli* and *Agrobacterium* cells have been used as microprojectiles to deliver foreign DNA and obtain stable transformants *(8)*, opening the door to deliver large genomes or entire chromosomes into recipient cells.

1.1. Chloroplast Transformation

Several approaches have been used to manipulate chloroplasts and chloroplast genes including generation of chloroplast mutants, protoplast fusion, organelle inactivation and chloroplast recombination *(9)*. Daniell and McFadden *(10)* reported the first expression of foreign genes by isolated chloroplasts. Reports of the introduction of intact chloroplasts into albino protoplasts to produce variegated progeny *(11)* suggests the very real opportunity to introduce transformed chloroplasts into a variety of nuclear backgrounds. Daniell et al. *(12,13)* also demonstrated the first transient expression of foreign genes in plastids of higher plants using the chloramphenicol acetyltransferase (*cat*), β-glucuronidase genes and the gene gun. We have reported conditions for optimal DNA delivery into plastids of cultured tobacco cells using an improved gene gun *(14)*. Daniell et al. also *(13)* have reported expression of β-glucuronidase in different cellular compartments following biolistic delivery of chloroplast or nuclear vectors into wheat leaves and calli. These observations paved the way for stable complementation of tobacco chloroplast genome using a modified chloroplast 16S rDNA gene conferring antibiotic resistance *(15)* and high frequency plastid transformation in tobacco by selection for a chimeric *aadA* gene *(16)*. In these studies, high copy number expres-

Table 1
Transgenic Plants Recovered Through Microprojectile Bombardment[a]

Plant species	Explant utilized for transformation	Ref.
Alfalfa	Cali from petiole, stem sections	*54*
Arabidopsis	Root sections	*27*
Banana	Embryogenic suspension cells	*56*
Barley	Embryogenic callus;	*28*
	immature embryos	*29,30*
Bean	Meristems	*25*
Citrus	Embryogenic cells	*63*
Cotton	Embryogenic	*31*
	suspensions; meristems	*32*
Cranberry	Stem sections	*33*
Cucumber	Cotyledons	*59*
Dendrobium orchid	Protocorms	*59*
Eucalyptus	Zygotic embryos	*65*
Gladiolus	Suspension cells and calli	*55*
Grape	Embryogenic suspension cells	*71*
Maize	Embryogenic	*34*
	suspensions;	*35*
	immature embryos	*36*
Oat	Embryogenic	*37*
Papaya	Zygotic/somatic embryos; hypocotyls	*38*
Pasture grass	Embryogenic calli	*64*
Peach	Embryo derived calli	*57*
Peanut	Meristems	*39*
Poplar	Embryogenic	*40*
Rice	Zygotic embryos	*41,42*
Rubber	Anther derived calli	*69*
Ryegrass	Cell suspension culture	*68*
Sorghum	Zygotic embryos	*43*
Soybean	Meristems	*44,45*
Spruce	Embryogenic callus	*46*
Sugarcane	Embryogenic callus	*47*
Sunflower	Apical meristems;	*48*
	half shoot apices	*67*
Sweet potato	Leaves, petioles	*60*
Tobacco	Leaves; suspension culture	*49*
Tomato	Cell culture	*66*
Tulip	Flower stem explants	*61*
Turfgrass	Embryogenic suspension culture	*70*
Wheat	Embryogenic callus;	*50*
	zygotic embryos;	*51*
	immature embryos	*52,53*
Yam	Cultured suspension cells	*20*

[a]Revised Table 1.

sion has been accomplished by introducing one or two copies of the foreign gene into each chloroplast genome of tobacco. Each plant cell contains 5000–10,000 copies of chloroplast genomes per cell and, therefore, it has been possible to express thousands of copies of the introduced foreign gene per cell. Biolistic transformation may be the only reliable and reproducible method to date for delivering foreign DNA into chloroplasts of higher plants. *Agrobacterium*-mediated chloroplast transformation *(17)* has met with limited success; PEG-mediated chloroplast transformation *(18)* may not be efficient because no transient expression was observed using this method by other groups *(14)*.

1.2. Principle

In a gunpowder-driven PDS-1000 biolistic device (DuPont, Wilmington, DE), a gunpowder charge is used to drive a plastic plunger (macroprojectile) down a barrel to accelerate DNA-coated metal particles (microprojectiles). The macroprojectile hits a stopping plate, letting microprojectiles continue to travel through a small hole present in the stopping plate into a partially evacuated chamber. Microprojectiles traveling at high speed forcefully enter the target cells or tissues which are placed at a given height inside the bombardment chamber. The gas and debris from the gunpowder charge is vented into a filter housing unit and is eventually trapped in the vacuum oil.

Sanford and coworkers, in collaboration with DuPont have developed a retrofit kit for the PDS-1000, providing the user with a gas-driven particle delivery system, the PDS-1000/He and this is now marketed by Bio-Rad (Hercules, CA). The PDS-1000/He, uses high pressure He instead of gunpowder as the driving force for the macrocarriers. In this method, a small chamber sealed at one end with rupturable membrane(s) is filled with He to high pressure. A solenoid-driven lance then ruptures the membrane releasing a shock wave, which enters the launch assembly device that accommodates a removable sleeve. There are removable rings inside the sleeve to facilitate launching of microprojectiles. A nylon mesh is locked in place across the axis of the sleeve; DNA-coated microprojectiles are loaded directly onto the center of the mesh. The He shock wave atomizes and accelerates the microprojectiles as it passes through the mesh. In yet another modification called the "flying disk" method, a plastic membrane is loosely held in the same position as the nylon mesh. Particles are dried on its surface. The disk is accelerated down the sleeve 1 cm, upon firing, where it impacts against a stopping screen releasing microprojectiles to hit the target cells or tissues.

Several other biolistic devices, including a microtargeting particle accelerator, an air-gun apparatus, and instruments based on electric discharge are in use and these are described in detail in a recent volume *(7)*.

2. Materials

2.1. Materials for Bombardment

Costar bottle top filters (100 mL, 0.2 μm), sterile disposable pipets, autoclaved Whatman filter paper disks (55 mm in diameter, autoclave the filter disks in a glass Petri dish covered with aluminum foil), sterile 60 × 15 mm petri plates, tweezers (autoclave; flame before each use), ethanol (95% and 70%), Pipetman and autoclaved tips (low binding, Marsh Biomedicals, MA), sterile plant cell culture media, Kimwipes (wrap in aluminum foil and autoclave), 3–4-d-old cultured plant cells or appropriate tissues, sterile macroprojectiles (wash in 100% ethanol in microtitre plates), sterile stopping plates (wash in 100% ethanol), gun cleaning patches, gun cleaning rod, 2.5*M* calcium chloride (weigh 1.84 g of $CaCl_2$ and dissolve in 5 mL of water, filter sterilize; make fresh just before use; may be stored at 4°C for short periods; do not freeze), sterile tungsten or gold particles, sterile Eppendorf tubes (low binding), sterile 1*M* spermidine-free base (highly hygroscopic; therefore, take a 1-g unopened bottle of spermidine stored at 4°C and add 6.8 mL of sterile water and filter sterilize; store as 25-μL aliquots at –20°C).

2.2. Materials for Plant Cell and Tissue Culture

2.2.1. Sugarbeet (SB-18-1R)

100 mL 10X sugarbeet medium: 70 mL water (autoclaved), 4.3 g MS salts, 100 mg I-inositol, 5 mg pyridoxine HCl, 5 mg nicotinic acid, 30 g sucrose (molecular biology grade, protease, nuclease free), 1 mg thiamine HCl (100 μL from a stock solution of 10 mg/mL).

Make up volume to 100 mL and freeze at –20°C. In order to make the 1X medium:

1. Thaw 10X stock solution and dilute to 1X using autoclaved water.
2. Adjust pH to 5.8 with KOH (initial pH will be about 4.4).
3. Autoclave for 20 min.

2.2.2. Tobacco (NT1)

100 mL 10X tobacco medium: 70 mL water (autoclaved), 4.3 g MS salts, 100 mg I-inositol, 100 mg KH_2PO_4, 30 g sucrose (molecular biology grade), 1 mg thiamine-HC1 (use 100 μL of 10 mg/mL stock), 0.2 mg 2,4 dichlorophenoxy acetic acid (use 200 μL from 1 mg/mL in 50% ethanol).

Store medium at –20°C. In order to make 1X medium:

1. Thaw 10X stock solution and dilute to 1X.
2. Adjust pH to 5.8 with KOH.
3. Autoclave for 20 min.

2.2.3. Maize (BMS)

In order to make 1 L medium: 900 mL water (autoclaved), MS salts 4.3 g, 200 mg myo-inositol, 4 mL 2,4 dichlorophenoxy acetic acid (from 0.5 mg/mL stock), 20 g sucrose, 130 mg asparagine, 1 mL BMS vitamin stock.

In order to prepare BMS vitamins, mix 25 mg each of pyridoxine HCl, calcium pantothenate, thiamine HCl, and 130 mg of nicotinic acid in 100 mL autoclaved water, stir thoroughly; and store aliquots at –20°C.

The glassware, after washing in chromic acid, should be autoclaved separately before using for media preparation. Autoclave flasks and caps separately. After the growth medium has cooled down to room temperature, add 3 mL of suspension culture (6–7-d-old cells) to 12 mL of sterile 1X medium. Grow cells at 26°C under continuous shaking (150 rpm) in the light. Sieve cultures using 500-µ mesh after every two subcultures to get rid of cell clumps or regenerated calli. Subculture cells every week and maintain at least three batches of each culture. Always check cultures for contamination by streaking cells on LB-agar plates and incubating at 37°C until next subculture.

2.2.4. Solid Culture

For long term storage of cultured plant cells solid cultures are necessary. Strictly follow sterile conditions during transfer of cells or calli.

1. Make 1X medium by diluting a 10X stock.
2. Adjust pH to 5.8 using KOH.
3. Add 1.5 g of phytagar to 100 mL of medium.
4. Autoclave for 20 min.
5. Pour media either into Petri plates or culture dishes and let it solidify in the hood.
6. After solidification, streak a few drops of cells on their respective medium and cover.
7. Sugarbeet cells should have foil wrapped around the plate to prevent differentiation, whereas tobacco cells may be left in the light. Cultures should be kept in the growth chamber (set at 26°C).
8. Transfer calli to fresh Petri plates or culture dishes once a month.
9. Divide into two batches of three plates each and transfer once every month.
10. Remember to check for bacterial or fungal contamination by streaking cells on LB plates.

3. Methods

3.1. Preparation for Bombardment

3.1.1. Preparation of Tissues for Bombardment

For bombardment of tissues, plants should be grown aseptically from seeds on MS medium supplemented with 2% sucrose and solidified with 0.8% agar.

In our experiments on bombardment of wheat leaves or calli, we used plants that were regenerated from anther culture or immature embryos. When plants were at the 2–3 leaf stage, they were transplanted aseptically into test tubes containing half strength MS medium supplemented with 2% sucrose and solidified with 0.8% agar. Calli were generated from immature embryos by a modified method of Sears and Deckard *(13)*. Immature embryos (10–12 d old) of wheat were placed in basal MS medium containing only half the amount of 2,4-dichlorophenoxy acetic acid (2,4- D, 1 mg/mL); calli were left in this medium until hard, white embryonic tissue developed. This tissue usually started forming after 2 mo when calli were transferred to fresh medium every 3–4 wk. Calli rich in this embryonic tissue were transferred to fresh medium and used for bombardment.

Tobacco plants (*Nicotiana tabacum*, cv Petit Havana) were aseptically grown from surface-sterilized seeds on agar supplemented with MS salts, B5 vitamins (*see* Section 5.) and sucrose (30 g/L). Leaves (3–5 cm in length) were excised from plants and placed abaxial side up on RMO medium (for composition, *see* Section 3.3.) in a Petri dish and were bombarded as described below.

3.1.2. Preparation of Cells for Bombardment

1. Spray the hood with 70% ethanol and wipe with sterile wipes.
2. Turn on the UV light in the laminar flow hood for a few minutes.
3. After the UV light has been turned off, place culture flasks in the hood and spray ethanol over flasks.
4. Wear gloves and spray ethanol over gloved hands.
5. Label flasks with date, strain, type of medium and the user's initials.
6. Flame the neck of the culture flask and inoculate 5 mL of desired plant cells (to be bombarded) into 20 mL fresh medium using a sterile disposable pipet.
7. Take 100 μL of the growth medium from the mother and daughter cultures and streak it on a LB plate (which is already partitioned and numbered at the bottom).
8. Place the streaked LB plates at 37°C and the culture flasks at 26°C in the shaker. Leave caps loose enough to facilitate aeration while flasks are on the shaker.
9. Grow cells for 4 d or until the midlog phase.

3.1.3. Preparation of Microprojectiles

1. The preparation of microprojectiles should be done the day before bombardment.
2. Place 60 mg of tungsten (M-10, 0.7 μm, Biorad PN# 165-2266) particles into an Eppendorf tube.
3. Add 1 mL of 100% ethanol and vortex vigorously for 2 min or sonicate (*see* Section 5.).
4. Let this sit at room temperature overnight or store at –20°C indefinitely.
5. Spin the Eppendorf tube containing particles for 2.5 min at maximum speed in a microcentrifuge kept in the hood and remove the supernatant.

6. Wash thrice with sterile water. Finger vortex between each wash. When washing with water the tungsten particles will be very loose; it is acceptable to leave some water in the first two washes, but in the third wash try to aspirate off as much water as possible.
7. Add 1 mL of 50% glycerol (sterile) and leave at room temperature.
8. This can be stored up to 1 wk; longer storage may result in oxidation of the metal.

3.2. Bombardment of Cells or Tissues

3.2.1. Gun Powder Method (PDS-1000)

1. Keep a stand in the hood and affix a bottle to which the Costar sterile filter unit is tightly screwed. Place a 5.5 cm sterile filter paper disk and soak in 4 mL of the appropriate growth medium.
2. In a 5.5-cm plastic Petri dish, place two 5.5-cm filter disks and add 1 mL of the growth medium.
3. Add 2 mL of culture and filter out the growth medium by applying mild suction.
4. Scrape and weigh cells. Use microscale balance and a boat to weigh cells. Discard these cells after determining fresh weight since they are no longer sterile.
5. Add 4 mL of the growth medium to the Costar filter unit. Add enough volume of culture to get 300 mg of cells, using a sterile pipet.
6. Using the same pipet disperse cells uniformly on the filter paper.
7. Apply a gentle vacuum until all the medium is drained off.
8. Using a small sterile tweezer, transfer the filter disk containing cells to a Petri plate that has two 5.5-cm filter disks moistened with the growth medium.
9. Binding of DNA to microprojectiles: washed particles 25 µL, DNA (1 µg/uL in TE pH 8.0) 5 µL, 2.5*M* $CaCl_2$ 25 µL (prepared fresh), 0.1*M* spermidine 5 µL (stored at –20°C). Finger vortex after each addition and finally allow to sit for 10 min. Spin down in an Eppendorf centrifuge for 10 s using the pulse button and discard 40 µL of the supernatant.
10. Vortex vigorously for 1 min. Take out a macroprojectile from the microtiter plate using the pin loader. Load 5 µL of the DNA-tungsten mixture onto the macroprojectile.
11. Push down the macroprojectile using the pin loader into the barrel, with the DNA-tungsten mixture pointing downward, toward the sample.
12. Even if mixed thoroughly, the third aliquot may not contain the same amount of tungsten particles as the first two. Therefore, use this sample to load in an agarose gel to check the binding efficiency of DNA onto the tungsten particles.
13. At the time of loading the sample on the macroprojectile, vortex the DNA-tungsten mixture and immediately load the sample on the macroprojectile.
14. The bombardment chamber should be sprayed with 70% ethanol and wiped with sterile tissue after each bombardment. The barrel should be cleaned using gun cleaning patches and the rod.
15. Keeping the sample in a petri plate (cells on filter paper or tissues on agar plates) at the fourth level, bombard under partial vacuum (0.07–0.1 atmosphere).

16. Each sample may be bombarded twice. Bombard at least three plates for each treatment or plasmid.
17. After bombardment, add 1 mL of the culture medium to each plate.
18. Transient foreign gene expression studies are usually done 36, 48, or 72 h after bombardment. Grow cells after bombardment in the light at 26°C in the growth chamber.

3.2.2. Helium Method (PDS-1000/HE)

1. After 10 min of incubation of DNA-tungsten suspension, the particles pelleted down by a pulse centrifugation in a microfuge should be washed once with 70 µL of 100% ethanol.
2. Resuspend the pellet in 30 µL of 100% ethanol.
3. Spread 8 µL of tungsten/DNA suspension and dry on the center of a 1 in. disk (25 mm) made of 2 mil (0.05 mm) plastic membrane (for the flying disk method, Macrocarrier, Biorad PN# 165-2335) or on a 1 in. disk of 94 µm nylon mesh (for helium entrainment method).
4. Load the flying disk into a brass launch ring which is then screwed into a sleeve with a metal screen (stopping screen, Biorad PN# 165-2336) on a retainer ring 1.3 cm below the brass launch ring.
5. For the helium entrainment method, place the nylon mesh between two brass rings and screw them into the sleeve and place the sleeve into the sleeve holder at desired height.
6. Seal the rupture end of the pressure chamber with an appropriate number of isopropanol soaked 2-mil plastic rupture disks (3 layers for 900 psi 4 for 1200 psi, and 5 for 1500 psi; most useful is 1100 psi, Biorad PN# 165-2329).
7. Load the target sample at a chosen platform in the sample chamber and bombard under partial vacuum (28 inches Hg; set the He tank regulation 200 psi over the rating of the rupture disk).

3.3. Stable Expression

Stable expression may be studied by bombarding chloroplast or nuclear vectors into cultured cells and growing them in the presence of appropriate selection agents. Following bombardment, cells should be washed from the filter paper with 6 mL of the growth medium into a 10 × 2 cm-deep Petri dish. The Petri dish should be sealed and incubated in the dark on a gyratory shaker (75–80 rpm) at 26°C. After 2 d, 6 mL of fresh medium containing appropriate antibiotic (e.g., 200 µg/mL kanamycin) should be added to each Petri dish and half the volume should be transferred to a fresh dish. After another week, cells should be again diluted with an equal volume of antibiotic (e.g., 100 µg/mL kanamycin) containing medium. At the same time, contents of each dish should be divided into two dishes and returned to the shaker. After 1 wk cultures should be centrifuged at low speed to collect cells, which should be pipeted on the surface of solid growth medium containing antibiotic and agar in Petri dishes. Stably transformed calli should be selected by their resistance to appropriate antibiotic on solid medium. Antibiotic resistant calli should develop after 4–5 wk and should be maintained as separate lines on growth medium containing antibiotic.

In order to obtain transgenic plants, plants should be aseptically grown from surface-sterilized seeds on agar supplemented with MS salts and sucrose (30 g/L). Leaves (3–5 cm in length) should be excised from plants and placed abaxial side up in RMO medium (MS salts medium containing in 1 L: 2 mg IAA, 0.5 mg BA, 1 mg thiamine, 100 mg inositol, 7 g agarose, 30 g sucrose, final pH 5.8) in a Petri dish and should be bombarded as described (*see* Section 3.2.). Seven days after bombardment, leaf tissue should be transferred to growth medium containing appropriate antibiotic (e.g., 100 mg/L kanamycin). Antibiotic resistant calli should be selected and transgenic plants should be regenerated from transformed leaf-derived calli. For optimal conditions for regeneration of different plant species, *see* references listed in Table 1. For chloroplast transformation *see* Section 5. Plantlets regenerated from antibiotic containing medium may be examined for foreign gene expression and stable integration of introduced foreign genes as described (Section 3.4.).

3.4. Molecular Analyses of Transformants

Chloroplast (ct) and nuclear DNA should be isolated from transgenic plants. Ethidium bromide stained gels of restriction digested ctDNA preparations may be examined to detect additional ctDNA fragments; insertion of foreign gene fragments from chloroplast vectors into ct genomes should introduce additional restriction sites into ct genomes of transgenic plants. These additional restriction sites are useful restriction fragment length polymorphism (RFLP) markers to detect homoplasmy or heteroplasmy of ct genomes of transgenic plants. In addition, ct or nuclear DNA digested with suitable restriction enzymes may be separated by electrophoresis in agarose gels and blotted onto nylon membranes. A fragment containing the foreign gene coding sequence may be used as the probe. All transgenic lines should be tested for the presence of the foreign gene coding sequence in chloroplast or nuclear genomes. Northern blots may be performed to verify the presence of foreign gene transcripts in order to confirm expression of the introduced gene. Western blots should reveal the presence of the foreign gene product.

3.5. Genetic Analyses of Transformants

Resistant clones obtained through chloroplast or nuclear engineering may be regenerated to mature fertile plants. Seeds obtained from transgenic plants and control untransformed plants may be germinated in the presence of appropriate concentrations of antibiotic or selection agent. Expression of the foreign gene product should be assessed in different tissues. By self-fertilizing the regenerated plants, the genetic transmission pattern of the first generation may be examined. A typical 3:1 ratio of resistant to sensitive seeds will indicate that the introduced foreign gene is inherited as a dominant Mendelian factor. However, if the resistance is not transmitted through the pollen, a typical maternal inheritance will be observed suggesting that the chloroplast genome has been altered. In several reports of stable genetic transformation of the tobacco chloroplast genome, the seed progeny obtained after selfing were resistant to appropriate antibiotics; lack of segregation indicated that the regenerated plants were homoplasmic for the antibiotic resistant marker *(15)*.

3.6. Assays for Reporter Gene Expression

3.6.1. Chloramphenicol Acetyltransferase

1. Transfer bombarded cells to corex tubes and wash once with 10 mL of TE buffer (250 m*M* Tris-HCl, 10 m*M* EDTA, pH 7.8). Pellet cells at 8000*g* for 10 min and transfer the cell pellet to 2-mL Eppendorf tubes and resuspend in 1 mL of TE buffer containing 2 m*M* phenylmethyl sulfonyl fluoride.
2. Sonicate cells for 20 s, twice using a probe sonicator.
3. After 15 min centrifugation at 4°C, transfer the supernatant to a new Eppendorf tube and assay for CAT activity.
4. To 500 μL of extract add 20 μL of D-threo (dichloro acetyl-1-^{14}C) chloramphenicol (1.85 MBg/m*M*, 0.5 μCi) and acetyl coenzyme A (0.5 mM).
5. After incubation for 30 min at 37°C, add 1 mL of ethyl acetate, vortex vigorously and spin briefly to separate the aqueous phase from the organic phase.
6. Transfer the organic phase to a new tube and evaporate in a speed vacuum concentrator.
7. Dissolve the acetylated products in 25 μL of ethyl acetate, spot the extract on silica gel thin layer plates and separate in chloroform-methanol (95:5 ascending).
8. After autoradiography, scrape the acetylated forms of chloramphenicol and count in a suitable scintillator or use a suitable image analysis system to quantify expression.

3.6.2. β-Glucuronidase

1. Prepare GUS buffer (final concentrations given in parenthesis): NaH_2PO_4 (100 m*M*), 1.38 g EDTA (10 m*M*) 372 mg, $K_4Fe(CN)_6$ (0.5 m*M*) 21 mg. Make up volume to 100 mL (autoclaved H_2O) and adjust pH to 7.0 with NaOH. Filter sterilize and store at 4°C.
2. Prepare substrate solution for the reaction (for 10 plates): Dissolve 5 mg of X-Gluc (5-bromo-4-chloro-3-indoyl-β-d-glucuronic acid, 0.5 μg/mL) in 100 μL of DMSO and then add to 10 mL of GUS buffer. Add 10 μL of 100% Triton X-100 (0.1%). Filter sterilize solution.
3. Add 0.5 mL to each plate containing the bombarded sample, preferably on the callus or the leaf or cells in the bombarded area. Keep all stock solutions and containers sterile to avoid contamination. Incubate at 37°C overnight.

3.7. Construction of Expression Vectors

3.7.1. Chloroplast Expression Vectors

Work in my laboratory for the past several years has involved construction of chloroplast expression vectors either for stable integration of foreign genes into the chloroplast genome or for transient/stable expression and autonomous replication of introduced plasmids in chloroplasts. As a first step toward achieving this, I have constructed a series of chloroplast expression vectors, using the promoter selection vector pKK232-8 (Pharmacia, Uppsala, Sweden), which is a pBR322 derivative containing a promoterless *cat* gene. A multiple cloning site (MCS) has been placed 5' proximal to the *cat* gene to facilitate insertion and analysis of promoter fragments. Transcription/translation of *cat* can be used to quantify the strength of promoters inserted into the MCS of pKK232-8. The plasmid contains the ribosomal RNA T1 and T2 terminators distal to the *cat* gene to allow cloning of strong promoters and three stop codons between MCS and the AUG of the *cat* gene to prohibit translational read-through into the *cat* gene. Restriction fragments of ctDNA containing the entire promoter region and 5'-untranslated region of the *psb*A gene from spinach (pHD306) or pea (pHD312) or the *rbc*L and *atp*B promoter region from maize (pHD103) were individually inserted into the MCS site; colonies were screened on LB plates containing chloramphenicol. Plasmids containing chloroplast promoter fragments have been investigated by analyzing transient expression of *cat* in cucumber etioplasts using the methodology of Daniell and McFadden *(10)*. The spinach or pea psbA promoter was found to be the strongest among all the promoters tested. The chloroplast expression vector pHD203 (Fig. 1A) contains a double *psb*A promoter fragment, in opposite orientation to facilitate insertion of additional genes. While one *psb*A promoter region would drive the

Fig. 1. *(facing page)* Chloroplast and nuclear expression vectors. **(A)** The chloroplast expression vector pHD203 contains two *psb*A promoter fragments inserted in opposite orientations to facilitate simultaneous transcription of two promoterless genes. **(B)** *E. coli uid*A gene coding for β-glucuronidase has been inserted into the MCS of pHD203 at *Pst*I-SmaI sites resulting in pHD203-GUS. **(C)** Pea ctDNA replication origin has been inserted into a chloroplast expression vector pHD312 which contains a *cat* gene driven by the pea *psb*A promoter resulting in the construct pHD407. **(D)** Stable chloroplast expression vector pHD-CG-*npt*II; the *npt*II gene present within chloroplast borders is driven by the *psb*A promoter and the unique SmaI site present between the *npt*II gene and the T1 terminator may be used to insert additional genes of interest. **(E)** The nuclear expression vector pBI121 contains a selectable marker gene (*npt*II) driven by the *nos* promoter and a reporter gene (*uid*A) driven by the CaMV promoter and are flanked by *nos* terminators.

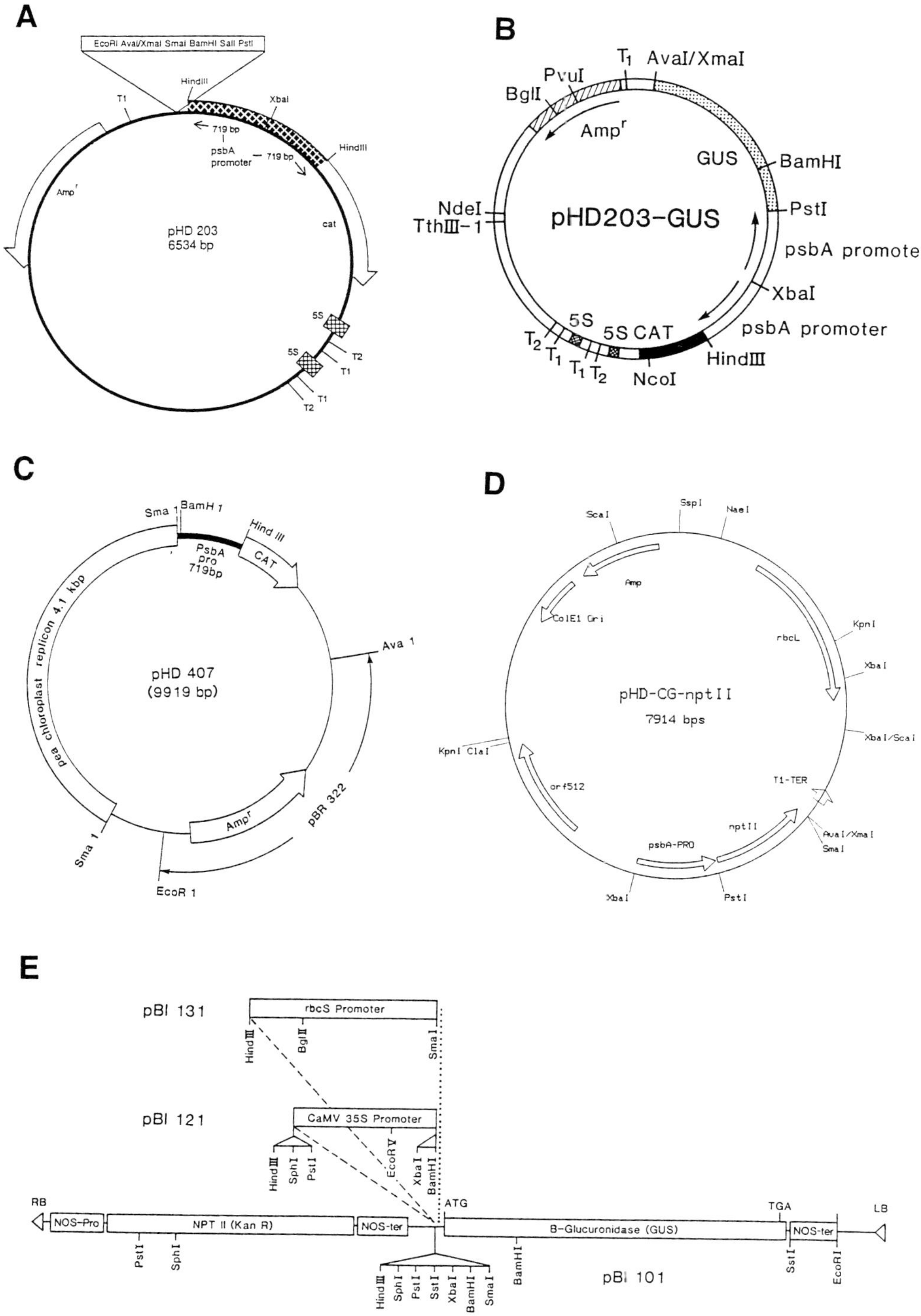
A
EcoRI AvaI/XmaI SmaI BamHI SalI PstI
HindIII
T1
XbaI
719 bp
psbA promoter
719 bp
HindIII
Amp^r
pHD 203
6534 bp
cat
5S
T2
5S
T1
T1
T2
B
T1
AvaI/XmaI
PvuI
BglI
Amp^r
GUS
BamHI
NdeI
TthIII-1
pHD203-GUS
PstI
psbA promoter
XbaI
5S
5S CAT
psbA promoter
T2
T1
T1
T2
NcoI
HindIII
C
Sma 1
BamH 1
Hind III
PsbA pro 719bp
CAT
pea chloroplast replicon 4.1 kbp
pHD 407
(9919 bp)
Ava 1
Amp^r
pBR 322
Sma 1
EcoR 1
D
ScaI
SspI
NaeI
Amp
ColE1 Ori
rbcL
KpnI
XbaI
pHD-CG-nptII
7914 bps
XbaI/ScaI
KpnI ClaI
orf512
T1-TER
nptII
AvaI/XmaI
SmaI
psbA-PRO
XbaI
PstI
E
pBI 131
rbcS Promoter
HindIII
BglII
SmaI
pBI 121
CaMV 35S Promoter
HindIII
SphI
PstI
EcoRV
XbaI
BamHI
RB
NOS-Pro
NPT II (Kan R)
NOS-ter
ATG
B-Glucuronidase (GUS)
TGA
NOS-ter
LB
PstI
SphI
HindIII
SphI
PstI
SstI
XbaI
BamHI
SmaI
BamHI
pBI 101
SstI
EcoRI

cat gene, the second promoter fragment is placed upstream of a MCS containing sites for several unique restriction enzymes. There is a ribosomal RNA T1 terminator distal to the MCS that would facilitate subcloning genes driven by strong promoters. There are convenient *Eco*RI and *Pst*I sites within *cat* and b-lactamase genes, respectively to screen for partial digestion of pHD203. *E. coli uid*A gene coding for β-glucuronidase has been inserted into the MCS of pHD203 at PstI-SmaI sites (Fig. 1B).

3.7.2. Insertion of Chloroplast Origin of Replication into Chloroplast Expression Vectors

In order to increase the copy number of the introduced plasmid, origin of replication sequences from plastid genomes may be included in chloroplast vectors. Several pea chloroplast DNA fragments *(21)* containing one or both replication origins have been identified and characterized. A well defined chloroplast replicon has been inserted into the chloroplast expression vector pHD312, which contains the pea *psb*A promoter 5' proximal to the promoterless *cat* gene, resulting in the construction of pHD407 (Fig 1C).

3.7.3. Insertion of Borders for Stable Integration into Chloroplast Genomes

Foreign genes should be integrated into chloroplast genomes at proper sites so that other genome functions are not disrupted. Stable chloroplast vectors also need selectable marker genes to screen for chloroplasts that contain foreign gene products. Methodology for construction of a typical stable chloroplast expression vector is described here. The chloroplast expression vector pHD203 (Fig. 1A) containing chloroplast *psb*A promoters in opposite orientations may be digested with a suitable enzyme, dephosphorylated and ligated with a gene of interest. Appropriate chloroplast border sequences may be inserted on either side of the foreign gene. For example, for constructing the vector pHD-CG-*npt*II, first tobacco chloroplast genomic sequences corresponding to *rbc*L and *orf*512 coding regions (57750–60593) were subcloned into the bluescript KS^+ vector as a SstI-EcoRV fragment resulting in the construct pBS-JB-ctBorders. The spacer region between these two coding regions was selected for inserting the selectable marker genes (*npt*II gene for kanamycin selection of transformed plastids) *(22)* in order to accomplish a high frequency of transformation *(16)*. All these genes were first independently cloned into pHD203 vector (Fig. 1A) which provides the chloroplast *psb*A promoter (5') and a strong T1 terminator (3'). The pBS-JB-ctBorders DNA was digested with XbaI, Klenow filled, and dephosphorylated. The *npt*II insert was taken out of pHD203-*npt*II as a XbaI-ScaI fragment by electroelution and was ligated with the vector DNA (pBS-JB-ct-Borders). The resultant construct (pHD-CG-nptII, Fig. 1D) contained the *npt*II gene within chloroplast border sequences. This vector may be further digested at a unique SmaI site (present between the nptII gene and the T1 terminator) and dephosphorylated. Any foreign gene of interest, driven by a chloroplast promoter may be inserted into the unique SmaI site in pHD-CG-nptII (after dephosphorylation of the vector DNA), provided the restriction fragment contains blunt ends; if one or both ends are sticky, restriction fragments may be inserted into the SmaI site after Klenow filling of sticky ends and ligation with the dephosphorylated vector (pHD-CG-*npt*II).

3.7.4. Nuclear Expression Vectors

The nuclear expression vector pPBI121 (Fig. 1E) carries a *uid*A gene driven by a CaMV-35S promoter and flanked at the 3' end by a *nos* polyA fragment *(23)*. The nuclear expression vector pUC8 CaMV CAT N is a 4.2 kbp plasmid containing a *cat* gene driven by a 35S CAMV promoter, flanked by a 3' *nos* PstI poly A fragment. For more information on nuclear expression vectors, refer to the previous chapter of this volume. For negative controls, pUC19 DNA or appropriate vector DNA should be used in all bombardments.

3.8. Evaluation of Results

3.8.1. β-Glucuronidase Expression in Anther-Derived Albino Plants

It was of interest to study GUS expression in anther-derived albino plants. Certainly, it was anticipated that the blue GUS product might be especially easy to visualize. Figure 2A shows the expression of GUS in the albino leaf bombarded with pHD203-GUS (left) but not in that bombarded with control pUC19 (right). The product of the *uidA* gene, β-glucuronidase, when present, cleaves glucuronic acid from the substrate X-gluc to produce an insoluble indigo dye following oxidative dimerization. Expression of GUS in albino leaves bombarded with pHD203-GUS (Fig. 2A) suggests the presence of a functional protein synthetic machinery in albino plastids. Chloroplast specific expression of GUS by pHD203-GUS is discussed in the next section.

3.8.2. Compartmentalized β-Glucuronidase Expression

Green plants derived from anther culture were preferred for studies on gene expression because the results were comparable to field grown plants but at the same time plants were free of bacteria since they were grown under totally sterile conditions. Though the tungsten particles were seen in samples bombarded with pUC19, no GUS expression was observed; this shows that the bombarded tissues were free of bacterial contamination. On the other hand, it was evident from samples that were bombarded with pPBI121 and pHD203-GUS that β-glucuronidase, when present, cleaved glucuronic acid from the

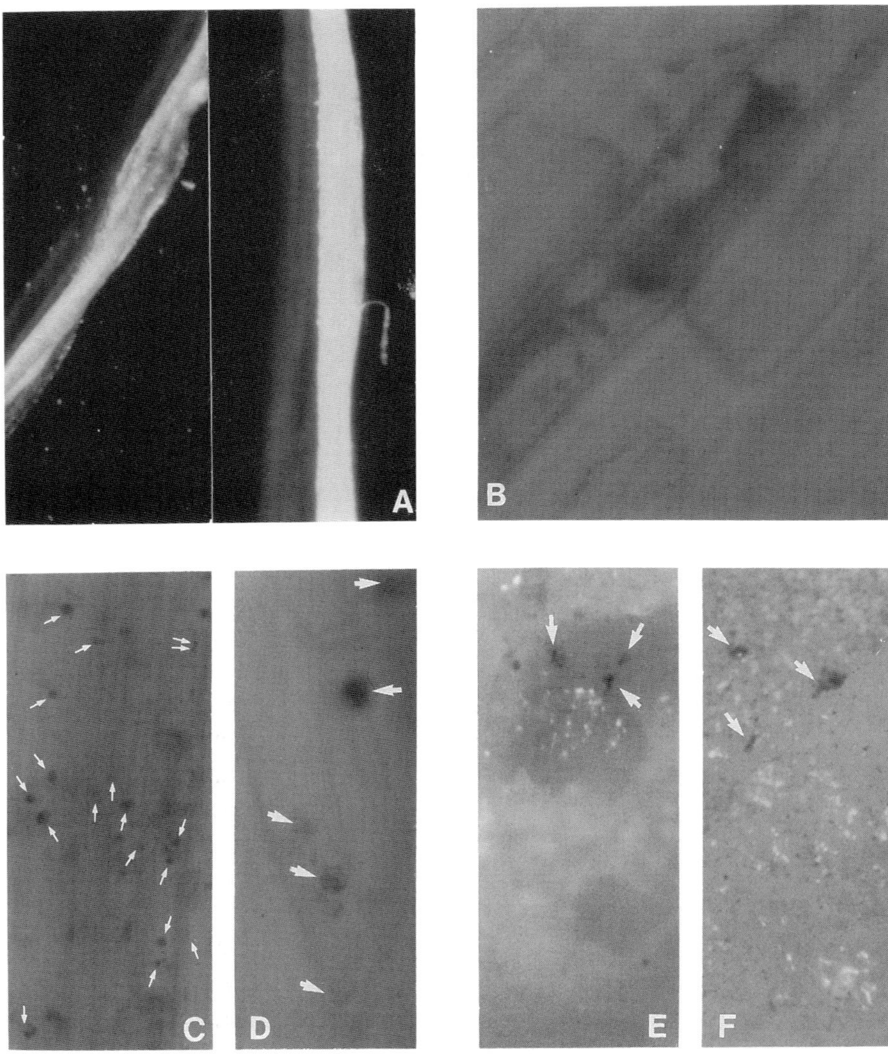

Fig. 2. (A) Expression of GUS in the albino wheat leaf bombarded with pHD203-GUS (left) and lack of expression in the leaf bombarded with pUC19 (right). Magnification: ×64 in the dissecting microscope (Zeiss, Steml-SV8). **(B)** Microscopic observation of GUS expression in cells from a green leaf bombarded with pBl121. Note the presence of GUS derived product spread evenly throughout the cytosol. Magnification: ×1411 viewed under oil immersion in Zeiss Axioplan. **(C,D)** Subcellular localization of GUS expression in cells from a green leaf bombarded with pH203-GUS. Arrows indicate plastids that expressed GUS. (C) Magnification: ×640, (D) Magnification ×1411, viewed under oil immersion in Zeiss Axioplan. **(E,F)** Calli were bombarded with (E) pHD203-GUS, (F) pUC19. Note the tungsten-DNA clumps

(continued)

substrate X-gluc to produce an insoluble indigo dye. In order to locate the compartment in which gene products from pPBI121 or pHD203-GUS function, bombarded leaves from anther-derived green plants were examined under the microscope. It is evident from Fig. 2B that the β-glucuronidase-derived product was present evenly throughout the cytosol when the nuclear expression vector pPBI121 had been used to bombard wheat leaves. On the other hand, when chloroplast expression vector pHD203-GUS was used for bombardments, the indigo dye was subcellularly localized within wheat cells (Fig. 2C,D). Chloroplasts noticeably lost their green color after the addition of GUS substrate, probably because the substrate contained organic solvents and detergents that destabilize pigment protein complexes. These results also show that a dicot chloroplast promoter (pea *psb*A) can indeed function efficiently in a monocot (wheat) chloroplast; thus chloroplast promoters are interchangeable among monocots and dicots; the *cat* gene driven by the maize *rbc*L promoter also functioned in tobacco chloroplasts *(12)*.

3.8.3. β-Glucuronidase Expression in Calli Derived from Immature Embryos

While anther-derived albino and green plants are ideal to study transient expression of foreign genes, regeneration of wheat plants from bombarded tissues may be a formidable challenge. Therefore, calli rich in embryonic tissue were generated from immature embryos of wheat. Figure 2E shows the expression of GUS in regenerable calli derived from immature embryos. When bombarded with foreign DNA, callus clumps were shattered upon impact of tungsten particles; however, this did not affect their subsequent gene expression. No background indigo dye was detected in negative controls, bombarded with pUC19, after incubation with the GUS substrate, indicating again that the tissues were free of bacterial contamination (Fig. 2F). A number of blue areas may be seen in the callus bombarded with pHD203-GUS (Fig. 2E), indicating that chloroplasts in a number of targeted cells were transformed. The data in Fig. 2 has been reprinted from Daniell et al. *(13)*.

3.8.4. Expression of cat *in Cultured Tobacco Cells*

Cultured NTI tobacco cells collected on filter papers were bombarded with tungsten particles coated with pUC118 (negative control), 35S CAT (nuclear expression vector), pHD312 (repliconless chloroplast expression vector), and

---

Fig. 2. *(continued)* (arrows) at the sites of bombardment and the scattered blue areas outside these sites in the callus bombarded with pHD 203-GUS. Magnification ×64 in the dissecting microscope (Zeiss, Steml-SV8).

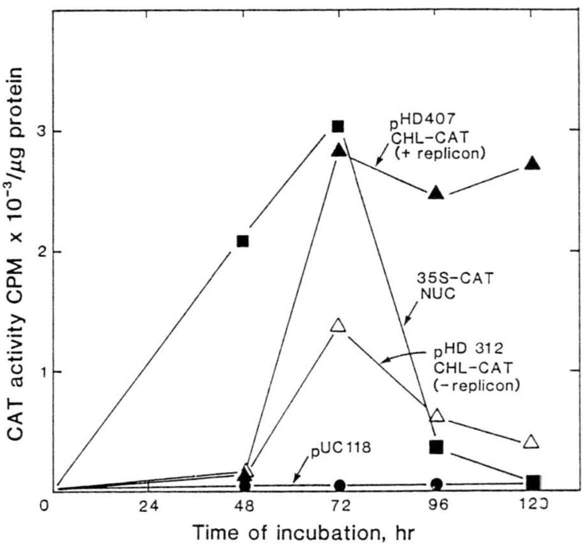

Fig. 3. Quantitative analysis of *cat* expression in tobacco NT1 suspension cells. Cultured cells were bombarded with pUC118 (negative control), 35S-CAT (nuclear expression vector), pHD312 (chloroplast expression vector without a replicon), and pHD407 (chloroplast expression vector containing pea ctDNA replicon). CAT was assayed by thin layer chromatography of [14C] chloramphenicol and its faster migrating acetylated products; after autoradiography of the separated acetylated chloramphenicol forms for 4 h, spots were scraped and radioactivity was counted. Silica gel had a background of 2203 cpm; the cpm ranges for different samples were as follows: pUC118: 2705–6556; 35S-CAT: 3993–220,353; pHD312: 2410–133,240; pHD407: 7484–267,364. Reprinted ref. *12* with permission.

pHD407 (chloroplast expression vector with replicon). Sonic extracts of cells bombarded with pUC118 showed no detectable cat activity in the autoradiograms (Fig. 3). Nuclear expression of *cat* was maximal 72 h after bombardment. Cells bombarded with chloroplast expression vectors showed a low level of expression until 48 h of incubation. An increase in the expression of *cat* was observed at 72 h in samples bombarded with pHD407; the repliconless vector pHD312 showed about 50% of this maximal activity. Although the expression of nuclear *cat* and the repliconless chloroplast vector decreased after 72 h, a high level of chloroplast *cat* expression was maintained in cells bombarded with pHD407. Replication of chloroplast expression vectors should have resulted in increased copy number, thereby increasing *cat* expression. These experiments paved the way for obtaining transgenic tobacco plants using autonomously replicating chloroplast vectors *(24)*. Organelle-specific expression of *cat* in appropriate compartments was checked by introducing various plasmid constructions into tobacco protoplasts by electroporation *(12)*. It is known that electroporation of protoplasts results in DNA delivery into the cytosol and not inside organelles.

3.8.5. Comparison of Different Biolistic Devices

We have used different biolistic systems to study the expression of foreign genes in various cellular compartments of plant cells *(14)*. Chloroplast transformation efficiencies increased dramatically (about 200-fold) using the He-driven biolistic device as compared to the more commonly used gunpowder charge driven device (Table 2). Using *uid*A as a reporter gene and the improved biolistic device, optimal bombardment conditions were established, consistently producing several hundred chloroplast transformants per Petri plate. Chloroplast transformation efficiency was found to be increased further (20-fold) with supplemental osmoticum in the bombardment and incubation media. Russell et al. *(25)* also have reported lower transformation frequencies in experiments that utilized the gunpowder device and tungsten particles as compared to the He device and gold particles owing to physical damage to cells from the gas blast and acoustic shock generated by the gunpowder device and toxicity of the tungsten particles. Tables 2 *(14)* and 3 *(26)* show a comparison of transformation efficiencies using different biolistic delivery systems.

4. Notes

Strictly follow the order of additions for coating tungsten particles with DNA; any change would result in a dramatic decrease in DNA binding. Make sure the stopping plate is kept in place and the vacuum is at least 28 in. of Hg. The degree of vacuum in the sample chamber during bombardment has a dramatic effect on cell viability and subsequent expression of the introduced foreign gene; maximal transformation was observed at 30 in. of Hg, whereas no transformation was observed at 10 in. of Hg *(19)*. Spray the gun chamber with 70% ethanol and wipe it with sterile tissues after each bombardment. Do not force the bullet into the barrel. DNA should be free of protein; otherwise it will form clumps with tungsten particles. DNA free of proteins may be obtained by repeated phenol-chloroform extractions or proteinase treatment followed by ethanol precipitation.

In order to avoid clumps during bombardment, which greatly reduces transformation efficiency, it is good practice to prepare a fresh stock of tungsten/DNA mixture and bombard as described below:

1. Suspend 50 mg of tungsten particles in 1 mL of absolute ethanol.
2. Sonicate suspension at maximum power for 10 min, three times (this can be stored at –20°C indefinitely).

Table 2
Effect of Configuration, Helium Pressure, Sample Level and Sleeve Level in NT1 Chloroplast Transformation Efficiency[a]

Configuration	Pressure (psi)	Sample level,[b] cm	Sleeve level, [c]cm	Efficiency[d]
He entrainment	900 (61 atm)	6.1	1.6	174
		6.1	0.55	134
		8.1	1.0	124
	1200 (82 atm)	8.1	1.6	163
		4.1	1.6	60
		8.1	0.55	36
		6.1	1.0	9
	1500 (102 atm)	6.1	1.6	59
		6.1	0.55	46
		4.1	1.0	2
		8.1	1.0	227
Flying disk	900 (61 atm)	6.1	1.6	18
		6.1	0.55	52
		4.1	1.0	27
		8.1	1.0	9
	1200 (82 atm)	8.1	1.6	2
		4.1	1.6	5
		4.1	0.55	4
		8.1	0.55	4
		6.1	1.0	1
	1500 (102 atm)	6.1	1.6	2
		6.1	0.55	3
		4.1	1.0	3
		8.1	1.0	2

[a]Modified from ref. *14*.
[b]cm distance from particle launch point to target cells.
[c]cm distance from rupture membrane to flying disc or nylon mesh.
[d]Efficiency is expressed as the number of blue spots per plate. Each treatment contained six replicates.

3. Take 250 µL of tungsten stock suspension and centrifuge for 5 s.
4. Remove ethanol and wash three times with sterile distilled water, centrifuging 3 min between washings.
5. Resuspend tungsten in 250 µL sterile distilled H_2O and aliquot 50 µL into each Eppendorf tube.
6. Into each Eppendorf tube add sequentially: 10 µL DNA (1 µg/µL), 50 µL of 2.5*M* $CaCl_2$, 20 µL of 0.1*M* spermidine-free base; vortex at 4°C for 20 min; add 200 µL of absolute ethanol to each tube and spin at 3000 rpm in a microfuge for 2 s

Table 3
Summary of Numerous Experiments Comparing Transformation Efficiency in Gunpowder-Driven vs He-Driven Systems[a]

	Number of Transformants		
Target	Gunpowder	He	Fold Increase
Yeast (colonies)	81	1792	22
Bacillus megaterium (colonies)	0.8	265	331
Escherichia coli (colonies)	3.8	380	100
Tobacco (NT1) (blue spots)	540	3115	6
Tobacco (NT1) (Km' calli)	3.6	23	6
Chicken Myotubes (luc activity)	112 k	1920 k	11
Mouse Skin (luc activity)	300	1543	5
Mouse Ear (luc activity)	1312	5563	4
Mouse Liver (luc activity)	0	309	—

[a]Modified from ref. *26*.

(tungsten sediments very easily in ethanol; therefore tungsten/DNA ethanol spins must be very brief; otherwise resulting pellet will not breakup; remove supernatant and rinse pellet in absolute ethanol four more times, centrifuging at 3000 rpm for 2 s) after each wash; resuspend pellet in 30 µL absolute ethanol and store mixture on ice (DNA should be used for bombardment within 2 h after preparation).

7. Pipet 5 µL of tungsten/DNA mixture onto the center of the macrocarrier (sterilized in 100% ethanol, dried in desiccant) which should spread evenly (any chunk will result in cell death).
8. Place the macrocarrier with tungsten/DNA mixture into a dish of desiccant and cover to aid drying.
9. Place a rupture disk in the holder ring and tighten ring to the helium barrel (sterilized in 70% ethanol).
10. Place a stopping screen and the macrocarrier with tungsten/DNA (in holder) into the retaining assembly and screw down.
11. Place assembly into the vacuum chamber at level 2 (second from top).
12. Place sample at the 5th level (from top).
13. Bombard as described in Section 3.2.2.

For better growth rate, plants may be aseptically grown on phytagar (0.6%) containing MS salts (4.3 g/L), sucrose (30 g/L), and B5 vitamin stock (100 mg/mL myo-inositol, 10 mg/mL thiamine-HCl, 1 mg/mL nicotinic acid and 1 mg/mL pyrodioxine-HCl). Add 1 mL of B5 vitamin stock for 1 L of the growth medium. Plants generated from apical meristems or axillary buds grow much faster than those germinated from seeds.

For chloroplast transformation, it is better to bombard and regenerate tobacco plants in RMOP medium containing: BA or BAP 1 mg/L, 1 NAA 0.1 mg/L, thiamine-HCl 1 mg/L, myo-insitol 100 mg/L, MS salts 4.3 g/L, phytagar 6 g/L, and sucrose 30 g/L. The duration of gene expression before the initiation of selection process may vary based on antibiotic or the selectable marker. It is advisable to choose a lower concentration of antibiotic (e.g., 50 µg/mL of kanamycin for chloroplast as opposed to 100 µg/mL for nuclear transformation).

Transgenic plants from approximately 30 commercially important cultivars of soybean and cotton plants from all major elite varieties have been obtained using a variety independent gene transfer procedures *(5)*. In soybean, transgenic plants were recovered through an organogenic process that resulted in chimeric as well as clonal plants, whereas in cotton almost all transgenic plants recovered to date have been the result of direct germination of the bombarded explant followed by pruning of nontransformed sectors on the plants to allow development of axils that have been clonally transformed *(5)*. This clearly demonstrates that the selection of the target tissue is a crucial step of utmost importance in the process of developing a stable transformation system. Explants utilized for transformation for different plant species is shown in Table 1. Another important aspect of stable expression studies is the choice of the selection agent. For example, there are numerous reports of a lack of correlation between transient β-glucuronidase (GUS) expression and stable *npt*II expression based on kanamycin selection; however, Tor et al. *(20)* have reported a 100% correlation between geneticin (G418) resistance and GUS expression. Geneticin is, therefore, a potent selection agent for the *npt*II gene.

Acknowledgments

The author acknowledges fruitful collaborations with L. Bogorad, J. C. Sanford, G. N. Ye, B. A. McFadden, B. L. Nielsen, and K. K. Tewari. The results presented here were supported by the following grants to HD: NSF Grant 8902065, State Board of Education Grants 88-022, 89-022, 89-010, and Washington Technology Center grant 312-206, USDA NRICGP grants 92-37301-7782, 93-37311-9455, 95-02770 U. S. Army, Natick RD & E contract DAAK60-93-C-0094, and NIH grant GM 16551-01.

References

1. Binns, A. N. (1990) Agrobacterium mediated gene delivery and the biology of host range limitations. *Physiol. Plant.* **79,** 135–139.

2. Jenes, B., Moore, H., Cao, J., Zhang, W., and Wu, R. (1993) *Techniques* for gene transfer in transgenic plants, in *Engineering and Utilization, vol. I* (Kung, S. and Wu, R., eds.), Academic, New York, pp. 125–146.
3. Fromm, M., Callis, J., Taylor, L. P., and Walbot, V. (1987) Electroporation of DNA and mRNA into plant protoplasts. *Methods Enzymol.* **153,** 351–366.
4. Neuhans, G. and Spangenberg, G. (1990) Plant transformation by microinjection techniques. *Physiol. Plant.* **79,** 213–217.
5. Christou, P. (1994) Application to plants: Agricultural Biotechnology and Crop Improvement, in *Particle Bombardment Technology for Gene Transfer* (Yang, N. and Christou, P., eds.), Oxford Univ. Press, Oxford, pp. 71–99.
6. Sanford, J. C. (1990) Biolistic plant transformation. *Physiol. Plant.* **79,** 206–209.
7. Yang, N. S. and Christou, P. (1994) Particle bombardment technology for gene transfer, UWBC Biotechnical Resources Series, Oxford Univ. Press, Oxford.
8. Rasmussen, J. L., Kikkert, J. R., Roy, M. K., and Sanford, J. C. (1994) Biolistic transformation of tobacco and maize suspension cells using bacterial cells as micro-projectiles. *Plant Cells Reports* **13,** 212–217.
9. Daniell, H. (1993) Foreign gene expression in chloroplasts of higher plants mediated by tungsten particle bombardment. *Methods Enzymol.* **217,** 536–556.
10. Daniell, H. and McFadden, B. A. (1987) Uptake and expression of bacterial and cyanobacterial genes by isolated cucumber etioplasts. *Proc. Natl. Acad. Sci. USA* **84,** 6349–6353.
11. Eigel, L., Oelmuller, R., and Koop, H. U. (1991) Transfer of defined number of chloroplasts into albino protoplasts using an improved subprotoplast/protoplast microfusion procedure:transfer of only two chloroplasts leads to variegated progeny. *Mol. Gen. Genet.* **227,** 446–451.
12. Daniell, H., Vivekananda, J., Nielsen, B., Ye, G. N., Tewari, K. K., and Sanford, J. C. (1990) Transient foreign gene expression in chloroplasts of cultured tobacco cells following biolistic delivery of chloroplast vectors. *Proc. Natl. Acad. Sci. USA.* **87,** 88–92.
13. Daniell, H., Krishnan, M., and McFadden, B. A. (1991) Expression of β-glucuronidase gene in different cellular compartments following biolistic delivery of foreign DNA into wheat leaves and calli. *Plant Cell Rep.* **9,** 615–619.
14. Ye, G. N., Daniell, H., and Sanford, J. C. (1990) Optimization of delivery of foreign DNA into higher plant chloroplasts. *Plant Mol. Biol.* **15,** 809–820.
15. Svab, Z., Hajdukiewitz, P., and Maliga, P. (1990) Stable transformation of plastids in higher plants. *Proc. Natl. Acad. Sci. USA.* **87,** 8526–8530.
16. Svab, Z. and Maliga, P. (1993) High frequency plastid transformation in tobacco by selection for a chimeric aadA gene. *Proc. Natl. Acad. Sci. USA* **90,** 913–917.
17. Block, M. de, Shell, J., and Montagu, M. van (1985) Chlorolast transformation by *Agrobacterium tumefaciens*. *EMBO J.* **4,** 1367.
18. Sporlein, B., Streubel, M., Dahlfeld, G., Westhoff, P., and Koop, H. U. (1991) PEG- mediated plastid transformation. A new system for transient gene expression assays in chloroplasts. *Theor. Appl. Genet.* **82,** 717–722.

19. Klein, T. M., Gradziel, T., Fromm, M. E., and Sanford, J. C. (1988) Factors influencing gene delivery into Zea mays cells by high-velocity microprojectiles. *Bio/Technology* **6,** 559–563.
20. Tor, M., Ainsworth, C., and Matell, S. H. (1993) Stable transformation of the food yam *Dioscorea alata* L. by particle bombardment. *Plant Cell Rep.* **12,** 468–473.
21. Reddy, M. K., Chondhry, N. R., Kumar, D., Mukherjee, S. K., and Tewari, K. K. (1994) Characterization and mode of in vitro replication of pca chloroplast ori A sequence *Eur. J. Biochem.* **220,** 993–941.
22. Carrer H., Hockenberry, T. N., Svab, Z., and Maliga, P. (1993) Kanamycin resistance as a selectable marker for plastid transformation in tobacco. *Mol. Gen. Genet.* **241,** 49–56.
23. Jefferson, R. A. (1987) Assaying chimeric genes in plants: the GUS gene fusion system. *Plant Mol. Biol. Rep.* **5,** 387–405.
24. Staub, J. M. and Maliga, P. (1994) Extrachromosomal elements in tobacco plastids. *Proc. Natl. Acad. Sci. USA* **91,** 7468–7472.
25. Russell, D. R., Wallace, K. M., Bathe, J. H., Martinell, B. J., and McCabe, D. E. (1993) Stable transformation of *Phaseolus vulgaris* via electric discharge-mediated particle acceleration. *Plant Cell Rep.* **12,** 165–169.
26. Sanford, J. C., Devit, M. J., Russell, J. A., Smith, F. D., Harpending, P. R., Roy, M. K. and Johnston, S. A. (1991) An improved, helium driven biolistic device. *Technique* **3,** 3–16.
27. Seki, M., Shigemoto, N., Komeda, Y., Imamura, J., Yamada, Y., and Morikawa, H. (1991) Transgenic *Arabidopsis thaliana* plants obtained by particle bombardment mediated transformation. *Appl. Microb. Biotech.* **36,** 228–230.
28. Wan, Y. and Lemaux, P. G. (1994) Generation of large numbers of independently transformed fertile barley plants. *Plant Physiol.* **104,** 37–48.
29. Ritala, A., Mannonen, L., Aspegren, K., Salmenkallio-Martila, M., Kurten, U., Hannus, R., Lozano, J. M., Teeri, T. H., and Kauppinen, V. (1993). Stable transformation of barley tissue culture by particle bombardment. *Plant Cell Rep.* **12,** 435–440.
30. Ritala, A., Aspergen, K., Kurten, U., Salmenkallio-Marttila, M., Mannonen, L., Hannus, R., Kauppinen, V., Teeri, T. H., and Enari, T.-M. (1994) Fertile transgenic barley by particle bombardment of immature embryos. *Plant Mol. Biol.* **24,** 317–325.
31. Finer, J. J. and McMullen, M. D. (1990) Transformation of cotton (*Gossypium hirsutum* L.) via particle bombardment. *Plant Cell Rep.* **8,** 586–589.
32. McCabe, D. E. and Martinell, B. J. (1991) Particle gun transformation applied to cotton, in *Molecular Biology of Plant Growth and Development.* ISPMB 3rd International Congress. (Hallick, R. B., ed.), Univ. of Arizona Press, Tucson.
33. Serres, R., Stang, E., McCabe, Russell, D., Mahr, D., and McCown B. (1992) Gene transfer using electric discharge particle bombardment and recovery of transformed cranberry plant. *J. Amer. Soc. Hort. Sci.* **117,** 174–180.
34. Gordon-Kamm, W. J., Spencer, T. M., Mangano, M. L., Adams, T. R., Daines, R. J., Start, W. G., O'Brien, J. V., Chambers, S. A., Adams, W. R., Willetts, N. G., Rice, T. B.,

Mackey, C. J., Krueger, R. W., Kausch, A. P., and Lemaux, P. G. (1990) Transformation of maize cells and regeneration of fertile transgenic plants. *Plant Cell* **2,** 603–618.

35. Fromm, M. E., Morrish, F., Armstrong, C., Williams, R., Thomas, J., and Klein, T. M. (1990). Inheritance and expression of chimeric genes in the progeny of transgenic maize plants. *Bio/Technology* **8,** 833–844.
36. Koziel, M. G., Beland, G. L., Bowman, C., Carozzi, N. B., Crenshaw, R., Crossland, L., Dawson, J., Desai, N., Hill, M., Kadwell, S., Launis, K., Lewis K., Maddox, D., McPherson, K., Megjki, M. R., Merlin, E., Rhodes, R., Warren, G. W., Wright, M., and Evola, E. V. (1993) Field performance of elite transgenic maize plants expressing an insecticidal protein derived from *Bacillus thuringiensis*. *Bio/Technology* **11,** 194–200.
37. Somers, D. A., Rines, H. W., Gu, W., Kaeppler, H. F., and Bushnell, W. R. (1992) Fertile, transgenic oat plants. *Bio/Technology* **10,** 1589–1594.
38. Fitch, M. M. M., Manshardt, R. M. Gonsalves, D., Slightom, J. L., and Sanford, J. C. (1990) Stable transformation of papaya via microprojectile bombardment. *Plant Cell Rep.* **9,** 189–194.
39. Brar, G. S., Cohen, B. A., and Vick, C. L. (1992) Germline transformation of peanut (*Arachis hypogaea* L.) utilizing electric discharge particle acceleration (Acell Tm) technology. Proceedings of the American Peanut Research and Education Soc., Norfolk, Virginia, **24,** 21.
40. Dayton-Wilde, H., Meagher, R. B., and Merkle, S. A. (1992) Expression of foreign genes in transgenic yellow-poplar plants. *Plant Physiol.* **98,** 114–120.
41. Christou, P., Ford, T., Kofron, M. (1991) Production of transgenic rice (*Oryza sativa* L.) plants from agronomically important indica and japonica varieties via electric particle accelertion of exogenous DNA into immature zygotic embryos. *Bio/Technology* **9,** 957–962.
42. Christou, P., Ford, T. L., and Kofron, M. (1992) The development of a variety-independent gene transfer method for rice. *Trends Biotech.* **10,** 239–246.
43. Casas, A. M., Kononwicz, A. K., Zehr, U. B., Tomes, D. T., Axtell, J. D. Butler, L. B., Bressan, R. A., and Hasegawa, P. M. (1993) Transgenic sorghum plants via microprojectile bombardment. *Proc. Natl. Acad. Sci. USA.* **90,** 11,212–11,216.
44. McCabe, D. E., Seain, W. F., Martinell, B. J., and Christous, P. (1988) Stable trans- formation of soybean (Glycine max) by particle acceleration. *Bio/Technology* **6,** 923–926.
45. Christou, P., McCabe, D. E., Martinell, B. J., and Swain, W. F. (1990) Soybean genetic engineering-commercial production of transgenic plants. *Trends Biotech.* **8,** 145–151.
46. Ellis, D. D., McCabe, D. E. McInnis, S., Ramachandran, R., Russell, D. R., Wallace, K. M., Martinell, B. J., Roberts, D. R., Raffa, K. F., and McCown, B. H. (1993) Stable transformation of *Picea glauca* by particle accelaeration—A model system for conifer transformation. *Bio/Technology* **11,** 84–89.
47. Bower, R. and Birch, R. G. (1992) Transgenic sugarcane plants via microprojectile bombardment. *Plant J.* **2,** 409–416.

48. Bidney, D., Scelonge, C., Martich, J., Burrus, M., Sims, L., and Huffman, G. (1992) Microprojectile bombardment of plant tissues increases transformation frequency by Agrobacterium tumefaciens. *Plant Mol. Biol.* **18,** 301–313.
49. Tomes, D. T., Weissinger, A. K., Ross, M., Higgins, R., Drummond, B. J., Schaaf, S., Malone-Schoneberg, J., Staebell, M., Flynn, P., Anderson, J., and Howard, J. (1990) Transgenic tobacco plants and their progeny drived from microprojectile bombardment of tobacco leaves. *Plant Mol. Biol.* **14,** 261–268.
50. Vasil, V., Castillo, A. M., Fromm, M. E., and Vasil, I. K (1992) Herbicide resistant fertile tansgenic wheat plants obtained by microprojectile bombardment of regenerable embryogenic callus. *Bio/Technology* **10,** 667–674.
51. Weeks, T. J., Anderson, O. D., and Blechl, A. E. (1993) Rapid production of multiple independent lines of fertile transgenic wheat *(Triticum aestivum). Plant Physiol.* **102,** 1077–1084.
52. Vasil, V., Srivastava, V., Castillo, A. M., Fromm, M. E., and Vasil, I. K. (1993) Rapid production of transgenic wheat plants by direct bombardment of cultured immature embryos. *Bio/Technology* **11,** 1553–1558.
53. Becker, D., Brettschneider, R., and Lorz, H. (1994) Fertile transgenic wheat from microprojectile bombardment of scutellar tissue. *Plant J.* **5,** 299–307.
54. Pereira, L. F. and Erickson, L. (1995) Stable transformation of alfalfa *(Medicago sative L.)* by partide bombardment. *Plant Cell Rep.* **14,** 290–293.
55. Kamo, K., Blowers, A., Smith, F., Van Eck, J., and Lawson, R. (1995) Stable transformation of gladioulus using suspension cells and callus. *J. Am. Soc. Hort. Sci.* **120,** 347–352.
56. Sagi, L., Panis, B., Remy, S., Schoofs, H., De Smet, K., Swennen, R., and Cammue, B. P. A. (1995) Genetic transformation of banana and plantain (*Musa* spp.) via partide bombardment. *Bio/Technology* **13,** 481–485.
57. Ye, X. J., Brown, S. K., Scorza, R., Cordts, J., and Sanford, J. C. (1994) Genetic transformation of peach tissues by partide bombardment. *J. Am. Soc. Hort. Sci.* **119,** 367–373.
58. Cabrera-Ponce, J. L., Vegas-Garcia, A., and Herrera-Estrella, L. (1995) Herbicide resistant transgenic papaya plants produced by and efficient particle bombardment transformation method. *Plant Cell Rep.* **15,** 1–7.
59. Kodama, H., Irifune, K., Kamada, H., and Morikawa, H. (1993) Transgenic roots produced by introducing R1-rol genes into cucumber cotyledons by particle bombardment. *Transgenic Res.* **2,** 147–152.
60. Prakash, C. S. and Varadarajan, V. (1992) Genetic transformation of sweet potato by particle bombardment. *Plant Cell Rep.* **11,** 53–57.
61. Wilmink, A., Van de Ven, B. C. E., and Dons, J. J. M. (1992) Expression of the gus-gene in the moncot tulip after introduction by particle bombardment and *Agrobacterium. Plant Cell Rep.* **11,** 76–80.
62. Kuehnle, A. R. and Sugil, N. (1992) Transformation of *Dendrobium* orchid using particle bombardment of protocorms. *Plant Cell Rep.* **11,** 484–488.

63. Yao, J., Wu, J., Gleave, A. P., Morris, B. A. M., Yao, J. L., and Wu, J. H. (1996) Transformation of citrus embryogenic cells using particle bombardment and production of transgenic embryos. *Plant Sci. Limerick.* **113,** 175–183.
64. Ross, A. H., Manners, J. M., and Birch, R. G. (1995) Embroyogenic callus production, plant regeneration and transcient gene expression following particle bombardment in the pasture grass, *Cenchrus ciliaris*. *Aust. J. Bot.* **43,** 192–199.
65. Rochange, T., Serrano, L., Marque, C., Teulieres, C., Boudet, A. M. (1995) DNA delivery into *Eucalyptus globulus* zygotic embryos through biolistics: optimization of the biological and physical parameters of bombardment for two different particle guns. *Plant Cell Rep.* **14,** 674–678.
66. Van-Eck, J. M., Blowers, A. D., and Earle, E. D. (1995) Stable transformation of tomato cell cultures after bombardment with plasmid and YAC DNA. *Plant Cell Rep.* **14,** 329–334.
67. Krittel, N., Gruber, V., Hahne, G., and Lence, P. (1994) Transformation of sunflower *(Heliarthus annus)*: a reliable protocol. *Plant Cell Rep.* **14,** 81–86.
68. VanderMass, H. M., de Jong, E. R., Reub, S., Hensgens, L. A. M., and Krens, F. A. (1994) Stable transformation and long term expression of the gusA reporter gene in callus lines of perennial ryegrass *(Lolium perenne)*. *Plant Mol. Biol.* **24,** 401–405.
69. Arokiaraj, P., Jones, H., Cheong, K. F., Coomber, S., and Charlwood, B. V. (1994) Gene insertion into *Hevea brasiliensis*. *Plant Cell Rep.* **13,** 425–431.
70. Hartman, C. L., Lee, L., Day, P. R., and Tumer, N. E. (1994) Herbicide resistant turfgrass *(Agrostis palustris)* by biolistic transformation. *Bio/Technology* **12,** 919–923.
71. Hebert, D., Kikkert, J. R., Smith, F. D., and Reisch, B. I. (1993) Optimization of biolistic transformation of embryogenic grape cell suspensions. *Plant Cell Rep.* **12,** 585–589.

36

Transformation of the Cereals Using *Agrobacterium*

Jean Gould

1. Introduction

The evolved gene transfer mechanism of *Agrobacterium tumefaciens* is the transformation method of choice. Its use is simple, gene transfers are precise and result in permanent genetic changes. The range of species known to be infected by this organism has grown from an original group of dicot plants known to produce galls after infection *(1)*, to include many other species that exhibit little or no gall formation.

The cereals were the last group of plants to be transformed using *Agrobacterium*. This economically important group was thought to be refractory to *Agrobacterium*-mediated transformation *(2,3)*, and based on this understanding, efforts to genetically modify the cereals focused on methods of direct gene transfer. With development of the microprojectile DNA delivery method in the 1980s, transformation of corn and rice became possible *(4–11)*. These and other direct DNA transfer methods represented a significant technological breakthrough; however, serious drawbacks have limited widespread application. Transformation is limited to genotypes that can be regenerated in culture from embryogenic tissue. The yield of transformed plants produced by direct DNA transfer is low (0.001–0.1%); however, the yield of plants transformed in the germline permitting inheritance by subsequent generations, is much lower. In comparison, *Agrobacterium*-mediated transformations have a 10–30% transformation rate, and the genes become permanent genomic incorporations and are inherited. Other drawbacks are: genetic damage (somaclonal variation) induced in culture *(12)* when plant tissues are taken through a dedifferentiation step *(13)*, transfer of multiple copies of the gene of interest (10–100 or more) which can produce problems in gene expression, and instability of the transferred genes in the plant's genome.

From: *Methods in Molecular Biology, vol. 62: Recombinant Gene Expression Protocols*
Edited by: R. Tuan Humana Press Inc., Totowa, NJ

The key to achieving transformation in the cereals with *Agrobacterium tumefaciens* was the realization that the transformation-competent cells were not the dividing dedifferentiating wound or callus cells known to be competent in tobacco and other dicot species, but were the dividing meristematic cells in cereals. This serendipity allowed development of the procedure described here, which is both simple and general. It relies on the natural competence of dividing meristematic cells in the apex of cereal embryos and/or seedlings, supervirulence factors in *A. tumefaciens* and the native regenerative drive of the shoot apex. The transformed apex develops directly into a transformed plant using a simple regeneration protocol *(17,33,34)* that works well with all maize, wheat, sorghum, and rice genotypes tested (*17,24,25*; Gould nonpublished).

1.1. Agrobacterium*-Mediated Transformation in Cereals*

Graves and Goldman *(14)* were the first to report that the meristematic regions of maize embryos and seedlings were susceptible to *A. tumefaciens* infection. Later, Grimsley's group using Agroinfection, demonstrated this aspect of *Agrobacterium* infectivity further *(15,16)*. Our work with maize *(17)* was the first to show that transformation followed inoculation of seedling apical meristems with *A. tumefaciens* and the plants generated produced progeny containing the transferred genes. Recent studies reporting *Agrobacterium*-mediated transformation of maize *(18,19)* and rice *(20–22)* substantiate these findings and demonstrate the effectiveness of *Agrobacterium* in cereal transformation.

1.2. Problems Inherent in Callus-Based Plant Regeneration Methods

In plant tissue culture, genotype fidelity is often lost during passage through a callus intermediate, as in shoot organogenesis from leaf disk, or embryogenesis from seedling tissues. The resulting mutations, or somaclonal variation, range from obvious gross phenotypic changes to imperceptible changes that can ultimately impact yield, as reported in barley *(11)*, and perhaps other qualitative traits as well. Gross chromosomal rearrangements and loss of chromosome material was observed in abnormal cotton plants regenerated in culture through embryogenesis *(12)*. A recently discovered mechanism, activation of latent retrotransposons in the genome of plants during extreme stress, infection, and dedifferentiation in tissue culture *(13)*, produces permanent genetic mutations. This mechanism may account for much culture-induced somaclonal variation.

1.3. Advantages to Regeneration from the Shoot Apical Meristem

Genetic fidelity is most closely maintained in the meristems of plants. In cereals, as in other plant species, the dividing cells in the apical meristem contain a competent cell type that accepts and sustains transformation by *Agrobacterium (16,17,23)*. Inoculation of a meristem allows direct transformation of an existing shoot and subsequent plant regeneration that is direct and rapid. Dedifferentiation does not occur and time in culture is minimal. Meristem or shoot apex culture was originally developed to remove the viral load from propagated germ plasm and is the method of choice in the nursery industry for production of clonal lines that are true to type.

This method of *Agrobacterium*-mediated shoot apex transformation was first demonstrated in maize *(17)* with production of normal fertile plants and inheritance of the transferred genes. We have subsequently applied this method to wheat, sorghum (*24,25*; Gould nonpublished), and rice (Gould nonpublished). Other successful variations of this method, employ the meristematic cells of the immature embryo axis of maize *(19)* and rice *(21,22)*.

2. Materials

1. Autoclave or other method for sterilization of media and equipment.
2. Distilled or reverse osmosis water.
3. Incubator or lighted shelving for tissue cultures, maintained at approx 26–29°C, light intensity of 70–100 µE/ms using a 16:8 h photoperiod.
4. *Agrobacterium tumefaciens* EHA101–kanamycin resistant *(26)*, or EHA105–kanamycin sensitive *(27)*, containing the "super-vir" virulence locus, pTiBo542 *(28)*.
5. Binary vector, such as the pBI101 series (Clonetech, Palo Alto, CA).
6. Commercial bleach (5.25% sodium hypoclorite).
7. Seeds: Obtain commercially grown seed. We have used: *Zea mays* L., B7 Funk's G90, or Pioneer varieties; *Triticum asetivum* L., Chinese Spring, or other spring wheat; *Sorghum bicolor* L., RTX 430 *(24,25)*; *Oryza sativa* L., Rexmont, or Lemont. Note: It is important to obtain high quality seed in good condition, grown under cool and dry conditions. Seed that has developed under hot and humid conditions will be internally contaminated and difficult to disinfect.
8. Media: Most components can be purchased through Sigma (St. Louis, MO.) Murashige and Skoog (MS) inorganic salt formulation *(31)* (Sigma), containing sucrose, thiamin-HCl, and kinetin. Agar: Difco Bacto Agar.

   Antibiotics: Kanamycin (Sigma); and carbenicillin 500 mg/L (Sigma) or Calvamox; 250 mg/L. (Smith, Klein, Beecham), or a penicillinase-resistant penicillin can be used.

   a. LB Medium (LB): 11 g/L Bacto agar *(32)*, 50 mg/L kanamycin and/or appropriate antibiotic.
   b. Seed Germination Medium (SG): 7 g agar/L distilled water, autoclaved.
   c. Inoculation medium (MS + kin): MS inorganic salts plus 0.1 mg/L, kinetin, 30 g/L sucrose, 0.1 mg/L thiamine-HCl, 8 g/L agar, pH 5.7.
   d. Selection medium (MS + K + C): MS inorganic salts plus 0.1 mg/L, kinetin, 30 g/L sucrose, 0.1 mg/L thiamine-HCl, 8 g/L agar, pH 5.7, carbenicillin 500 mg/L (or Clavamox, 250 mg/L), a selective antibiotic or herbicide, i.e., kana-

mycin (7.5–15 mg/L kanamycin if the *nos* promoter is used with *neo*, as in NPTII). Antibiotics are added as filter-sterilized solutions of 1–3 mL each, to autoclaved medium that has been allowed to cool so that it is not hot to the touch.

e. Rooting medium (MS + C): MS inorganic salts plus 30 g/L sucrose, 0.1 mg/L thiamin-HCl, carbenicillin 500 mg/L, agar 8 g/L, pH 5.7.

f. Embryo germination medium (EM): This medium is identical to rooting medium (MS + C) above; however, carbenicillin is omitted.

9. Option: Bacterial activation solution (ACS): 75 m*M* MES pH 5.4 (Wayne Barnes personal communication), acetosyringone, 30–100 μ*M*, octopine 100 mg/L *(20,27,37)*. Allow bacteria to become induced in the solution, 0–2 h.
10. Promoters and Genes: Kanamycin is effective in cereal meristem transformation; however, geneticin (Sigma) can be more effective. The visual marker gene GUS (beta-glucuronidase, *uid*A or *gus*A) has been in use since 1987 and many reliable methods for use of this reporter gene are available *(29)*. The standard gene constructions developed for use in tobacco and other dicot species, i.e., NPTII containing kanamycin resistance driven by the nos promoter, and GUS driven by the CaMV 35S promoter can be used with the cereals. However, expression of these reporter genes can be low. In wheat for example, GUS expression driven by the CaMV 35S promoter is limited to the tips of seedling roots (Fig. 1). Both antibiotic and visual marker genes should be driven by promoters that are effective in cereals, i.e., the rice actin 1 (*act* 1), the maize ubiquitin (*ubi*1), the nopaline synthase *(nos)*, and rice ribulose-1,5 biphosphate carboxylase/oxygenase (*rubc*S) promoters *(30)*. Except for *nos*, these cereal promoters are not available in binary vectors for use in *Agrobacterium* at this writing, but are available in small plasmids for electroporation or microparticle direct plant transformation from the CAMBIA, Molecular Genetic Resource Service, Canberra, Australia.

3. Method

3.1. Seed Germination

As with many procedures, there are seemingly minor steps involved that can make a difference in how well the procedure works. For this procedure, four elements are crucial: induction of virulence functions in *A. tumefaciens* (*see* Note 1), induction of competency in cells of the plant apex (*see* Note 2), combining both agrobacteria and plant cells that are in transformation competent states (*see* Note 3), and a selection process that is tailored to the level of promoter activity in the inoculated apical meristem (*see* Note 4).

1. Rinse seeds in running water for 15–30 min.
2. Surface sterilize seeds in 20% household bleach for 15 min.
3. Rinse 3X with autoclaved sterile water.
4. Place seeds on water/agar (Bacto) solidified medium (8 g/L) to germinate. Incubate in light at 27°C, or room temperature, for 7 d.

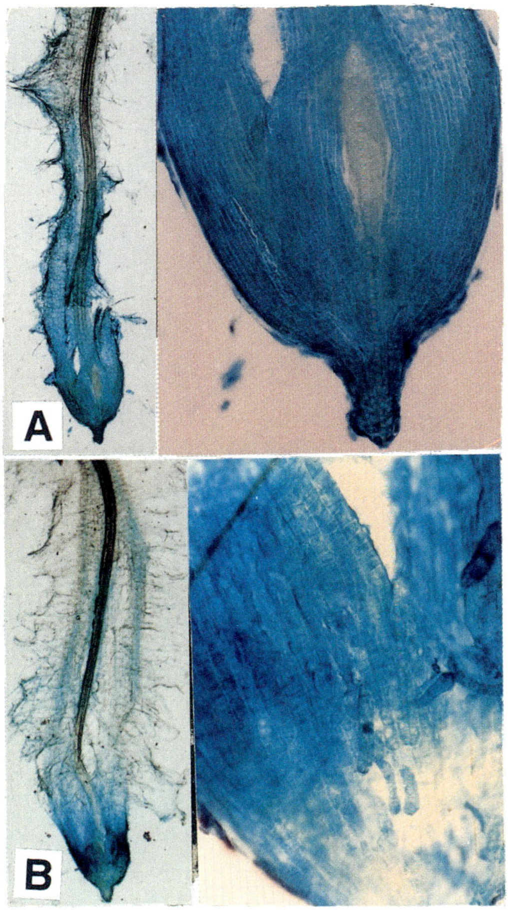

Fig. 1. GUS expression in the F_1 progeny of wheat: Micrographs of roots taken from aseptically germinated seedling progeny of a regenerated wheat plant X52 stained for GUS activity using X-GLUC. Shown are two seedlings: X52-8 **(A)** and X52-7 **(B)**. GUS activity was never detected in roots of control wheat seedlings. The blue X-GLUC precipitate is homogenous throughout the cytoplasm of the cells and can be seen in the nucleus of a group of cells in the foreground of B. This pattern of expression was observed in all the GUS positive wheat seedlings assayed. Debris near roots are detached plant cells and cuticle.

3.2. Shoot Apex Isolation

Shoot apex isolation can be complex but it does not have to be. For more background *see* refs. *18,33,34*. The apex can be inoculated from the shoot base described in ref. *17*, or from the side, which is described here. Common to both procedures is opening a facet of the apical meristem and inoculation of the super-virulent *Agrobacterium* at this site.

1. Remove the shoot from the seed and remove excess leaf tissue, a process very much like cleaning green onions. The shoot apex is embedded in the base of the seedling near the seed. Remove the shoot from the seed and roots without damaging the apex, using a scalpel or beveled hypodermic needle (23–25 5/8-gage). The isolated shoot is color-coded, white at the base and green at the shoot end. Trim 1/3 to 1/2 from the leafy shoot; this region contains only rolled leaves. This step produces a cylindrical "log" of tissue. The location of the shoot apex can sometimes be seen as a bulge within this log. In the B73 variety of corn, some of the epidermal cells in this region are red.
2. Open up a side of the shoot apex. Stand the "log" on end, green side up and cut down through the center of the leaves to bisect the apical meristem from the top. Leaf primordia and innermost elongating leaves are not removed, but left intact with the apex to aid survival and growth of the shoot on the simple MS based medium.
3. Locate the apex and inoculate the exposed side of the meristem. Trim away the excess stem tissue from beneath the meristem leaving a base of approx 1 mm. Culture lengthwise with inoculated region facing up. Inoculate this wound with *Agrobacterium*. The method is forgiving.

3.3. Inoculation and Culture

1. Inoculate *Agrobacterium tumefaciens* on to a fresh plate of LB medium containing the appropriate antibiotic, grow at room temperature, 25–27°C.
2. Scrape the new growth of bacteria from culture plate (2–4 d growth). If using induction solution mix 50:50 v/v with 50–100 µL induction medium.
3. Introduce onto the wound made the shoot meristem.
4. Cocultivate inoculated shoots 2–3 d on an MS *(31)* agar-solidified medium containing thiamine-HCl at 0.1 mg/L, and kinetin at 0.1 mg/L at room temperature.
5. Transfer shoots to hormone-free medium containing 500 mg/L carbenicillin, or 250 mg/L Clavamox, for 7 d to remove bacteria.
6. Transfer shoots to selection medium containing 7.5–15 mg/L kanamycin and 500 mg/L carbenicillin, or 250 mg/L Clavamox. Shoots remain on this medium for up to 4 wk or until roots form. The procedure is outlined in Table 1.

3.4. Transfer to Soil

Shoots surviving antibiotic selection root spontaneously after 3–4 wk in culture and are ready to transfer to soil 4–6 wk after isolation. Flowering begins 8 wk after transfer to soil, and senescence begins after 12 wk.

3.5. Progeny Germination and Assays

Germinate mature embryos on an MS-based, hormone-free medium containing thiamine-HCl, 0.1 mg/L, and sucrose, 20 g/L. After 1–2 wk, seedlings

Table 1
Inoculation and Culture Schedule for Cereals: *Triticum asetivum* L., Chinese Spring[a]

Procedure	Time	Medium
3.1. Seed germination	5-7 d in light	Seed germination medium (SG)
3.2. Shoot isolation		MS + 0.1 mg/L kinetin (MS + kin)
3.3. Shoot inoculation	3–7 d	Scrape new growth of bacteria from plate. Inoculate apical meristem (Optional: mix bacteria 1:1 with activator solution: 75 m*M* MES pH 5.4 (Wayne Barnes, personal communication), acetosyringone (acetophenone) 30–100 μ*M* *(27,37)*, octopine 100 mg/L *(20)*
Reculture shoots	7 d	MS + carbenecillin 500 mg/L
Reculture shoots	14 d	MS + kanamycin 7.5 mg/L + carbenecillin 500mg/L
3.4. Transfer regenerated plants to soil	5–8 wk	
Anthesis	4 wk	
3.5. Progeny germination and assays	2 wk	MS (+ sucrose 20 g/L) (a test for *Agrobacterium* contamination in seedling progeny)
3.6. DNA analyses: PCR amplification for the transferred genes	~100 d from transfer of R_0 plants to soil	
Transfer progeny to soil		

[a]Shoot isolation, inoculation, plant regeneration, and production of F_1 progeny (example: *Triticum aestivum* L., Chinese Spring). All of the grain cereals follow a similar schedule. Rooted plants are obtained approx 30 d from the time of shoot excision and inoculation. In wheat and corn, seedling progeny can be obtained approx 100 d later.

are ready for transfer to soil. Prior to transfer, remove and assay leaf tissue and root tips for assays of reporter gene activity, i.e., GUS (Fig. 1) using the X-GLUC colorimetric or MUG fluorimetric assay *(29)*.

3.6. DNA Analyses

1. Isolate DNA using the method of Dellaporta et al. *(35)*, and estimate concentration fluorimetrically using the TKO 100 (Hoeffer Scientific, San Francisco CA).
2. PCR oligonucleotide primers and cycle parameters for the amplification of a 1 kb fragment spanning *nos* and *neo* sequences from 200 pg target plant DNA in each

amplification, have been described previously *(17)*. Amplify a fragment in the coding region of *uidA* using a 5' oligomer CTC GAC GGCCTG TGG GCA TTC AGT-40 b downstream from the ATG initiation codon, and a 3' oligomer-TAA CCT TCA CCC GGT TGC CAG AGG-approx 680 b downstream from the 5' oligomer. Separate amplified DNA by gel electrophoresis, blot gel to a membrane, and hybridize with radiolabeled *uid*A or *neo* probes as above.

3. Genomic DNA hybridization: Digest total plant DNA with HindIII or EcoRI, separate by agarose gel electrophoresis, blot, and hybridize with radiolabeled *uid*A or *neo* probe. The procedures used with wheat are the most sensitive and are described in detail elsewhere *(36)*. Following hybridization, wash membranes 0.5–1.0 h in 0.1X SSC and 0.1% SDS at 65°C. Expose blots to film at –80°C for 7 d.
4. Chimeras: The regenerated plants are often chimeric for the transferred genes and because of low copy, activity and signal can be low. However, the F_1 progeny of these plants are not chimeric and usually the transgenes are easier to detect in progeny than in the original regenerated plant.

4. Notes

1. Bacterial virulence: The hypervirulence and broad host-range of *A. tumefaciens* EHA101 and EHA105 are essential for successful transformation. However, *Agrobacterium* are soil organisms and the *vir* plasmid (and associated virulence functions) is lost when incubated over 29°C. These temperatures are easily met and exceeded in incubators set at 28°C, and in plant tissue culture plates under high direct lighting. To protect virulence functions, it is best to grow cultures of *Agrobacterium* at room temperature and cultures of inoculated shoots at room temperature under indirect lighting during the period of cocultivation. Virulence can be induced by a variety of methods. We use acetosyrinone (Aldrich, Milwaukee, WI), nopaline (Sigma), or octopine (Sigma) *(20,27,37)* in MES buffer.
2. Transformation competency–apical meristem: In the procedure outlined here, the isolated and inoculated seedling shoot apices are immediately cultured onto a medium containing kinetin to promote cell division and aid regeneration of the apical meristem *(17,33,34)*. Addition of a cytokinin (kinetin or benzyl adenine) and perhaps the apex isolation procedure itself, may ensure transformation competency in the meristem.
3. Timing: Care is usually taken to insure induction of *Agrobacterium* virulence prior to inoculation. In this method *both* plant and bacterial systems should be in a competent state.
4. Selection: At this time, a permissive selection protocol is used because the efficiency of the promoter(s) used to drive antibiotic resistance in the cereal meristem is unclear. Ideally, the promoter used with a resistance gene should be active in the meristem for antibiotic selection to function properly; however, the promoters that are usually available with resistance genes have often been chosen in tobacco callus-based procedures and are not suited for use in a meristem-based method. In addition, plant generation is dependent on vitality of the meristem. If

cells in the meristem die, the organization inherent in the meristem is destroyed, and regeneration is abolished. In the NPTII construction, the *nos* promoter drives *neo*. If this construction is used, the activity of this promoter in cereal meristem is usually sufficient to provide protection to 7.5–10 mg/L of kanamycin.

Acknowledgments

The author is indebted to Toshio Murashige for the training and research philosophy that proved to be essential to carry out this project; Michael Devey for contributing to the initiation of these experiments and valuable assistance in later stages; Stanton Gelvin and Wayne Barnes for insight into *A. tumefaciens* virulence; Osamu Hasegawa, Eugenio Ulian, and Tsay-Souk Ko for tremendous effort, insightful discussion, and friendship. This research was supported by The Texas Agricultural Experiment Station and by Nisshinbo, Inc. of Tokyo.

References

1. DeCleene, M. (1979) The susceptibility of monocotyledons to *Agrobacterium tumefaciens*. *Phytopathol. Z.* **113,** 81–89.
2. Potrykus, I. (1990) Gene transfer to cereals, an assessment. *Bio/Technologyogy* **8,** 535–542.
3. Chilton, M.-D. (1993) Progress on *Agrobacterium* transformation of cereals. *Proc. Natl. Acad. Sci. USA* **90,** 3549–3553.
4. Klein, T., Wolf, E., Wu, R., and Stanford, J. (1987) High velocity microprojectiles for delivering nucleic acids into living cells. *Nature* **327,** 70–73.
5. Fromm, M., Morrish, F., Armstrong, C., Williams, R., Thomas, J., and Klein, T. (1990) Inheritance and expression of chimeric genes in the progeny of transgenic maize plants. *Bio/Technology* **8,** 833–839.
6. Gordon-Kamm, W., Spencer, M., Mangano, M., Adams, T., Daines, R., Start, G., O'Brien, J., Chambers, S., Adams, W., Willetts, N., Rice, T., Mackey, C., Krueger, R., Kausch, A., and Lemaux, P. (1990) Transformation of maize cells and regeneration of fertile transgenic plants. *Plant Cell* **2,** 603–618.
7. Luo, Z.-X. and Wu, R. (1989) A simple method for the transformation of rice via the pollen-tube pathway. *Plant Mol. Biol. Rep.* **7,** 69–77.
8. Christou, P., Ford, T., and Kofron, M. (1991) Production of transgenic rice (*Oryza sativa* L.) plants from agronomically important indica and japonica varieties via electric discharge particle acceleration of exogenous DNA into immature zygotic embryos. *Bio/Technology* **9,** 957–962.
9. Vasil, V., Castillo, A., Fromme, M., and Vasil, I. (1992) Herbicide resistant fertile transgenic wheat plants obtained by microprojectile bombardment of regenerable embryogenic callus. *Bio/Technology* **10,** 667–678.
10. Weeks, J., Anderson, O., and Blechl, A. (1993) Rapid production of multiple independent lines of fertile transgenic wheat *(Triticum aestivum)*. *Plant Physiol.* **102,** 1077–1084.

11. Lemaux, P., Wan, Y., Bregitzer, P., Halbert, S., Waytt, S., Cho, M.-J., Marx, G., and Buchanan, B. (1995) Molecular breeding of barley. Plant Genome III, San Diego CA, Jan. 18, 1995, 142.
12. Li, R., Stelly, D., and Trolinder, N. (1989) *Genome* **32,** 1128–1134.
13. Hirochika, H. (1993) Activation of tobacco retrotransposons during tissue culture. *EMBO J.* **12,** 2521–2528.
14. Graves, A. and Goldman, S. (1986) The transformation of *Zea mays* seedlings with *Agrobacterium tumefaciens. Plant Mol. Biol.* **43,** 50.
15. Grimsley, N., Hohn, T., Davis, J., and Hohn, B. (1987) *Agrobacterium* mediated delivery of infectious maize streak virus into maize plants. *Nature* **325,** 177–179.
16. Schlappi, M. and Hohn, B. (1992) Competence of immature maize embryos for *Agrobacterium*-mediated gene transfer. *Plant Cell* **4,** 7–16.
17. Gould, J., Devey, M., Hasegawa, O., Ulian, E., Peterson, G., and Smith, R. (1991) Transformation of *Zea mays* L., using *Agrobacterium tumefaciens* and the shoot apex. *Plant Physiol.* **95,** 424–426.
18. Hansen, G., Das, A., and Chilton, M.-D. (1994) Constitutive expression of the virulence genes improves the efficiency of plant transformation by *Agrobacterium* to tobacco and corn. *PNAS USA* **91,** 7603–7607.
19. Ishida, Y., Satio, H., Otha, S., Hiei, Y., and Komari, T. (1995) *Agrobacterium*-mediated transformation of maize. *Plant Physiol. Suppl.* **108: 152,** (Suppl.) 801.
20. Rainieri, D., Bottino, P., Gordon, M., and Neste,r E. (1990) *Agrobacterium* transformation of rice (*Oryza sativa* L.) *Bio/Technology* **8,** 33–38.
21. Chan, M.-T., Chang, H.-H., Ho, S.-L., Tong, W.-F., and Yu, S.-M. (1993) *Agrobacterium*-mediated production of transgenic rice plants expressing a chimeric alpha-amylase promoter/beta-glucuronidase gene. *Plant Mol. Biol.* **22,** 491–506.
22. Hiei, Y., Ohta, S., Komari, T., and Kumashiro, T. (1994) Effecient transformation of rice (*Oryza sativa* L.) mediated by *Agrobacterium* and sequence analysis of the boundaries of the T-DNA. *Plant J.* **6,** 271–275.
23. Ulian, E., Smith, R., Gould, J., and McKnight, T. (1988) Transformation of plants via the shoot apex. *Vitro, Cell Devl. Biol.* **24,** 951–954.
24. Gould, J., Devey, M., Hasegawa, O., Ko, T.-S., Villalon, D., Rigoldi, M., Ulian, E., Peterson, G., and Smith, R. (1991) Transformation of the Graminae by *Agrobacterium tumefaciens*. ISPMB, Tucson AZ.
25. Gould, J., Devey, M., Ko, T.-S., Peterson, G., Hasegawa, O., and Smith, R. (1992) Transformation of Graminae using *Agrobacterium tumefaciens. J. Cell. Biochem.* Keystone Symposia on Molecular & Cellular Biology, Supplement 16F, p. 207.
26. Hood, E., Helmer, G., and Fraley, R. (1986) The hypervirulence of *Agrobacterium tumefaciens* A281 is encoded in a region of pTBo542 outside of T-DNA. *J. Bact.* **168,** 1291–1301.
27. Li, X.-Q., Liu, C.-N., Ritchie, S., Peng, Y., Gelvin, S., and Hodges, T. (1992) Factors influencing *Agrobacterium*-mediated transient expression of gusA in rice. *Plant Mol. Biol.* **20,** 1037–1048.

28. Hood, E., Jen, G., Kayes, L., Kramer, J., Fraley, R., and Chilton, M. D. (1990) Restriction endonuclease map of pTiBo542, a potential Ti plasmid vector for genetic engineering of plants. *Bio/Technology*. **2,** 702–709.
29. Gallagher, S. (1992) *GUS Protocols*. Academic, New York, p.221.
30. McElroy, D., Chamberlain, D., Moon, E., and Wilson, K. (1995) Development of *gus*A reporter gene constructions for cereal transformation. *Mol. Breeding* **1,** 27–37.
31. Murashige, T. and Skoog, F. (1962) A revised medium for rapid growth of and bioassays with tob acco tissue cultures. *Physiol. Plant* **15,** 473–497.
32. Sambrook, J., Frisch, E., and Maniatis, T. (1989) *Molecular Cloning. A Laboratory Manual*, 2nd ed. Cold Spring Harbor Laboratory, Cold Spring Harbor, NY, Appendix, A.1-A.6.
33. Shabde, M. and Murashige, T. (1977) Hormonal requirements of excised *Dianthus caryophyllus* L. Shoot apical meristem in vitro. *Amer. J. Bot.* **64,** 443–448.
34. Irish, E. and Nelson, T. (1988) Development of maize plants from cultured shoot apices. *Planta* **175,** 9–12.
35. Dellaporta, S., Wood, J. and Hicks J. (1985) Maize DNA miniprep, in *Molecular Biology of Plants* (Malmberg, R., et al., eds.), Cold Spring Harbor Laboratory, Cold Spring Harbor, NY, pp. 36–37.
36. Devey, M. and Hart, G. (1993) Chromasomal localization of intergenomic RFPL loci in hexaploid wheat. *Genome* **36,** 913–918.
37. Veluthambi, K., Krishnan, M., Gould, J. H., Smith, R. H., and Gelvin, S. B. (1989) Opines stimulate induction of the vir genes of *Agrobacterium tumefaciens* Ti plasmid. *J. Bact.* **171,** 3696–3703.

37

Transformation of Microalgae Using Silicon Carbide Whiskers

Terri G. Dunahay, Sally A. Adler, and Jonathan W. Jarvik

1. Introduction

Microalgae are single-celled or colonial photosynthetic organisms that are receiving increasing attention for use their potential usefulness in a number of industrial applications, including bioremediation and the production of high-value specialty chemicals and fuels. The ability to genetically engineer these organisms, in order to study and eventually manipulate their metabolic pathways, would greatly enhance the utility of microalgae as scientifically and industrially important species.

A significant barrier to genetic transformation in microalgae, as for all plant cells, is the introduction of foreign DNA into the host cell through the cell wall. For many species of higher plants, protoplasts can be formed by enzymatic removal of the cell wall. DNA can be introduced into the wall-less cells by chemically induced direct uptake *(1)* or by electroporation *(2)*. However, because of the complex and diverse composition of algal cell walls, protocols to produce viable protoplasts have not been developed for most microalgal strains.

The only microalgae for which reproducible genetic transformation systems exist are the green flagellate *Chlamydomonas reinhardtii* and a related colonial species, *Volvox carteri (3,4)*. Development of genetic transformation techniques for *C. reinhardtii* was facilitated by the availability of the mutant strain *cw*-15, which lacks a cell wall. Kindle *(5)* demonstrated that DNA could be introduced into these wall-less strains by agitating the cells in the presence of glass beads and polyethylene glycol (PEG). The disadvantage of this method is that the requirement for a specific genetic trait in the host cells to achieve efficient transformation (e.g., the *cw* mutation) limits the applicability. An

From: *Methods in Molecular Biology, vol. 62: Recombinant Gene Expression Protocols*
Edited by: R. Tuan Humana Press Inc., Totowa, NJ

alternative approach that allowed the use of any *C. reinhardtii* strain as a transformation target was the production of wall-deficient cells using autolysin, a species-specific wall-degrading enzyme produced during gametogenesis *(5)*. Walled *C. reinhardtii* cells were also successfully transformed using the glass bead technique, but at a greatly reduced frequency.

The methods discussed above are specifically applicable to *C. reinhardtii*. A device that theoretically can be used to introduce DNA into any walled cell was developed several years ago by Sanford *(6)*. The microprojectile accelerator uses a rapid burst of pressure to inject DNA-coated tungsten or gold beads into cells. This technique, called biolistics, has been used successfully to transform walled cells of higher plants, bacteria, yeast, and at least two species of microalgae. The primary disadvantage of this technique is the expense of purchasing or leasing the biolistic device.

We were interested in developing a simple, inexpensive technique for introducing DNA into walled algal cells. In 1990, Kaeppler *(7)* reported using silicon carbide (SiC) whiskers to mediate the introduction of DNA into plant cells. We have developed a technique based on the SiC method of Kaeppler and the glass bead technique of Kindle *(5)* that uses SiC whiskers to transform microalgae *(8)*. We have demonstrated this technique using walled cells of *C. reinhardtii*, as this is the only unicellular alga for which good genetic transformation markers are readily available. The SiC method produces transformants at an efficiency of up to 10^{-5} per cell for walled cells and up to 10^{-4} per cell for *cw*-15 strains. In contrast to agitation of the cells with glass beads, agitating the cells with SiC whiskers for up to 10 min results in little loss in cell viability. The gentle nature of the SiC protocol may allow more flexibility in the development of protocols for the transformation of other microalgal strains.

Efficient nuclear transformation of *C. reinhardtii* has been accomplished thus far only by the use of homologous genes as selectable markers *(3)*. The protocol outlined below has been developed using the wild-type ARG7 gene, which encodes the enzyme argininosuccinate lyase (ASL), to complement arg7 auxotrophs *(9)*. We have also used SiC whiskers to mediate the transformation of a nitrate reductase-deficient strain of *C. reinhardtii* with the wild-type gene NIT1 *(10)*. Other selectable marker systems for *C. reinhardtii* nuclear transformation are discussed in *(3)*. Many *C. reinhardtii* strains and plasmids, including the plasmids carrying the NIT1 or ARG7 genes, can be obtained from Dr. Elizabeth Harris at the Chlamydomonas Genetics Center at Duke University, Durham, NC.

2. Materials

1. Silicon carbide whiskers: SiC whiskers are believed to be an inhalation hazard, similar to asbestos fibers. Care should be taken in handling and disposing of the

material. Gloves and a dust mask should be worn when weighing the whiskers, and residual powder should be disposed of as a hazardous waste. The material can be handled safely once the whiskers are suspended in water. Tubes containing suspensions of SiC whiskers should be tightly sealed before agitation to avoid the release of aerosols.

2. An actively dividing culture of the appropriate strain of *C. reinhardtii*. Most of the work described here used strain CC-1861, which contains a lesion in the ARG7 gene and requires exogenous arginine for growth. The cells are generally grown in HSA medium *(11)* supplemented with 13 m*M* arginine, in flasks bubbled with air. Optimal transformation is obtained with cells in late exponential stage (approx 5×10^6 cells/mL).
3. Sterile liquid algal growth medium HSA.
4. 100-mm Petri plates containing HSA, solidified with 1.5% agar.
5. Plating agar: HSA containing 1.2% agar. Plating agar should be melted and maintained at 45°–50°C in a water bath. For each transformation 2.5 mL is needed.
6. Plasmid DNA: The plasmid pARG7.8, received from the laboratory of Dr. Paul Lefebvre, consists of the *Sal*I-*Sal*I-*Bam*HI (i.e., *Sal*I partial digest) genomic fragment containing the full-length ARG7 gene from wild-type *C. reinhardtii (9)* ligated into the *Sal*I and *Bam*HI sites of the vector pBR329 (Saul Purton, personal communication). The plasmids can be isolated from *E. coli* by any standard plasmid purification protocol (i.e., alkaline lysis followed by extraction with phenol:chloroform *[12]*) or by using a commercial plasmid purification kit, such as the Maxiprep Plasmid Isolation Kit (Qiagen, Chatsworth, CA). For optimal transformation efficiencies, linearize the plasmid with restriction endonuclease *Xmn*I (this enzyme cuts the plasmid within the vector sequence), then inactivate the enzyme by heating the mixture to 65°C for 20 min prior to use. For transformation, the plasmid should be dissolved in sterile, distilled water or TE buffer (10 m*M* Tris pH 8.0, 1 m*M* EDTA) at a concentration of 1.0 µg/µL. Use 20 µg DNA per transformation.
7. Polyethylene glycol, mol wt 8000: a 20% (w/v) solution in distilled water, filter sterilized using a 0.22 µm filter and stored in a dark container.
8. A benchtop vortex mixer, such as the Fisher Vortex Genie 2 or the Lab Line Super Mixer.

3. Methods

1. Sterilize 50 mg aliquots of the SiC whiskers by suspension in 70% ethanol for 15–20 min, followed by several washes with sterile distilled water. Alternatively, the SiC can be sterilized by autoclaving. Place 50 mg whiskers in a 1.5-mL microfuge tube, puncture a small hole in the tube, cover with foil, and autoclave. Suspend the treated whiskers in 1 mL of sterile, distilled water.
2. Before harvesting the cells for transformation, remove the cells from the bubbler and allow the flask to sit without aeration for approx 4 h. We have observed that the addition of this step increases the efficiency of transformation up to 20-fold. This step also leads to an enhancement of transformation frequency using the

glass bead protocol (Paul Lefebvre, personal communication). The physiological basis of this effect is unclear. However, after 4 h without aeration, the cells appear spherical (rather than ovoid) and are more susceptible to lysis by 0.5% Nonidet P-40, much like wall-less cells. Thus, we suspect that their enhanced competence for transformation is due to an alteration in the cell wall.

3. Count the algal cells using a hemocytometer. Using sterile centrifuge tubes, pellet the cells by centrifugation at 4000*g* for 10 min. Suspend in HSA liquid medium to a concentration of approx 2.5×10^8 cells/mL.
4. For each transformation, aliquot 40 μL sterile SiC whiskers (50 mg/mL) into a 15-mL disposable plastic centrifuge tube using a wide bore pipet tip. (An ordinary pipet tip can be converted by cutting off the distal 5 mm with a sterile razor blade.) Add 220 μL HSA liquid medium. Add sequentially: 200 μL cells (5×10^7), 20 μL linearized DNA (1.0 μg/μL), and 160 μL PEG (20% PEG 8000). The final volume of the transformation mix should be 640 μL. (Note that the final volume of HSA plus DNA should equal 240 μL, so if you use a different volume of DNA, adjust the volume of HSA accordingly.)
5. Agitate the sample using a vortex mixer at top speed for 2 min, stopping briefly every 10 s.
6. Immediately after vortexing, add 1.86 mL HSA to bring the volume to 2.5 mL. This dilutes the PEG, which is toxic to the algae, and is important for achieving high transformation rates. Add 2.5 mL plating agar to the tube, mix by gentle inversion, and pour the entire volume onto a 100-mm Petri dish containing HSA agar.
7. Allow the agar to cool and harden. Wrap the edges of the plates with Parafilm (Marathon PaperMills, Rothschild, WI) to prevent desiccation, and incubate the plates under normal growth conditions. Transformed colonies should be visible in 7–10 d.

4. Notes

1. Until recently, a drawback to this technique has been the difficulty in obtaining a reliable source of SiC whiskers. SiC whiskers are used primarily in the production of metal and ceramic composites, and are not generally commercially available in small quantities. As mentioned above, SiC whiskers are considered to be an inhalation hazard, and most manufacturers we contacted were reluctant to send out small samples. The original experiments (described in ref. *8*) were performed using whiskers designated TWS100 (0.3 μm to 0.6 μm wide × 15 μm long), obtained as a gift from Tokai Carbon America, Inc., NY. A second sample from Tokai, designated TWS100C, was very hydrophobic, and formed large clumps in water. Tokai Carbon will no longer give out small samples of their whiskers, although you can purchase the whiskers in 2-kg quantities for approx $2000. The experiment described in this article used primarily Silar SC-9 whiskers (0.6 μm wide × 10–80 μm long) obtained as a gift from Dr. David Somers, University of Minnesota. These whiskers were produced by Advanced Composite Materials. Advanced Composites has recently begun selling 5-g quantities of Biograde

Whiskers (similar to Silar SC-9) for \$250. To our knowledge, the only other supplier of research amounts of SiC whiskers is Alfa Chemicals. These whiskers are 1.1 μm × 15 μm, and can be ordered in 50-g quantities for approx \$100 through the Johnson Matthey Company, Ward Hill, MA. Of the four types of SiC tested, the Silar SC-9 produced significantly more transformants in *C. reinhardtii*, up to 500 per 5×10^7 cells. The other SiC whiskers tested produced between 25–200 transformants per 5×10^7 cells. The Silar SC-9 whiskers were also found to work better than SiC whiskers from several other sources (including Tokai Carbon and Alfa) for transformation of maize suspension culture cells *(13)*. It is unclear what properties of the whiskers (purity, size, size variation within a sample) affect transformation efficiency.

2. Preliminary experiments testing the level of transformation as a function of whisker concentration indicate that the whiskers can be reduced up to eightfold (i.e., use only 5 μL of the 50 mg/mL whisker suspension) with little decrease in transformation efficiency. Reducing the amount of whiskers further resulted in substantial reductions in the numbers of transformants produced.
3. The presence of PEG in the transformation mix is important, as the number of transformants obtained decreased by 80–90% when PEG was omitted. A final concentration of PEG of 4.5–5.0% worked well; a range of PEG concentrations was not tested.
4. There are a number of growth media available for *C. reinhardtii (11)*. The particular medium used does not seem to be important for transformation, as long as the medium used for selection has the appropriate properties (i.e., no arginine for selection using the ARG7 gene). However, it is critical not to use media significantly different from the normal growth medium for transformation, i.e., do not use HSA to grow cells and perform transformations in SGII *(14)*.
5. Plating and manipulation of cell cultures may be performed in a sterile transfer hood to minimize contamination. If bacterial contamination is a problem, ampicillin (50 μg/mL) can be added to the plates.
6. In the experiments described here, the cells were plated in soft agar. Survival of the cells following agitation with SiC was similar for cells plated in soft agar or for cells spread directly onto SGII plates; however, it is not clear if plating the cells in agar increases the numbers of transformants obtained.
7. We have also demonstrated transformation using the NIT1 gene to transform *C. reinhardtii* strain CC-2453 nit1-305. These cells carry a mutation in the gene for nitrate reductase and thus cannot utilize nitrate as their sole nitrogen source. The cells grow well in a medium containing ammonium or urea; for example, SGII *(14)*. Cells transformed with the wild-type nitrate reductase gene can be identified by their ability to grow on SGII/NO_3, which is SGII medium in which the NH_4NO_3 is replaced by an equimolar amount of KNO_3. When selecting for NIT1 transformants, it is important to solidify selectable media with *washed* agar, as commercially available agar may contain residual NH_4^+ that can promote growth in the nitrate reductase-deficient cells. The agar can be prepared by extensively

washing the agar powder with distilled water before mixing with the growth medium for autoclaving. Alternatively, washed agar is available from Sigma (St. Louis, MO).

8. SiC whiskers can also be used to transform cell wall-deficient strains of *C. reinhardtii*. As was seen for walled strains, agitating *cw*-15 strains with SiC did not significantly affect cell viability. Transformation rates were found to be 10- to 50-fold higher when *cw*-15 strains were used, producing up to 5000 transformed colonies per plate.
9. It has been reported that *C. reinhardtii* cells treated simultaneously with two plasmids, one containing a selectable marker gene and one containing a second nonselectable gene, will often integrate and express both transgenes *(4,10,15)*. We have also used SiC whiskers successfully to cotransform *C. reinhardtii* with two genes on different plasmids. When the second, nonselectable plasmid is used in a twofold molar excess relative to pARG7.8, we typically observe that 30–50% of the cells isolated as arginine auxotrophs are also transformed with the second plasmid.
10. The experiments described here demonstrate nuclear transformation of *C. reinhardtii* mediated by SiC whiskers. We have not carefully investigated the possibility of using SiC to mediate chloroplast transformation; one preliminary attempt was unsuccessful. Kindle *(16)* has shown that the glass bead transformation technique can be used for chloroplast transformation, but at a much lower efficiency than can be achieved using biolistics.

References

1. Paszkowski, J., Shillito, R. D., Saul, M., Mandák, V., Hohn, T., Hohn, B., and Potrykus, I. (1984) Direct gene transfer to plants. *EMBO J.* **3,** 2717–2722.
2. Joersbo, M. and Brunstedt, J. (1991) Electroporation: mechanism and transient expression, stable transformation and biological effects in plant protoplasts. *Physiol. Plant.* **81,** 256–264.
3. Kindle, K. L. and Sodeinde, O. A. (1994) Nuclear and chloroplast transformation in *Chlamydomonas reinhardtii*: strategies for genetic manipulation and gene expression. *J. Appl. Phycol.* **6,** 231–238.
4. Schiedlmeier, B., Schmitt, R., Mueller, W., Kirk, M. M., Gruber, H., Mages, W., and Kirk, D. L. (1994) Nuclear transformation of *Volvox carteri*. *Proc. Natl. Acad. Sci. USA* **91,** 5080–5084.
5. Kindle, K. L. (1990) High frequency nuclear transformation of *Chlamydomonas reinhardtii*. *Proc. Natl. Acad. Sci. USA* **87,** 1228–1232.
6. Sanford, J. C., Smith, F. D., and Russell, J. A. (1993) Optimizing the biolistic process for different biological applications. *Methods Enzymol.* **217,** 483–509.
7. Kaeppler, H. F., Gu., W., Somers, D. A., Rines, H W., and Cockburn, A. F. (1990) Silicon carbide fiber-mediated DNA delivery into plant cells. *Plant Cell Rep.* **9,** 415–418.
8. Dunahay, T. G. (1993) Transformation of *Chlamydomonas reinhardtii* with silicon carbide whiskers. *BioTechniques* **15,** 452–460.

9. Debuchy, R. L., Purton, S., and Rochaix, J.-D. (1989) The argininosuccinate lyase gene of *Chlamydomonas reinhardtii*: an important tool for nuclear transformation and for correlating the genetic and molecular maps of the ARG7 locus. *EMBO J.* **8,** 2803–2809.
10. Kindle, K. L., Schnell, R. A., Fernández, E., and Lefebvre, P. A. (1989) Stable nuclear transformation of *Chlamydomonas* using the *Chlamydomonas* gene for nitrate reductase. *J. Cell. Biol.* **109,** 2589–2601.
11. Harris, E. H. (1989) *The Chlamydomonas Sourcebook.* Academic, San Diego, CA.
12. Sambrook, J., Fritsch, E. F., and Maniatis, T. (1989) *Molecular Cloning: A Laboratory Manual.* Cold Spring Harbor Laboratory Press, Cold Spring Harbor, NY.
13. Wang, K., Drayton, P., Frame, B., Dunwell, J., and Thompson, J. (1995) Whisker-mediated plant transformation: an alternative technology. *In Vitro Cell. Dev. Biol. Plant* **31,** 101–104.
14. Sager, R. and Granick, S. (1953) Nutritional studies with *Chlamydomonas reinhardtii. Ann. N. Y. Acad. Sci.* **56,** 831–838.
15. Nelson, J. A., Savereide, P. B., and Lefebvre, P. A. (1994) The *CRY1* in *Chlamydomonas reinhardtii*: structure and use as a dominant selectable marker for nuclear transformation. *Mol. Cell. Biol.* **14,** 4011–4019.
16. Kindle, K. L., Richards, K. L., and Stern, D. B. (1991) Engineering the chloroplast genome: Techniques and capabilities for chloroplast transformation in *Chlamydomonas reinhardtii. Proc. Natl. Acad. Sci. USA* **88,** 1721–1725.

Index